Super-Resolution Imaging in Biomedicine

Series in Cellular and Clinical Imaging

Series Editor
Ammasi Periasamy

PUBLISHED

Coherent Raman Scattering Microscopy
edited by Ji-Xin Cheng and Xiaoliang Sunney Xie

Imaging in Cellular and Tissue Engineering
edited by Hanry Yu and Nur Aida Abdul Rahim

Second Harmonic Generation Imaging
edited by Francesco S. Pavone and Paul J. Campagnola

The Fluorescent Protein Revolution
edited by Richard N. Day and Michael W. Davidson

Natural Biomarkers for Cellular Metabolism:
Biology, Techniques, and Applications
edited by Vladimir V. Gukasyan and Ahmed A. Heikal

Optical Probes in Biology
edited by Jin Zhang, Sohum Mehta, and Carsten Schultz

Super-Resolution Imaging in Biomedicine
edited by Alberto Diaspro and Marc A. M. J. van Zandvoort

SERIES IN CELLULAR AND CLINICAL IMAGING
AMMASI PERIASAMY, SERIES EDITOR

Super-Resolution Imaging in Biomedicine

Edited by
Alberto Diaspro
Marc A. M. J. van Zandvoort

CRC Press
Taylor & Francis Group
Boca Raton London New York

CRC Press is an imprint of the
Taylor & Francis Group, an **informa** business

Cover Image: Courtesy of Paolo Bianchini, Nanoscopy, IIT and LAMBS

CRC Press
Taylor & Francis Group
6000 Broken Sound Parkway NW, Suite 300
Boca Raton, FL 33487-2742

Printed on acid-free paper
Version Date: 20160915

International Standard Book Number-13: 978-1-4822-4434-2 (Hardback)

Visit the Taylor & Francis Web site at
http://www.taylorandfrancis.com

and the CRC Press Web site at
http://www.crcpress.com

To Teresa, Claudia, and Irene

To Eline, Rosa, and Myrthe

Contents

Series Preface xi

Preface xiii

Editors xv

Contributors xvii

Section I Overview of Super-Resolution Technologies

1 In the Realm of Super-Resolved Fluorescence Microscopy 3

Alberto Diaspro, Paolo Bianchini, Giuseppe Vicidomini, Colin J.R. Sheppard, Luca Lanzanò, Francesca Cella Zanacchi, and Lauretta Galeno

2 Super-Resolution in a Standard Microscope: From Fast Fluorescence Imaging to Molecular Diffusion Laws in Live Cells 19

Carmine Di Rienzo, Enrico Gratton, Fabio Beltram, and Francesco Cardarelli

3 Photoswitching Fluorophores in Super-Resolution Fluorescence Microscopy 49

Ulrike Endesfelder

4 Image Analysis for Single-Molecule Localization Microscopy 79

Martin Ovesný, Pavel Křížek, Josef Borkovec, Zdeněk Švindrych, and Guy M. Hagen

5 Deconvolution of Nanoscopic Images 99
Hans van der Voort

Section II The Methods

6 Re-Scan Confocal Microscopy 125
Giulia M.R. De Luca, Ronald M.P. Breedlijk, and Erik M.M. Manders

7 Structured Illumination Microscopy 137
Sara Abrahamsson, Graeme Ball, Kai Wicker, Rainer Heintzmann, and Lothar Schermelleh

8 Correlative Nanoscopy: AFM Super-Resolution (STED/STORM) 149
Jenu Chacko, Francesca Cella Zanacchi, Benjamin Harke, Alberto Diaspro, and Claudio Canale

9 Stochastic Optical Fluctuation Imaging 173
Simon Christoph Stein, Anja Huss, Ingo Gregor, and Jörg Enderlein

10 Super-Resolution Two-Photon Excitation Microscopy Utilizing Transmissive Liquid Crystal Devices 189
Kohei Otomo, Terumasa Hibi, Yuichi Kozawa, Sari Ipponjima, Shunichi Sato, and Tomomi Nemoto

11 Fluorescent Nanodiamonds in Biological and Biomedical Imaging and Sensing 215
Metin Kayci, Flavio M. Mor, and Aleksandra Radenovic

Section III Applications in Biology and Medicine

12 Optical Nanoscopy with SOFI 241
Xuanze Chen, Zhiping Zeng, Xi Zhang, Pingyong Xu, and Peng Xi

13 Super-Resolution Fluorescence Microscopy of the Nanoscale Organization of RNAP and DNA in *E. coli* 259
Christoph Spahn, Ulrike Endesfelder, and Mike Heilemann

14 Correlative Live-Cell and Super-Resolution Microscopy and Its Biological Applications 281
Melike Lakadamyali

15 SAX Microscopy and Its Application to Imaging of 3D-Cultured Cells 301

Katsumasa Fujita, Kentaro Mochizuki, and Nicholas I. Smith

16 Quantitative Super-Resolution Microscopy for Cancer Biology and Medicine 321

Alec Peters, Andrew Nickerson, Li-Jung Lin, and Xiaolin Nan

17 Distribution, Organization, and Dynamics of EGF Receptor in the Plasma Membrane Studied by Super-Resolution Imaging 357

Susana Rocha, Herlinde De Keersmaecker, Eduard Fron, Mitsuhiro Abe, Asami Makino, Hiroshi Uji-i, Johan Hofkens, Toshihide Kobayashi, Atsushi Miyawaki, and Hideaki Mizuno

18 Super-Resolution Microscopy: The Dawn of a New Era in Cardiovascular Research 383

Marc A.M.J. van Zandvoort

Index 415

Series Preface

A picture is worth a thousand words.

This proverb says everything. Imaging began in 1021 with the use of a pinhole lens in a camera in Iraq; later in 1550, the pinhole was replaced by a biconvex lens developed in Italy. This mechanical-imaging technology migrated to chemical-based photography in 1826 with the first successful sunlight picture made in France. Today, digital technology counts the number of light photons falling directly on a chip to produce an image at the focal plane; this image may then be manipulated in countless ways using additional algorithms and software. The process of taking pictures ("imaging") now includes a multitude of options—it may be either invasive or noninvasive, and the target and details may include monitoring signals in two, three, or four dimensions.

Microscopes are an essential tool in imaging used to observe and describe protozoa; bacteria; spermatozoa; and any kind of cells, tissues, or whole organisms. Pioneered by Antonie van Leeuwenhoek in the 1670s and later commercialized by Carl Zeiss in 1846 in Jena, Germany, microscopes have enabled scientists to better grasp the often misunderstood relationship between microscopic and macroscopic behavior, by allowing for the study of the development, organization, and function of unicellular and higher organisms, as well as structures and mechanisms at the microscopic level. Further, the imaging function preserves temporal and spatial relationships that are frequently lost in traditional biochemical techniques and gives two- or three-dimensional resolution that other laboratory methods cannot. For example, the inherent specificity and sensitivity of fluorescence, the high temporal, spatial, and three-dimensional resolution that is possible, and the enhancement of contrast resulting from the detection of an absolute rather than relative signal (i.e., unlabeled features do not emit) are several advantages of fluorescence techniques. In addition, the plethora of well-described spectroscopic techniques providing different types of information and the commercial availability of fluorescent probes such as visible fluorescent proteins (many of which exhibit an environment- or analytic-sensitive response) increase the range of possible applications such as the development of biosensors for basic and clinical research. Recent advancements in optics; light sources; digital imaging systems; data acquisition methods; and image enhancement, analysis, and display methods have further broadened the applications in which fluorescence microscopy can be applied successfully.

Another development has been the establishment of multiphoton microscopy as a three-dimensional imaging method of choice for studying biomedical specimens from single cells to whole animals with submicron resolution. Multiphoton microscopy methods utilize naturally available endogenous fluorophores—including NADH, TRP, and FAD—whose autofluorescent properties provide a label-free approach. Researchers may then image various functions and organelles at molecular levels using two-photon and fluorescence lifetime imaging (FLIM) microscopy to distinguish normal from cancerous conditions. Other widely used nonlabeled imaging methods are coherent anti-Stokes Raman scattering spectroscopy (CARS) and stimulated Raman scattering (SRS) microscopy, which allow imaging of molecular function using the molecular vibrations in cells, tissues, and whole organisms. These techniques have been widely used in gene therapy, single molecule imaging, tissue engineering, and stem cell research. Another nonlabeled method is harmonic generation (SHG and THG), which is also widely used in clinical imaging, tissue engineering, and stem cell research. There are many more advanced technologies developed for cellular and clinical imaging including multiphoton tomography, thermal imaging in animals, and ion imaging (calcium, pH) in cells.

The goal of this series is to highlight these seminal advances and the wide range of approaches currently used in cellular and clinical imaging. Its purpose is to promote education and new research across a broad spectrum of disciplines. The series emphasizes practical aspects, with each volume focusing on a particular theme that may cross various imaging modalities. Each title covers basic to advanced imaging methods, as well as detailed discussions dealing with interpretations of these studies. The series also provides cohesive, complete state-of-the-art, and cross-modality overviews of the most important and timely areas within cellular and clinical imaging.

Since my graduate student days, I have been involved and interested in multimodal imaging techniques applied to cellular and clinical imaging. I have pioneered and developed many imaging modalities throughout my research career. The series manager, Ms Luna Han, recognized my genuine enthusiasm and interest to develop a new book series on Cellular and Clinical Imaging. This project would not have been possible without the support of Luna. I am sure that all the volume editors, chapter authors, and myself have benefited greatly from her continuous input and guidance to make this series a success.

Equally important, I personally thank the volume editors and the chapter authors. This has been an incredible experience working with colleagues who demonstrate such a high level of interest in educational projects, even though they are all fully occupied with their own academic activities. Their work and intellectual contributions based on their deep knowledge of the subject matter will be appreciated by everyone who reads this book series.

Ammasi Periasamy
University of Virginia

Preface

Super-resolution microscopy is leading to important new advances in biomedicine and in those fields dealing with the study of biological systems. In many ways, it is the heir to two-photon microscopy, which itself represented a big revolution over a decade ago. We have put together this book to showcase these new advances and indicated how they are enabling increased spatial resolution in subnanometer scale images of fixed and living cells and tissues. This allows us to obtain new views and new information about molecules, pathways, and dynamics in biological systems. Among the key techniques in this book one can find stimulated emission depletion (STED), structured illumination microscopy (SSIM), photoactivated localization microscopy (PALM), and stochastic optical reconstruction microscopy (STORM).

Currently there are many commercialized super-resolution imaging techniques, opening avenues for practical biomedical applications. In spite of this, biomedical applications are only slowly developing, most probably due to unawareness of the enormous potential or uncertainty about probes that can be used. Furthermore, it is not so easy to decipher what is going on in nature, switching from diffraction limited information to information on the molecular scale, having the possibility to go toward "unlimited" spatial resolution. This book is intended to break this preconceptual barrier and to push researchers to utilize the new microscopes available today, highlighting the advantages and disadvantages of each of the techniques in order to empower readers to make informed choices about which technique to employ for their specific problem. It also gives examples of biomedical applications using commercial microscopes. The unifying theme of this book is *how to bridge the resolution gap with electron microscopy for a biological question using a tunable spatial resolution*. We stress the idea that a key step in getting super-resolution lies, more than in the past, in sample preparation procedures.

The first part of the book starts with a general introduction to super-resolution microscopy, followed by a chapter demonstrating that, even in a standard fluorescence microscope, subdiffraction information can be obtained. Next, the needed fluorophores are discussed, after which two important chapters on image processing follow. The second part describes various current super-resolution techniques, but also discusses correlative and combined techniques, such as

AFM and super-resolution; two-photon and super-resolution; and fiber optical endoscopy. Finally, the third part covers applications of super-resolution microscopy in current biomedical research.

We organized and edited this book to offer an effective support toward educational and decisional aspects both to the beginners and to the advanced researchers. We hope that after finishing this book, the reader will at least take back home a new idea. We also worked to answer the central question, "Which is the best super-resolved microscopy technique to answer a specific scientific question?" Of course, we realize that each new finding creates many more new questions. That is what science is all about. The future will teach us what all the freshly generated data mean and how to interpret them, but it is clear that new roads lie ahead. Despite receiving the Nobel Prize, super-resolution techniques are still very young and plenty of excellent scientific work has to be done.

Alberto Diaspro
Istituto Italiano di Tecnologia

Marc A.M.J. van Zandvoort
Maastricht University

MATLAB® is a registered trademark of The MathWorks, Inc. For product information, please contact:

The MathWorks, Inc.
3 Apple Hill Drive
Natick, MA 01760-2098 USA
Tel: +1-508-647-7000
Fax: +1-508-647-7001
E-mail: info@mathworks.com
Web: www.mathworks.com

Editors

Alberto Diaspro is the director of the Department of Nanophysics at the Istituto Italiano di Tecnologia (IIT), Genoa, Italy, deputy director of IIT, and chair of the Nikon Imaging Center at IIT (www.nic.iit.it). He is a professor of applied physics and supervisor for the PhD courses in the bioengineering and robotics and physics programs in the Department of Physics, the University of Genoa, Italy. He was the president of Optics with Life Sciences (OWLS) and European Biophysical Societies' Association (EBSA) and vice president of International Commission of Optics (ICO). He is a founder of the Nanoscale Biophysics Subgroup of the Biophysical Society. During the 1990s, he carried out part of his research activity at Drexel University, Philadelphia, Pennsylvania; Universidad Autonoma de Madrid, Spain; and Czech Academy of Sciences, Prague, Czech Republic. He also coordinated a research program (2004–2012) at the IFOM-IEO Campus in Milano, Italy, on Biomedical Research and is currently associated with the Institute of Biophysics of the National Research Council (CNR) since 2006. He founded Laboratory for Advanced Microscopy, Bioimaging and Spectroscopy (LAMBS; www.lambs.it) in 2003. He constructed a hybrid artificial "nanobiorobot" as part of European Union and national research projects (2000–2005), and designed and developed the first Italian multiphoton microscope with a research grant of the National Institute of Physics of Matter (1998). He directed the design and development of the first Italian nanoscopy architecture at the Neuroscience and Brain Technologies Department of IIT in 2008. At present, he coordinates the Nanobiophotonics IIT Research Program and is PI of the Nanoscopy Research Team. He is the coordinator of several EU and national research programs, and published more than 300 international peer reviewed papers, 7000 citations, H = 39 (source Google Scholar). He is the editor-in-chief of the Wiley international journal *Microscopy Research and Technique* and an active member of international editorial boards and societies (SIOF, SIF, SISM, SIBPA, BS, EBSA, OWLS, IEEE, SPIE, and OSA). He is an IEEE senior member, OSA senior member, and SPIE fellow. He received the Emily M. Gray Award in 2014. He is the president of the Scientific Council of "Festival of Science" (www.festivalscienza.it).

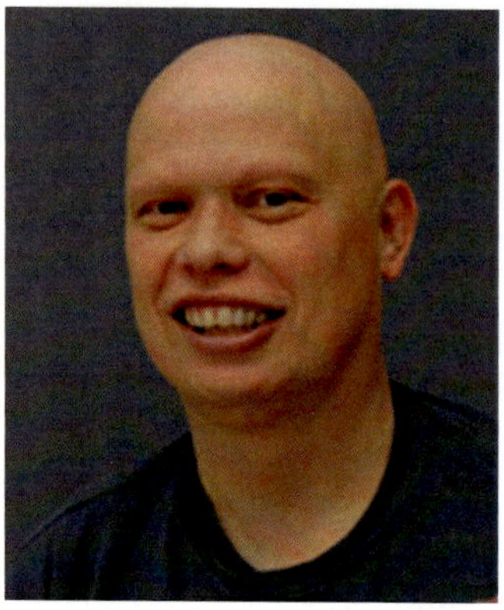

Marc A. M. J. van Zandvoort earned his PhD in biophysics at Utrecht University, Netherlands, in 1994. He then became a postdoc at Universitá di Bologna, Italy, and subsequently obtained a grant from the European Community for a postdoctoral position at LENS Institute, Florence, Italy. In 1996, he returned to the Netherlands, where he started as a postoctoral position at Utrecht University and at Maastricht University, Netherlands, in 1998. In 2000, he joined the Department of Biophysics, Maastricht University as an assistant professor and became an associate professor in the Department of Biomedical Engineering at the same university in 2007. In 2009, he obtained a part-time position at the Universitätsklinikum Aachen, Germany, as a professor of biophysics of microscopy and the leader of the Two-Photon Core Facility, a position he still holds. In 2012, he moved to the Department of Genetics and Cell Biology at Maastricht University. Currently, he is the leader of the Two-Photon Core Facility in Aachen and expert in Advanced Optical Microscopy at the CORE lab in Maastricht. He is the editor of *PLoS One* and a regular reviewer of several journals. He organized conferences such as the Congress Focus on Microscopy 2013 and the yearly Triple M symposium on Advanced Microscopy in Maastricht. He is a representative for Maastricht in the NL-BioImaging-Advanced Microscopy initiative. His recent expression of interest for EuroBioImaging was granted the qualification "highest priority," partly because of the strong international network and collaboration between Maastricht and Aachen. He is also a member of the Enabling Technologies project group, a commercial initiative supported by Maastricht University, DSM, and the Province of Limburg, to create an advanced imaging platform at Maastricht University. Finally, he is actively involved in various German microscopic DFG proposals and networks. He has published more than 100 papers in both physics and biomedical journals and in addition gives speeches regularly (over 50 in the past years) at conferences and courses. His research currently focuses on the applications of advanced microscopy (multiphoton, high-resolution STED microscopy, and correlative light electron microscopy [iCORR]) to biomedical cardiovascular, metabolic, and brain research questions. He has recently developed a method to measure plaque microvessel functionality using *in vivo* multiphoton microscopy in mouse carotid arteries.

Contributors

Mitsuhiro Abe
Lipid Biology Laboratory
RIKEN Advanced Science Institute
Saitama, Japan

Sara Abrahamsson
Lulu and Anthony Wang Laboratory of Neural Circuits and Behavior
The Rockefeller University
New York, New York

Graeme Ball
College of Life Sciences
University of Dundee
Dundee, United Kingdom

Fabio Beltram
Center for Nanotechnology Innovation @NEST
Istituto Italiano di Tecnologia
and
NEST
Scuola Normale Superiore and Istituto Nanoscienze—CNR
Pisa, Italy

Paolo Bianchini
Nanoscopy and Nikon Imaging Center
Istituto Italiano di Tecnologia
Genoa, Italy

Josef Borkovec
First Faculty of Medicine
Institute of Cellular Biology and Pathology
Charles University in Prague
Prague, Czech Republic

Ronald M.P. Breedlijk
Van Leeuwenhoek Centre for Advanced Microscopy
Molecular Cytology
Swammerdam Institute for Life Sciences
University of Amsterdam
Amsterdam, the Netherlands

Claudio Canale
Nanophysics
Istituto Italiano di Tecnologia
Genoa, Italy

Francesco Cardarelli
Center for Nanotechnology Innovation @NEST
Istituto Italiano di Tecnologia
Pisa, Italy

Jenu Chacko
Nanophysics
Istituto Italiano di Tecnologia
Genoa, Italy

and

Department of Biomedical Engineering
University of California, Irvine
Irvine, California

Xuanze Chen
Department of Biomedical Engineering, College of Engineering
Bejing University
Beijing, China

Herlinde De Keersmaecker
Molecular Imaging and Photonics
Department of Chemistry
Katholieke Universiteit Leuven
Heverlee, Belgium

Giulia M.R. De Luca
Van Leeuwenhoek Centre for Advanced Microscopy
Molecular Cytology
Swammerdam Institute for Life Sciences
University of Amsterdam
Amsterdam, the Netherlands

Carmine Di Rienzo
Center for Nanotechnology Innovation @NEST
Istituto Italiano di Tecnologia
and
NEST
Scuola Normale Superiore and Istituto Nanoscienze—CNR
Pisa, Italy

Alberto Diaspro
Nanoscopy and Nikon Imaging Center
Istituto Italiano di Tecnologia
and
Department of Physics
University of Genoa
Genoa, Italy

Jörg Enderlein
Institute of Physics
Georg August University
and
Bernstein Center for Computational Neurobiology
and
DFG Research Center 'Nanoscale Microscopy and Molecular Physiology of the Brain'
Göttingen, Germany

Ulrike Endesfelder
Max Planck Institute for Terrestrial Microbiology
Department of Systems and Synthetic Microbiology
and
LOEWE Center for Synthetic Microbiology
Marburg, Germany

Eduard Fron
Molecular Imaging and Photonics
Department of Chemistry
Katholieke Universiteit Leuven
Heverlee, Belgium

Katsumasa Fujita
Department of Applied Physics
Osaka University
Osaka, Japan

Lauretta Galeno
Nanoscopy and Nikon Imaging Center
Istituto Italiano di Tecnologia
Genoa, Italy

Enrico Gratton
Laboratory for Fluorescence Dynamics
Department of Biomedical Engineering
University of California, Irvine
Irvine, California

Ingo Gregor
Institute of Physics
Georg August University
Göttingen, Germany

Guy M. Hagen
UCCS Center for the BioFrontiers Institute
University of Colorado at Colorado Springs
Colorado Springs, Colorado

Benjamin Harke
Abberior Instruments GmbH
Göttingen, Germany

Mike Heilemann
Institute of Physical and Theoretical Chemistry
Goethe-University Frankfurt
Frankfurt, Germany

Rainer Heintzmann
Institute of Physical Chemistry
and
Institute of Photonic Technology
Microscopy Research Department
Friedrich Schiller University Jena
Jena, Germany

Terumasa Hibi
Research Institute for Electronic Science
and
Graduate School of Information Science and Technology
Hokkaido University
Sapporo, Japan

Johan Hofkens
Molecular Imaging and Photonics
Department of Chemistry
Katholieke Universiteit Leuven
Heverlee, Belgium

Anja Huss
Institute of Physics
Georg August University
Göttingen, Germany

Sari Ipponjima
Graduate School of Information Science and Technology
Hokkaido University
Sapporo, Japan

Metin Kayci
Laboratory of Nanoscale Biology
Institute of Bioengineering
School of Engineering
Ecole Polytechnique Fédérale de Lausanne
Lausanne, Switzerland

Toshihide Kobayashi
Lipid Biology Laboratory
RIKEN Advanced Science Institute
Saitama, Japan

Yuichi Kozawa
Institute of Multidisciplinary Research for Advanced Materials
Tohoku University
Sendai, Japan

Pavel Křížek
First Faculty of Medicine
Institute of Cellular Biology and Pathology
Charles University in Prague
Prague, Czech Republic

Melike Lakadamyali
ICFO—Institut de Ciencies Fotoniques
Barcelona, Spain

Luca Lanzanò
Nanoscopy
Istituto Italiano di Tecnologia
Genoa, Italy

Li-Jung Lin
Department of Biomedical Engineering
Knight Cancer Institute
and
OHSU Center for Spatial Systems Biomedicine
Oregon Health and Science University
Portland, Oregon

Asami Makino
Lipid Biology Laboratory
RIKEN Advanced Science Institute
Saitama, Japan

Erik M.M. Manders
Van Leeuwenhoek Centre for Advanced Microscopy
Molecular Cytology
Swammerdam Institute for Life Sciences
and
Nikon Centre of Excellence on Super Resolution Microscopy Development
University of Amsterdam
Amsterdam, the Netherlands

Atsushi Miyawaki
Laboratory for Cell Function Dynamics
RIKEN Brain Science Institute
Saitama, Japan

Hideaki Mizuno
Laboratory of Biomolecular Network Dynamics
Biochemistry, Molecular and Structural Biology Section
Department of Chemistry
Katholieke Universiteit Leuven
Heverlee, Belgium

Kentaro Mochizuki
Graduate School of Life Sciences
Tohoku University
Sendai, Japan

and

Osaka University
Osaka, Japan

Flavio M. Mor
Laboratory of Nanoscale Biology
Institute of Bioengineering
School of Engineering
Ecole Polytechnique Fédérale de Lausanne
Lausanne, Switzerland

Xiaolin Nan
Department of Biomedical Engineering
Knight Cancer Institute
and
OHSU Center for Spatial Systems Biomedicine
Oregon Health and Science University
Portland, Oregon

Tomomi Nemoto
Research Institute for Electronic Science
and
Graduate School of Information Science and Technology
Hokkaido University
Sapporo, Japan

Andrew Nickerson
Department of Biomedical Engineering
Knight Cancer Institute
and
OHSU Center for Spatial Systems Biomedicine
Oregon Health and Science University
Portland, Oregon

Kohei Otomo
Research Institute for Electronic Science
Hokkaido University
Sapporo, Japan

Martin Ovesný
First Faculty of Medicine
Institute of Cellular Biology and Pathology
Charles University in Prague
Prague, Czech Republic

Alec Peters
Department of Biomedical Engineering
Knight Cancer Institute
and
OHSU Center for Spatial Systems Biomedicine
Oregon Health and Science University
Portland, Oregon

Aleksandra Radenovic
Laboratory of Nanoscale Biology
Institute of Bioengineering
School of Engineering
Ecole Polytechnique Fédérale de Lausanne
Lausanne, Switzerland

Susana Rocha
Molecular Imaging and Photonics
Department of Chemistry
Katholieke Universiteit Leuven
Heverlee, Belgium

Shunichi Sato
Institute of Multidisciplinary Research for Advanced Materials
Tohoku University
Sendai, Japan

Lothar Schermelleh
Department of Biochemistry
University of Oxford
Oxford, United Kingdom

Colin J.R. Sheppard
Nanoscopy and
Molecular Microscopy and Spectroscopy
Istituto Italiano di Tecnologia
Genoa, Italy

Nicholas I. Smith
Biophotonics Laboratory
Immunology Frontier Research Center
Osaka University, Suita, Japan

Christoph Spahn
Institute of Physical and Theoretical Chemistry
Goethe-University Frankfurt
Frankfurt, Germany

Simon Christoph Stein
Institute of Physics
Georg August University
Göttingen, Germany

Zdeněk Švindrych
Department of Biology
W. M. Keck Center for Cellular Imaging
University of Virginia Charlottesville
Charlottesville, Virginia

Hiroshi Uji-i
Molecular Imaging and Photonics
Department of Chemistry
Katholieke Universiteit Leuven
Heverlee, Belgium

Hans van der Voort
Scientific Volume Imaging
Hilversum, the Netherlands

Marc A.M.J. van Zandvoort
School for Cardiovascular Diseases CARIM
Department of Genetics and Cell Biology Section
Molecular Cell Biology
Maastricht University
Maastricht, the Netherlands

and

Institute for Molecular Cardiovascular Research
Universitatsklinikum Aachen
Aachen, Germany

Giuseppe Vicidomini
Molecular Microscopy and Spectroscopy
and
Nikon Imaging Center
Istituto Italiano di Tecnologia
Genoa, Italy

Kai Wicker
Institute of Physical Chemistry
Friedrich Schiller University Jena
Jena, Germany

Peng Xi
Department of Biomedical Engineering
Bejing University Hospital
Bejing, China

Pingyong Xu
Institute of Biophysics
Chinese Academy of Sciences
Beijing, China

Francesca Cella Zanacchi
Nanoscopy and Nikon Imaging Center
Istituto Italiano di Tecnologia
Genoa, Italy

and

The Institute of Photonics Sciences
Barcelona, Spain

Zhiping Zeng
Institute of Biophysics
Chinese Academy of Sciences
Beijing, China

Xi Zhang
Institute of Biophysics
Chinese Academy of Sciences
Beijing, China

Section I

Overview of Super-Resolution Technologies

1

In the Realm of Super-Resolved Fluorescence Microscopy

Alberto Diaspro, Paolo Bianchini, Giuseppe Vicidomini, Colin J.R. Sheppard, Luca Lanzanò, Francesca Cella Zanacchi, and Lauretta Galeno

1.1 Introduction
1.2 Optical Microscopy and Spatial Resolution
1.3 Super-Resolution
1.4 Circumventing the Diffraction Barrier
1.5 Super-Resolved Ensemble Fluorophore Microscopy: Targeted Read-Out Approach
1.6 Super-Resolved Single Fluorophore Microscopy—Stochastic "Read-Out" Approach
1.7 Nanoscopium Nominare Libuit
Acknowledgments
References

"Don't believe what your eyes are telling you. All they show is limitation. Look with your understanding."

Richard Bach
Jonathan Livingston Seagull

1.1 Introduction

As brilliantly discussed by Richard Feynman in his seminal lecture at an American Physical Society meeting at Caltech on December 29, 1959, "There is a plenty of room at the bottom." "[...] in no other field than biology are scientists making more rapid progress. [...] and to answer questions relating to the most central and fundamental problems of biology [...] we should make the ... microscope 100 times better" [1]. The tremendous progress in the biological sciences over recent decades has been somewhat facilitated through the development of "light" technologies that we can consider as bio-imaging oriented, as in the case of confocal and two-photon excitation microscopy [2]. In particular, in the field of fluorescence microscopy there is continuing parallel progress [3,4]. So, light technology in biology is more than just optics. We aim to see the sharpest details in living systems starting from our "obsession" for those advances relating to spatial and temporal resolution coupled with the intrinsic property of visible light to produce moderate and controllable perturbation on biological systems, say, under the microscope.

At least, since the publication of the seminal works by Gabor [5] and Giuliano Toraldo di Francia [6] we know that resolution is not a fundamental limit, but instead the fundamental limit is set by concepts of information theory. The term super-resolution is, however, somewhat confused. It is not the aim of this communication to clarify this issue that has been brilliantly discussed in a comparatively recent paper [7–9]. Here, we are pleased to introduce some general aspects relating to the topic "the development of super-resolved fluorescence microscopy," for which the Royal Swedish Academy of Sciences awarded Eric Betzig, Stefan W. Hell, and William E. Moerner the Nobel Prize in Chemistry 2014 [10,11].

The impact of such a development is extremely relevant for scientists who are now able to address important biological and biophysical questions at the single molecule level in living cells. For instance, the way individual molecules form synapses across the brain neural network, or proteins aggregate when developing Parkinson's, Huntington's, or Alzheimer's diseases, or nuclear pore complexes and chromatin–DNA spatially organize themselves when developing a cancerous disease are only a few of the new case studies experienced under a new perspective.

In the past 40 years, optical fluorescence microscopy, due to its inherent ability of imaging living systems during their temporal evolution, had a continuous update on three main tracks, namely, three-dimensional (3D) imaging, penetration depth at low perturbation, and resolution improvements. However, computational optical sectioning, confocal laser scanning, two-photon excitation, and super-resolved methods can be considered as milestones in optical microscopy.

It is evident that fluorescence microscopy has become an indispensable tool in cell biology because of its unique advantages: it is a largely noninvasive technique and it can probe the deeper layers of a specimen at ambient conditions and enables spectroscopic diagnosis with chemical sensitivity. The use of fluorescence is a key process: molecules of interest, properly labeled, produce the main signal at the basis of the image formation process (Figure 1.1) [3,12].

More generally, optical microscopy had to tackle a physical limitation: using glass lenses the spatial resolution is set to approximately half the wavelength of light being used. Demonstration of the possibility of circumventing such a limit to study biological systems has been given by the Nobel laureates in Chemistry 2014 (www.nobelprize.org) utilizing fluorescent molecules. This avant-garde result has driven optical microscopy to the nanoscale dimension [4].

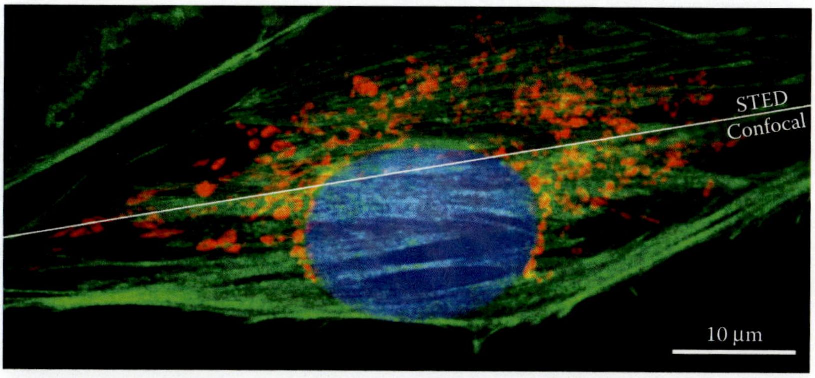

Figure 1.1

Representative fluorescence image of a biological cell. Due to high biochemical affinity for compartments, blue, red, and green are related to DNA, mitochondria, and actin filaments, respectively. In this image, the STED and confocal modes are reported.

1.2 Optical Microscopy and Spatial Resolution

The optical microscope, using "round-shaped glasses," allows seeing better details than the ones one can get with the naked eye. Even if its use is still controversial, the Nimrud lens or Layard lens is a 3000-year piece of rock crystal, discovered at the Assyrian palace of Nimrud, that may have been used as a magnifying glass, or as a burning glass also being considered as a piece of decorative inlay. However, both Seneca the Younger (3 BCE–65) and Gajus Plinius (23–79 CE) described the magnification properties of a water-filled spherical glass container. Later, the English philosopher Roger Bacon in 1267, also influenced by al-Kindī (801–873) known as "the Philosopher of the Arabs," argued how "[We] may number the smallest particles of dust and sand by reason of the greatness of the angle under which we may see them." In 1538, the Italian physician Girolamo Fracastoro reported that "If anyone should look through two spectacle glasses, one being superimposed on the other, he will see everything much larger." Around 1590 three Dutch opticians—Hans and Zacharias Jansen, and Hans Lippershey—invented the compound microscope. Anyway, Johannes Faber, fellow of the Accademia dei Lincei, in 1625, in a letter to Federico Cesi, conceived the term microscope writing the milestone sentence *microscopium nominare libuit* referred to Galileo Galilei's "small eye-glass" (*occhialino*). At some time before 1668, Antonie van Leeuwenhoek learned to grind lenses, made simple microscopes, and began observing with them (Figure 1.2). In addition, a quantitative leap in optical microscopy came in 1873, when the physics of lens construction was examined by German physicist Ernst Abbe. The most popular relationship to address the spatial resolution performances of an optical microscope in the observation plane (x–y) is the so-called Abbe's formula that sets as the spatial resolution limit one half the wavelength and the wavelength for the lateral and axial one, respectively.

A more precise and formal approach for spatial resolution can be based on the point spread function (PSF) that for a space-invariant linear system is related to

Figure 1.2

Replica of the Antonie van Leeuwenhoek microscope within the scenario of a modern microscopy lab.

Dirac's impulse response. Similarly, we could switch to a description in terms of communication channels dealing with spatial frequencies in the Fourier space [13].

The cut-off frequency is set with the inverse of the Abbe limit:

$$w_{\text{cut-off}} = 2\pi/\text{d} = 2\pi(2n\sin\alpha)/\lambda \tag{1.1}$$

where:

d is the minimum recognizable separation distance between two points
λ is the wavelength
n is the refractive index of the medium between the lens and the object
α is the semi-aperture angle of the lens being used

As a consequence of the convolution between the point source and the PSF, the Fourier transform of the image transmitted by the microscope is band-limited. The lack of high frequencies in the formed image is the reason for the limited resolution. It is worth noting that since Abbe considered a fine grating placed on an optical microscope for his lens study, the Abbe formula brings a resolution definition referred to the object's spatial frequencies that fits well with the Rayleigh definition.

1.3 Super-Resolution

Now, resolution is not a fundamental limit. The limit is set by concepts of information theory [6,8,14,15]. The term "super-resolution" is sometimes utilized outside rigorous treatments. Lukosz proposed that all super-resolution approaches can be treated using the theorem of invariance of the number of degrees of freedom of an image formation system. Cox and Sheppard combined Felgett and Linfoot and Lukosz results in terms of information capacity [16–18]. The key point lies in the fact that, considering confocal imaging, the case of super-resolution by analytic continuation was treated. Super-resolution can be used to describe a number of fundamentally different schemes. One class is related to sharpening up the PSF while the spatial bandwidth does not change. This is the Toraldo di Francia [19] approach related to super-gain antennas and super-resolving pupils. Resolution is increased at the expense of the field of view. This class includes digital deconvolution. Another class, called restricted super-resolution, is based on modulation/demodulation (coding/decoding) and multiplexing. A third interesting class, unrestricted super-resolution, allows increasing the cut-off frequency without a fundamental restriction. Today, we use the term super-resolution in a broad sense to indicate the ability of discerning details below the Abbe limit and in the very special case of fluorescence as predicted by McCutchen in 1967 [20]. The revolutionary aspect lies in the fact that, with the methods developed by Eric Betzig, Willy Moerner, and Stefan Hell, there is theoretically no limit and at the same time there is the possibility of tuning the spatial resolution according to the scientific question posed [4,21].

1.4 Circumventing the Diffraction Barrier

Since the advent of confocal and two-photon excitation laser scanning microscopy [2], merging the concept of reducing the field of view and of improving the signal-to-noise ratio, improvements in spatial resolution have been achieved at the expense of time. Resolution improvements have been made with confocal [22,23] and two-photon excitation microscopy [24–26]. These methods and related advances have been a relevant scientific and experimental basis for the development of super-resolution approaches.

In general, the image formed by the modern super-resolved fluorescence microscopes does not come from a "multipoint" one-shot picture recorded on a photosensor but is something like a pointillist painting made using a "super fine" brush. The key to success comes from the merging of optics with photophysics/chemistry of fluorescent molecules. In the far field, resolution improvements have been achieved with confocal and two-photon excitation microscopy. The suppression of fluorescence background, since the very first attempts in spectroscopy, is a common aspect and is a key factor especially when imaging a real 3D specimen from the single cell to organs and tissues. In order to improve resolution and to make imaging spatially isotropic, important improvements in the axial resolution have been realized by means of 4PI and I5M [27].

Structured illumination microscopy (SIM) allowed a further improvement of Abbe's limit by using interference patterns at low photon budget for the specimens. Figure 1.3 shows an example.

However, in their initial implementations such methods remained confined by Abbe's limit of resolution rule. In the near field, resolution has been successfully attacked by different implementations with the main drawback of being limited to surfaces [28].

The main idea behind the implementation of super-resolved fluorescence microscopes is based on preventing the simultaneous emission of adjacent spectrally identical fluorophores. The most successful approaches for getting super-resolution in fluorescence optical microscopy have been classified as "super-resolved ensemble fluorophore microscopy" or targeted readout methods and "super-resolved single fluorophore microscopy" or stochastic readout method, respectively. Now, the implementation of the first principle, related to the S. W. Hell research, for getting super-resolution in the far field, dates back to the original work by Hell and Wichmann [29]. It is known as STED (stimulated emission depletion) and utilizes the general concept of RESOLFT

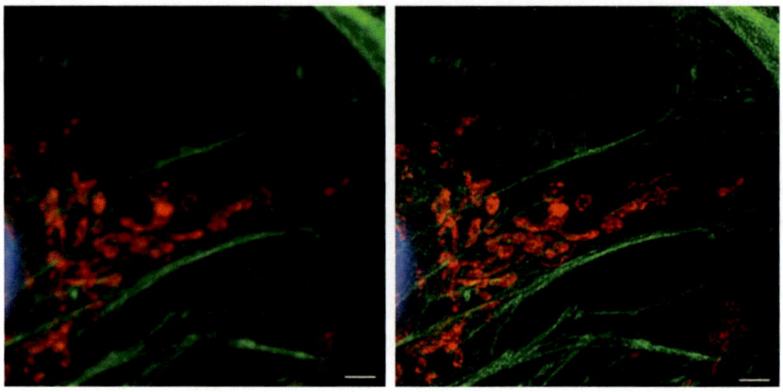

(a) (b)

Figure 1.3

Comparison between conventional wide-field microscopy and 3D-SIM taken using a N-SIM Nikon super-resolution microscope at NIC@IIT. Imaging of BPAE cell with Mitotracker Red CMXRos (red), Alexa Fluor 488 phalloidin (green), and DAPI nucleus (blue) with wide-field microscopy (a) and 3D-SIM (b). The 3D-SIM image has been collected using the 3D-SIM mode which allows for optical sectioning. Nucleus structure in both panels has been collected in conventional wide-field mode. Scale bar 5 μm.

(reversible saturable optical linear fluorescence transitions) [30]. Considering an ensemble of molecules excited by a diffraction-limited light beam, STED of fluorescence affects all molecules in the sample except those in a very small controlled region. Such a region can be arbitrary smaller than the diffraction-limited excitation spot by saturating stimulated emission. An image, as for the confocal or two-photon excitation microscopy case, can be formed by scanning the light spot defining the fluorescing region across the sample. Gustafsson [31] implemented this principle in the so-called saturable SIM (SSIM). His implementation is very bright and effective and allows the production of super-resolved 3D images of living systems. Both methods, STED and SSIM, produce spatial frequencies higher than Abbe's cut-off one. The *a priori* information linked to the STED method lies in the knowledge of the position of the scanning beam.

The second principle, related to the E. Betzig and W. E. Moerner research, is implemented considering the *a priori* information that photons detected from a sample at a certain time come from individual fluorescent molecules that are spatially located in a sparse way, that is, are spatially separated by distances higher than the resolution limit. This fact allows the determination with a very high precision of the position of the individual molecules. It is self-evident how, in both cases, the spatial resolution is improved at the expense of time, as in confocal and two-photon excitation laser scanning microscopy. It is worth noting that super-resolution with SSIM techniques have been demonstrated by Gustafsson [31].

1.5 Super-Resolved Ensemble Fluorophore Microscopy: Targeted Read-Out Approach

At the beginning of the 1990s, Stefan Hell quantitatively demonstrated the experimental conditions for the novel concept of STED microscopy [29,32]. Then he developed a fluorescence microscope to provide the experimental proof-of-principle of the functioning of STED microscopy.

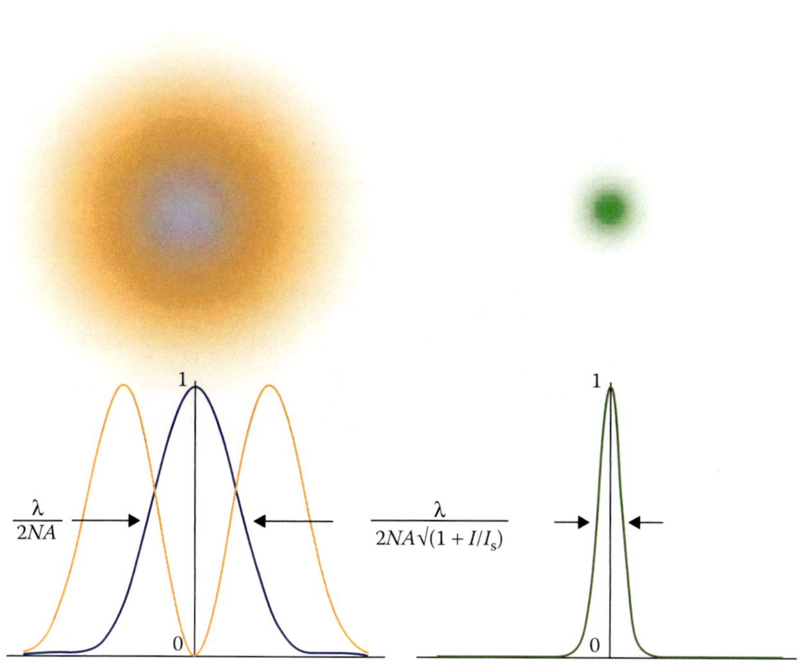

Figure 1.4

The depletion process performed through the doughnut beam induces a sharpening of the PSF.

Two laser beams are used in a scanning scheme. The former is used to excite fluorophores in the sample within a diffraction-limited spatial region (circular shape), the latter is typically red-shifted with respect to the first beam and provides a high-intensity profile endowed with a zero intensity minimum in the focal region that grows in all directions from the focus (donut shape). The role of the second beam, usually called STED beam, is to rapidly "push" fluorophores that have been excited by the first beam to the ground state, passing for a high vibrational energy state of the electronic ground state. By optimizing beam characteristics and pulse sequences for the excitation and STED beams, fluorescence excited by the first beam is extinguished everywhere except in the region close to the zero of the STED beam. Increasing the STED beam intensity (I) makes this region shrink considerably as the probability of depletion grows (Figure 1.4).

The resulting extension of the fluorescing region in the lateral plane scales like

$$d = \lambda/(2n\sin\alpha\ \sqrt{1 + I/I_{sat}}) \tag{1.2}$$

which can be regarded as an extension of Abbe's equation, which approximates the diffraction limit. It follows from this expression that Abbe's limit hampering resolution has been canceled. One can see the sharpening of the fluorescing region as an approaching Dirac's function of the illuminated region.

The diffraction-limited focal spot can be made infinitely small as driven by the ratio between the intensity of the STED laser beam and it is a photophysical property of the molecules under investigation. I_{sat} is the intensity needed for depopulating 50% of the excited state for a specific fluorescent molecule. It is clear that

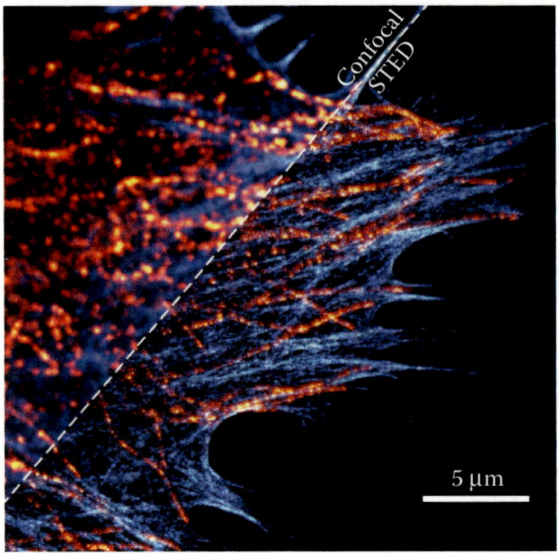

Figure 1.5

Image of a 3T3 cell in which we have labeled actin filaments with Alexa Fluor 488 phalloidin (cyan) and tubulin with Abberior Star 635p (red). The image has been collected sequentially by two Hybrid detectors in the spectral range 500–550 nm and 650–720 nm, respectively. We have used an HCX PL APO CS 100x 1.4 NA oil objective (Leica Microsystems, Mannheim, Germany) and a scan speed of 1400 Hz for 2048 pixel per line with a 64-line average. Confocal resolution (top left) and STED image (bottom right) shows filaments for actin and tubulin. The microscope used is a customized commercial STED set-up, constituted by a commercial Leica TCS SP5 gated STED-CW whereby a pulsed laser at 766 nm (Picoquant LDH-P-FA-766; Picoquant, Berlin, Germany) has been coupled through the Leica multiphoton port, the beam profile has been made into a doughnut shape by a vortex phase plate, and the pulses have been electronically synchronized with the supercontinuum laser source used in the Leica microscope for excitation. In this configuration, we combine two different STED approaches, namely the gCW-STED with pulsed STED implementation. Such a set-up allows multicolor STED nanoscopy with green and far-red emitting fluorescent molecules.

there may be other types of problems, such as photodamage to biological tissues or photobleaching of the fluorescent molecules, due to the utilization of intense STED beams. To alleviate this problem, one can, for example, use the more general concept of RESOLFT by introducing other ground-state depletion mechanisms than stimulated emission, which do not require such high intensity [32]. Moreover, two-photon excitation approaches [33] and time-based concepts for keeping the fluorescing region small at reduced STED beam intensity [34,35] have been recently used. Moreover, 3D super-resolution and a parallelization of the concept have been implemented [36]. Figures 1.5 and 1.6 show two implementations of the STED principles.

It is worth noting that original approaches are growing within the targeted read-out approach; in particular, approaches exploiting the temporal information connected to the process that can alleviate some problems related to the beam intensities needed for getting spatial super-resolution. This can be done in several ways, among them one can use deconvolution schemes [37], temporal photophysical characteristics of the labeling fluorophores [38,39], or more comfortable mechanisms of "depletion" as the ones used in RESOLFT [40]. Alternative schemes are moving to label free super-resolution approaches [41].

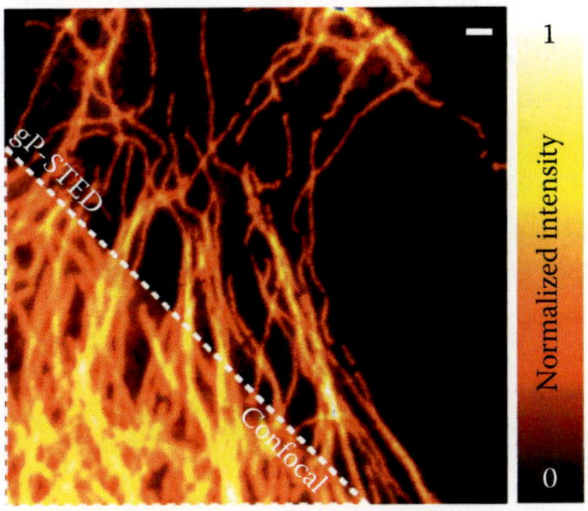

Figure 1.6

Comparison of confocal and gated pulsed STED (gP-STED) imaging with sub-nanosecond STED beam. Imaging of beta-tubulin ATTO647 *N* labeled cell (STED beam power = 39 mW at 80 MHz). Images clearly reveal the enhancement of spatial resolution obtained with STED microscopy. Since, the pulse width (~600 ps) of the sub-nanosecond STED beam is comparable with the fluorescence lifetime (~4 ns), the fluorescence emission generated during the depletion STED is not negligible, and must be filtered by time gating detection to fully explore the resolution and contrast boost. Scale bar 1 μm.

1.6 Super-Resolved Single Fluorophore Microscopy—Stochastic "Read-Out" Approach

To date, starting from Einstein considerations on molecules and their behavior until pioneering development in the so-called fluorescence correlation spectroscopy (FCS) domain explored in the W. W. Webb and R. Rigler Labs, the detection of single molecules in a crowded environment was an important research topic toward super-resolved fluorescence microscopy [42]. The first measurements on a single fluorophore in a dense medium were made in Moerner's Lab [43]. Moerner's Lab results on pentacene molecules decisively inspired the single-molecule field to applications in microscopy [44] by demonstrating that it was possible to deal with single fluorescent molecules both in terms of absorption and emission. His experimental approach allowed the measurement of the fluorescence of a single molecule by taking advantage of the better signal-to-noise ratio improved by absorption-based observations. Other results were reported in those years at room temperature and in liquid medium using confocal microscopy or pulsed excitation coupled to delayed photon detection for background radiation annihilation, respectively. Betzig started his race to study single fluorescent molecules using scanning near-field microscopy on an air-dried surface environment. He achieved super-resolution observations in the near-field [28]. Both single-molecule spectroscopy and single-molecule microscopy were carried out in the successive years including single-fluorescent-molecule tracking studies [44].

Super-resolved single-fluorophore microscopy can be understood as a class of techniques for which spatial super-resolution is achieved by the possibility of localizing with theoretical unlimited precision individual molecules considered as point sources of photons. The most important keyword is connected with the knowledge of the fact that emission stems from single molecules.

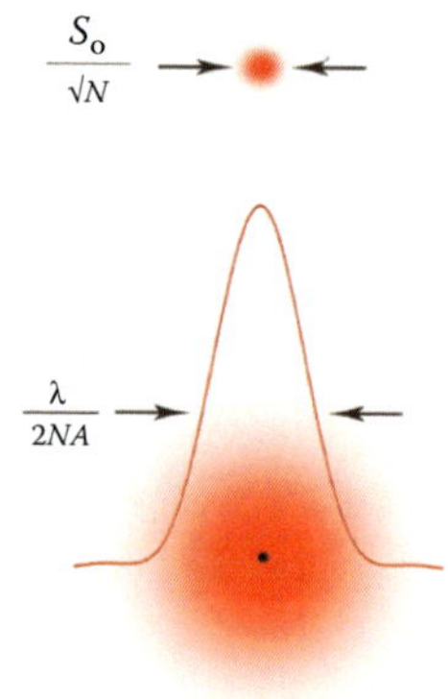

Figure 1.7

Increasing localization precision in a single molecule detection scheme.

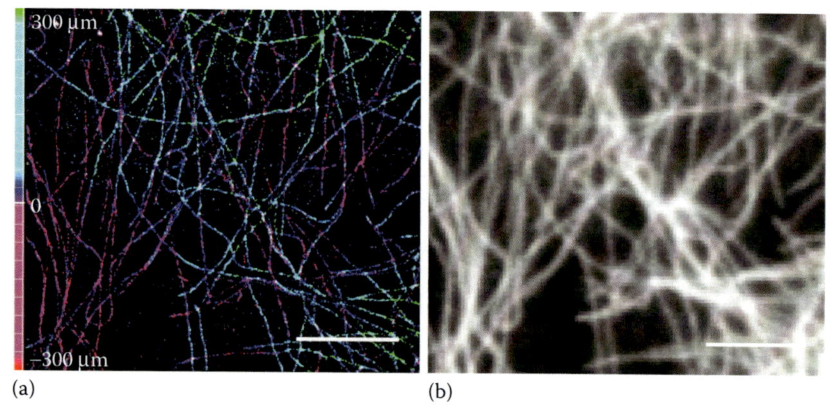

Figure 1.8

Image of cytoskeletal structures (microtubules labeled with Alexa 647) in HeLa cells. (a) 3D storm image of microtubules. (b) Corresponding wide field image. Localization precision: lateral 20 nm and axial 65 nm. Scale bar 5 μm.

Within this framework, the microscope detecting system receives photons from the single molecules embedded in a dense environment with a spatial probability density driven by the microscope PSF. Now, neglecting background and pixelation and considering fluorescent molecules one by one, the center of the PSF can be estimated with an error along both the x and y coordinates, that scales with a factor like the square root of N with respect to the Abbe's resolution limit [45], where N is the total number of photons registered by the detector from a certain single emitting molecule (Figure 1.7).

So far, the ability to localize a point source is improved (Figure 1.8). In fact, this means that with 10,000 detected photons from a single fluorescent molecule one gets a factor of 100 smaller than Abbe's limit and with increasing numbers of detected photons there is no limit for a possible spatial resolution improvement.

Betzig [46] suggested two steps, namely, (1) independent determination of the PSF of each spectral class and (2) super-resolved estimation of the centers of their PSFs with the precision of Equation 1.5. Then, merging the positions of all classes together, a super-resolution image could be realized. Brakenhoff and collaborators [47] implemented a nonoptimized version of this concept. However, the advent of the GFP (green fluorescent protein) pushed the study of the behavior of GFP mutants at single-molecule level. Interesting results were obtained at ambient temperature with the protein embedded in an aerated aqueous gel [48]: a "blinking" behavior was discovered leading the molecules to a final dark stable state from which fluorescence could be restored by irradiation at a different wavelength. Dark and bright states of the molecules could be controlled by light. Patterson and Lippincott-Swartz performed a key experiment demonstrating the photo properties of photoactivatable (PA) GFP variants [49].

The molecules, initially fluorescent inactive, could be activated by proper irradiation and then were able to emit fluorescence when excited at a certain wavelength; as all the fluorescent molecules, such a PA-GFP, could be permanently inactivated by photobleaching. This can be also achieved under two-photon excitation regime, taking advantage of its intrinsic 3D confinement of the process [50].

Betzig realized that such photo properties could be used to optimize the process of having a sparse distribution of fluorescent molecules within a dense population: using a very low photoactivation intensity one has a poor probability of random photoactivation; thus, it causes the generation of a sparse subset of activated molecules. A subsequent excitation of the activated sparse cluster enables the localization of the positions of the activated fluorescent molecules according to the single molecule localization improvement reported in Figure 1.7 [51]. A number of a sparse set of molecules can be activated and permanently inactivated by bleaching until "all" subsets from the dense population of molecules have been sampled and used to form a super-resolved image.

In 2006, a fundamental paper was published [52] reporting how a super-resolved fluorescence microscopy image modality could be applied to a thin (100 nm) section of a fixed mammalian lysosome. The method was named PALM (photoactivated localization microscopy). During the very same year both Zhuang and collaborators [53] and Hess and collaborators [54] reported about super-resolved fluorescence microscopy. The respective approaches were called STORM (stochastic optical reconstruction microscopy) and fPALM (fluorescence-PALM). Moreover, a growing number of papers came out within a comparatively very short time [55]. Imaging in thick samples became an interesting topic as recently noted by Moerner [44] that indicated using a light sheet illumination scheme as a promising method. As an example, IML–SPIM (individual molecule localization—selective plane illumination microscopy) allowed for the first time super-resolved single-fluorophore microscopy applied to a 200 μm thick tumoral spheroid sample [56].

1.7 Nanoscopium Nominare Libuit

The limiting role of diffraction for microscopy is extensively discussed in the famous work by Ernst Abbe in 1873 as reported in [57]. Today we can say that the super-resolved fluorescence microscopy revolution lies in the fact that there is no longer any theoretical limitation to the study of the smallest details in biological cells and in any other sample. As well, this possibility is tunable and can be linked to the biological question requiring a variable increase of spatial resolution to be answered. Super-resolved fluorescence methods are growing in terms of dissemination and continuing evolution [58–60]. A road map and trend can be outlined as function of demand mainly comes from biology, medicine, and biophysics [61,62]. Here, we would like to mention two directions in terms of development. A brand new one, dealing with a string role in terms of sample treatment and preparation is expansion microscopy [63,64] that has the high potential for being efficiently coupled to the "classic" super-resolution methods reported here. Such a combination could provide incredible results. Similarly, the development of a correlative approach coupling super-resolved fluorescence microscopy with atomic force microscopy has the potential of opening new research windows, for example, in nanomechanical explorations [65,66]. Correlative nanoscopy has already found interesting applications in biophysics [67–69].

Many other approaches are under development or developed within the realm of super-resolved fluorescence microscopy. In the style of Johannes Faber, we conclude using the term "nanoscopium nominare libuit" for the super-resolution fluorescence microscope that has become a nanoscope [10].

Acknowledgments

This chapter is adapted and partially reproduced with permission from Ref. [10] Società Italiana di Fisica. The authors are indebted to Eileen Sheppard for proofreading.

References

1. Feynman, R.P., March 1, 1992. There's plenty of room at the bottom (data storage). *J. Microelectromech. S.*, 1(1):60–66. doi:10.1109/84.128057. A reprint of the talk (1959). Free adapted sentence in the text.
2. Diaspro, A. 2001. *Confocal and Two-Photon Microscopy: Foundations, Applications, and Advances*. Wiley-Liss, Wilmington, DE.
3. Mondal, P.P., and A. Diaspro. 2014. *Fundamentals of Fluorescence Microscopy. Exploring Life with Light*. Springer, Dordrecht, the Netherlands.
4. Diaspro, A. 2010. *Nanoscopy and Multidimensional Optical Fluorescence Microscopy*. Chapman & Hall/CRC Press, London.
5. Gabor, D. 1946. Theory of communication. Part 1: The analysis of information. *J. Inst. Electr. Eng. III Radio Comm. Eng.*, 93(26):429. doi:10.1049/ji-3-2.1946.0074.
6. Toraldo di Francia, G. 1955. Resolving power and information. *J. Opt. Soc. Am.*, 45:497.
7. Mertz, J. 2010. *Introduction to Optical Microscopy*. Roberts, Greenwood Village, CO, Chapter 18.
8. Sheppard, C.J.R., 2007. Fundamentals of superresolution. *Micron*, 38:165–169.
9. Bechhoefer, J. 2015. What is superresolution microscopy? *Am. J. Phys.*, 83:22–29.
10. Diaspro, A. 2014. Circumventing the diffraction limit. *Il Nuovo Saggiatore*, 30(5–6):45–51.
11. Möckl, L. D.C. Lamb, and C. Bräuchle. 2014. Super-resolved fluorescence microscopy: Nobel Prize in chemistry 2014 for Eric Betzig, Stefan Hell, and William E. Moerner. *Angew. Chem. Int. Ed. Engl.*, 53:13972–13977.
12. Jameson, D.M., 2014. *Introduction to Fluorescence*. CRC Press, Boca Raton, FL.

13. Goodman, J.W., 1968. *Introduction to Fourier Optics*. McGraw-Hill, New York.
14. Cox, I.J., C.J.R. Sheppard, T. Wilson, 1982. Super-resolution by confocal fluorescent microscopy. *Optik*, 60:391–396.
15. Sheppard, C.J.R., 1984. Optical resolution and the spatial frequency cut-off. *Optik*, 66:311–315.
16. Cox, I.J., and C. Sheppard. 1986. Information capacity and resolution in an optical system. *JOSA A*, 3(8):1152–1158.
17. Felgett, P.B., and E.H. Linfoot. 1955. On the assessment of optical images. *Proc. R. Soc. London Ser. A*, 247:369–407.
18. Lukosz, W., 1966. Optical systems with resolving powers exceeding the classical limit, Part 1. *J. Opt. Soc. Am.*, 56:1463–1472.
19. Toraldo di Francia, G., 1955. Super-resolution. *Opt. Acta.*, 2:5–8.
20. McCutchen, C.W., 1967. Superresolution in microscopy and the Abbe resolution limit. *J. Opt. Soc. Am.*, 57:1190–1192.
21. Schermelleh, L., R. Heintzmann, and H. Leonhardt. 2010. A guide to super-resolution fluorescence microscopy. *J. Cell Biol.*, 190:165.
22. Cremer, C., and T. Cremer. 1978. Considerations on a laser-scanning-microscope with high resolution and depth of field. *Microsc. Acta*, 81:31.
23. Sheppard, C.J., and T. Wilson. 1981. The theory of the direct-view confocal microscope. *J. Microsc.*, 124:107.
24. Sheppard, C.J., and R. Kompfner. 1978. Resonant scanning optical microscope. *Appl. Opt.*, 17:2879.
25. Denk, W., J.H. Strickler, and W.W. Webb. 1990. Two-photon laser scanning fluorescence microscopy. *Science*, 248:73.
26. Hell, S.W., and E.H.K. Stelzer. 1992. Fundamental improvement of resolution with a 4Pi-confocal fluorescence microscope using two-photon excitation. *Opt. Commun.*, 93:277.
27. Gustafsson, M.G.L., D.A. Agard, and J.W. Sedat. 1995. Sevenfold improvement of axial resolution in 3D widefield microscopy using two objective lenses. *Proc. SPIE*, 2412:147.
28. Betzig, E., and J.K. Trautman. 1992. Near-field optics: Microscopy, spectroscopy, and surface modification beyond the diffraction limit. *Science*, 257:189.
29. Hell, S.W., and J. Wichman. 1994. Breaking the diffraction resolution limit by stimulated emission. *Opt. Lett.*, 19:780.
30. Hell, S.W., 2003. Toward fluorescence nanoscopy. *Nat. Biotechnol.*, 21:1347.
31. Gustafsson, M.G., 2005. Nonlinear structured-illumination microscopy: Wide-field fluorescence imaging with theoretically unlimited resolution. *Proc. Natl. Acad. Sci. U.S.A.*, 102:1301.
32. Hell, S.W., and M. Kroug. 1995. Ground-state-depletion fluorscence microscopy: A concept for breaking the diffraction resolution limit. *Appl. Phys. B.*, 60:495.
33. Bianchini, P., B. Harke, S. Galiani, G. Vicidomini, and A. Diaspro. 2012. Single-wavelength two-photon excitation-stimulated emission depletion (SW2PE-STED) superresolution imaging. *Proc. Natl. Acad. Sci. U.S.A.*, 109:6390.
34. Vicidomini, G., et al. 2014. Gated CW-STED microscopy: A versatile tool for biological nanometer scale investigation. *Methods*, 66:124.
35. Harke, B., et al. 2008. Three-dimensional nanoscopy of colloidal crystals. *Nano Lett.*, 8:1309.

36. Chmyrov, A., J. Keller, T. Grotjohann, M. Ratz, E. d'Este, S. Jakobs, C. Eggeling, S. W. Hell. 2013. Nanoscopy with more than 100,000 'doughnuts'. *Nat. Methods*, 10:737.
37. Castello, M., A. Diaspro, and G. Vicidomini. 2014. Multi-images deconvolution improves signal-to-noise ratio on gated stimulated emission depletion microscopy. *Appl. Phys. Lett.*, 105:234106–234115.
38. Hernández, I.C., M. Castello, L. Lanzanò, M. d'Amora, P. Bianchini, A. Diaspro, and G. Vicidomini. 2016. Two-photon excitation STED microscopy with time-gated detection. *Sci. Rep.*, 6:19419.
39. Lanzanò, L., I. Coto Hernández, M. Castello, E. Gratton, A. Diaspro, and G. Vicidomini. 2015. Encoding and decoding spatio-temporal information for super-resolution microscopy. *Nat. Comm.*, 6:1–9.
40. Testa, I., N.T. Urban, S. Jakobs, C. Eggeling, K.I. Willig, and S.W. Hell. 2012. Nanoscopy of living brain slices with low light levels. *Neuron*, 75:992–1000.
41. Liu, N., M. Kumbham, I. Pita, Y. Guo, P. Bianchini, A. Diaspro, S.A.M. Tofail, A. Peremans, C. Silien. 2016. Far-field subdiffraction imaging of semiconductors using nonlinear transient absorption differential microscopy. *ACS Photon.*, 3(3):478–485.
42. Hirschfeld, T., 1976. Optical microscopic observation of single small molecules. *Appl. Opt.*, 15:2965.
43. Moerner, W.E., and L. Kador. 1989. Optical detection and spectroscopy of single molecules in a solid. *Phys. Rev. Lett.*, 62:2535.
44. Sahl, S.J., and W.E. Moerner. 2013. Super-resolution fluorescence imaging with single molecules. *Current Opin. Struct. Biol.*, 23:778.
45. Thompson, R.E., D.R. Larson, and W.W. Webb. 2002. Precise nanometer localization analysis for individual fluorescent probes. *Biophys. J.*, 82:2775.
46. Betzig, E., 1995. Proposed method for molecular optical imaging. *Opt. Lett.*, 20:237.
47. van Oijen, A.M., J. Kohler, J. Schmidt, M. Muller, and G.J. Brakenhoff. 1998. 3-Dimensional super-resolution by spectrally selective imaging. *Chem. Phys. Lett.*, 292:183.
48. Dickson, R.M., A.B. Cubitt, R.Y. Tsien, and W.E. Moerner. 1997. On/off blinking and switching behaviour of single molecules of green fluorescent protein. *Nature*, 388:355.
49. Patterson, G.H., J. Lippincott-Schwartz. 2002. A photoactivatable GFP for selective photolabeling of proteins and cells. *Science*, 297:1873.
50. Schneider, M., S. Barozzi, I. Testa, M. Faretta, and A. Diaspro. 2005. Two-photon activation and excitation properties of PA-GFP in the 720-920-nm region. *Biophys. J.*, 89:1346.
51. Deschout, H., F. Cella Zanacchi, M. Mlodzianoski, A. Diaspro, J. Bewersdorf, S.T. Hess, and K. Braeckmans. 2014. Precisely and accurately localizing single emitters in fluorescence microscopy. *Nat. Methods*, 11:253–266.
52. Betzig, E., G.H. Patterson, R. Sougrat, O.W. Lindwasser, S. Olenych, J.S. Bonifacino, M.W. Davidson, J. Lippincott-Schwartz, and H.F. Hess. 2006. Imaging intracellular fluorescent proteins at nanometer resolution. *Science*, 313:1642.
53. Rust, M.J., M. Bates, and X. Zhuang. 2006. *Nat. Sub-diffraction-limit imaging by stochastic optical reconstruction microscopy (STORM). Methods*, 3:793.

54. Hess, S.T., T.P. Girirajan, and M.D. Mason. 2006. Ultra-high resolution imaging by fluorescence photoactivation. *Biophys. J.*, 91:4258.
55. Thompson, M.A., M.D. Lew, and W.E. Moerner. 2012. Extending microscopic resolution with single-molecule imaging and active control. *Ann. Rev. Biophys.*, 41:321.
56. Cella Zanacchi, F., Z. Lavagnino, M. Perrone Donnorso, A. Del Bue, L. Furia, M. Faretta, and A. Diaspro. 2011. Live-cell 3D super-resolution imaging in thick biological samples. *Nat. Methods*, 8:1047.
57. Lauterbach, M.A., 2012. Finding, defining and breaking the diffraction barrier in microscopy – A historical perspective. *Opt. Nanoscopy*, 1:1–8.
58. Eggeling, C., K.I. Willig, S.J. Sahl, and S.W. Hell. 2015. Lens-based fluorescence nanoscopy. *Quart. Rev. Biophys.*, 48:178–243.
59. Shao, D. Li, L., B.-C. Chen, X. Zhang, M. Zhang, B. Moses, D.E. Milkie et al. 2015. Extended-resolution structured illumination imaging of endocytic and cytoskeletal dynamics. *Science*, 349:aab3500.
60. Moerner, W.E., Y. Shechtman, and Q. Wang. 2015. Single-molecule spectroscopy and imaging over the decades, *Faraday Discuss*, 184:9–36. doi:10.1039/c5fd00149h, published online November 30, 2015.
61. Bianchini, P., C. Peres, M. Oneto, S. Galiani, G. Vicidomini, and A. Diaspro. 2015. STED nanoscopy: A glimpse into the future. *Cell Tissue Res.*, 360(1):143–150.
62. Hell, S.W., S.J. Sahl, M. Bates, X. Zhuang, R. Heintzmann, M.J. Booth, J. Bewersdorf et al. 2015. The 2015 super-resolution microscopy roadmap. *J. Phys. D Appl. Phys.*, 48:443001.
63. Chen, F., P.W. Tillberg, and E.S. Boyden. 2015. Expansion microscopy. *Science*, 347:543–548.
64. Chozinski, T.J., A.R. Halpern, H. Okawa, H.-J. Kim, G.J. Tremel, R.O.L. Wong, and J.C. Vaughan. 2016. Expansion microscopy with conventional antibodies and fluorescent proteins. *Nat. Methods*.
65. Harke, B., J.V. Chacko, H. Haschke, and C. Canale. 2012. A novel nanoscopic tool by combining AFM with STED microscopy. *Opt. Nanoscopy*, 1:1–6.
66. Smith. C., 2012. Microscopy: Two microscopes are better than one. *Nature*, 492:293–297.
67. Chacko, J.V., F.C. Zanacchi, and A. Diaspro. 2013. Probing cytoskeletal structures by coupling optical superresolution and AFM techniques for a correlative approach. *Cytoskeleton (Hoboken)*, 70:729–740.
68. Monserrate, A., S. Casado, and C. Flors. 2013. Correlative atomic force microscopy and localization-based super-resolution microscopy: Revealing labelling and image reconstruction artefacts. *ChemPhysChem*, 15:647–650.
69. Viero, G., L. Lunelli, A. Passerini, P. Bianchini, R.J. Gilbert, P. Bernabo, T. Tebaldi, A. Diaspro, C. Pederzolli, and A. Quattrone. 2015. Three distinct ribosome assemblies modulated by translation are the building blocks of polysomes. *J. Cell Biol.*, 208:581–596.

2

Super-Resolution in a Standard Microscope

From Fast Fluorescence Imaging to Molecular Diffusion Laws in Live Cells

Carmine Di Rienzo, Enrico Gratton, Fabio Beltram, and Francesco Cardarelli

2.1 Introduction
2.2 Spatiotemporal Image Correlation to Study Molecular Diffusion: A Theoretical Framework
2.3 From Theory to Experiments: Properties of the STICS-*i*MSD Method
2.4 Biological Applications: Data Presentation
2.5 Discussion
References

Living systems establish local steady-state conditions by maintaining a complex landscape of precisely regulated interactions, which govern the spatial distribution of their molecular constituents. In this regard, fluorescence fluctuation microscopy encompasses a diversified arsenal of analysis tools that provide a quantitative link between cell structural organization and the underlying molecular dynamics. A spatial or temporal analysis of the fluctuating fluorescence signal arising from suitably labeled reporters allows noninvasive measurements of molecular and biochemical parameters such as concentration, diffusion coefficients, binding constants, degree of oligomerization, and so on directly within nonperturbed, live samples. Thus, within the constraints imposed by the requirement of fluorescence labeling, fluctuation microscopy has the potential to add a dynamic molecular dimension to standard fluorescence imaging *in vivo*. In particular, we shall discuss a fluorescence fluctuation-based method that makes it possible to probe the actual molecular "diffusion law" directly from imaging, in the form of a mean square displacement (MSD) versus time-delay plot (image-derived MSD [*i*MSD]), with no need to preliminarily assume any interpretative model. Also, we shall show that the present method makes it possible to investigate the relevant diffusion law from a length scale of about 20 nm. This can be accomplished by fast imaging of the region of interest. For each time delay in the resulting stack of images, the average spatial correlation function is calculated. By fitting the series of correlation functions at each time delay, the molecular "diffusion law" is quantitatively derived. This approach is tested *in silico* through several simulated 2D diffusion conditions, and in live cells, through a series of fluorescently labeled benchmark molecules (both lipids and proteins). Notably, this approach does not require extraction of the individual molecular trajectories nor the use of particularly bright fluorophores and can be used also when the molecules under study are present at high density (e.g., GFP-tagged molecules transiently transfected into cells). We shall discuss the ability of this approach to resolve average molecular dynamic properties well below the limit imposed by diffraction. Overall, we shall show that this novel approach represents a powerful tool for the determination of kinetic and thermodynamic parameters over very wide spatial and temporal scales, thus unveiling dynamic molecular properties that can be used to build reliable physical models of cellular processes.

2.1 Introduction

A major challenge of present (and future) biophysics is to quantitatively study how molecular subensembles dynamically interact to fulfill their physiological roles in living cells and organisms. This is the classic research target of molecular cell biology and quantitative fluorescence microscopy: individual molecular components (proteins, lipids, etc.) are cloned/purified/synthesized, tagged, and analyzed. In this context, high-speed single-particle tracking (SPT) techniques play a crucial role and yield information at the nano–meso-scale on the dynamics and interactions of several molecular cellular components. Based on SPT, for instance, Kusumi and coworkers thoroughly investigated the compartmentalization of the fluid plasma membrane into submicron domains throughout the cell membrane and the hop diffusion of many relevant molecules (for a review see Kusumi et al. 2005a, 2005b; 2010). Widespread exploitation of the SPT approach is hindered, however, by at least four characteristics that, at present, appear unavoidable: (1) the experiment relies on production, purification, and labeling of a single molecule with a suitable marker particle (e.g., a gold colloidal particle or quantum dot); (2) the usable labels are rather large on the molecular scale and can induce cross-linking of target molecules or steric hindrance effects, thus affecting the biological function under study; (3) the size (colloidal gold particles typically used have diameters of 20–40 nm) and chemical nature of the label typically prevents the application of SPT measurements to intracellular molecules; (4) a large number of single-molecule trajectories must be recorded to fit statistical criteria, thus dramatically increasing the time required to gather and analyze SPT data. In regard to these SPT drawbacks, fluorescence correlation spectroscopy (FCS) is a very attractive alternative approach. In fact, thanks to its intrinsic single-molecule sensitivity even in the presence of many molecules, it can easily afford the required statistics in a reasonable amount of time. FCS can be performed well with genetically encoded FPs and, in general, with fluorophores not particularly bright. The basic principle of fluctuation analysis is to decrease the detection volume so that small numbers of fluorescent molecules are excited at the same time. Molecules stochastically cross the open detection volume defined by the laser beam, leading to a fluctuating occupancy that follows Poisson statistics (which in turn obeys the relation that the variance is proportional to the average number of observed molecules). The underlying molecular dynamics are extracted as characteristic decay times through fluctuation correlation analysis. In its classic view, FCS is used as a "local" measurement of the concentration and residency time of molecules present in the diffraction-limited focal area defined by the laser beam. Many efforts targeted the extension of the FCS principle to the spatial scale. For instance, the focal area was duplicated (Ries and Schwille 2006), moved in space in laser "scanning" microscopes (Berland et al. 1996; Ruan et al. 2004; Heinemann et al. 2012; Cardarelli et al. 2012a, 2012b; Di Rienzo et al. 2014b), or combined with fast cameras (Kannan et al. 2006; Unruh and Gratton 2008; Di Rienzo et al. 2013b, 2014a). Using these "spatiotemporal" correlation approaches, diffusion constant, concentration, and partitioning coefficient of several membrane components were measured on both model membranes and actual biological ones (Schwille et al. 1999; Weiss et al. 2003). An alternative way to sample both time and space is based on the size change of the focal area. For instance, it was demonstrated that molecular FCS-based diffusion laws can be recovered by performing fluctuation analysis at various spatial scales larger than the laser focal area (Wawrezinieck et al. 2005; Lenne et al. 2006). Based on these data, inferences were drawn about the dynamical organization of cell membrane components by extrapolation below the diffraction limit. Along this reasoning, similar studies can be performed by downsizing the focal area: recent reports successfully exploited the ~30 nm focal area attained by stimulated emission depletion (STED) (Hell and Wichmann 1994) to directly probe the nanoscale dynamics of membrane components in a living cell by fluctuation analysis (Eggeling et al. 2009; Mueller et al. 2011; Sezgin et al. 2012). In all the FCS experiments described, however, the size of the focal spot represents a limit in spatial resolution that cannot be overcome. In addition, the information collected always requires assumptions and models that describe the molecular dynamics under study. These models must then be translated into equations to be fitted to FCS measurements. In an effort to overcome these limitations an approach based on spatiotemporal image correlation spectroscopy (Hebert et al. 2005) method will be discussed here, that is suitable for the study of the dynamics of fluorescently tagged molecules in live cells with high temporal (up to 10^{-4} s for images and 10^{-6} for line scans) and spatial (well below the diffraction limit) resolution (schematic representation in Figure 2.1).

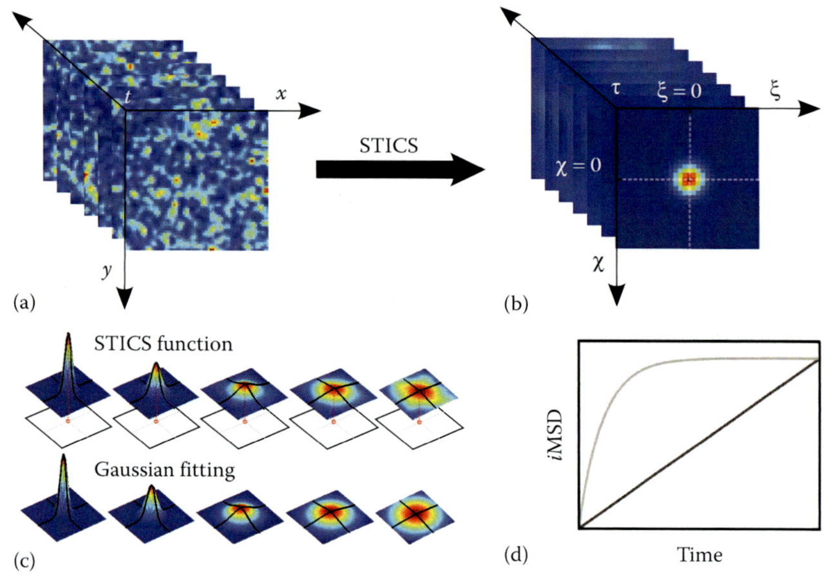

Figure 2.1

From STICS to *i*MSD to study protein diffusion on membranes. STICS operation converts (a) a stack of intensity, images $I(x,y,t)$ into (b) a stack of images representing the spatiotemporal evolution of correlation $G(\xi,\chi,\tau)$. (c) When particles dynamic is governed only by diffusion the maximum of correlation remains in the origin (see peak projection on Cartesian axis). It is possible to approximate $G(\xi,\chi,\tau)$ with a Gaussian function, whose variance corresponds to the particles average *i*MSD. (d) Plot of *i*MSD versus time may be used to distinguish, for example, between free and confined diffusion. (From Di Rienzo, C. et al., *Proc. Natl. Acad. Sci. U.S.A.*, 110, 12307–12312, 2013a. With permission.)

First, fast imaging of the region of interest is performed. For each time delay in the resulting stack of images the average of the spatial correlation function is calculated. By fitting the series of correlation functions, the actual protein "diffusion law" can be obtained directly from imaging, in the form of a MSD versus time-delay plot (*i*MSD). Thanks to the *i*MSD versus time plot, protein diffusion modes can be directly identified with no need for an interpretative model or assumptions about the spatial organization of the membrane. The method was put to test both *in silico*, with several 2D diffusion patterns, and in live cells, with several benchmark molecules diffusing on the plasma membrane. In particular, we shall discuss the present method in the case of a GFP-tagged variant of the transmembrane transferrin receptor (TfR), a well-known benchmark of confined diffusion (Sako and Kusumi 1994, 1995). We shall show how one can recover diffusion constants, confinement areas, confinement strength coefficients, partitioning coefficients, and so on over many microns in the sample, and their variation in response to drug treatments or temperature changes. The present strategy represents a useful asset for future research on membrane dynamic organization particularly when the labeled particles under study are too dim or too dense to be individually tracked.

2.2 Spatiotemporal Image Correlation to Study Molecular Diffusion: A Theoretical Framework

We define the fluorescence intensity at point $\mathbf{r}(\mathbf{r}=(x,y,z))$ and at time t as

$$I(\mathbf{r},t)=\varepsilon\int_V W(\mathbf{r}-\mathbf{r}')c(\mathbf{r},t)d\mathbf{r}', \tag{2.1}$$

where:

ε takes into account the total efficiency in photon production, collection, and measurement
$W(\mathbf{r})$ represents the point spread function (PSF) of the detection system
$c(\mathbf{r}, t)$ the concentration

The integral is calculated over all space V. We now introduce the spatiotemporal image correlation function as

$$G(\rho,\tau)=\frac{\langle I(\mathbf{r},t)\cdot I(\mathbf{r}+\rho,t+\tau)\rangle}{\langle I(\mathbf{r},t)\rangle^2}-1, \tag{2.2}$$

where:

$\rho=(\xi,\psi,\zeta)$ represents the spatial lag in x, y, and z directions
τ is the temporal lag
$\langle\ldots\rangle$ indicates the average over space and time variables

If we now substitute the definition of intensity in Equation 2.2 and Fourier transform we obtain

$$G(\rho,\tau)=\frac{\int e^{-i\rho\mathbf{k}}W(\mathbf{k})^2 S(\mathbf{k},\tau)d\mathbf{k}}{N\int W(\mathbf{k})^2 d\mathbf{k}}, \tag{2.3}$$

where:

N is the average number of fluorophore in the observation volume
$W(\mathbf{k})$ is the spatial Fourier transform of the instrumental PSF
$S(\mathbf{k},\tau)=\frac{1}{\langle c\rangle}\int d\rho e^{-i\rho\mathbf{k}}\langle\delta c(\mathbf{r},t)\cdot\delta c(\mathbf{r}+\rho,t+\tau)\rangle$ is the intermediate scattering function

If we consider the limit of dilute solution, the motion of each single fluorophore is independent from the others and the intermediate scattering function can be approximated by the characteristic function of particle displacement (the incoherent intermediate scattering function) as

$$S(\mathbf{k},\tau)\approx P(\mathbf{k},\tau)=\left\langle e^{-i\mathbf{k}\Delta\mathbf{R}(\tau)}\right\rangle, \tag{2.4}$$

where $\Delta\mathbf{R}(\tau)$ is a random variable that describes the 3D displacement of the single particle after a time delay τ. Substituting Equation 2.4 in Equation 2.3 we can recognize that the dividend in Equation 2.3 is the Fourier transform of a product of characteristic function and thus is a probability density function. So, we can easily conclude that

$$G(\rho,\tau) \propto P(\rho,\tau) \otimes \big(W(\rho) \otimes W(\rho)\big), \tag{2.5}$$

where $P(\rho,\tau)$ is the probability density function of particle displacement. This result points out that $G(\rho,\tau)$ is proportional to the probability distribution function of the distance (in time and space) between two photons collected from the same molecule. Thus $G(\rho,\tau)$ can be used to measure the properties of particle displacement independently of particle motion and instrumental PSF. As one example, here we shall solve Equation 2.5 for the case of a Gaussian PSF defined as

$$W(\rho) = W(\xi,\chi,\zeta) = \frac{1}{\sigma_0^2\sigma_z(2\pi)^{3/2}} \exp\left(-2\frac{\xi^2+\chi^2}{\sigma_0^2} - 2\frac{\zeta^2}{z_0^2}\right), \tag{2.6}$$

where σ_0^2 and σ_z^2 are classically defined as the waist of the PSF in the radial and in the axial direction, respectively. We further simplify the calculation assuming Gaussian transport and defining

$$P(\xi,\chi,\zeta,\tau) = \left(\frac{d}{2\pi\delta r^2(\tau)}\right)^{d/2} \exp\left(-d\frac{\xi^2+\chi^2+\zeta^2}{2\delta r^2(\tau)}\right), \tag{2.7}$$

where:

$\delta r^2(\tau)$ is the MSD of the particle
d is the dimensionality of the motion

Thus, we can solve Equation 2.5 as

$$G(\xi,\chi,\zeta,\tau) = \frac{1}{N\sigma_r^2(\tau)\sigma_a(\tau)} \exp\left(-\frac{\xi^2+\chi^2}{\sigma_r^2(\tau)} - \frac{\zeta^2}{\sigma_a^2(\tau)}\right), \tag{2.8}$$

where $\sigma_r^2(\tau) = [2\delta r^2(\tau)/d + \sigma_0^2]$ and $\sigma_a^2(\tau) = [2\delta r^2(\tau)/d] + z_0^2$ are named *i*MSD in the radial and in the axial direction, respectively. These definitions reflect directly the contribution of the laser beam waist and the particle MSD $[2\delta r^2(\tau)/d]$ to the correlation function and merge together the contribution from the imaging of the two subsequent photons in space and time on the particle displacement. Here we continue with the case of 2D motion and thus Equation 2.8 is simplified to

$$G_D(\xi,\chi,\tau) = g_D(\tau)\exp\left(-\frac{\xi^2+\chi^2}{\sigma_r^2(\tau)}\right), \tag{2.9}$$

with:

$$g_D(\tau) = \frac{1}{N\sigma_r^2(\tau)} \tag{2.10}$$

$G(\xi, \chi, \tau)$ in Equation 2.9 is a Gaussian function centered at the origin ($\xi = 0$, $\chi = 0$) with variance that increases with delay time and amplitude that decreases with delay time. These quantities (amplitude and waist of the Gaussian) can be used to measure particle diffusive dynamics.

2.2.1 The Contribution of Particle Size

When the particle under study has more than one fluorophore and its size is not negligible (e.g., large protein clusters, vesicles, etc.), $G(\xi, \chi, \tau)$ must include also the spatial extension of the particle itself. In the general case, the size of the clustering region, the number of labels, and the distribution of the labels affect the correlation function (Sengupta et al. 2011). In fact, the distance between the two collected photons will include a new random variable that takes into account the distance between the fluorophore that produced the photons and the center of the particle ($\mathbf{\Delta S}$). If we consider that the number of labels is high, and considering $\mathbf{\Delta S}$ uncorrelated to both particle displacement and photon detection uncertainty, we can include the effect of particle size rewriting Equation 2.5 as

$$G(\rho, \tau) \propto P(\rho, \tau) \otimes \left(W(\rho) \otimes W(\rho)\right) \otimes \left(S(\rho) \otimes S(\rho)\right), \tag{2.11}$$

where $S(\boldsymbol{\rho})$ represents the probability distribution function of $\mathbf{\Delta S}$. As a first approximation, we can consider $S(\boldsymbol{\rho})$ as a Gaussian function with variance σ_p^2. This approximation allows us to consider the effect of the particle size as an apparent increase in the instrumental PSF size and thus we can write:

$$\sigma_p^2 = \frac{\sigma_r^2(0) - \sigma_0^2}{2}, \tag{2.12}$$

where the factor of two originates from the correlation between a pair of photons (thus the variance due to the particle is added twice). Thus, $\sigma_r^2(0)$ can be used to verify the presence of large diffusing aggregates and estimate their size.

2.2.2 Interpretation of the *i*MSD Plot

For free diffusing particles, the variance of particle displacement increase linearly in time and then the *i*MSD plot is linear according to

$$\sigma_r^2(\tau) = 4D\tau + \sigma_0^2, \tag{2.13}$$

where D is the diffusion coefficient.

For particles diffusing in a confined area, the particle displacement is limited by the size of the confining region and the *i*MSD reaches an asymptotic limit. In this latter case, assuming a square confinement (Kusumi et al. 1993), the *i*MSD can be expressed as

$$\sigma_r^2(\tau) \cong \frac{L^2}{3}\left(1 - \exp\left(-\frac{\tau}{\tau_c}\right)\right) + \sigma_0^2, \tag{2.14}$$

where:

L defines the linear size of the confinement area

τ_c is an index of the time scale of confinement

As an intermediate case, that is, when a particle is only transiently confined to a meshwork or by spatially organized large obstacles, it shows a faster local diffusivity (D_{micro}) that dominates the faster timescale and a slower, long-range diffusivity (D_{macro}) that is an effect of the transient confinement and will dominate the longer timescale. In this case, $\sigma_r^2(\tau)$ can be approximated as

$$\sigma_r^2(\tau) \cong \frac{L^2}{3}\left(1-\exp\left(-\frac{\tau}{\tau_c}\right)\right)+4D_{\text{macro}}\tau+\sigma_0^2. \tag{2.15}$$

Finally, following the treatment by Wawrezinieck et al. (2005), we shall introduce the "confinement strength" parameter, defined as $S_{\text{conf}} = D_{\text{micro}}/D_{\text{macro}}$. This parameter directly quantifies the impact of confinement on molecular long-range diffusivity. If we consider an active movement, the variance of particle displacement in time will be superlinear and then we can define the *i*MSD as

$$\sigma_r^2(\tau)=(K_\alpha\tau)^\alpha+\sigma_0^2, \tag{2.16}$$

where K_α takes into account the speed of the process and $1 \le \alpha \le 2$ the superlinearity of the displacements. In particular, when $\alpha = 1$ the motion is Brownian, Equation 2.15 become Equation 2.13, and $K_\alpha = 4D$. When $\alpha = 2$ the particle experiences 2D randomly oriented flux and $K_\alpha = v$, where v is the flux speed. The theoretical framework presented so far can describe both isotropic diffusion and transient confinement by an extended meshwork. On the contrary, dynamic partitioning of molecules within isolated (nano) domains requires further considerations. In this system, molecules can experience two conditions: they can freely diffuse without entering the nanodomain or they can be trapped in a fixed nanodomain much smaller than the PSF and only after a certain time (trapping time, τ_T) they freely diffuse away. In the simplest case, molecules are trapped for a time significantly longer than the characteristic time of diffusion $\tau_D = \sigma_r^2(0)/4D$ so that the observed dynamics is the result of two separate components. For the trapping component we shall consider particles as blocked inside a domain and consequently treat the correlation function as simple binding. Accordingly, the probability to find the particle inside the domain does not depend on spatial variables but is a function of time only (Digman and Gratton 2009):

$$p_T(\tau)=\exp\left(-\frac{\tau}{\tau_T}\right). \tag{2.17}$$

Equation 2.5 can be recast into

$$G_T(\xi,\chi,\tau)=g_T(\tau)\cdot\exp\left(-\frac{\xi^2+\chi^2}{\sigma_T^2}\right), \tag{2.18}$$

with

$$g_T(\tau)=\frac{\gamma_T}{N_T}\exp\left(-\frac{\tau}{\tau_T}\right), \tag{2.19}$$

where γ_T and N_T indicate the "contrast" between the populated and unpopulated domains and the average number of domains, respectively. Note that the *i*MSD for particles bound to domains (σ_T^2) is flat because it does not depend on τ and cannot be resolved from the instrumental waist in the case of very small domains. Larger domains, that may contain more than one particle, will contribute to the *i*MSD as an offset (see below). The correlation function for particles freely diffusing off the domains is described by Equation 2.9. Overall, the correlation function for dynamic partitioning systems can be written as

$$G(\xi,\chi,\tau)=g_T(\tau)\cdot\exp\left(-\frac{\xi^2+\chi^2}{\sigma_T^2}\right)+g_D(\tau)\cdot\exp\left(-\frac{\xi^2+\chi^2}{\sigma_r^2(\tau)}\right) \quad (2.20)$$

In the case of binding times close to the diffusive time, however, the two components (binding and diffusion) are not completely distinguishable. In this case a (two-component) Gaussian approximation may become unsatisfactory since the characteristic temporal evolution of particle spatial distribution for binding and diffusion are partially overlapped. Finally, an obvious limit case is that of a binding time much faster than the diffusion time. In this case, dynamic partitioning of molecules within nanodomains becomes indistinguishable from their free diffusion, that is, a one-component Gaussian model yields a satisfactory description of the system.

2.3 From Theory to Experiments: Properties of the STICS-*i*MSD Method

2.3.1 Simulations: A Guideline to Recognize the Particle Mode of Motion

In order to verify the main properties of the proposed analysis, five sets of guided simulations for point particles in a 2D box will be presented that correspond to peculiar molecular diffusion patterns found in live cells. Briefly, simulations were run in MATLAB® R14 (The MathWorks, Natick, MA), using the optimization and image processing toolboxes. The same sampling conditions of real experiments are imposed. All populations are set to have the same total number of particles (density of 10 particles/μm^2), with equal quantum yield, and randomly seeded in a 64×64 pixels squared matrix (0.1 μm/pixel). All the simulated acquisitions are 6×10^3 frames long, with a time step of 10 ms/frame and beam radius set to 300 nm. Each x and y coordinate was changed separately by adding a random number drawn from a normal distribution with a mean of zero, and a standard deviation σ defined as

$$\sigma=\sqrt{2D\Delta_t}, \quad (2.21)$$

where:

D is the diffusion coefficient

Δ_t is the sampling time between sequential images

In the transient confinement simulation, barriers are simulated as infinitely thin lines that particles can cross with a probability "P" independent of time. The recovered particle-position matrix was then convolved with a Gaussian filter to simulate the diffraction-limited acquisition. Periodic boundary conditions for both particle movement and convolution are applied. The contribution of background and counting noise was introduced following the approach described by Kolin et al. (2006). The spatial extension of particles was simulated convolving particle position with a Gaussian function having a standard deviation equal to the particle size. Averaging of multiple repetitions of the simulations was used to account for longer exposure times. We shall discuss (1) pure isotropic diffusion, (2) confined diffusion due to a meshwork, (3) partially confined diffusion, (4) guided motion, and (5) dynamic partitioning (Figure 2.2).

Notably, the *i*MSD plot makes it possible to distinguish these different behaviors very easily. In simulation (1), for instance, particles are diffusing isotropically with diffusivity "D" ($\mu m^2/s$) (Figure 2.2a): the calculated *i*MSD shows a linear trend in time with a characteristic slope that depends on the D value (Figure 2.2b). The plot in Figure 2.2c shows the accordance between the simulated diffusion coefficient (D_{theor}) and that retrieved from fitting to Equation 2.13 (D_{exp}). In the second set of simulations, particles are diffusing in a continuous meshwork where adjacent domains of size "L" are defined by impenetrable barriers (Figure 2.2d). In such a configuration, particles can diffuse freely within each domain (with diffusivity "D_{micro}") but are not able to jump (or "hop") to the adjacent domains ($P = 0$). Accordingly, the *i*MSD shows an initial linear behavior (that reflects molecular diffusion within the mesh) and then saturates to an asymptotic value that depends on the confinement area (Figure 2.2e). Again, accordance between fixed (L_{theor}) and retrieved (L_{exp}) confinement-length values is shown in Figure 2.2f. We point out that the *i*MSD approach is able to measure confinement values well below the diffraction limit (in the present example, $L = 150$ nm as compared to a simulated PSF of 300 nm). In simulation (3) a non-null probability to jump between adjacent meshes is introduced ($P > 0$): a new "long-range" diffusivity appears, D_{macro} (Figure 2.2g). Accordingly, the *i*MSD yields the expected saturation trend when P approaches 0 ($D_{macro} \rightarrow 0$) and, by contrast, a linear trend when P is close to 1 ($D_{micro} = D_{macro}$) (Figure 2.2h). For all the intermediate P values, the *i*MSD shows an initial slope that depends on D_{micro}, a subsequent inflection mirroring the partial confinement, and a long-range linear behavior that depends on P (Figure 2.2h). Fitting the *i*MSD plot to Equation 2.15 yields the confinement strength (S_{conf}) parameter that in turn decreases by increasing the hopping probability P (Figure 2.2i). In simulation (4), we consider the case of a mesh of randomly oriented straight paths (Figure 2.2j). In more detail, the particle is free to diffuse with diffusivity D_{out} when in the path with a certain probability (P_{in}). Here, the particle diffuses slower ($D_{in}/D_{out} < 1$) and is dragged by a flux of speed v in the direction of the path before spontaneously coming off it (P_{out}). Moreover, when two paths cross, the particle randomly continues along the previous path or follows the new one. We set $P_{in} = P_{out} = 1$, $D_{out} = 1\ \mu m^2\ s^{-1}$ and in order to modulate the residency time of the particle on the path we vary D_{in}/D_{out} ratio: the smaller the D_{in}/D_{out} ratio, the more time the particle will spend within the path before leaving it. Thus, as shown in Figure 2.2k, if $D_{in}/D_{out} << 1$ the particle will move dragged by a flux leading to a quadratic dependence of the *i*MSD on time ($\alpha \sim 2$). On the other hand, if $D_{in}/D_{out} \sim 1$ the mesh has a negligible effect on particle motion and the *i*MSD becomes linear as expected for free diffusion. Finally, for intermediate values of the D_{in}/D_{out} ratio the shape of the *i*MSD is well approximated by Equation 2.15 and, as expected,

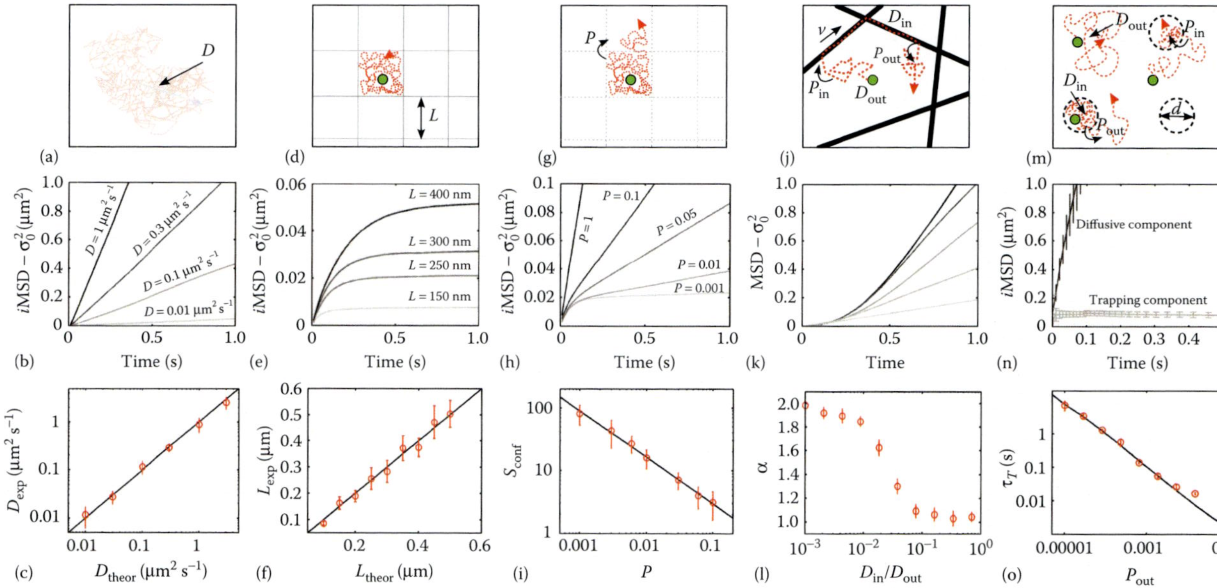

Figure 2.2

*i*MSD analysis on simulated 2D diffusion. (a) Simulated condition: 2D isotropic diffusion, with diffusivity D. (b) In this case the *i*MSD is linear, with a higher slope for increasing D values. (c) Accordance between the theoretical D value (imposed in the simulation) and that recovered from the analysis. (d) Simulated condition: 2D isotropic diffusion in a meshwork of impenetrable barriers (particles have probability $P = 0$ to overcome the barrier). (e) The *i*MSD plot starts linear and then reaches a plateau that identifies the confinement area and the corresponding characteristic linear size L. (f) Accordance between the theoretical L value and that recovered from the analysis. (g) Simulated condition: 2D isotropic diffusion in a meshwork of penetrable barriers. Particles diffuse within the mesh, and have probability $P > 0$ to overcome the barrier, thus generating a "hop" diffusion component. (h) The *i*MSD plot starts linear (with a slope dependent on D_{micro}) and then deviates toward a lower slope which depends on P. (i) Calculated S_{conf} as a function of the imposed P. (j) Simulated condition: particle net flux. Particles isotropically diffuse in the medium and have a probability P_{in} to get bound to a structure where they exert net flux with speed v, and probability P_{out} to return in the medium. (k) If the ratio $D_{in}/D_{out} << 1$, the particle will move dragged by a flux and we expect a quadratic dependence of the *i*MSD in time ($\alpha \sim 2$). On the other hand, if $D_{in}/D_{out} \sim 1$ the mesh has a negligible effect on particle motion and the *i*MSD become linear as we expect for the free diffusion. Finally, for intermediate values of D_{in}/D_{out} ratio the shape of the *i*MSD is well approximated by Equation 2.15. (l) As expected, the recovered α varies between 1 and 2 as a function of P_{out}. (m) Simulated condition: dynamic partitioning. Particles diffuse freely outside and inside the domain with diffusivity D_{out} and D_{in}, respectively, and have probability P_{in} and P_{out} to enter or exit the domain, respectively. (n) Two characteristic *i*MSD traces in dynamic partitioning: diffusion (black) and trapping (gray). Diffusion shows linear *i*MSD, whereas trapping yields a constant *i*MSD. (o) Calculated τ_T as a function of the imposed P_{out} (red dots) with respect to the trapping time directly calculated from molecular trajectories (black line). (Adapted from Di Rienzo, C. et al., *Proc. Natl. Acad. Sci. U.S.A.*, 110, 12307–12312, 2013a. With permission.)

the recovered α varies between 1 and 2 as a function of P_{out} (Figure 2.2l). In the last set of simulations, particles are allowed to diffuse freely outside and inside the domain with diffusivity D_{out} and D_{in}, respectively, while having a probability P_{in} and P_{out} to enter or exit the domain, respectively (Figure 2.2m). This configuration corresponds to the case of dynamic partitioning. Accordingly, two characteristic *i*MSD traces are found: a linear one, corresponding to free diffusion (black) and a constant one, corresponding to trapping (gray) (Figure 2.2n). The trapping time (τ_T) can be calculated as a function of the simulated P_{out} (red dots) and compared with the trapping time derived from molecular trajectories (black line) (Figure 2.2o).

2.3.2 The Correlation Amplitude $g(\tau)$

It is worth mentioning that, as previously demonstrated by Shusterman et al. (2008) for single-point FCS, $g(\tau)$ can be used to measure the *i*MSD directly by Equation 2.10. In fact, the *i*MSD trace calculated from the width of the Gaussian is superimposable to the one obtained by Equation 2.10. Thus, fitting the amplitude shall represent an independent strategy to measure particle diffusivity (Figure 2.3), particularly useful in the case of multicomponent systems. However, it must be noted that the amplitude of the Gaussian strongly depends on the number of particles. Thus, photophysical phenomena (e.g., reversible and irreversible photobleaching) will inevitably contribute to generate the correlation function amplitude, affecting the measurement of diffusion. By contrast, the *i*MSD measured from the variance is not affected by time-dependent effects and consequently yields a more reliable estimate of molecular diffusivity. Along this reasoning, it shall also appear clear that the comparison between the *i*MSD measured from the Gaussian width and the inverse of $g(\tau)$ can represent a powerful tool to distinguish between fast diffusion and blinking.

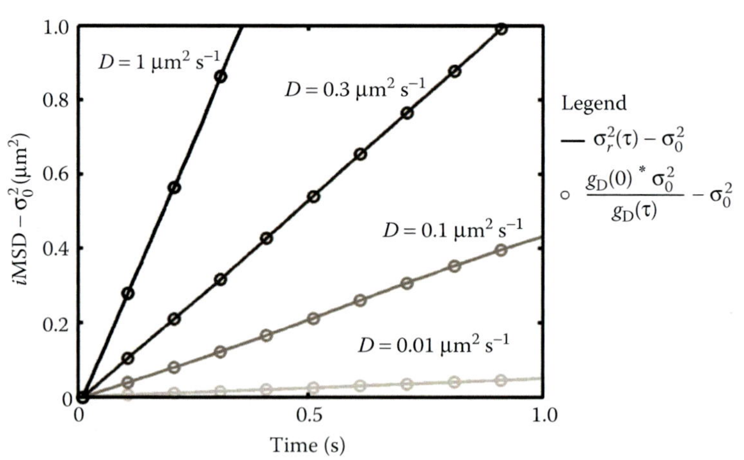

Figure 2.3

For diffusive particles the temporal correlation function $g_D(\tau)$ could represent an independent strategy to measure the *i*MSD. Correspondence between the *i*MSD obtained by separately fitting the height and waist of $G_D(\xi, \chi, \tau)$ for a simulation of free-diffusing particles. (From Di Rienzo, C. et al., *Proc. Natl. Acad. Sci. U.S.A.*, 110, 12307–12312, 2013a. With permission.)

2.3.3 σ_0, Particle Size, and the Role of Sampling Speed

Thus far, we presented only *i*MSD traces obtained after subtraction of the σ_0 value (extracted from data fitting). However, σ_0 has the potential to provide quantitative information about the size of the diffusing particle. To prove this, we shall show here a set of simulations in which particles of increasing size (indicated by the particle radius, *R*) are able to freely diffuse in a box (Figure 2.4a). As expected, changing the particle size does not affect the overall *i*MSD trend (linear in this case), but impacts on the σ_0 intercept value. In particular, σ_0 increases at increasing *R* values. The difference between the experimental σ_0 and an independently calibrated experimental waist value (in the present example PSF = 300 nm) yields the actual particle size. Figure 2.4b shows the accordance between the imposed particle size and the one retrieved from the *i*MSD plot in a rather broad range of particle sizes, from few tens to several hundreds of nanometers. Of note, in a wide-field acquisition the calculated spatial correlation $G(\tau = 0)$ contains a "broadening" effect compared to the extrapolated σ_0 value due to the extent of particle diffusion during exposure time (Figure 2.5). This in turn implies that

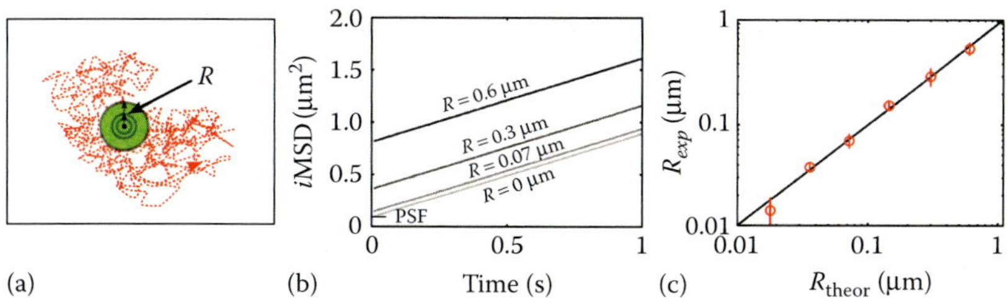

Figure 2.4

Effect of particle size on σ_0. (a) Simulated condition: 2D isotropic diffusion, with an increasing particle size (*R*). (b) As expected, the *i*MSD is linear with increasing σ_0 values for increasing *R* values. (c) Accordance between the theoretical *R* value (imposed in the simulation) and that recovered from the analysis. (From Di Rienzo, C. et al., *Proc. Natl. Acad. Sci. U.S.A.*, 110, 12307–12312, 2013a. With permission.)

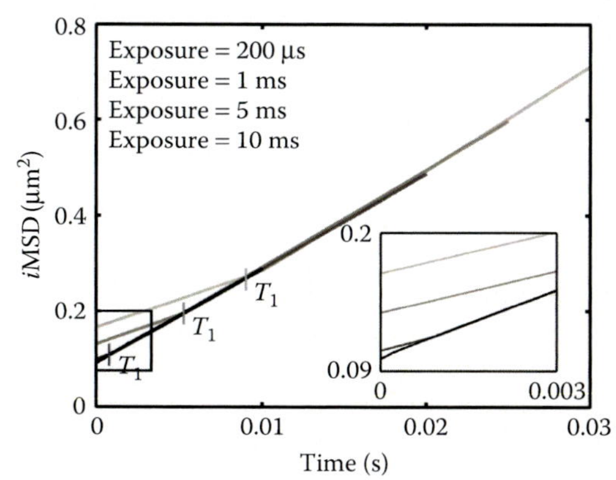

Figure 2.5

The effect of exposure time on spatial correlation. *i*MSD plot for a fast diffusing particle acquired at different exposure times. The spatial correlation $G(\xi, \chi, 0)$ shows a "broadening" effect due to particle diffusion during the exposure time. (From Di Rienzo, C. et al., *Proc. Natl. Acad. Sci. U.S.A.*, 110, 12307–12312, 2013a. With permission.)

fast-diffusing particles will appear larger in $G(\tau = 0)$ if the rate of frame acquisition is longer than the time a particle takes to diffuse a fraction of the PSF.

2.3.4 An "average" Single-Molecule Measurement

As already stated above, FCS has emerged as a very powerful method to address the motion of molecules in both the interior and exterior of a cell. In particular, it provides information at the single-molecule level by averaging the behavior of many molecules, and thus yields very good statistics in a limited amount of time. This unique property, together with the possibility to sample fluctuations in space (i.e., through FCS spatial variants, discussed above) has the potential to fill the gap between FCS and SPT technologies. This is better illustrated here by examining the effect of particle density, signal-to-background ratio, and number of photons per particle on the applicability of *i*MSD approach to measure particle "average" displacements (evaluated by comparing the expected parameter values with those retrieved from the analysis of simulated data) (Figure 2.6). Overall,

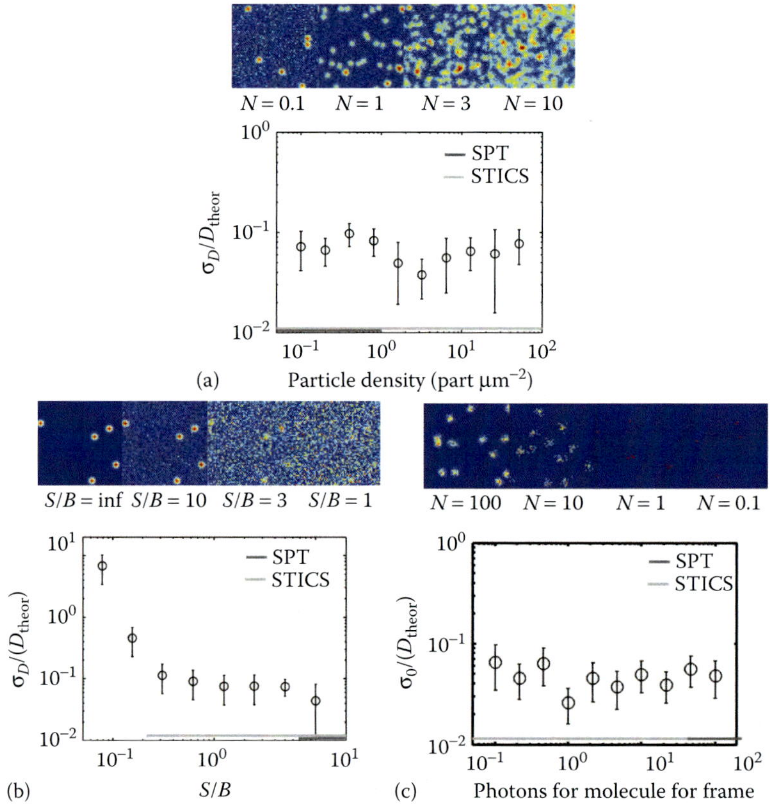

Figure 2.6

Comparison of the *i*MSD approach with SPT. The effect of particle density (a), signal-to-background ratio (b), and number of photons (c) on the applicability of our approach to measure particle *i*MSD was evaluated by comparing the expected parameter values with those retrieved from the analysis of simulated data (see SI). Overall, the range of feasibility of our approach proves to be wider compared with that of a typical SPT measurement. Parameter values are expressed as mean ± standard deviation. (Adapted from Di Rienzo, C. et al., *Proc. Natl. Acad. Sci. U.S.A.*, 110, 12307–12312, 2013a. With permission.)

the range of feasibility of the *i*MSD analysis proves to be wider compared with that of a typical SPT measurement. In particular, SPT, being based on particle localization, requires a high number of photons per particle per frame: in ideal conditions, more than 100 photons per particle per frame are required to reach a localization resolution better than 30 nm. The presented approach, instead, is feasible also when the label produces an average of less than 1 photon per particle per frame (Figure 2.6c). Thus, given a labeling technology, a 3-order or magnitude faster timescale can be explored by the present method. This property allows considering the *i*MSD approach not only an "average" single-molecule technique but also an "average" single-photon technique.

2.4 Biological Applications: Data Presentation

In this section, few exemplary applications of the STICS-*i*MSD method to relevant biological cases will be presented and discussed. More in detail, particles diffusing in a 2D biological system (i.e., the plasma membrane of live cells) will be used, in analogy with the simulations presented above. Despite the large efforts made in the last forty years of research (Singer and Nicolson 1972; Kusumi et al. 2012), our knowledge of the organization of molecules on the plasma membrane of live cells is still rudimentary. Current models include the notion that membranes are crowded environments (Ryan et al. 1988) with a complex spatial organization (or "heterogeneity"), that they interact with the cytoskeletal components, and contain microdomains (or "lipid rafts") of different size and lipid/protein composition (Edidin 2003; Kusumi et al. 2004, Simons and Vaz 2004). Using electron tomography, the 3D structure of the membrane skeleton on the cytoplasmic face of the plasma membrane was clarified (Morone et al. 2006). It was found that the cytoplasmic face of the membrane is covered by a meshwork of actin filaments, which are closely associated with the membrane surface (~0.83 nm estimated distance). Since transmembrane proteins protrude into the cytoplasm, their intracellular domains may collide with these actin filaments that can act in turn as "fences," inducing temporary confinement of the protein within mesh domains. Transmembrane proteins are also assumed to hop between these domains, whenever there is space between the membrane and the actin filament owing to membrane structural fluctuations and/or when the actin filament temporarily dissociates (Kusumi et al. 2005b). This complex behavior is thought to be a master regulator of specific molecular interactions, such as the rapid assembly or disassembly of specific multiprotein/lipid complexes (Capps et al. 2004; Kodippili et al. 2009). At the same time, cholesterol-mediated lipid interactions are thought to have a functional role in many membrane-associated processes such as signaling events (Simons and Ikonen 1997; Brown and London 2000; Fielding 2006; Jacobson et al. 2007; Hanzal-Bayer and Hancock 2007). Although several experiments indicate their existence, lipid nanodomains ("rafts") remain controversial owing to their putative size of 5–200 nm, that spans the range between the extent of a protein complex and the resolution limit of optical microscopy. Overall, a major challenge in the study of these dynamic processes/structures is the requirement of high-temporal and spatial resolution observations, capable of monitoring the behavior of the different molecular components at the nanoscale.

2.4.1 The TfR Transient Confinement

The TfR is used as an exemplary case of transmembrane protein confined by the cytoskeleton meshwork (Sako and Kusumi 1995) (schematic representation in Figure 2.7a). In particular, a GFP-tagged variant of the TfR protein is transiently transfected into CHO-K1 cells (Figure 2.7b). In order to selectively excite the fluorophores on the basal membrane of living cells we use total internal reflection (TIR) illumination, by means of a standard TIR fluorescence (TIRF) microscope (Leica AF6000). To collect the fluorescence signal, a high magnification objective (100× *NA* 1.47) and an electron multiplied charge coupled device (EMCCD) camera (physical size of the pixel on the chip: 16 μm) are used. Thus, a membrane region of ~50 μm^2 is selected (red box in Figure 2.7b) and imaged repeatedly at ~100 frames per second. As expected, STICS analysis shows that TfR molecules disperse on the membrane in a symmetric fashion due to diffusion only, thus broadening the correlation Gaussian in every direction (Figure 2.7c). As shown in Figure 2.7d, the total acquisition time is long enough to fully describe the characteristic correlation curve of TfR molecules. Notably, the *i*MSD plot presented in Figure 2.7e (red curve) allows concluding that GFP-TfR experiences transient confinement within the analyzed membrane region. Also, by fitting the *i*MSD curve to Equation 2.15, several relevant biological parameters can be calculated, as follows: GFP-TfR diffusion within the mesh domains $D_{micro} = 0.15\ \mu m^2\, s^{-1}$ (see also the inset in Figure 2.7e); long-range "hop" diffusion $D_{macro} = 0.03\ \mu m^2\, s^{-1}$; confinement length $L = 400$ nm. Of note, this analysis is correct as far as the region of interest considered is much larger than the average space explored by the molecule in the time window of the measurement: here, GFP-TfR displacement is analyzed within a 1-second time window, during which the protein diffuses an average of less than 1 μm in linear length ($D = 0.2\ \mu m^2\, s^{-1}$), as compared with a region linear size of 6.4 μm. Consequently, less than 0.3% of TfR molecules will travel up to a distance of 2 μm from their initial position in 1 s. As control, we treated cells with Latrunculin-B, a well-known actin-perturbing agent (Figure 2.7f). Under these conditions, the GFP-TfR *i*MSD exhibits a dramatic shift toward an overall linear trend, suggesting that meshwork-induced confinement of the protein is almost completely eliminated (green curve in Figure 2.7e). As a further experiment, the temperature dependence of GFP-TfR diffusion is evaluated: the *i*MSD curves obtained for the two limit temperatures (i.e., 20°C and 40°C) are shown in Figure 2.8a. The D_{macro} value increases with the temperature, confirming that GFP-TfR "hop" diffusion is favored at higher temperatures (Figure 2.8b). This is not surprising, considering that diffusion in the presence of integer cytoskeleton architecture is an energy-requiring process. Indeed, by a linear fit of D_{macro} data in an Arrhenius plot, an activation energy of 1.4 ± 0.5 eV is estimated, a value comparable to that previously determined for thermally activated cytoskeleton rearrangements (Sunyer et al. 2009). The average σ_0 value retrieved for GFP-TfR in physiological conditions is 0.098 μm^2, a value significantly higher than the experimental waist (0.075 μm^2): this may indicate the presence of large protein aggregates, or it can be the result of a slow sampling rate. Figure 2.9a shows this concept by comparison of the GFP-TfR *i*MSD calculated at 10 ms repetition time (light red line) with that obtained at 200 ms repetition time (red dots). To tackle this uncertainty, TfR dynamics at a much faster timescale is explored. In the case of large aggregates, in fact, σ_0 would remain unchanged at a faster timescale. By contrast, in the case of isolated proteins, σ_0 should match the waist value.

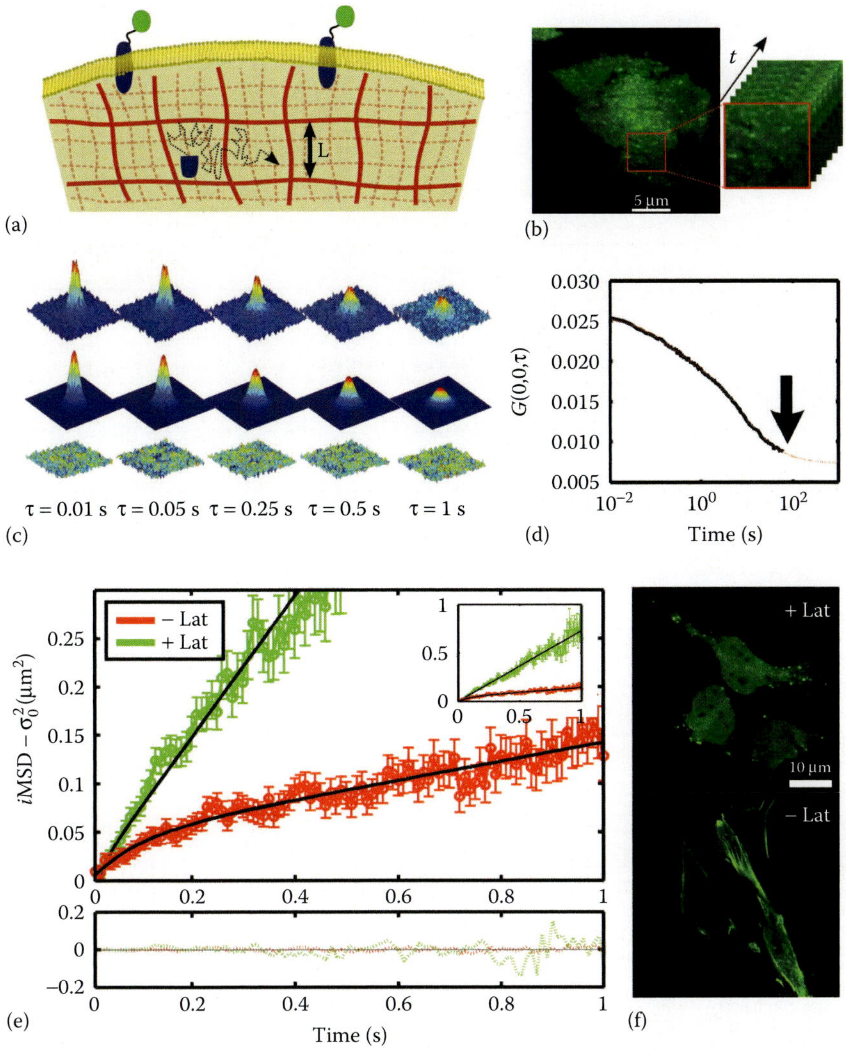

Figure 2.7

Analysis of TfR dynamics in living cells. (a) Schematic representation of a GFP-tagged TfR diffusing in the cytoskeleton hierarchical meshwork, with particular emphasis on the spatial size accessible in the tens-of-millisecond timescale. (b) TIRF microscopy image of a CHO cell expressing GFP-tagged TfR (left) and the detail of the membrane-patch used for the measurement (right). (c) Correlation function temporal evolution with the corresponding Gaussian fit and residues. (d) The autocorrelation plot shows that the characteristic time of the fluctuations is shorter than the total length of the measurement (arrow). Thus, immobile fraction removal is a safe operation (see main text). (e) *i*MSD versus time plot for GFP-TfR in physiological conditions (red curve) and after 30 min of Latrunculin-B treatment (green curve). The inset shows the *i*MSD trend at a short timescale. (f) Fluorescence images of cells transfected with actin-GFP that show the effect of Latrunculin-B on the integrity of actin filaments after 30 min of treatment. (From Di Rienzo, C. et al., *Proc. Natl. Acad. Sci. U.S.A.*, 110, 12307–12312, 2013a. With permission.)

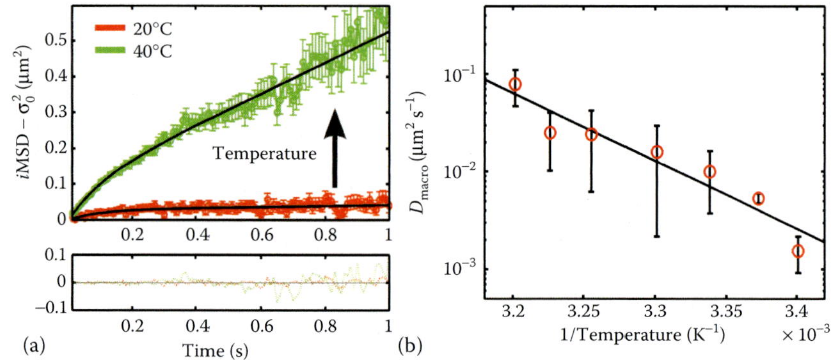

Figure 2.8

Temperature dependence of TfR dynamics. (a) *i*MSD for GFP-tagged TfR in CHO cells at 20°C (red curve) and at 40°C (green curve). An increased slope at longer times indicates that hop diffusivity increases with temperature. (b) Arrhenius plot of D_{macro}. (From Di Rienzo, C. et al., *Proc. Natl. Acad. Sci. U.S.A.*, 110, 12307–12312, 2013a. With permission.)

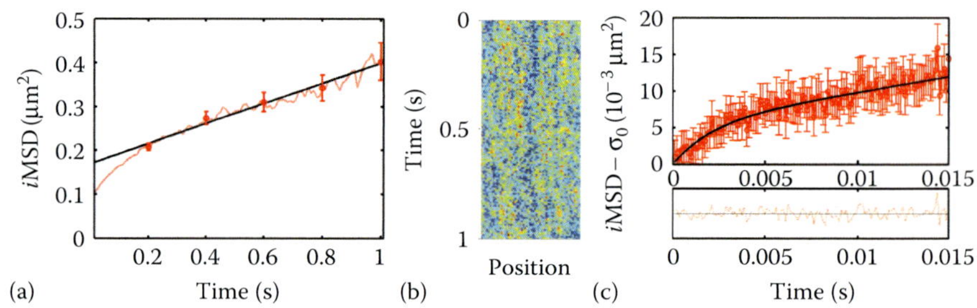

Figure 2.9

Effect of sampling frequency on σ_0: faster acquisition is needed to reveal subtle dynamics. (a) Comparison of the GFP-TfR *i*MSD calculated at 10 ms repetition time (light red line) with that at 200 ms (red dots). The black line represents the best linear fit of red dots. (b) Representative GFP-TfR intensity carpet obtained by raster scanning of a line of 6 μm. (c) *i*MSD derived from the carpet in (b). (From Di Rienzo, C. et al., *Proc. Natl. Acad. Sci. U.S.A.*, 110, 12307–12312, 2013a. With permission.)

Thus, a ~100-times faster acquisition is achieved by line scanning with a confocal microscope (Figure 2.9b). Notably, we are able to reconstruct the TfR diffusion law at a short timescale (125 μs–10 ms): we obtain a transient confinement with a linear size of 140 nm ($D_{micro} = 0.7\ \mu m^2\ s^{-1}$ and $D_{macro} = 0.2\ \mu m^2\ s^{-1}$), far below the diffraction limit and in keeping with SPT measurements conducted on a similar timescale (Fujiwara et al. 2002). Most importantly, the newly obtained σ_0 value matches the instrumental PSF ($\sigma_0 \sim$ PSF ~ 240 nm), thus excluding the presence of large diffusing protein aggregates.

2.4.2 The GFP-GPI Dynamic Partitioning

Available data demonstrate that glycosylphosphatidylinositol- and myristoyl-tagged GFP-variant fusion proteins are targeted to the plasma membrane (Lenne et al. 2006; Rhee et al. 2006). Also, data suggest that this lipid-modified fusion protein is dynamically targeted to distinct lipid-enriched membrane compartments, also known as "lipid rafts" (Eggeling et al. 2009). The mechanism regulating GFP-GPI dynamics and localization in live cells is typically referred to as "dynamic partitioning." As shown above, simulations demonstrate that the *i*MSD approach has the ability to discriminate between transiently confined diffusion in a meshwork and dynamic partitioning into membrane nanodomains (i.e., lipid rafts). In fact the latter needs two Gaussian components for a satisfactory fitting. This reflects the presence of two segregated molecular diffusive behaviors: (1) pure isotropic diffusion outside of the nanodomains and (2) confined diffusion within the nanodomains. Of note, in this case the *i*MSD trace calculated from the variance does not coincide with the *i*MSD calculated from the temporal correlation (fitting the amplitudes). This is expected for a multicomponent system, as already discussed. Concerning experiments, Figure 2.10a shows a representative TIRF image of GFP-GPI transiently transfected into live CHO-K1 cells. The time evolution of GFP-GPI correlation function (Figure 2.10b, first line) and the corresponding fitting by Equation 2.9 (Figure 2.10b, second line) show that a one-component Gaussian model is not enough for a satisfactory description of the system. By contrast, fitting with a two-Gaussian model (Figure 2.10b, fourth line) well describes the experimental function, producing random residuals (Figure 2.10b, fifth line). The experimental *i*MSD curves are qualitatively keeping with the theoretical predictions, in

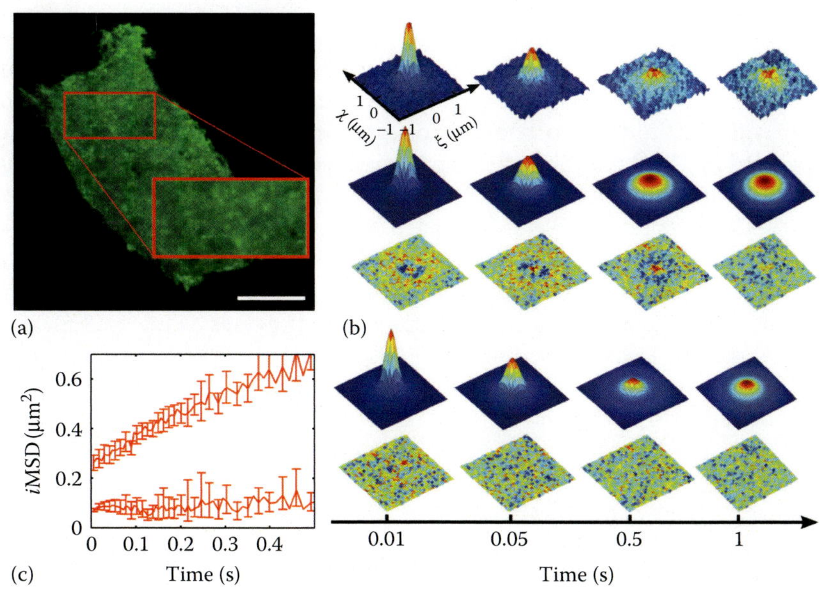

Figure 2.10

GFP-GPI correlation function shows the typical pattern of dynamic partitioning. (a) A representative TIRF image of GFP-GPI. (b) Time evolution of GFP-GPI correlation function (first line) and the corresponding fitting by Equation 2.4 (second line). Residuals (third line) show that a one-component Gaussian model does not afford a satisfactory description of the system. Fitting with a two Gaussian model (fourth line) describes better the experimental function, producing random residuals (fifth line). (c) The calculated *i*MSD curves are qualitative in keeping with the theoretical predictions, in the sense that they show an almost constant *i*MSD component (trapped molecules) and a linearly increasing one (diffusing molecules). (From Di Rienzo, C. et al., *Proc. Natl. Acad. Sci. U.S.A.*, 110, 12307–12312, 2013a. With permission.)

the sense that they show an almost constant *i*MSD component (trapped molecules) and a linearly increasing one (diffusing molecules) (Figure 2.10c).

2.4.3 The Transient Receptor Vanilloid 1 Super Diffusion

Transient Receptor Potential Vanilloid 1 (TRPV1) is a nonselective cation channel that integrates several stimuli into nociception and neurogenic inflammation. It has been previously demonstrated that, in static condition, TRPV1 and β-tubulin yield a well-detectable fluorescence resonance energy transfer (FRET) signal, thereby demonstrating the existence of a complex (here named TRPV1-T). This interaction is here analyzed from a spatiotemporal point of view by the *i*MSD approach, but adopted in FRET modality, that is, using EGFP-labeled TRPV1 and TagRFP-labeled β-tubulin. Remarkably, in this case, the *i*MSD of TRPV1-T clearly shows a hyperbolic trend in time (Figure 2.11). This "super-diffusive" behavior can be reasonably attributed to guided/active mobility (Dix et al. 2006), and is characterized by a diffusion constant of $D = (9.0 \pm 2.6) \times 10^{-3}\ \mu m^2\ s^{-1}$ and $\alpha = 1.5 \pm 0.3$. Interestingly, we find the radius of the super-diffusive TRPV1-T species (σ_0) to be around 400 nm. This finding suggests that the interaction between the TRPV1 and microtubules occurs across a large submicron domain of cell membrane. Consistently, a highly patterned signal is visible also in the static FRET image, where several fluorescent spots with analogous sizes are clearly visible (see example in Figure 2.11c).

2.4.4 Lipids Diffusion on Cell Membranes

Differently from transmembrane proteins or lipid-anchored proteins, membrane lipids are expected to diffuse freely on the plasma membrane (Eggeling et al. 2009; Mueller et al. 2011). As a paradigmatic example, we shall use a fluorescently tagged palmitoylphosphoethanolamine (PPE). In this case, we introduce the labeled molecule into the plasma membrane of a living cell by means of fusogenic liposomes obtained by hydration (procedure described in detail in Kleusch et al. 2012). A representative TIRF image of the basal membrane of a CHO cell labeled with Atto488-PPE is presented in Figure 2.12. As expected, the measured *i*MSD (Figure 2.12c) is linear in time, indicating a mostly free 2D diffusion of this molecule on the membrane, in keeping with previous STED-FCS measurements (Eggeling et al. 2009; Mueller et al. 2011). It is worth mentioning that, thanks to the very high speed achievable by the used camera-based setup, all the measured lipid displacements are well below the diffraction limit. This observation demonstrates the ability of the *i*MSD approach to resolve average molecular displacements down to few tens of nanometers, also for fast-diffusing membrane components, such as lipids.

2.5 Discussion

2.5.1 Why Super-Resolution?

The ability of the *i*MSD approach to super-resolve the average dynamic behavior of molecules in live cells has been particularly stressed thus far. At this point, it is legitimate to ask whether the same level of information may be accessed by recently developed technologies capable of combining subdiffraction spatial resolution with the high temporal resolution needed to track molecules in 2D (or 3D) environments. Some are based on localization precision, such as stochastic optical reconstruction microscopy (STORM) (Rust et al. 2006; Jones et al. 2011),

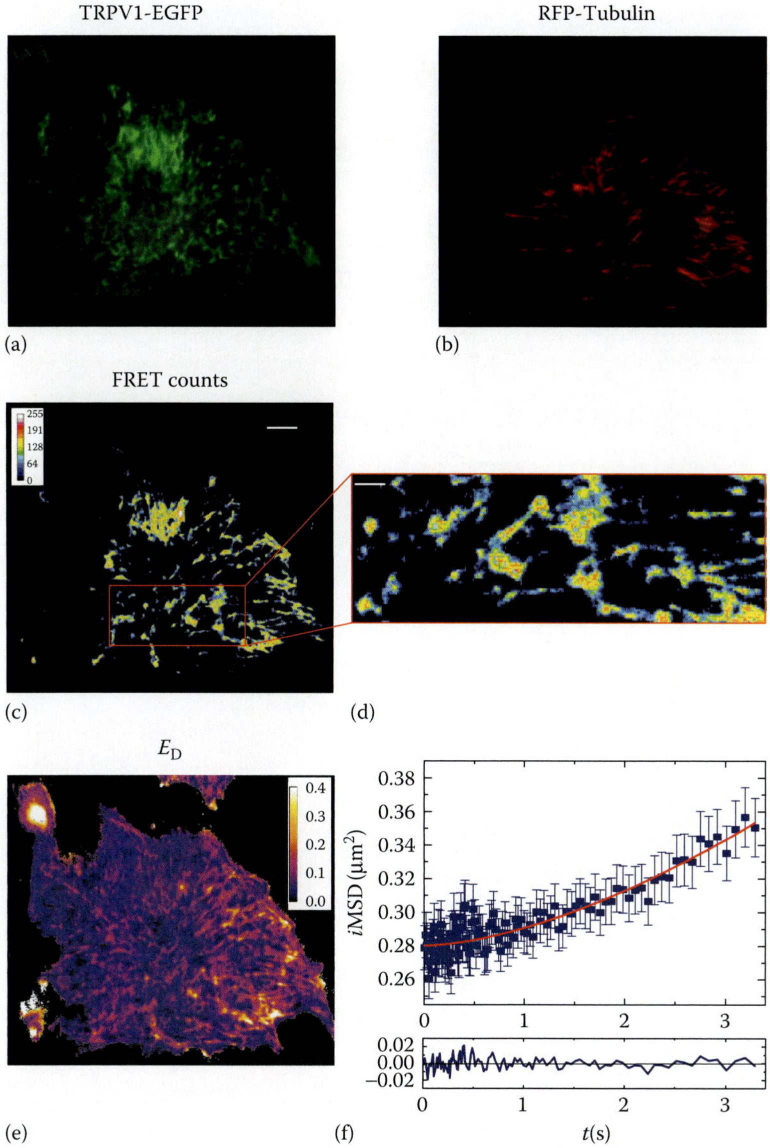

Figure 2.11

*i*MSD-FRET analysis of TRPV1–microtubules interaction (TRPV1-T). (a) Donor emission image, (b) acceptor emission image, (c) FRET counts image (scale bar: 5 μm), (d) zoom of the cellular region used for the STICS-FRET measurement (scale bar: 1 μm), (e) FRET efficiency image FRET counts, and (f) *i*MSD versus time plot for FRET signal between TRPV1-EGFP and RFP tubulin in physiological condition. The fitting shows that TRPV1-T undergoes super-diffusion in the cellular membrane. (Adapted from Storti, B. et al. *PLoS One*, doi:10.1371/journal.pone.0116900.)

photoactivated localization microscopy (PALM) (Betzig et al. 2006), fluorescence PALM (FPALM) (Hess et al. 2006), and SPT PALM (sptPALM) (Manley et al. 2008): the relatively large amount of photons required at each snapshot, however, limits the time resolution of these methods to at least several milliseconds, thus hampering their effectiveness in applications like those reported here. Of note, a

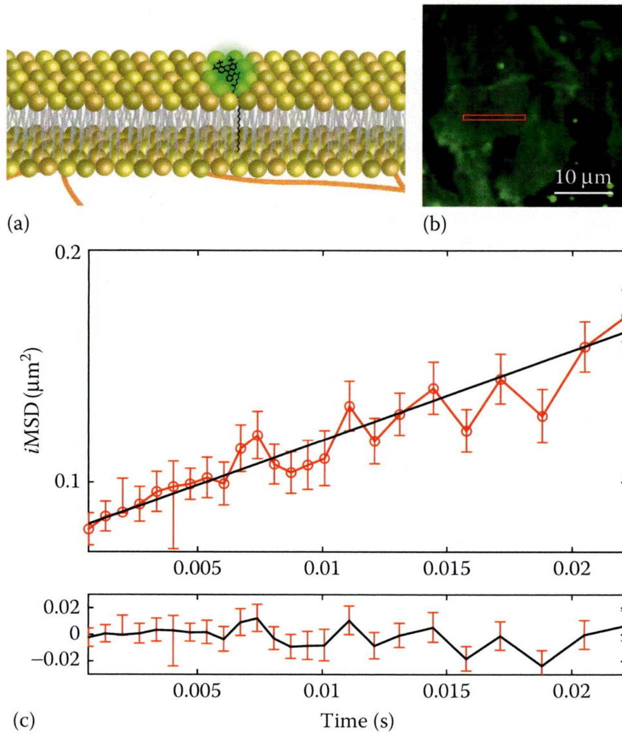

Figure 2.12

ATTO488-PPE diffusion law in live cell membranes. (a) Schematic representation of 534 ATTO488-PPE insertion into the cell membrane. (b) TIRF image of CHO basal membrane labeled with ATTO488-PPE: a ROI (red box) is selected in a mostly uniform part of the cell, avoiding cell border and highly fluorescent spots. (c) The *i*MSD measured in the selected ROI shows a linear behavior confirming a free diffusion model for this component. (Adapted from Di Rienzo, C. et al., *J. Vis. Exp.*, 92, 2014a.)

promising alternative has been opened by spatially modulating the fluorescence emission with STED (or reversible saturable optical linear fluorescence transitions [RESOLFT]) (Klar and Hell 1999; Hell 2007). These approaches combine the possibility to shape the observation volume well below the diffraction limit with the use of fast scanning microscopes and detection systems. In particular, in a STED-FCS measurement, the average transit time of molecules for decreasing observation volumes is measured by fluctuation analysis of the fluorescence signal. This allows obtaining a local measurement of molecular diffusion directly below the diffraction limit, as demonstrated by addressing the spatiotemporal dynamics of lipids and proteins in live cell membranes (Eggeling et al. 2009; Hedde et al. 2013). By contrast, in the approach presented here, diffusion is measured as an average of all the particles' behavior in the region of interest, by means of a standard, diffraction-limited, system. Reported results, however, demonstrate that this method is not limited by diffraction, but only by the temporal resolution available. In fact, analogously to what is done in other super-resolution techniques, such as PALM and STORM, molecular displacements well below the diffraction limit can be measured. Contrary to STED-FCS, this approach can be easily applied to a wide range of standard microscopy setups, such as raster scanning microscopes or widefield camera-based microscopes. Also, contrary to STED-FCS, the measurement presented here does not require a fluorophore-dependent calibration of the size of the instrumental waist. The actual resolution in the measurement of particle displacements by the *i*MSD approach depends on how accurately the correlation function is calculated. This property recalls the SPT case, where the resolution strictly depends on how accurately the particle "image" is reconstructed, and well explains that the *i*MSD method is not inherently limited by diffraction. To measure a significant correlation in less than a minute, few photons per molecule per frame (usually <10 photons) are enough. In fact, the contribution of all the observed particles is averaged together during correlation function calculation, even if particles are not isolated. This is an intrinsic property of fluctuation spectroscopy methods and makes it possible to use relatively dim and dense labels, such as fluorescent proteins transfected in live cells. In this light, it shall appear clear that the minimum measurable displacement depends on the diffusivity of the particle and on the time resolution of the measurement. The experiments showed here demonstrate, for instance, that a time resolution of approximately 10^{-4} s is needed to measure an average displacement of 50 nm on the plasma membrane, where the maximum measured diffusivity for proteins and lipids is around 5 μm^2 s^{-1}. An important additional requirement for this method to accurately describe molecular displacements is an appropriate spatial sampling: in particular, a pixel size smaller than the experimental waist is typically needed for a satisfactory fit of the correlation function. In most standard microscopes, the PSF waist varies in the 200–500 nm range (variability depends on the numerical aperture of the selected objective and on the wavelength used) and can be easily calibrated by using nanosized fluorescent particles. The pixel size can be adapted to the system under study taking into account a simple rule: lower the pixel size, higher the accuracy in the description of the correlation function: in general, a pixel size of 70–150 nm (2–3 times lower than the instrumental waist) can be enough. Furthermore, the minimum size of the region of interest shall be at least three times larger than the maximum displacement of interest (plus the instrumental waist): this helps to reach a statistically significant sampling of molecular displacements for the fitting algorithm to properly work.

2.5.2 Experimental Requirements

From an experimental point of view, the presented approach requires only access to a microscope equipped with a fast acquisition module (e.g., in the TIRF or SPIM configuration [Di Rienzo et al. 2014a; Hedde et al. 2014]). As discussed above, a time resolution below 1 ms is needed to properly describe the dynamics of fast membrane lipids at a spatial scale below 100 nm. In order to reach this temporal resolution, a region of interest smaller than the whole chip of the camera shall be selected. In this way, the camera will record from a reduced number of lines, thus increasing the time resolution. Typically, however, several milliseconds are inevitably lost to shift the charges from the "exposure" to the "readout" chip of the camera. To beat this limit, an emerging technology allows shifting the recorded lines only instead of the whole frame, with an effective reduction of the exposed chip size (for this configuration to be effective, the chip outside of the region of interest must be covered by slits mounted in the optical path). Thanks to this technology, a time resolution down to 10^{-4} s can be achieved to acquire an entire image. Almost the same time resolution can be achieved among adjacent lines with a Leica SP5 confocal microscope in the resonant-scan mode (e.g., a line of 128 pixel, 6.4 μm long, is acquired at a frequency of 8000 Hz, that is 125 μs per line).

Few details about the detectors to be used are worthy of note. In general, the requirements are the same of FCS-based measurements. Single-photon counter detectors, like Avalanche photodiodes (APDs), are the most common choice for these measurements. APDs provide good single-photon sensitivity and a very fast response, although they have a limited dynamic range. On the other hand, classical PMTs, particularly at high-gain voltage, display very good single-photon sensitivity, but measure an average intensity that is not directly proportional to the collected light (due to the presence of an offset). Although this offset is low compared to the dynamic range of the camera (few hundred compared to 2^{16} in 16 bit readout) and negligible in experiments where many photons are collected, it has to be carefully taken into account to achieve a correct normalization of the correlation function. On the other hand, the offset can be used as a reference to identify the amount of signal collected (e.g., in low-light conditions). Of note, in order to estimate the average amount of photons that are collected during the acquisition, the average digital level associated to each collected photon has to be measured. This quantity can be retrieved by exposing the detector to a very low light intensity (e.g., the background light in the room); under these conditions, it can be reasonably assumed that just single photons reach the camera, that is, the measured intensity can be related to zero or one detected photon only (Dalal et al. 2008). Commercial detector arrays suitable for FCS include the EMCCDs, to whom all the considerations made for PMTs can be applied. An additional alternative is represented by sCMOSs camera. These devices have been successfully applied to FCS measurements for very bright particles (Singh et al. 2013). Yet, the characteristic noise level of these systems severely limits their potential use in combination with relatively dim molecules such as fluorescent proteins. In the context of the emerging imaging technologies, an interesting approach is based on the possibility to produce very thin light sheets (1–2 μm) through the sample (Voie et al. 1993). The light sheet allows selective illumination of a single plane (single plane illumination microscopy, SPIM) in the sample and, combined with a camera-based acquisition system, fast optical sectioning in 3D (Huisken et al. 2004). Because of these characteristics, SPIM has been successfully conjugated with FCS (Wohland et al. 2010) and could represent a valid tool to extend the presented analysis to 3D environments, such as the cytoplasm or the nucleus of living cells (Hedde et al. 2014).

2.5.3 Potential Impact

The aim here is to revisit the spatial distribution and dynamics of proteins and lipids in live cells using a novel FCS-based approach which provides conceptually the same physical quantities of classical SPT techniques but using small and not particularly bright molecular labels (e.g., GFP). Methods based on spatiotemporal image fluctuation analysis preserve single-molecule sensitivity in the presence of many similarly labeled molecules, with no need to extract the trajectory of each molecule. This in turn allows handling large data sets and determining a number of average kinetic and thermodynamic parameters on wide spatial and temporal scales directly from imaging, with no need to preliminarily assume any interpretative model. Such a quantitative access to the thermodynamic landscape underlying molecular diffusion and interactions in live cells will represent the basis to successfully address the different intra- and extracellular dynamic organization of molecules under selected experimental conditions (e.g., physiological vs. pathological condition). Based on these considerations, we envision that the presented strategy shall represent a useful asset to guide future research in biomedicine.

References

Berland, K. M., P. T. So, Y. Chen, W. W. Mantulin, and E. Gratton. 1996. Scanning two-photon fluctuation correlation spectroscopy: Particle counting measurements for detection of molecular aggregation. *Biophysical Journal* 71(1):410–420. doi:10.1016/S0006-3495(96)79242-1.

Betzig, E., G. H. Patterson, R. Sougrat, O. W. Lindwasser, S. Olenych, J. S. Bonifacino, M. W. Davidson, J. Lippincott-Schwartz, and H. F. Hess. 2006. Imaging intracellular fluorescent proteins at nanometer resolution. *Science* 313(5793):1642–1645. doi:10.1126/science.1127344.

Brown, D. A. and E. London. 2000. Structure and function of sphingolipid- and cholesterol-rich membrane rafts. *The Journal of Biological Chemistry* 275(23):17221–17224.

Capps, G. G., S. Pine, M. Edidin, and M. C. Zuniga. 2004. Short class I major histocompatibility complex cytoplasmic tails differing in charge detect arbiters of lateral diffusion in the plasma membrane. *Biophysical Journal* 86(5):2896–2909.

Cardarelli, F., L. Lanzano, and E. Gratton. 2012a. Capturing directed molecular motion in the nuclear pore complex of live cells. *Proceedings of the National Academy of Sciences of the United States of America* 109(25):9863–9868. doi:10.1073/pnas.1200486109.

Cardarelli, F., L. Lanzano, and E. Gratton. 2012b. Fluorescence correlation spectroscopy of intact nuclear pore complexes. *Biophysical Journal* 101(4):L27–L29. doi:10.1016/j.bpj.2011.04.057.

Dalal, R. B., M. A. Digman, A. F. Horwitz, V. Vetri, and E. Gratton. 2008. Determination of particle number and brightness using a laser scanning confocal microscope operating in the analog mode. *Microscopy Research and Technique* 71(1):69–81. doi:10.1002/Jemt.20526.

Di Rienzo, C., E. Gratton, F. Beltram, and F. Cardarelli. 2013a. Fast spatiotemporal correlation spectroscopy to determine protein lateral diffusion laws in live cell membranes. *Proceedings of the National Academy of Sciences of the United States of America* 110(30):12307–12312. doi:10.1073/pnas.1222097110.

Di Rienzo, C., E. Gratton, F. Beltram, and F. Cardarelli. 2014a. From fast fluorescence imaging to molecular diffusion law on live cell membranes in a commercial microscope. *Journal of Visualized Experiments* 92. doi:10.3791/51994.

Di Rienzo, C., E. Gratton, V. Piazza, F. Beltram, and F. Cardarelli. 2014b. Probing short-range protein Brownian motion in the cytoplasm of living cells. *Nature Communications*. doi:10.1038/ncomms6891.

Di Rienzo, C., E. Jacchetti, F. Cardarelli, R. Bizzarri, F. Beltram, and M. Cecchini. 2013b. Unveiling LOX-1 receptor interplay with nanotopography: Mechanotransduction and atherosclerosis onset. *Scientific Reports* 3:1141. doi:10.1038/srep01141.

Digman, M. A. and E. Gratton. 2009. Analysis of diffusion and binding in cells using the RICS approach. *Microscopy Research and Technique* 72(4):323–332.

Dix, J. A., E. F. Y. Hom, and A. S. Verkman. 2006. Fluorescence correlation spectroscopy simulations of photophysical phenomena and molecular interactions: a molecular dynamics/monte carlo approach. *The Journal of Physical Chemistry B* 110(4):1896–1906.

Edidin, M. 2003. The state of lipid rafts: from model membranes to cells. *Annual Review of Biophysics and Biomolecular Structure* 32:257–283.

Eggeling, C., C. Ringemann, R. Medda, G. Schwarzmann, K. Sandhoff, S. Polyakova, V. N. Belov, et al. 2009. Direct observation of the nanoscale dynamics of membrane lipids in a living cell. *Nature* 457(7233):1159–1162. doi:10.1038/nature07596.

Fielding, C. J. 2006. *Lipid Rafts and Caveolae: From Membrane Biophysics to Cell Biology.*

Fujiwara, T., K. Ritchie, H. Murakoshi, K. Jacobson, and A. Kusumi. 2002. Phospholipids undergo hop diffusion in compartmentalized cell membrane. *The Journal of Cell Biology* 157(6):1071–1081.

Hanzal-Bayer, M. F. and J. F. Hancock. 2007. Lipid rafts and membrane traffic. *FEBS Letters* 581(11):2098–2104.

Hebert, B., S. Costantino, and P. W. Wiseman. 2005. Spatiotemporal image correlation spectroscopy (STICS) theory, verification, and application to protein velocity mapping in living CHO cells. *Biophysical Journal* 88(5):3601–3614. doi:10.1529/biophysj.104.054874.

Hedde, P. N., R. M. Dorlich, R. Blomley, D. Gradl, E. Oppong, A. C. Cato, and G. U. Nienhaus. 2013. Stimulated emission depletion-based raster image correlation spectroscopy reveals biomolecular dynamics in live cells. *Nature Communications* 4:2093. doi:10.1038/ncomms3093.

Hedde, P. N., M. Stakic, and E. Gratton. 2014. Rapid measurement of molecular transport and interaction inside living cells using single plane illumination. *Scientific Reports* 4. doi:10.1038/srep07048.

Heinemann, F., V. Betaneli, F. A. Thomas, and P. Schwille. 2012. Quantifying lipid diffusion by fluorescence correlation spectroscopy: A critical treatise. *Langmuir* 28(37):13395–13404. doi:10.1021/la302596h.

Hell, S. W. 2007. Far-field optical nanoscopy. *Science* 316(5828):1153–1158. doi:10.1126/science.1137395.

Hell, S. W. and J. Wichmann. 1994. Breaking the diffraction resolution limit by stimulated emission: stimulated-emission-depletion fluorescence microscopy. *Optics Letters* 19(11):780–782.

Hess, S. T., T. P. Girirajan, and M. D. Mason. 2006. Ultra-high resolution imaging by fluorescence photoactivation localization microscopy. *Biophysical Journal* 91(11):4258–4272. doi:10.1529/biophysj.106.091116.

Huisken, J., J. Swoger, F. Del Bene, J. Wittbrodt, and E. H. Stelzer. 2004. Optical sectioning deep inside live embryos by selective plane illumination microscopy. *Science* 305(5686):1007–1009. doi:10.1126/science.1100035.

Jacobson, K., O. G. Mouritsen, and R. G. W. Anderson. 2007. Lipid rafts: At a crossroad between cell biology and physics. *Nature Cell Biology* 9(1):7–14.

Jones, S. A., S. H. Shim, J. He, and X. Zhuang. 2011. Fast, three-dimensional super-resolution imaging of live cells. *Nature Methods* 8(6):499–508. doi:10.1038/nmeth.1605.

Kannan, B., J. Y. Har, P. Liu, I. Maruyama, J. L. Ding, and T. Wohland. 2006. Electron multiplying charge-coupled device camera based fluorescence correlation spectroscopy. *Analytical Chemistry* 78(10):3444–3451. doi:10.1021/ac0600959.

Klar, T. A. and S. W. Hell. 1999. Subdiffraction resolution in far-field fluorescence microscopy. *Optics Letters* 24(14):954–956. doi:10.1364/OL.24.000954.

Kleusch, C., N. Hersch, B. Hoffmann, R. Merkel, and A. Csiszár. 2012. Fluorescent lipids: Functional parts of fusogenic liposomes and tools for cell membrane labeling and visualization. *Molecules* 17(1):1055–1073.

Kodippili, G. C., J. Spector, C. Sullivan, F. A. Kuypers, R. Labotka, P. G. Gallagher, K. Ritchie, and P. S. Low. 2009. Imaging of the diffusion of single band 3 molecules on normal and mutant erythrocytes. *Blood* 113(24):6237–6245.

Kolin, D. L., S. Costantino, and P. W. Wiseman. 2006. Sampling effects, noise, and photobleaching in temporal image correlation spectroscopy. *Biophysical Journal* 90(2):628–639. doi:10.1529/biophysj.105.072322.

Kusumi, A., T. K. Fujiwara, R. Chadda, M. Xie, T. A. Tsunoyama, Z. Kalay, R. S. Kasai, and K. G. Suzuki. 2012. Dynamic organizing principles of the plasma membrane that regulate signal transduction: Commemorating the fortieth anniversary of Singer and Nicolson's fluid-mosaic model. *Annual Review of Cell and Developmental Biology* 28:215–250.

Kusumi, A., H. Ike, C. Nakada, K. Murase, and T. Fujiwara. 2005a. Single-molecule tracking of membrane molecules: Plasma membrane compartmentalization and dynamic assembly of raft-philic signaling molecules. *Seminars in Immunology* 17(1):3–21. doi:10.1016/j.smim.2004.09.004.

Kusumi, A., I. Koyama-Honda, and K. Suzuki. 2004. Molecular dynamics and interactions for creation of stimulation-induced stabilized rafts from small unstable steady-state rafts. *Traffic* 5(4):213–230.

Kusumi, A., C. Nakada, K. Ritchie, K. Murase, K. Suzuki, H. Murakoshi, R. S. Kasai, J. Kondo, and T. Fujiwara. 2005b. Paradigm shift of the plasma membrane concept from the two-dimensional continuum fluid to the partitioned fluid: High-speed single-molecule tracking of membrane molecules. *Annual Review of Biophysics and Biomolecular Structure* 34:351–378. doi:10.1146/annurev.biophys.34.040204.144637.

Kusumi, A., Y. Sako, and M. Yamamoto. 1993. Confined lateral diffusion of membrane receptors as studied by single particle tracking (nanovid microscopy). Effects of calcium-induced differentiation in cultured epithelial cells. *Biophysical Journal* 65(5):2021–2040. doi:10.1016/S0006-3495(93)81253-0.

Kusumi, A., Y. M. Shirai, I. Koyama-Honda, K. G. Suzuki, and T. K. Fujiwara. 2010. Hierarchical organization of the plasma membrane: Investigations by single-molecule tracking vs. fluorescence correlation spectroscopy. *FEBS Letters* 584(9):1814–1823. doi:10.1016/j.febslet.2010.02.047.

Lenne, P. F., L. Wawrezinieck, F. Conchonaud, O. Wurtz, A. Boned, X. J. Guo, H. Rigneault, H. T. He, and D. Marguet. 2006. Dynamic molecular confinement in the plasma membrane by microdomains and the cytoskeleton meshwork. *EMBO Journal* 25(14):3245–3256. doi:10.1038/sj.emboj.7601214.

Manley, S., J. M. Gillette, G. H. Patterson, H. Shroff, H. F. Hess, E. Betzig, and J. Lippincott-Schwartz. 2008. High-density mapping of single-molecule trajectories with photoactivated localization microscopy. *Nature Methods* 5(2):155–157. doi:10.1038/nmeth.1176.

Morone, N., T. Fujiwara, K. Murase, R. S. Kasai, H. Ike, S. Yuasa, J. Usukura, and A. Kusumi. 2006. Three-dimensional reconstruction of the membrane skeleton at the plasma membrane interface by electron tomography. *The Journal of Cell Biology* 174(6):851–862.

Mueller, V., C. Ringemann, A. Honigmann, G. Schwarzmann, R. Medda, M. Leutenegger, S. Polyakova, V. N. Belov, S. W. Hell, and C. Eggeling. 2011. STED nanoscopy reveals molecular details of cholesterol- and cytoskeleton-modulated lipid interactions in living cells. *Biophysical Journal* 101(7):1651–1660. doi:10.1016/j.bpj.2011.09.006.

Rhee, J. M., M. K. Pirity, C. S. Lackan, J. Z. Long, G. Kondoh, J. Takeda, and A.-K. Hadjantonakis. 2006. In vivo imaging and differential localization of lipid-modified GFP-variant fusions in embryonic stem cells and mice. *Genesis (New York, NY: 2000)* 44(4):202–218.

Ries, J. and P. Schwille. 2006. Studying slow membrane dynamics with continuous wave scanning fluorescence correlation spectroscopy. *Biophysical Journal* 91(5):1915–1924. doi:10.1529/biophysj.106.082297.

Ruan, Q., M. A. Cheng, M. Levi, E. Gratton, and W. W. Mantulin. 2004. Spatial-temporal studies of membrane dynamics: Scanning fluorescence correlation spectroscopy (SFCS). *Biophysical Journal* 87(2):1260–1267. doi:10.1529/biophysj.103.036483.

Rust, M. J., M. Bates, and X. Zhuang. 2006. Sub-diffraction-limit imaging by stochastic optical reconstruction microscopy (STORM). *Nature Methods* 3(10):793–795. doi:10.1038/nmeth929.

Ryan, T. A., J. Myers, D. Holowka, B. Baird, and W. W. Webb. 1988. Molecular crowding on the cell surface. *Science (New York, NY)* 239(4835):61–64.

Sako, Y. and A. Kusumi. 1994. Compartmentalized structure of the plasma membrane for receptor movements as revealed by a nanometer-level motion analysis. *Journal of Cell Biology* 125(6):1251–1264.

Sako, Y. and A. Kusumi. 1995. Barriers for lateral diffusion of transferrin receptor in the plasma membrane as characterized by receptor dragging by laser tweezers: fence versus tether. *Journal of Cell Biology* 129(6):1559–1574.

Schwille, P., J. Korlach, and W. W. Webb. 1999. Fluorescence correlation spectroscopy with single-molecule sensitivity on cell and model membranes. *Cytometry* 36(3):176–182. doi:10.1002/(SICI)1097-0320(19990701)36:3%3C176::AID-CYTO5%3E3.0.CO;2-F.

Sengupta, P., T. Jovanovic-Talisman, D. Skoko, M. Renz, S. L. Veatch, and J. Lippincott-Schwartz. 2011. Probing protein heterogeneity in the plasma membrane using PALM and pair correlation analysis. *Nature Methods* 8(11):969–975. doi:10.1038/Nmeth.1704.

Sezgin, E., I. Levental, M. Grzybek, G. Schwarzmann, V. Mueller, A. Honigmann, V. N. Belov, C. Eggeling, U. Coskun, K. Simons, and P. Schwille. 2012. Partitioning, diffusion, and ligand binding of raft lipid analogs in model and cellular plasma membranes. *Biochimica et Biophysica Acta* 1818(7):1777–1784.

Shusterman, R., T. Gavrinyov, and O. Krichevsky. 2008. Internal dynamics of superhelical DNA. *Physical Review Letters* 100(9):098102.

Simons, K. and E. Ikonen. 1997. Functional rafts in cell membranes. *Nature* 387(6633):569–572.

Simons, K. and W. L. Vaz. 2004. Model systems, lipid rafts, and cell membranes. *Annual Review of Biophysics and Biomolecular Structure* 33:269–295.

Singer, S. J. and G. L. Nicolson. 1972. The fluid mosaic model of the structure of cell membranes. *Science* 175(4023):720–731. doi:10.1126/science.175.4023.720.

Singh, A. P., J. W. Krieger, J. Buchholz, E. Charbon, J. Langowski, and T. Wohland. 2013. The performance of 2D array detectors for light sheet based fluorescence correlation spectroscopy. *Optics Express* 21(7):8652–8668.

Storti, B., C. Di Rienzo, F. Cardarelli, F. Beltram, and R. Bizzarri. 2015. Unveiling TRPV1 spatio-temporal organization in live cell membranes. *PLoS One*. doi:10.1371/journal.pone.0116900.

Sunyer, R., F. Ritort, R. Farre, and D. Navajas. 2009. Thermal activation and ATP dependence of the cytoskeleton remodeling dynamics. *Physical Review E, Statistical, Nonlinear, and Soft Matter Physics* 79(5 Pt 1):051920.

Unruh, J. R. and E. Gratton. 2008. Analysis of molecular concentration and brightness from fluorescence fluctuation data with an electron multiplied CCD camera. *Biophysical Journal* 95(11):5385–5398. doi:10.1529/biophysj.108.130310.

Voie, A. H., D. H. Burns, and F. A. Spelman. 1993. Orthogonal-plane fluorescence optical sectioning: Three-dimensional imaging of macroscopic biological specimens. *Journal of Microscopy (Oxford)* 170(Pt 3):229–236. doi:10.1111/j.1365-2818.1993.tb03346.x.

Wawrezinieck, L., H. Rigneault, D. Marguet, and P. F. Lenne. 2005. Fluorescence correlation spectroscopy diffusion laws to probe the submicron cell membrane organization. *Biophysical Journal* 89(6):4029–4042. doi:10.1529/biophysj.105.067959.

Weiss, M., H. Hashimoto, and T. Nilsson. 2003. Anomalous protein diffusion in living cells as seen by fluorescence correlation spectroscopy. *Biophysical Journal* 84(6):4043–4052. doi:10.1016/S0006-3495(03)75130-3.

Wohland, T., X. Shi, J. Sankaran, and E. H. Stelzer. 2010. Single plane illumination fluorescence correlation spectroscopy (SPIM-FCS) probes inhomogeneous three-dimensional environments. *Optics Express* 18(10):10627–10641. doi:10.1364/OE.18.010627.

3

Photoswitching Fluorophores in Super-Resolution Fluorescence Microscopy

Ulrike Endesfelder

3.1 Short History of Microscopy
3.2 Choosing a Suitable Fluorophore
3.3 Mechanisms behind the Photoswitching of Fluorophores for Super-Resolution Microscopy
3.4 Super-Resolution Microscopy without the Need of Photoswitching
3.5 Outlook
References

3.1 Short History of Microscopy

3.1.1 Getting Close to the Limits: Improving Contrast and Resolution

For resolving biological samples in greater detail than naturally possible by the discrimination power of the human eye, scientists first used technical aids like water drops or eye glasses. In the late sixteenth century, finally the first compound microscopes with combinations of carefully cut lenses, resulting in an objective and ocular system, were built (Singer 1914). In 1873, when the physicists' discussion on whether the nature of light is better described as a particle or by a wave-like behavior was at its climax, Ernst Abbe then introduced the term "numerical aperture" of a microscopic system, a dimensionless number that characterizes the range of angles over which the system can collect light

$$NA = n \sin \alpha$$

Here n is the index of refraction (air 1.00, water 1.33, typical immersion oil 1.52, or typical optical component glass 1.52 [Schott glass]), and α is the half-angle of the maximum light cone that can be collected by the lens system. Abbe further states that the finest detail that can be resolved is proportional to $\lambda/2$ NA, with λ being the wavelength of light (Abbe 1873).

The lens-based microscopic systems now quickly improved: The astronomer Joseph von Fraunhofer developed the first achromatic system to correct for chromatic aberration, Ernst Abbe and Otto Schott worked on the refinement of the quality of optical glass and August Köhler invented the "Köhler illumination," which reduces image artifacts and provides high contrast by an even, maximal defocused illumination in the sample plane. Also, scientists grasped the concept of the electromagnetic wave spectrum and its properties. Consequently, new techniques emerged: On the one hand using shorter wavelengths (e.g., X-ray [Hildenbrand 1958]) and on the other hand using other light properties for contrast enhancement (e.g., the ultramicroscope, where samples are illuminated orthogonally to the detection path [Siedentopf and Zsigmondy 1902, Nobel Prize in Chemistry, R. Zsigmondy, 1925], phase contrast microscopes [Zernike 1942; Nobel Prize in Physics, F. Zernike, 1953], or differential interference contrast [DIC] microscopes [Allen et al. 1969]). Next to the light microscopy methods, the wave-like behavior of matter itself was described and used to develop electron microscopy (e.g., electron microscopy [EM], [Knoll and Ruska 1932, Nobel Prize in Physics, E. Ruska, 1986]). Further, scanning methods such as the scanning tunneling microscope (STM) (Binnig and Rohrer 1987), Nobel Prize in Physics, G. Binnig and H. Rohrer, 1986) or the atomic force microscope (AFM) (Binnig et al. 1986) evolved.

The phenomenon of naturally occurring luminescence of objects like crystals, involving the effect of fluorescence as well as phosphorescence, was first described by Sir George Gabriel Stokes, who defined the shift to longer wavelengths between the absorption of the illumination light and the re-emitted fluorescence light by the then so-called Stokes shift (Stokes 1852). Scientists now started to use the naturally occurring autofluorescence of whole biological objects or of intracellular components like chlorophyll and vitamins and the first fluorescence microscopes were built (1904 August Köhler [Köhler 1904], 1911 Carl Reichert and Oskar Heimstädt [Heimstädt 1911], 1913 Carl Zeiss and Heinrich Lehmann [Lehmann 1913]). Max Haitinger generalized the technique by introducing fluorescent probes as specific markers in the 1930s. Following the staining procedures of Haitinger (1937) and others, different nonautofluorescent structures and organelles within cells could now be specifically labeled with differently colored fluorophores (first called "fluorochromes") and thus were accessible by fluorescence microscopy. Labeling strategies further improved in the 1950s, when Albert H. Coons and Melvin H. Kaplan invented the staining of tissues by antigens that are bound by fluorophore-labeled antibodies (Coons et al. 1942; Coons and Kaplan 1950).

Engineering-improved organic dyes that are, for example, more photostable, brighter, or cell-membrane permeable is of on-going research up to today (Mason 1999). The invention of the confocal microscope (Minsky 1961) and the use of lasers as monochromatic light sources (Nobel Prize in Physics, C. Townes, N. Basov, and A. Prokhorov, 1964) led to the development of the confocal laser scanning microscope (CLSM)—a big renaissance for fluorescence microscopy in an era when EM was popular. In 1992, Douglas Prasher managed to extract the complementary DNA (cDNA) of the so-called green fluorescent protein (GFP), a native fluorescent protein from the jellyfish *Aequorea victoria* (Prasher et al. 1992) that was then demonstrated to co-express to virtually any target protein ([Chalfie et al. 1994], Nobel Prize in Chemistry, O. Shimomura, M. Chalfie and R. Y. Tsien, 2008). GFP and other native fluorescent proteins (e.g., *Discosoma striata* RFP1) were soon modified by selecting for various, properties and now span a large gallery of fluorescent proteins in the visible light range (Shaner et al. 2005). The inherent specific contrast of single molecules combined with the ability of live-cell imaging makes fluorescence microscopy one of the most powerful tools in modern microscopy for life sciences. Today, several fluorescence techniques (microscopic and non-microscopic) are used to study single-cell dynamics, by total internal reflection fluorescence microscopes (TIRFMs) (Axelrod 1981) combined with fluorescence resonance energy transfer (FRET) (Foerster 1948; Stryer 1978) methods, fluorescence correlation spectroscopy (FCS) (Magde et al. 1972; Lakowicz 2007), fluorescence recovery after photobleaching (FRAP) (Axelrod et al. 1976; Soumpasis 1983), fluorescence *in situ* hybridization (FISH) (Gall and Pardue 1969; John et al. 1969; Bauman et al. 1980), or fluorescence lifetime imaging microscopy (FLIM) (Lakowicz et al. 1992).

3.1.2 Fluorescence Super-Resolution Microscopy Techniques

The resolution today is often introduced by the Rayleigh criterion. In 1896, Lord Rayleigh defined the maximal resolution of a system by the minimal distance between two point-like objects in which they can still be separated as individual sources. He regarded two point sources of equal strength as just discernible when the main diffraction maximum of one image coincides with the first minimum of the other. For an epifluorescence microscope with a circular aperture where the light is collected with the same objective this yields

$$d = 0.61\ \lambda\ \mathrm{NA}$$

with λ being the wavelength of light and NA being the numerical aperture (Rayleigh 1896).

Nevertheless, as already demonstrated by Zsigmondy using his ultramicroscope (Siedentopf and Zsigmondy 1902), particles with dimensions below the resolution limit of visible light can be resolved. Since then several light microscopy techniques improved the obtained image resolution. A confocal or multiphoton fluorescence approach reduces the out-of-focus fluorescence and allows for 3D imaging by optical sectioning. The numerical aperture can be further increased by the use of two opposing objectives and thus the larger combined collection angle which is realized in scanning 4Pi and widefield I^5M (Stelzer and Hell 1992; Gustafsson et al. 1999). Near-field scanning optical microscopy (NSOM) makes use of the fact that the diffraction-limit applies only for light that has propagated for a distance sufficiently larger than its wavelength. By placing a nanometer-sized excitation and detection tip near the sample, the resolution is only limited by the aperture of the tip. This principle was already presented by Edward Hutchinson Synge in 1928 (in tradition of Zsigmondy addressed by "resolution into the ultra-microscopic region") (Synge 1928). Experimentally, a resolution below 20 nm was demonstrated by Betzig et al. (1991).

Starting in 1994 (the same year when GFP was first co-expressed), stimulated-emission depletion (STED) microscopy was the first proposed super-resolution far-field excitation technique and many new far-field methods that circumvent the diffraction barrier by taking advantage of spatially or temporally confined

Table 3.1 Timeline of the Invention of Different Super-Resolution Fluorescence Techniques

1990	Two-photon microscopy	Denk et al. (1990)
1991	Near-field scanning optical microscopy (NSOM)	Betzig et al. (1991)
1992	4Pi microscopy	Stelzer and Hell (1992)
1994	Stimulated emission depletion (STED) microscopy	Hell and Wichmann (1994)
1999	Image-interference and incoherent-interference-illumination microscopy (I^5M)	Gustafsson et al. (1999)
2000	Structured illumination microscopy (SIM)	Gustafsson (2000)
2004	Selective plane illumination microscopy (SPIM)	Huisken et al. (2004)
	Nanometer-localized multiple single-molecule (NALMS) fluorescence microscopy	Qu et al. (2004)
2006	Photoactivated localization microscopy (PALM)	Betzig et al. (2006)
	Stochastic optical reconstruction microscopy (STORM)	Rust et al. (2006)
	Fluorescence photoactivation localization (FPALM)	Hess et al. (2006)
	Points accumulation for imaging in nanoscale topography (PAINT)	Sharonov and Hochstrasser (2006)
2007	PALM with independently running acquisition (PALMIRA)	Geisler et al. (2007)
2008	*Direct* STORM (*d*STORM)	Heilemann et al. (2008)
	Ground state depletion followed by individual molecule	Folling et al. (2008)
	return (GSDIM)	Steinhauer et al. (2008)
	Blink microscopy	Lemmer et al. (2008)
	Spectral precision distance microscopy (SPDM)	
2009	Super-resolution optical fluctuation imaging (SOFI)	Dertinger et al. (2009)
	Quantum dot triexciton imaging (QDTI)	Hennig et al. (2009)
	Reversible photobleaching microscopy (RPM)	Baddeley et al. (2009)
2010	Universal PAINT	Giannone et al. (2010)
2011	Superresolution by power-dependent active intermittency	Lew et al. (2011)
	PAINT (SPRAIPAINT)	Burnette et al. (2011)
	Bleaching/blinking-assisted localization microscopy (BaLM)	
2014	DNA and exchange PAINT	Jungmann et al. (2014)

fluorescence evolved (see Table 3.1 and the introducing chapters for the techniques covered in this book). In this context, two sorts of super-resolved far-field methods can be distinguished: The first one concentrates on the incident excitation light pattern; the second kind on the modulation of the detected emission light in time. Further, the first group makes use of structured illumination schemes which spatially modulate the fluorescence of molecules such that not all of them emit light simultaneously. STED (Hell and Wichmann 1994) employs a donut-shaped depletion beam that forces molecules into their energy ground state which leaves only the fluorophores at the subdiffraction-sized center of the donut fluorescent and scans the samples with nanometer precision, structured illumination microscopy (SIM) (Lukosz and Marchand 1963; Gustafsson 2000) uses high-spatial frequency interference patterns to produce detectable and computable low-frequency moiré patterns in the sample. The second technique relies on single-molecule imaging, using stochastic photomodulation of individual fluorophores to separate the fluorescence emission in time. These localization-based techniques, such as photoactivated localization microscopy (PALM) (Betzig et al. 2006), (*direct*) stochastic optical reconstruction microscopy, (*d*)STORM (Rust et al. 2006; Heilemann et al. 2008), ground state depletion followed by individual molecule return (GSDIM) (Folling et al. 2008), and many others (see detailed review Moerner [2012]), are today commonly summarized using the term single-molecule localization microscopy (SMLM). They all require the tight control of the photoswitching of individual fluorophores, but also strongly rely on the performance of the post-processing algorithms used to generate the super-resolved data (see review Small and Stahlheber [2014]).

Today, all super-resolution techniques have proven to be live-cell compatible, to allow for multiple color staining and to be suitable for obtaining 3D information. Structured, scanning or surface illumination techniques naturally bring optical sectioning with them, for epifluorescence illumination schemes several optical effects, such as interference of beams (Shtengel et al. 2009), double-helically arranged point spread functions (Pavani et al. 2009), biplane alignment (Juette et al. 2008), or astigmatism (Huang et al. 2008), have been used to provide the 3D information.

3.2 Choosing a Suitable Fluorophore

When choosing suitable fluorophores for specific experiments some general requirements should be considered. There are different types of fluorophores; organic dyes like the carbocyanines, rhodamines, and oxazines; fluorescent proteins which most (except some of the near infrared fluorophores that descend from bacterial phytochromes [Shu et al. 2009]) possess a 11-stranded β-barrel structure like GFP; or also semiconductor crystal quantum dots (see Figure 3.1 and [Burns et al. 2006; Giepmans et al. 2006; Wang et al. 2006; Lakowicz 2007; Shcherbo et al. 2010; Toseland 2013]).

The chromophore backbones of the organic dyes and fluorescent proteins are conjugated electronic π-systems. These can consist of aromatic rings as well as C=C, C=O, or N=N bonds. The spectral properties depend on the extension of the conjugated electronic system, on the number of electrons, and on substituents (Klessinger 1978). Usually, an elongation of the conjugated system will shift the absorption maximum to longer wavelengths as can be easily seen for the cyanine dyes in Figure 3.2 with their properties as listed in Table 3.2.

Thus, every fluorophore covers a unique excitation and emission spectrum which should match with the given microscopic system (illumination wavelength,

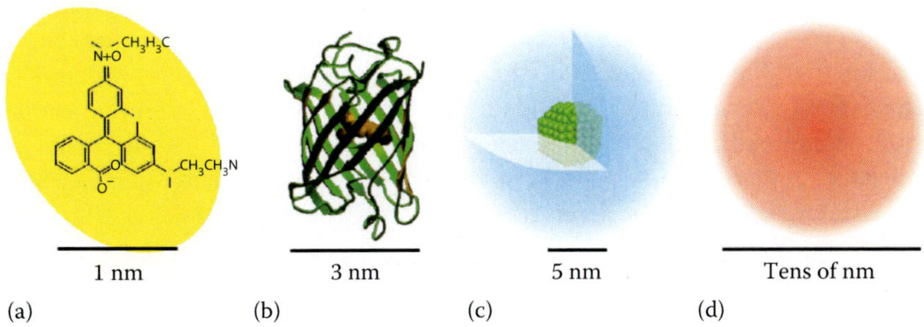

Figure 3.1

Common types of fluorophores in the life sciences are (a) organic dye molecules (tetramethylrhodamine [TMR] is displayed), (b) fluorescent proteins (GFP is displayed), (c) polymer-coated, water-soluble semiconductor quantum dots, and (d) dyed polymer particles. (From Burns, A. et al. *Chem. Soc. Rev.*, 35:1028–1042, 2006. Adapted by permission of The Royal Society of Chemistry.)

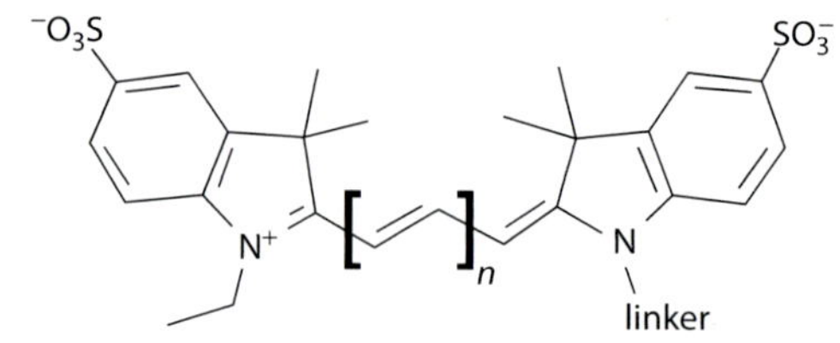

Figure 3.2

Cyanine dye derivatives.

Table 3.2 The Absorption and Emission Spectra of the Cyanine Fluorophores Scale with the Length of Their Polymethine Chain

	Polymethine Chain	λ_{abs} (nm)	λ_{em} (nm)
Cy3	$R_2N–C=C–C=C–C=N^+R_2$	550	562
Cy5	$R_2N–C=C–C=C–C=C–C=N^+R_2$	649	664
Cy7	$R_2N–C=C–C=C–C=C–C=C–C=N^+R_2$	743	774

filters and mirrors and sensitivity of the detector) and the specific type of experiment, for example, live cells are usually more sensitive for irradiation with shorter light wavelengths, certain spectral ranges exhibit pronounced autofluorescence and in case of a thick tissue sample, near-infrared fluorophores allow imaging at deeper areas inside the tissue. Favored emission and excitation spectra should be separated by a large enough Stokes shift, and, in multicolor experiments, the chosen set of fluorophores should exhibit sharp and defined spectra to yield minimal crosstalk in excitation and emission channels. Quantum dots possess convenient wide absorption but sharp, by their confined size, defined emission spectra for multicolor excitation schemes (Jaiswal and Simon 2004).

Next, a high brightness of the fluorophore provides sufficient signal above the background which is especially crucial for single-molecule imaging when the fluorescence of individual fluorophores is captured. Both, the extinction coefficient which quantifies the absorption of the incoming light as well as the quantum yield which specifies the ratio of emitted photons to the absorbed photons are important. Further, the fluorophore should exhibit a large photostability, which determines the temporal framework and applicable laser intensities of the experiment until the fluorophore then is irreversibly destroyed, namely photobleached. Here, molecular oxygen is an important factor. Oxygen is an efficient quencher of the triplet state of the fluorophores and thus supports their return back into the ground state. Fluorophores so yield higher fluorescence counts as the long residency times of the fluorophore in the "dark" triplet state are shortened. On the other hand, as the ground state of oxygen is a triplet state, excited fluorophores in the triplet state are very efficient generators of singlet oxygen. This then leads to the production of reactive oxygen species (ROS) like hydrogen peroxide, superoxide anions, or hydroxyl radicals, which then are the cause of strong irreversible photobleaching (DeRosa and Crutchley 2002). Thus, helpful "antifading" strategies which prolong stable measurement times usually prevent bleaching rates by removing the oxygen from the fluorophore environment but also include alternative triplet quenchers that replace the oxygen in its quencher role. These alternative quenchers, like the traditionally used β-mercaptoethanol (BME) (Kishino and Yanagida 1988) or mercaptoethylamine (MEA) (Heilemann et al. 2005; Widengren et al. 2007), alter the photophysics optimizing for a more stable fluorescence emission (see comparison of different strategies by Cordes et al. [2011]). The use of a reducing and oxidizing system (ROXS), usually a combination of methyl viologen (MV) and ascorbic acid (AA) and oxygen removal, is another popular formulation to minimize photobleaching rates and blinking, although there are exceptional fluorophores that still show blinking behavior under these conditions (Vogelsang et al. 2008). Also, fluorophore photophysics differ between solvents. For example, the use of heavy water (D_2O) can result in a higher fluorescence quantum yield as compared to normal water (Lee et al. 2013; Klehs et al. 2014).

Further, when exciting fluorophores with polarized light, a suitable fluorescence lifetime allows observing molecule orientations and interactions by fluorescence anisotropy experiments (Weber 1952; Lakowicz 2007; Sauer et al. 2010). Other factors to include into the design of an experiment can be environmental dependencies on temperature, pH, or special surrounding components. For fluorescent proteins, folding and maturation efficiency as well as velocity are important and can also differ due to environment, for example, the presence of molecular O_2 is mostly needed for final chromophore maturation (Heim et al. 1994; Evdokimov et al. 2006; Zhang et al. 2006; Subach et al. 2009a). In case of photoconvertible fluorophores the low-fluorescent dark state of the molecule should exhibit only neglectable signal when compared to the high-fluorescent on-state of the fluorophore; a quantity that improved a lot after developing the first photoactivatable protein paGFP (Patterson and Lippincott-Schwartz 2002). What is seen during the measurement is, in the end, an experimental contrast value determined by the fluorophore and the microscope parameters together.

As mentioned above, the fluorophore generally should be insensitive to environmental factors to allow a quantitative interpretation of the results. For some fluorophores, a quantified variability for one property was designed on purpose. Examples are pH-sensitive probes (e.g., review of Kneen et al. [1998] or indicating exocytosis [Matsuyama et al. 2000]), ratiometric probes (e.g., determining pH [Miesenbock et al. 1998; Bizzarri et al. 2009], measuring protein dynamics [Matsuda et al. 2008] or protein age [Khmelinskii et al. 2012]), indicator probes (e.g., sensing halides [Galietta et al. 2001], Ca^{2+} [Nakai et al. 2001], channel gating voltage [Ataka and Pieribone 2002]), and many others (see, e.g., Ibraheem and Campbell [2010]; You et al. [2015]).

The available strategies to tag the biomolecule of interest are crucial. Different labeling techniques for proteins, lipids, or nucleic acids result in very different tag sizes. For example, a fluorescent protein is ~3 nm large, a quantum dot has ~10–30 nm core size dependent on its spectral characteristics and an itself small organic dye (<1 nm) immunolabeled via a classical antibody sandwich (typically a primary full and secondary F[ab] fragment antibody) is ~22 nm large (Hinterdorfer and Van Oijen 2009; Endesfelder et al. 2014). Especially at resolving powers typical in super-resolution fluorescence microscopy, it is important to emphasize that it is always the label which is visualized and not the tagged molecule of interest itself. Thus, the size and orientation (e.g., concerning polarization effects) of the fluorophore and the achievable labeling density directly influence the obtainable resolution. High density labeling, which means having an at least average nearest-neighbor distance of below twice the sampling rate, is necessary (Shannon 1949).

Otherwise, the density of fluorophores limits the effective resolution. In case of a genetic label, the fluorophore or protein tag (e.g., SNAP, Halo, or Clip [Keppler et al. 2003; Los et al. 2008; Gautier et al. 2008]) should be codon optimized for the chosen system, express efficiently and should possess neglectable oligomerization tendencies. Alternatively, organic dyes are extrinsic labels and have to be brought into the cell which can be challenging, especially for live cell labeling. Only few dyes are membrane permeable (Grimm et al. 2015), for others and for *in vitro* pre-labeled targets like purified proteins or oligonucleotides, electroporation, bead loading, scraping, membrane transfer, or nanoinjection techniques have to be envisaged (Taylor and Wang 1978; Neumann et al. 1982; Mcneil and Warder 1987; McNeil 1988; Clarke and Mcneil 1992; Barber et al. 1996; Bruckbauer et al. 2007; Sakon and Weninger 2010). All labeling strategies should be checked for physiology, for example, they can be toxic for the cell (Alford et al. 2009) or lead to artifacts in cellular protein functions, for example, by influencing the protein distribution or by causing steric hindrances (El-Sagheer and Brown 2010; Swulius and Jensen 2012).

Last but not least, the different super-resolution techniques covered in this book have special demands. In case of STED, the molecules are constantly forced from the excited state into the electronic ground state via stimulated emission. Fluorophores that have a large extinction coefficient and a high stimulated emission cross section as well as a high quantum yield are favorable, as this helps contrast and detection of the fluorescence of the few fluorophores left in the center of the excitation pattern. The rate of the stimulated depletion of the excited state scales with the applied depletion energy, thus fluorophores chosen for STED further have to be exceptionally photostable. Further, the depletion wavelength should be carefully chosen as when fluorophores re-enter the electronic ground state, they can also be re-excited. For SIM, the most crucial parameter is the photostability. As the technique works by measuring the fluorescence response of a defined patterned excitation, variations in fluorescence not caused by the illumination strongly interfere. Therefore, the most photostable fluorophores and the use of an effective antifading agent is common practice. Finally, for SMLM techniques, the rigid control of the photoswitching is most crucial. The dark time, τ_{OFF}, must be long enough to guarantee single fluorophore distinction in the sample at any time during the experiment. Even when applying algorithms that can handle a high number of fluorescent molecules at a time, the techniques easily get impaired when the molecule density is too high (Holden et al. 2011; Wolter et al. 2011).

To conclude, more complex imaging experiments usually demand for more careful considerations and generally, also further additional characteristics such as, for example, choosing the spectral overlap between donor emission and acceptor excitation for FRET or deciding on strategies for optimal multiphoton absorbance can be listed. In this context, different photoswitching strategies, as evaluated in detail below, require individual optimizations, for example, customized specific photoconversion efficiencies for convertible fluorophores or tailored imaging buffers for selected organic dyes.

3.3 Mechanisms behind the Photoswitching of Fluorophores for Super-Resolution Microscopy

The key for many super-resolution microscopy techniques is the tight control of the photophysics of the fluorophores which are actively manipulated using light. These "photoswitches" are fluorophores that exhibit at least two stable and selectively addressable states, which can be divided into fluorescent and nonfluorescent states and which can be conveyed into another upon irradiation with light. For super-resolution microscopy, the molecules have to reside for a long time in a nonfluorescent dark state. How and which dark nonfluorescent state is achieved differs between the techniques and the type of fluorophore used. Further, the photoswitching process(es) can be reversible state transitions, like *cis–trans* isomerization, but also irreversible transformations, for example, the photocleavage of a part of the molecule. A typical model scheme is displayed in Figure 3.3. Some fluorophore photophysics are mainly described by the occupancy and transition rates between two reversible states, for other fluorophores more states have to be taken into account when modeling the switching behavior between several states properly.

3.3.1 Nonfluorescent Dark States of Fluorophores

What is utile as a nonfluorescent dark state highly depends on the specific molecular states a fluorophore possesses and their occupation probabilities and transition rates. Regarding electronic transitions, as summarized in Figure 3.4, fluorescence occurs from the excited singlet state with a characteristic lifetime in the nanosecond range; the triplet state occupation upon spin conversion of the excited electron typically lasts several microseconds before recovery into the electronic ground state. Upon laser illumination, the fluorescence emission of a fluorophore over time can therefore be described as an almost constant fluorescence signal where the fluorophore is cycled between the ground and first excited singlet state, which is interrupted from time to time by occupancies of the triplet state (dependent on the intersystem crossing probability) until the fluorophore is finally photobleached. As a typical temporal resolution of, for example, current EMCCD cameras for super-resolution acquisitions is in the millisecond regime, these transitions are usually too rapid to be followed in the accumulated fluorescence emissions images and are more important to consider when fast dynamics techniques like FCS or

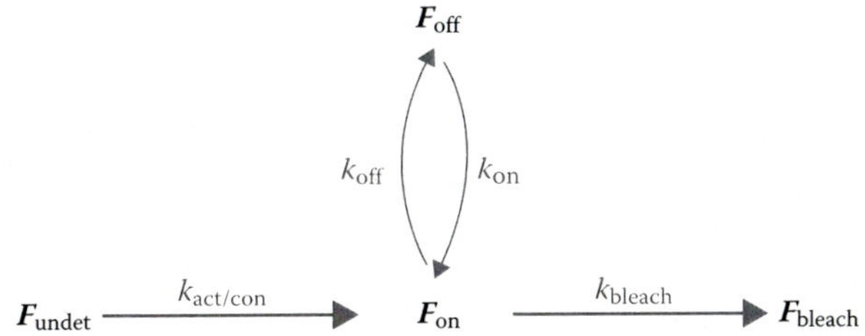

Figure 3.3

Exemplary photoswitching model of a fluorophore. Depicted is an irreversible photomodulation process with a rate $k_{act/con}$ from an undetected fluorophore state F_{undet} to the fluorescent state F_{on}. Here, the F_{undet} state can be a nonfluorescent dark state or a fluorescent state that is conversed to a second, usually red-shifted, fluorescent state used as F_{on} state. Second, an irreversible photobleaching step with a bleaching rate k_{bleach} transfers the fluorophore from F_{on} to the irreversible bleached state F_{bleach}. Further, reversible photoswitching with the switching rate k_{off} from the fluorescent state F_{on} to the nonfluorescent dark state F_{off} and the recovering rate k_{on} for the reverse transition from the nonfluorescent dark state F_{off} back to the fluorescent state F_{on} is displayed. In general, bleaching can also occur from the states F_{off} and F_{undet} but these are neglected in this model.

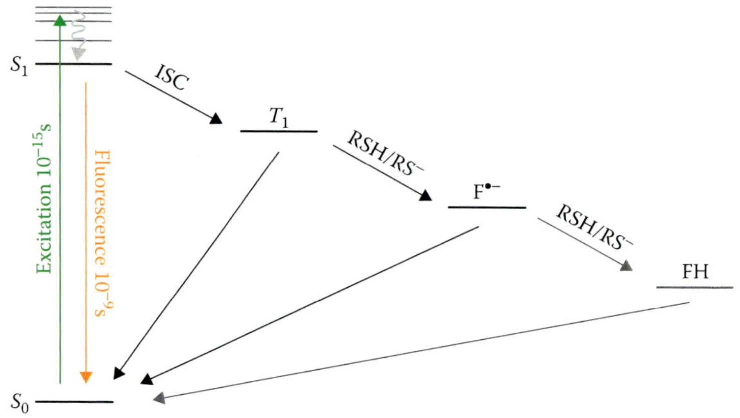

Figure 3.4

Upon irradiation, the fluorophore emits fluorescence photons and is cycled between its singlet ground and excited state S_0 and S_1, interrupted by triplet state (T_1) occupation via intersystem crossing upon spin flip of the excited electron with a rate k_{isc}. The excited fluorophore in the triplet state can react with molecular oxygen to return to the electronic ground state or can be reduced by the reductant with the rate k_{red} to a radical state (F•). This radical anion can be oxidized back into S_0 with rate k_{ox} or, as several radicals of rhodamine and oxazine dyes possess an absorption band around ~400 nm, they can also recover via UV illumination. Fluorophores such as ATTO 655 and ATTO 680 accept a second electron to form a fully reduced leuco-form (FH). Oxidation of FH also recovers fluorescence. (From van de Linde, S. et al., *Nat. Protoc.* 6, 991–1009, 2011b. With permission.)

FLIM are applied. Nevertheless, in other imaging schemes, also the electronic ground state can be a highly occupied dark state when, in the case of STED, the molecules are forced into it by stimulated emission.

When the molecules get reduced into a semi-radical state or fully reduced leuco form like shown in Figure 3.4, they can stably remain in this condition for several minutes up to several hours if the environmental conditions are carefully chosen ([Heilemann et al. 2009; Vogelsang et al. 2010; van de Linde et al. 2011a; van de Linde and Sauer 2014]; e.g., if oxygen depletion yields long residency times in a dark state). Another option to keep fluorophores primarily dark is to use caged systems where a photoreactive group quenches the fluorescence until it is cleaved by UV light (Mitchison et al. 1998).

Dark states of endogenous fluorescent proteins can be obtained by photoactivating the final maturation process of a prior uncompleted chromophore, by photocleaving the chromophore or by intramolecular dynamics, for example, by a *cis–trans* conformational change (Figure 3.5). Nevertheless, these mostly irreversible transformable fluorophores can also undergo reversible blinking triggered by imaging buffers which can force the fluorescent proteins into reversible dark states similar to when switching organic fluorophores (Endesfelder et al. 2011; Winterflood and Ewers 2014).

3.3.2 Reversible Photoswitching of Fluorophores

Reversible photoswitching is described by the ideal model scheme in which a fluorophore is switched only between a dark state F_{off} and a fluorescent state F_{on}. It is the major scheme for STED, where the highly photostable and fast switchable fluorophores should not bleach away while they are forced into the electronic

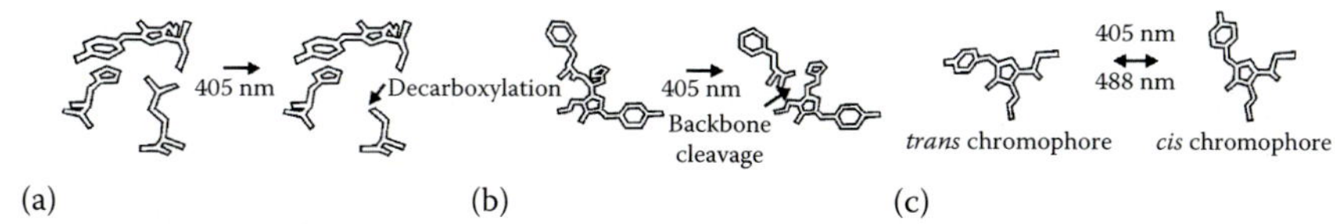

Figure 3.5

Switching mechanisms of fluorescent proteins. (a) Photoactivation of fluorophores like paGFP or PAmCherry1 by decarboxylation, (b) photoconversion of fluorophores like EosFP or Dendra via backbone cleavage, and (c) reversible photoswitching by conformational changes between *cis* and *trans* isoforms of chormophores like Dronpa.

ground state but rather should be able to stably fluoresce when they get into the center of the STED beam. For SMLM, most of the fluorophores should reside in the dark state to enable single-molecule signal detection, which means that the k_{off} rate should be much larger than the recovering rate k_{on}.

Most of the reversible photoswitchable fluorescent proteins (detailed reviews are [Shcherbakova et al. 2014; Adam 2014]), like, for example, already the first photochromic asFP595 (Lukyanov et al. 2000; Schäfer et al. 2007; Schäfer et al. 2008), undergo conformational changes between *cis* and *trans* isoforms which usually can be controlled by an off- and another on-switching laser wavelength. The mechanism behind the switching with the two different wavelengths is explained by the associated reversible protonation of the chromophore between a protonated *trans* and anionic *cis* form that possess different absorption spectra. Usually, the proteins are nonfluorescent when protonated and fluorescent when anionic. The majority of the photoswitchable fluorescent proteins can be assigned into two groups: The larger group consists of so-called negative photoswitching fluorescent proteins that are switched into their dark form by the excitation light that is used for read-out of the fluorescence signal. For example, the first green photoswitchable fluorescent protein Dronpa (Ando et al. 2004) and the fast switcher rsFastLime (Andresen et al. 2008) belong to this family. The other, mechanically more complex, group are the positive photoswitching fluorescent proteins like, for example, rsCherry (Stiel et al. 2008) or Padron (Andresen et al. 2008), a positive switching variant of Dronpa. Unlike in the negative switching mechanism, the stable conformation of the positive photoswitching fluorescent proteins is the (but also nonfluorescent) *trans*-chromophore. Thus, the switching is controlled counterintuitively by applying the fluorescence excitation laser for switching the fluorophore into its fluorescent form.

A fluorophore which is special as it possesses a unique switching mechanism is the fluorescent protein Dreiklang (Brakemann et al. 2011). The switching was shown to be achieved by a reversible hydration of the chromophore and, remarkably, is uncoupled from the fluorescence read-out wavelength. Particular are also the proteins IrisFP (Adam et al. 2008) and NijiFP (Adam et al. 2011), both obtained by a single mutation from their irreversibly photoconvertible parents EosFP and Dendra (see Section 3.3.3.), respectively. They still can be photoconverted from green to red but now also possess negative reversible photoswitching properties into dark forms for both colors.

Most of the organic dyes, which are used by the techniques STORM, *d*STORM, and GSDIM are switched by using chemical additives in a special imaging buffer or by applying special mounting media. In this way, the photophysics of the fluorophores are influenced by the strongly defining environments and the state occupancies and transition rates are altered. As already discussed in Section 3.2, molecular oxygen plays an ambivalent role as it is a common source for photobleaching via singlet oxygen and at the same time an efficient triplet-state quencher yielding higher fluorescence count rates (DeRosa and Crutchley 2002). Alternative triplet-state quenchers can assume this role when depleting the oxygen with an, for example, enzymatic, oxygen scavenger system.

Here, oxygen scavenging has been successfully accomplished by different strategies. Common are enzymatic systems like the buffer including glucose oxidase, glucose, and catalase (Benesch and Benesch 1953; Englander et al. 1987; Kishino and Yanagida 1988), the mix of protocatechuate-dioxygenase and protocatechuic acid (Patil and Ballou 2000; Aitken et al. 2008) or a system containing pyranose oxidase, glucose, and catalase (Swoboda et al. 2012). A fast depleting chemical system consists of methylene blue (MB) and MEA (Schäfer et al. 2013).

Successfully employed triplet-state quenchers and reductants are MEA, BME, dithiothreitol (DTT), glutathione (GSH), 6-hydroxy-2,5,7,8-tetramethylchroman-2-carboxylic acid (Trolox), cyclo-octatetraene (COT), and potassium iodide (KI) (Kishino and Yanagida 1988; Dijk et al. 2004; Heilemann et al. 2005; Rasnik et al. 2006; Widengren et al. 2007; Heilemann et al. 2009; Chmyrov et al. 2010). Further, antioxidants like *n*-propyl gallate (*n*PG) or AA (Giloh and Sedat 1982; Dijk et al. 2004; Steinhauer et al. 2008) and many anti-fading reagents like nitrobenzyl alcohol (NBA); paraphenylenediamine (PPD); 1,4-diazabicyclo[2.2.2] octane (DABCO); or commercial products like Vectashield, Fluor-stop, Mowiol, Slow-fade or Citifluor can retard photobleaching (e.g., Valnes and Brandtzaeg [1985]; Longin et al. [1993]; Olivier et al. [2013]; Dittrich and Schwille [2014]). It has further been shown that protective agents can also be covalently linked to the fluorophore (Altman et al. 2012).

Finally, for placing the dyes into long-lasting, but reversible dark states, the different photophysical and -chemical parameters influencing the molecular states of the fluorophore have to be in line with each other. Typical imaging buffers define the pH, include optimized concentrations of reducing agents and remove oxygen when favorable. Fluorophores are excited by sufficiently high laser irradiations to quickly drive them into the dark state and, for radical dark states, can often also be efficiently brought back into the electronic ground state via an UV absorption band. Several groups have explored this at first undesired photoblinking of different dyes in diverse environments and later, when this blinking behavior was suddenly favorable for super-resolution microscopy, have focused on the identification of their individual optimal photochemical photoswitching properties which have been unmasked (like for the oxazine Atto 655 and the carbocyanine Cy5 in Figure 3.6) being reduced dye radicals, fully reduced leuco forms or adducts of the fluorophore with the additive chemical (Zondervan et al. 2003; Heilemann et al. 2009; Vogelsang et al. 2009; van de Linde et al. 2011a, 2011b; Dempsey et al. 2009, 2011; Ha and Tinnefeld 2012; Vaughan et al. 2013; van de Linde and Sauer 2014). For fluorophore pairs, for example, for the pair Cy3–Cy5, a close proximity of the activator dye (Cy3) to the reporter dye (Cy5) can lead to fluorescence of the reporter upon excitation of the donor molecule (Rust et al. 2006; Bates et al. 2007).

Reversible fluorophore switching of organic dyes can also be obtained by photochromic compounds that undergo a photoinduced structural transformation

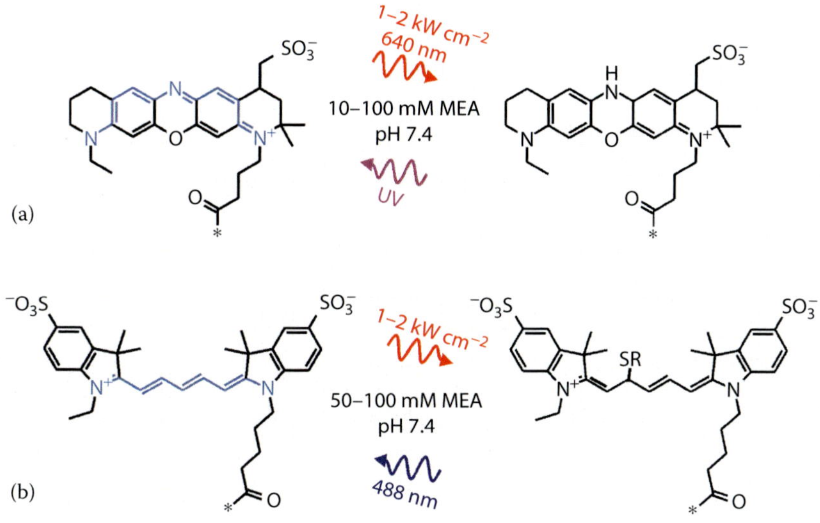

Figure 3.6

Reversible photoswitching of organic dyes in chemical thiol-containing imaging buffers. (a) ATTO655 upon irradiation at 640 and 405 nm in, for example, 10–100 mM MEA and (b) Cy5 upon irradiation at 640 and 488 nm in, for example, 50–100 mM MEA and oxygen depletion.

like, for example, photochromic rhodamines, spirophrans, or diarylethenes (Folling et al. 2007; Hu et al. 2008; Belov et al. 2009; Seefeldt et al. 2010; Uno et al. 2011; Montenegro et al. 2012). Recently, a spontaneously blinking fluorophore based on intramolecular spirocyclization has been designed (Uno et al. 2014).

3.3.3 Irreversible Photoactivation or -Conversion for SMLM

Irreversible photoactivation or -conversion is described by the ideal model scheme in which a fluorophore is transferred from an undetected fluorophore state F_{undet} to the fluorescent state F_{on}. Here, the F_{undet} state can be a nonfluorescent dark state or a first, non-observed fluorescent state that is conversed into a second, usually red-shifted, fluorescent state which is read-out as F_{on} state until the fluorophore is photobleached into the irreversible state F_{bleach}. This unidirectional mechanism is thus not suited for STED or SIM, but can be employed for single-molecule read-out in SMLM.

The fluorophore that was used in the first PALM experiments in Betzig et al. (2006) was the 4 years earlier engineered pAGFP (Patterson and Lippincott-Schwartz 2002). By illumination with UV light, the chromophore is decarboxylated and can then be excited by the now in the blue regime formed absorption band of its mature chromophore form. Today, also red photoactivatable fluorophore like PAmCherry1 (which possesses a similar decarboxylation activation mechanism like paGFP) or PAmKate are available (Subach et al. 2009a, 2009b; Gunewardene et al. 2011).

The green-to-red photoconversion of the photoconvertible fluorescent proteins like Kaede (Ando et al. 2002), mEos2 (McKinney et al. 2009), or Dendra2 (Gurskaya et al. 2006) relies on a backbone cleavage between two residues by near UV light which leads to an extension of the chromophore that now absorbs and emits photons at longer wavelengths. Recently, also an orange-to-red photoconvertible protein, PSmOrange2, was obtained (Subach et al. 2012). Finally, photoconversion processes are not limited to fluorophores that experience backbone cleavage upon UV irradiation. The protein psCFP2 undergoes a cyan-to-green photoconversion but is assumed to be based on a decarboxylation similar to the photoactivatable proteins (Chudakov et al. 2005; Lukyanov et al. 2005).

Organic dyes can be kept in stable nonfluorescent forms by close-by quencher molecules. Irreversible photoactivation schemes of organic dyes have thus been realized by attaching photocleavable quenching moieties to the fluorophore, so-called caged fluorophore systems (Mitchison et al. 1998; Lord et al. 2009; Belov et al. 2010; Maurel et al. 2010; Wysocki et al. 2011; Klan et al. 2013; Grimm et al. 2013; Belov et al. 2014). A chemical approach for obtaining kind of photoactivatable probes from standard fluorophores like cyanines or rhodamines is to reduce the fluorophores with sodium borohydride ($NaBH_4$). These stably nonfluorescent hydrogenated fluorophores can then be oxidized in small subsets into their fluorescent form via UV illumination (Kundu et al. 2009; Vaughan et al. 2012; Carlini et al. 2014).

3.3.4 Advanced Super-Resolution Applications

A super-resolution strategy for multiple targets advances the switching strategy from one fluorophore to several fluorophores which might demand for contradictory optimizations, as, for example, the switching chemical buffers used can differ between fluorophores and have strong impact on the photoswitching properties, for example, PAmcherry1, which needs oxygen to be activated should not be combined with buffers that remove it (Subach et al. 2009a; Endesfelder et al. 2011).

By an endogenous coexpression of a fluorescent protein tag, one literally can observe and count the individual molecules in the cell one-by-one in SMLM studies, but the absolute photoactivation efficiencies or fluorophore blinking characteristics should be comprised to correct for under- and overcounting effects (Annibale et al. 2011a, 2011b; Lando et al. 2012; Puchner et al. 2013; Durisic et al. 2014a, 2014b).

For highest structural resolutions, dense labeling and high photon yields for precise measurements have to be obtained. A high temporal resolution usually comes at the cost of a lower spatial resolution due to faster imaging routines. Further, when imaging living cells, the compromise between optimal imaging strategies and the health of the cells is critical: Photodamage of living cells due to laser irradiation is dependent on the wavelength, the irradiation dose, the cell fragility, and the cellular components. Although not fully investigated, DNA strand breakage, cell heating and the generation of ROS and free radicals (e.g., due to photochemistry of exciting or switching fluorophores) are documented as known causes (Hinterdorfer and Van Oijen 2009).

Correlative microscopy combines the opportunities of different methods, allows measuring different features of the exact same sample and can further be an important control to unmask artifacts or misinterpreted results of a single technique. For example, dynamical tracking studies combined with structural imaging yield a more complete image of the intracellular vesicle transport (Balint et al. 2013). Another prospective strategy is super-resolution microscopy correlated with another microscopic technique like, for example, EM, atomic-force microscopy, lightsheet microscopy, or high-throughput methods. Here, the labeling and preparation strategy, which is optimized for both techniques is one of the most challenging questions, as for example in correlative EM, cryo- or resin-covered environments impair the photophysics of the fluorophores. Driven by numerous developments in both techniques, correlative SMLM-EM today is at the edge to become an important tool to perform structural biology that is also capable to track function (review of Endesfelder [2015]).

3.4 Super-Resolution Microscopy without the Need of Photoswitching

Next to photoswitching fluorophores, other parameters can be used for discerning the single molecules. Discrimination also becomes possible by measuring the fluorescence lifetime directly (Heilemann et al. 2002) or indirectly by observing the drop of fluorescence intensity for intrinsically blinking and photobleaching molecules and therewith determining the molecule positions like in nanometer-localized multiple single-molecule (NALMS) fluorescence microscopy (Qu et al. 2004), subtracting patterns in defocused imaging to enhance the resolution (SPIDER) microscopy (Dedecker et al. 2009) or Bleaching/blinking-assisted localization microscopy (BaLM) (Burnette et al. 2011). Further, transient stochastic binding kinetics followed by dissociation or photobleaching can be utilized. The fluorophores are present at a low enough concentration in the solution surrounding the sample such that only a few binding events are observed at a given time ensuring a single-molecule detection. Techniques developed are the points accumulation for imaging in nanoscale topography (PAINT) technique (Sharonov and Hochstrasser 2006) and its further PAINT descendants (Rocha et al. 2009; Giannone et al. 2010; Jungmann et al. 2010, 2014; Lew et al. 2011), the method nanometer accuracy by stochastic catalytic reactions (NASCA) microscopy (Roeffaers et al. 2009), and binding-activated localization microscopy (BALM) (Schoen et al. 2011). In the case of some for Nile red used by Sharonov and Hochstrasser in their original PAINT work, the binding enhances the fluorescence properties of the previously in water mostly nonfluorescent substrate. In super-resolution optical fluctuation imaging (SOFI), temporal fluorescence fluctuations are analyzed without the need of single-molecule resolution (Dertinger et al. 2009).

3.5 Outlook

Current developments focus on more suitable fluorescent probes concerning all advanced imaging schemes sketched. Recent efforts concentrate on rational fluorophore design that builds on the understanding of the underlying processes to yield less blinking fluorophores for quantitative counting experiments, on soluble dyes that are membrane-permeable and live-cell compatible, on far-red and infrared fluorescent proteins that possess higher fluorescence quantum yields (currently around 5%) or on fluorophores optimized for correlative schemes (Adam et al. 2014; Lukinavicius et al. 2014; Chenxi et al. 2015; Paez-Segala et al. 2015; Grimm et al. 2015).

On the other hand, more and more sophisticated algorithms help to compensate for imperfect fluorophore behavior, for example, by tracking blinking events, by automating imaging routines and controlling high-throughput schemes or by performing multiemitter fitting to cope for high k_{on}/k_{off} ratios that lead to high fluorescent spot densities which are hard to discriminate or to allow for faster imaging speeds (Edelstein et al. 2014; Min et al. 2014). Analysis methods evolve for more localization-based analysis tools that determine the resolution, define protein clusters, or quantify protein interactions (Nieuwenhuizen et al. 2013; Sengupta et al. 2013; Diez et al. 2014; Vandenberg et al. 2015; Pengo et al. 2015). By modeling schemes, fluorophore photophysics as well as the dynamics of the targeted biomolecule can be clarified (Persson et al. 2013; Fricke et al. 2014; Rollins et al. 2015). Finally, particle averaging methods allow for structural biology studies of macromolecular complexes using super-resolution fluorescence microscopy that nowadays can complement EM and crystallography (Szymborska et al. 2013; Schucker et al. 2015).

References

Abbe, E. 1873. Beiträge zur Theorie des Mikroskops und der mikroskopischen Wahrnehmung. *Archiv für mikroskopische Anatomie* 9(1):413–418. doi: 10.1007/BF02956173.

Adam, V. 2014. Phototransformable fluorescent proteins: Which one for which application? *Histochem Cell Biol* 142(1):19–41. doi: 10.1007/s00418-014-1190-5.

Adam, V., R. Berardozzi, M. Byrdin, and D. Bourgeois. 2014. Phototransformable fluorescent proteins: Future challenges. *Curr Opin Chem Biol* 20:92–102. doi: 10.1016/j.cbpa.2014.05.016.

Adam, V., M. Lelimousin, S. Boehme, G. Desfonds, K. Nienhaus, M. J. Field, J. Wiedenmann, S. McSweeney, G. U. Nienhaus, and D. Bourgeois. 2008. Structural characterization of IrisFP, an optical highlighter undergoing multiple photo-induced transformations. *Proc Natl Acad Sci USA* 105(47):18343–18348. doi: 10.1073/pnas.0805949105.

Adam, V., B. Moeyaert, C. C. David, H. Mizuno, M. Lelimousin, P. Dedecker, R. Ando et al. 2011. Rational design of photoconvertible and biphotochromic fluorescent proteins for advanced microscopy applications. *Chem Biol* 18(10):1241–1251. doi: 10.1016/j.chembiol.2011.08.007.

Aitken, C. E., R. A. Marshall, and J. D. Puglisi. 2008. An oxygen scavenging system for improvement of dye stability in single-molecule fluorescence experiments. *Biophys J* 94(5):1826–1835. doi: 10.1529/biophysj.107.117689.

Alford, R., H. M. Simpson, J. Duberman, G. C. Hill, M. Ogawa, C. Regino, H. Kobayashi, and P. L. Choyke. 2009. Toxicity of organic fluorophores used in molecular imaging: Literature review. *Mol Imaging* 8(6):341–354.

Allen, R. D., G. B. David, and G. Nomarski. 1969. The Zeiss-Nomarski differential interference equipment for transmitted-light microscopy. *Z Wiss Mikrosk* 69(4):193–221.

Altman, R. B., Q. Zheng, Z. Zhou, D. S. Terry, J. D. Warren, and S. C. Blanchard. 2012. Enhanced photostability of cyanine fluorophores across the visible spectrum. *Nat Methods* 9(5):428–429. doi: 10.1038/nmeth.1988.

Ando, R., H. Hama, M. Yamamoto-Hino, H. Mizuno, and A. Miyawaki. 2002. An optical marker based on the UV-induced green-to-red photoconversion of a fluorescent protein. *Proc Natl Acad Sci USA* 99(20):12651–12656. doi: 10.1073/pnas.202320599.

Ando, R., H. Mizuno, and A. Miyawaki. 2004. Regulated fast nucleocytoplasmic shuttling observed by reversible protein highlighting. *Science* 306(5700):1370–1373. doi: 10.1126/science.1102506.

Andresen, M., A. C. Stiel, J. Folling, D. Wenzel, A. Schonle, A. Egner, C. Eggeling, S. W. Hell, and S. Jakobs. 2008. Photoswitchable fluorescent proteins enable monochromatic multilabel imaging and dual color fluorescence nanoscopy. *Nat Biotechnol* 26(9):1035–1040. doi: 10.1038/nbt.1493.

Annibale, P., S. Vanni, M. Scarselli, U. Rothlisberger, and A. Radenovic. 2011a. Identification of clustering artifacts in photoactivated localization microscopy. *Nat Methods* 8(7):527–528. doi: 10.1038/nmeth.1627.

Annibale, P., S. Vanni, M. Scarselli, U. Rothlisberger, and A. Radenovic. 2011b. Quantitative photo activated localization microscopy: Unraveling the effects of photoblinking. *PLoS One* 6(7):e22678. doi: 10.1371/journal.pone.0022678.

Ataka, K. and V. A. Pieribone. 2002. A genetically targetable fluorescent probe of channel gating with rapid kinetics. *Biophys J* 82(1 Pt 1):509–516. doi: 10.1016/S0006-3495(02)75415-5.

Axelrod, D. 1981. Cell-substrate contacts illuminated by total internal reflection fluorescence. *J Cell Biol* 89(1):141–145.

Axelrod, D., D. E. Koppel, J. Schlessinger, E. Elson, and W. W. Webb. 1976. Mobility measurement by analysis of fluorescence photobleaching recovery kinetics. *Biophys J* 16(9):1055–1069.

Baddeley, D., I. D. Jayasinghe, C. Cremer, M. B. Cannell, and C. Soeller. 2009. Light-induced dark states of organic fluochromes enable 30 nm resolution imaging in standard media. *Biophys J* 96(2):L22–L24. doi: 10.1016/j.bpj.2008.11.002.

Balint, S., I. Verdeny Vilanova, A. Sandoval Alvarez, and M. Lakadamyali. 2013. Correlative live-cell and superresolution microscopy reveals cargo transport dynamics at microtubule intersections. *Proc Natl Acad Sci USA* 110(9):3375–3380. doi: 10.1073/pnas.1219206110.

Barber, K., R. R. Mala, M. P. Lambert, R. Z. Qiu, R. C. MacDonald, and W. L. Klein. 1996. Delivery of membrane-impermeant fluorescent probes into living neural cell populations by lipotransfer. *Neurosci Lett* 207(1):17–20. doi: 10.1016/0304-3940(96)12497-6.

Bates, M., B. Huang, G. T. Dempsey, and X. Zhuang. 2007. Multicolor superresolution imaging with photo-switchable fluorescent probes. *Science* 317(5845):1749–1753. doi: 10.1126/science.1146598.

Bauman, J. G. J., J. Wiegant, P. Borst, and P. Vanduijn. 1980. A new method for fluorescence microscopical localization of specific DNA-sequences by in situ hybridization of fluorochrome-labeled RNA. *Exp Cell Res* 128(2):485–490. doi: 10.1016/0014-4827(80)90087-7.

Belov, V. N., M. L. Bossi, J. Fölling, V. P. Boyarskiy, and S. W. Hell. 2009. Rhodamine spiroamides for multicolor single-molecule switching fluorescent nanoscopy. *Chemistry* 15(41):10762–10776. doi: 10.1002/chem.200901333.

Belov, V. N., G. Y. Mitronova, M. L. Bossi, V. P. Boyarskiy, E. Hebisch, C. Geisler, K. Kolmakov, C. A. Wurm, K. I. Willig, and S. W. Hell. 2014. Masked rhodamine dyes of five principal colors revealed by photolysis of a 2-diazo-1-indanone caging group: synthesis, photophysics, and light microscopy applications. *Chemistry* 20(41):13162–13173. doi: 10.1002/chem.201403316.

Belov, V. N., C. A. Wurm, V. P. Boyarskiy, S. Jakobs, and S. W. Hell. 2010. Rhodamines NN: A novel class of caged fluorescent dyes. *Angew Chem Int Ed Engl* 49(20):3520–3523. doi: 10.1002/anie.201000150.

Benesch, R. E. and R. Benesch. 1953. Enzymatic removal of oxygen for polarography and related methods. *Science* 118(3068):447–448.

Betzig, E., G. H. Patterson, R. Sougrat, O. W. Lindwasser, S. Olenych, J. S. Bonifacino, M. W. Davidson, J. Lippincott-Schwartz, and H. F. Hess. 2006. Imaging intracellular fluorescent proteins at nanometer resolution. *Science* 313(5793):1642–1645. doi: 10.1126/science.1127344.

Betzig, E., J. K. Trautman, T. D. Harris, J. S. Weiner, and R. L. Kostelak. 1991. Breaking the diffraction barrier: Optical microscopy on a nanometric scale. *Science* 251(5000):1468–1470. doi: 10.1126/science.251.5000.1468.

Binnig, G. and H. Rohrer. 1987. Scanning tunneling microscopy—From birth to adolescence. *Angew Chem Int Ed Engl* 26(7):606–614. doi: 10.1002/anie.198706061.

Binnig, G., C. F. Quate, and Ch. Gerber. 1986. Atomic force microscope. *Phys Rev Lett* 56(9):930–933. doi: 10.1103/PhysRevLett.56.930.

Bizzarri, R., M. Serresi, S. Luin, and F. Beltram. 2009. Green fluorescent protein based pH indicators for in vivo use: A review. *Anal Bioanal Chem* 393(4):1107–1122. doi: 10.1007/s00216-008-2515-9.

Brakemann, T., A. C. Stiel, G. Weber, M. Andresen, I. Testa, T. Grotjohann, M. Leutenegger et al. 2011. A reversibly photoswitchable GFP-like protein with fluorescence excitation decoupled from switching. *Nat Biotechnol* 29(10):942–947. doi: 10.1038/nbt.1952.

Bruckbauer, A., P. James, D. J. Zhou, J. W. Yoon, D. Excell, Y. Korchev, R. Jones, and D. Klenerman. 2007. Nanopipette delivery of individual molecules to cellular compartments for single-molecule fluorescence tracking. *Biophys J* 93(9):3120–3131. doi: 10.1529/biophysj.107.104737.

Burnette, D. T., P. Sengupta, Y. Dai, J. Lippincott-Schwartz, and B. Kachar. 2011. Bleaching/blinking assisted localization microscopy for superresolution imaging using standard fluorescent molecules. *Proc Natl Acad Sci USA* 108(52):21081–21086. doi: 10.1073/pnas.1117430109.

Burns, A., H. Ow, and U. Wiesner. 2006. Fluorescent core-shell silica nanoparticles: towards "Lab on a Particle" architectures for nanobiotechnology. *Chem Soc Rev* 35(11):1028–1042. doi: 10.1039/b600562b.

Carlini, L., A. Benke, L. Reymond, G. Lukinavicius, and S. Manley. 2014. Reduced dyes enhance single-molecule localization density for live superresolution imaging. *ChemPhysChem* 15(4):750–755. doi: 10.1002/cphc.201301004.

Chalfie, M., Y. Tu, G. Euskirchen, W. W. Ward, and D. C. Prasher. 1994. Green fluorescent protein as a marker for gene expression. *Science* 263(5148):802–805.

Chenxi, D., B. Martin, K. Mariam El, H. Xavier, A. Virgile, and B. Dominique. 2015. Rational design of enhanced photoresistance in a photoswitchable fluorescent protein. *Methods Appl Fluoresc* 3(1):014004.

Chmyrov, A., T. Sanden, and J. Widengren. 2010. Iodide as a fluorescence quencher and promoter-mechanisms and possible implications. *J Phys Chem B* 114(34):11282–11291. doi: 10.1021/Jp103837f.

Chudakov, D. M., S. Lukyanov, and K. A. Lukyanov. 2005. Fluorescent proteins as a toolkit for in vivo imaging. *Trends Biotechnol* 23(12):605–613. doi: 10.1016/j.tibtech.2005.10.005.

Clarke, M. S. F. and P. L. Mcneil. 1992. Syringe loading introduces macromolecules into living mammalian-cell cytosol. *J Cell Sci* 102:533–541.

Coons, A. H., H. J. Creech, R. N. Jones, and E. Berliner. 1942. The demonstration of pneumococcal antigen in tissues by the use of fluorescent antibody. *J. Immunol* 45(3):159–170.

Coons, A. H. and M. H. Kaplan. 1950. Localization of antigen in tissue cells II. Improvements in a method for the detection of antigen by means of fluorescent antibody. *J Exp Med* 91(1):1–13.

Cordes, T., A. Maiser, C. Steinhauer, L. Schermelleh, and P. Tinnefeld. 2011. Mechanisms and advancement of antifading agents for fluorescence microscopy and single-molecule spectroscopy. *Phys Chem Chem Phys* 13(14):6699–6709. doi: 10.1039/c0cp01919d.

Dedecker, P., B. Muls, A. Deres, H. Uji-i, J.-I. Hotta, M. Sliwa, J-P. Soumillion, K. Müllen, J. Enderlein, and J. Hofkens. 2009. Defocused wide-field imaging unravels structural and temporal heterogeneity in complex systems. *Adv Mat* 21(10–11):1079–1090. doi: 10.1002/adma.200801873.

Dempsey, G. T., M. Bates, W. E. Kowtoniuk, D. R. Liu, R. Y. Tsien, and X. W. Zhuang. 2009. Photoswitching mechanism of cyanine dyes. *J Am Chem Soc* 131(51):18192–18193. doi: 10.1021/Ja904588g.

Dempsey, G. T., J. C. Vaughan, K. H. Chen, M. Bates, and X. Zhuang. 2011. Evaluation of fluorophores for optimal performance in localization-based super-resolution imaging. *Nat Methods* 8(12):1027–1036. doi: 10.1038/nmeth.1768.

Denk, W., J. H. Strickler, and W. W. Webb. 1990. Two-photon laser scanning fluorescence microscopy. *Science* 248(4951):73–76.

DeRosa, M. C. and R. J. Crutchley. 2002. Photosensitized singlet oxygen and its applications. *Coord Chem Rev* 233:351–371. doi: 10.1016/S0010-8545(02)00034-6.

Dertinger, T., R. Colyer, G. Iyer, S. Weiss, and J. Enderlein. 2009. Fast, background-free, 3D super-resolution optical fluctuation imaging (SOFI). *Proc Natl Acad Sci USA* 106(52):22287–22292. doi: 10.1073/pnas.0907866106.

Diez, L. T., C. Bonsch, S. Malkusch, Z. Truan, M. Munteanu, M. Heilemann, O. Hartley, U. Endesfelder, and A. Fürstenberg. 2014. Coordinate-based co-localization-mediated analysis of arrestin clustering upon stimulation of the C-C chemokine receptor 5 with RANTES/CCL5 analogues. *Histochem Cell Biol* 142(1):69–77. doi: 10.1007/s00418-014-1206-1.

Dijk, M. A., L. C. Kapitein, J. Mameren, C. F. Schmidt, and E. J. Peterman. 2004. Combining optical trapping and single-molecule fluorescence spectroscopy: Enhanced photobleaching of fluorophores. *J Phys Chem B* 108(20):6479–6484. doi: 10.1021/jp049805+.

Dittrich, P. S. and P. Schwille. 2014. Photobleaching and stabilization of fluorophores used for single-molecule analysis with one- and two-photon excitation. *App Phys B* 73(8):829–837. doi: 10.1007/s003400100737.

Durisic, N., L. L. Cuervo, and M. Lakadamyali. 2014a. Quantitative super-resolution microscopy: Pitfalls and strategies for image analysis. *Curr Opin Chem Biol* 20:22–28. doi: 10.1016/j.cbpa.2014.04.005.

Durisic, N., L. Laparra-Cuervo, A. Sandoval-Alvarez, J. S. Borbely, and M. Lakadamyali. 2014b. Single-molecule evaluation of fluorescent protein photoactivation efficiency using an in vivo nanotemplate. *Nat Methods* 11(2):156–162. doi: 10.1038/nmeth.2784.

Edelstein, A. D., M. A. Tsuchida, N. Amodaj, H. Pinkard, R. D. Vale, and N. Stuurman. 2014. Advanced methods of microscope control using μ Manager software. *J Biol Methods* 1(2). doi: 10.14440/jbm.2014.36.

El-Sagheer, A. H. and T. Brown. 2010. Click chemistry with DNA. *Chem Soc Rev* 39(4):1388–1405. doi: 10.1039/b901971p.

Endesfelder, U. 2015. Advances in correlative single-molecule localization microscopy and electron microscopy. *NanoBioImaging* 1(1). doi: 10.2478/nbi-2014-0002.

Endesfelder, U., S. Malkusch, B. Flottmann, J. Mondry, P. Liguzinski, P. J. Verveer, and M. Heilemann. 2011. Chemically induced photoswitching of fluorescent probes—A general concept for super-resolution microscopy. *Molecules* 16(4):3106–3118. doi: 10.3390/molecules16043106.

Endesfelder, U., S. Malkusch, F. Fricke, and M. Heilemann. 2014. A simple method to estimate the average localization precision of a single-molecule localization microscopy experiment. *Histochem Cell Biol* 141(6):629–638. doi: 10.1007/s00418-014-1192-3.

Englander, S. W., D. B. Calhoun, and J. J. Englander. 1987. Biochemistry without oxygen. *Anal Biochem* 161(2):300–306.

Evdokimov, A. G., M. E. Pokross, N. S. Egorov, A. G. Zaraisky, I. V. Yampolsky, E. M. Merzlyak, A. N. Shkoporov, I. Sander, K. A. Lukyanov, and D. M. Chudakov. 2006. Structural basis for the fast maturation of Arthropoda green fluorescent protein. *EMBO Rep* 7(10):1006–1012. doi: 10.1038/sj.embor.7400787.

Foerster, T. 1948. Zwischenmolekulare Energiewanderung und Fluoreszenz. *Annalen der Physik* 2(1–2):55–75.

Fölling, J., V. Belov, R. Kunetsky, R. Medda, A. Schönle, A. Egner, C. Eggeling, M. Bossi, and S. W. Hell. 2007. Photochromic rhodamines provide nanoscopy with optical sectioning. *Angew Chem Int Ed Engl* 46(33):6266–6270. doi: 10.1002/anie.200702167.

Fölling, J., M. Bossi, H. Bock, R. Medda, C. A. Wurm, B. Hein, S. Jakobs, C. Eggeling, and S. W. Hell. 2008. Fluorescence nanoscopy by ground-state depletion and single-molecule return. *Nat Methods* 5(11):943–945. doi: 10.1038/nmeth.1257.

Fricke, F., S. Malkusch, G. Wangorsch, J. F. Greiner, B. Kaltschmidt, C. Kaltschmidt, D. Widera, T. Dandekar, and M. Heilemann. 2014. Quantitative single-molecule localization microscopy combined with rule-based modeling reveals ligand-induced TNF-R1 reorganization toward higher-order oligomers. *Histochem Cell Biol* 142(1):91–101. doi: 10.1007/s00418-014-1195-0.

Galietta, L. J., P. M. Haggie, and A. S. Verkman. 2001. Green fluorescent protein-based halide indicators with improved chloride and iodide affinities. *FEBS Lett* 499(3):220–224.

Gall, J. G. and M. L. Pardue. 1969. Formation and detection of RNA-DNA hybrid molecules in cytological preparations. *Proc Natl Acad Sci USA* 63(2):378–383. doi: 10.1073/pnas.63.2.378.

Gautier, A., A. Juillerat, C. Heinis, I. R. Correa, Jr., M. Kindermann, F. Beaufils, and K. Johnsson. 2008. An engineered protein tag for multiprotein labeling in living cells. *Chem Biol* 15(2):128–136. doi: 10.1016/j.chembiol.2008.01.007.

Geisler, C., A. Schönle, C. von Middendorff, H. Bock, C. Eggeling, A. Egner, and S. W. Hell. 2007. Resolution of lambda/10 in fluorescence microscopy using fast single molecule photo-switching. *Appl Phys A Mater Sci Proces* 88(2):223–226. doi: 10.1007/s00339-007-4144-0.

Giannone, G., E. Hosy, F. Levet, A. Constals, K. Schulze, A. I. Sobolevsky, M. P. Rosconi et al. 2010. Dynamic superresolution imaging of endogenous proteins on living cells at ultra-high density. *Biophys J* 99(4):1303–1310. doi: 10.1016/j.bpj.2010.06.005.

Giepmans, B. N., S. R. Adams, M. H. Ellisman, and R. Y. Tsien. 2006. The fluorescent toolbox for assessing protein location and function. *Science* 312(5771):217–224. doi: 10.1126/science.1124618.

Giloh, H. and J. W. Sedat. 1982. Fluorescence microscopy: Reduced photobleaching of rhodamine and fluorescein protein conjugates by n-propyl gallate. *Science* 217(4566):1252–1255.

Grimm, J. B., B. P. English, J. Chen, J. P. Slaughter, Z. Zhang, A. Revyakin, R. Patel et al. 2015. A general method to improve fluorophores for live-cell and single-molecule microscopy. *Nat Methods* 12(3):244–250. doi: 10.1038/nmeth.3256.

Grimm, J. B., A. J. Sung, W. R. Legant, P. Hulamm, S. M. Matlosz, E. Betzig, and L. D. Lavis. 2013. Carbofluoresceins and carborhodamines as scaffolds for high-contrast fluorogenic probes. *ACS Chem Biol* 8(6):1303–1310. doi: 10.1021/cb4000822.

Gunewardene, M. S., F. V. Subach, T. J. Gould, G. P. Penoncello, M. V. Gudheti, V. V. Verkhusha, and S. T. Hess. 2011. Superresolution imaging of multiple fluorescent proteins with highly overlapping emission spectra in living cells. *Biophys J* 101(6):1522–1528. doi: 10.1016/j.bpj.2011.07.049.

Gurskaya, N. G., V. V. Verkhusha, A. S. Shcheglov, D. B. Staroverov, T. V. Chepurnykh, A. F. Fradkov, S. Lukyanov, and K. A. Lukyanov. 2006. Engineering of a monomeric green-to-red photoactivatable fluorescent protein induced by blue light. *Nat Biotechnol* 24(4):461–465. doi: 10.1038/nbt1191.

Gustafsson, M. G. 2000. Surpassing the lateral resolution limit by a factor of two using structured illumination microscopy. *J Microsc* 198(Pt 2):82–87.

Gustafsson, M. G., D. A. Agard, and J. W. Sedat. 1999. I5M: 3D widefield light microscopy with better than 100 nm axial resolution. *J Microsc* 195(Pt 1): 10–16.

Ha, T. and P. Tinnefeld. 2012. Photophysics of fluorescent probes for single-molecule biophysics and super-resolution imaging. *Annu Rev Phys Chem* 63:595–617. doi: 10.1146/annurev-physchem-032210-103340.

Haitinger, M. 1937. *Die Fluoreszenzanalyse in der Mikrochemie: Mit 4 Abbildungen und 7 Tabellen*. Vienna, Austria: E. Haim.

Heilemann, M., D. P. Herten, R. Heintzmann, C. Cremer, C. Muller, P. Tinnefeld, K. D. Weston, J. Wolfrum, and M. Sauer. 2002. High-resolution colocalization of single dye molecules by fluorescence lifetime imaging microscopy. *Anal Chem* 74(14):3511–3517.

Heilemann, M., E. Margeat, R. Kasper, M. Sauer, and P. Tinnefeld. 2005. Carbocyanine dyes as efficient reversible single-molecule optical switch. *J Am Chem Soc* 127(11):3801–3806. doi: 10.1021/ja044686x.

Heilemann, M., S. van de Linde, A. Mukherjee, and M. Sauer. 2009. Super-resolution imaging with small organic fluorophores. *Angew Chem Int Ed Engl* 48(37):6903–6908. doi: 10.1002/anie.200902073.

Heilemann, M., S. van de Linde, M. Schüttpelz, R. Kasper, B. Seefeldt, A. Mukherjee, P. Tinnefeld, and M. Sauer. 2008. Subdiffraction-resolution fluorescence imaging with conventional fluorescent probes. *Angew Chem Int Ed Engl* 47(33):6172–6176. doi: 10.1002/anie.200802376.

Heim, R., D. C. Prasher, and R. Y. Tsien. 1994. Wavelength mutations and posttranslational autoxidation of green fluorescent protein. *Proc Natl Acad Sci USA* 91(26):12501–12504.

Heimstädt, O. 1911. Das Fluoreszenzmikroskop. *Z Wiss Mikrosk* 28:330–337.

Hell, S. W. and J. Wichmann. 1994. Breaking the diffraction resolution limit by stimulated emission: Stimulated-emission-depletion fluorescence microscopy. *Opt Lett* 19(11):780–782.

Hennig, S., S. van de Linde, M. Heilemann, and M. Sauer. 2009. Quantum dot triexciton imaging with three-dimensional subdiffraction resolution. *Nano Lett* 9(6):2466–2470. doi: 10.1021/nl9012387.

Hess, S. T., T. P. Girirajan, and M. D. Mason. 2006. Ultra-high resolution imaging by fluorescence photoactivation localization microscopy. *Biophys J* 91(11):4258–4272. doi: 10.1529/biophysj.106.091116.

Hildenbrand, G. 1958. Grundlagen der Röntgenoptik und Röntgenmikroskopie. In *Ergebnisse der Exakten Naturwissenschaften*. Berlin/Heidelberg, Germany: Springer, pp. 1–133.

Hinterdorfer, P. and A. Van Oijen. 2009. *Handbook of Single-Molecule Biophysics*. New York: Springer Science + Business Media.

Holden, S. J., S. Uphoff, and A. N. Kapanidis. 2011. DAOSTORM: An algorithm for high-density super-resolution microscopy. *Nat Methods* 8(4):279–280. doi: 10.1038/nmeth0411-279.

Hu, D., Z. Tian, W. Wu, W. Wan, and A. D. Li. 2008. Photoswitchable nanoparticles enable high-resolution cell imaging: PULSAR microscopy. *J Am Chem Soc* 130(46):15279–15281. doi: 10.1021/ja805948u.

Huang, B., W. Wang, M. Bates, and X. Zhuang. 2008. Three-dimensional super-resolution imaging by stochastic optical reconstruction microscopy. *Science* 319(5864):810–813. doi: 10.1126/science.1153529.

Huisken, J., J. Swoger, F. Del Bene, J. Wittbrodt, and E. H. Stelzer. 2004. Optical sectioning deep inside live embryos by selective plane illumination microscopy. *Science* 305(5686):1007–1009. doi: 10.1126/science.1100035.

Ibraheem, A. and R. E. Campbell. 2010. Designs and applications of fluorescent protein-based biosensors. *Curr Opin Chem Biol* 14(1):30–36. doi: 10.1016/j.cbpa.2009.09.033.

Jaiswal, J. K. and S. M. Simon. 2004. Potentials and pitfalls of fluorescent quantum dots for biological imaging. *Trends Cell Biol* 14(9):497–504. doi: 10.1016/j.tcb.2004.07.012.

John, H. A., M. L. Birnstie, and K. W. Jones. 1969. RNA-DNA hybrids at cytological level. *Nature* 223(5206):582–587. doi: 10.1038/223582a0.

Juette, M. F., T. J. Gould, M. D. Lessard, M. J. Mlodzianoski, B. S. Nagpure, B. T. Bennett, S. T. Hess, and J. Bewersdorf. 2008. Three-dimensional sub-100 nm resolution fluorescence microscopy of thick samples. *Nat Methods* 5(6):527–529. doi: 10.1038/nmeth.1211.

Jungmann, R., M. S. Avendano, J. B. Woehrstein, M. Dai, W. M. Shih, and P. Yin. 2014. Multiplexed 3D cellular super-resolution imaging with DNA-PAINT and Exchange-PAINT. *Nat Methods* 11(3):313–318. doi: 10.1038/nmeth.2835.

Jungmann, R., C. Steinhauer, M. Scheible, A. Kuzyk, P. Tinnefeld, and F. C. Simmel. 2010. Single-molecule kinetics and super-resolution microscopy by fluorescence imaging of transient binding on DNA origami. *Nano Lett* 10(11):4756–4761. doi: 10.1021/nl103427w.

Keppler, A., S. Gendreizig, T. Gronemeyer, H. Pick, H. Vogel, and K. Johnsson. 2003. A general method for the covalent labeling of fusion proteins with small molecules in vivo. *Nat Biotechnol* 21(1):86–89. doi: 10.1038/Nbt765.

Khmelinskii, A., P. J. Keller, A. Bartosik, M. Meurer, J. D. Barry, B. R. Mardin, A. Kaufmann et al. 2012. Tandem fluorescent protein timers for in vivo analysis of protein dynamics. *Nat Biotechnol* 30(7):708–714. doi: 10.1038/nbt.2281.

Kishino, A. and T. Yanagida. 1988. Force measurements by micromanipulation of a single actin filament by glass needles. *Nature* 334(6177):74–76. doi: 10.1038/334074a0.

Klan, P., T. Solomek, C. G. Bochet, A. Blanc, R. Givens, M. Rubina, V. Popik, A. Kostikov, and J. Wirz. 2013. Photoremovable protecting groups in chemistry and biology: Reaction mechanisms and efficacy. *Chem Rev* 113(1): 119–191. doi: 10.1021/cr300177k.

Klehs, K., C. Spahn, U. Endesfelder, S. F. Lee, A. Fürstenberg, and M. Heilemann. 2014. Increasing the brightness of cyanine fluorophores for single-molecule and superresolution imaging. *ChemPhysChem* 15(4):637–641. doi: 10.1002/cphc.201300874.

Klessinger, M. 1978. Konstitution und Lichtabsorption organischer Farbstoffe. *Chemie in unserer Zeit* 12(1):1–11.

Kneen, M., J. Farinas, Y. X. Li, and A. S. Verkman. 1998. Green fluorescent protein as a noninvasive intracellular pH indicator. *Biophys J* 74(3):1591–1599.

Knoll, M. and E. Ruska. 1932. Das elektronenmikroskop. *Zeitschrift für Physik* 78(5–6):318–339. doi: 10.1007/BF01342199.

Köhler, A. 1904. *Mikrophotographische Untersuchungen mit ultraviolettem Licht.* Vol. 21, Germany: Zeitschrift für wissenschaftliche Mikroskopie und für mikroskopische Technik.

Kundu, K., S. F. Knight, N. Willett, S. Lee, W. R. Taylor, and N. Murthy. 2009. Hydrocyanines: A class of fluorescent sensors that can image reactive oxygen species in cell culture, tissue, and in vivo. *Angew Chem Int Ed Engl* 48(2):299–303. doi: 10.1002/anie.200804851.

Lakowicz, J. R. 2007. *Principles of Fluorescence Spectroscopy.* New York: Springer Science & Business Media.

Lakowicz, J. R., H. Szmacinski, K. Nowaczyk, K. W. Berndt, and M. Johnson. 1992. Fluorescence lifetime imaging. *Anal Biochem* 202(2):316–330. doi: http://dx.doi.org/10.1016/0003-2697(92)90112-K.

Lando, D., U. Endesfelder, H. Berger, L. Subramanian, P. D. Dunne, J. McColl, D. Klenerman et al. 2012. Quantitative single-molecule microscopy reveals that CENP-A(Cnp1) deposition occurs during G2 in fission yeast. *Open Biol* 2(7):120078. doi: 10.1098/rsob.120078.

Lee, S. F., Q. Verolet, and A. Furstenberg. 2013. Improved super-resolution microscopy with oxazine fluorophores in heavy water. *Angew Chem Int Ed Engl* 52(34):8948–8951. doi: 10.1002/anie.201302341.

Lehmann, H. 1913. Das Lumineszenz-Mikroskop seine Grundlagen und seine Anwendungen, *Z. wiss. Mikrosk* 30:417–470.

Lemmer, P., M. Gunkel, D. Baddeley, R. Kaufmann, A. Urich, Y. Weiland, J. Reymann, P. Muller, M. Hausmann, and C. Cremer. 2008. SPDM: Light microscopy with single-molecule resolution at the nanoscale. *Appl Phys B Lasers Opt* 93(1):1–12. doi: 10.1007/s00340-008-3152-x.

Lew, M. D., S. F. Lee, J. L. Ptacin, M. K. Lee, R. J. Twieg, L. Shapiro, and W. E. Moerner. 2011. Three-dimensional superresolution colocalization of intracellular protein superstructures and the cell surface in live Caulobacter crescentus. *Proc Natl Acad Sci USA* 108(46):E1102–E1110. doi: 10.1073/pnas.1114444108.

Longin, A., C. Souchier, M. Ffrench, and P. A. Bryon. 1993. Comparison of anti-fading agents used in fluorescence microscopy: Image analysis and laser confocal microscopy study. *J Histochem Cytochem* 41(12):1833–1840. doi: 10.1177/41.12.8245431.

Lord, S. J., N. R. Conley, H. L. Lee, S. Y. Nishimura, A. K. Pomerantz, K. A. Willets, Z. Lu et al. 2009. DCDHF fluorophores for single-molecule imaging in cells. *ChemPhysChem* 10(1):55–65. doi: 10.1002/cphc.200800581.

Los, G. V., L. P. Encell, M. G. McDougall, D. D. Hartzell, N. Karassina, C. Zimprich, M. G. Wood et al. 2008. HaloTag: A novel protein labeling technology for cell imaging and protein analysis. *ACS Chem Biol* 3(6):373–382. doi: 10.1021/cb800025k.

Lukinavicius, G., L. Reymond, E. D'Este, A. Masharina, F. Gottfert, H. Ta, A. Guther et al. 2014. Fluorogenic probes for live-cell imaging of the cytoskeleton. *Nat Methods* 11(7):731–733. doi: 10.1038/nmeth.2972.

Lukosz, W. and M. Marchand. 1963. Optischen Abbildung unter Überschreitung der beugungsbedingten Auflösungsgrenze. *Opt Acta Int J Opt* 10(3): 241–255. doi: 10.1080/713817795.

Lukyanov, K. A., D. M. Chudakov, S. Lukyanov, and V. V. Verkhusha. 2005. Innovation: Photoactivatable fluorescent proteins. *Nat Rev Mol Cell Biol* 6(11):885–891. doi: 10.1038/nrm1741.

Lukyanov, K. A., A. F. Fradkov, N. G. Gurskaya, M. V. Matz, Y. A. Labas, A. P. Savitsky, M. L. Markelov et al. 2000. Natural animal coloration can be determined by a nonfluorescent green fluorescent protein homolog. *J Biol Chem* 275(34):25879–25882. doi: 10.1074/jbc.C000338200.

Magde, D., W. W. Webb, and E. Elson. 1972. Thermodynamic fluctuations in a reacting system—Measurement by fluorescence correlation spectroscopy. *Phys Rev Lett* 29(11):705–708. doi: 10.1103/PhysRevLett.29.705.

Mason, W. T. 1999. *Fluorescent and Luminescent Probes for Biological Activity: A Practical Guide to Technology for Quantitative Real-Time Analysis.* San Diego, CA: Academic Press.

Matsuda, T., A. Miyawaki, and T. Nagai. 2008. Direct measurement of protein dynamics inside cells using a rationally designed photoconvertible protein. *Nat Methods* 5(4):339–345. doi: 10.1038/nmeth.1193.

Matsuyama, S., J. Llopis, Q. L. Deveraux, R. Y. Tsien, and J. C. Reed. 2000. Changes in intramitochondrial and cytosolic pH: Early events that modulate caspase activation during apoptosis. *Nat Cell Biol* 2(6):318–325. doi: 10.1038/35014006.

Maurel, D., S. Banala, T. Laroche, and K. Johnsson. 2010. Photoactivatable and photoconvertible fluorescent probes for protein labeling. *ACS Chem Biol* 5(5):507–516. doi: 10.1021/Cb1000229.

McKinney, S. A., C. S. Murphy, K. L. Hazelwood, M. W. Davidson, and L. L. Looger. 2009. A bright and photostable photoconvertible fluorescent protein. *Nat Methods* 6(2):131–133. doi: 10.1038/nmeth.1296.

McNeil, P. L. 1988. Incorporation of macromolecules into living cells. In *Methods in Cell Biology*. D. L. Taylor, Y.-L. Wang, and K. W. Jeon (eds.), San Diego, CA: Academic Press, pp. 153–173 (Chapter 10).

McNeil, P. L. and E. Warder. 1987. Glass-beads load macromolecules into living cells. *J Cell Sci* 88:669–678.

Miesenbock, G., D. A. De Angelis, and J. E. Rothman. 1998. Visualizing secretion and synaptic transmission with pH-sensitive green fluorescent proteins. *Nature* 394(6689):192–195. doi: 10.1038/28190.

Min, J., C. Vonesch, H. Kirshner, L. Carlini, N. Olivier, S. Holden, S. Manley, J. C. Ye, and M. Unser. 2014. FALCON: Fast and unbiased reconstruction of high-density super-resolution microscopy data. *Sci Rep* 4:4577. doi: 10.1038/srep04577.

Minsky, M. 1961. Microscopy apparatus. Patent number: US3013467 A.

Mitchison, T. J., K. E. Sawin, J. A. Theriot, K. Gee, and A. Mallavarapu. 1998. Caged fluorescent probes. In *Methods in Enzymology*. Marriott G. (ed.), San Diego, CA: Academic Press, pp. 63–78.

Moerner, W. E. 2012. Microscopy beyond the diffraction limit using actively controlled single molecules. *J Microsc* 246(3):213–220. doi: 10.1111/j.1365-2818.2012.03600.x.

Montenegro, H., M. Di Paolo, D. Capdevila, P. F. Aramendia, and M. L. Bossi. 2012. The mechanism of the photochromic transformation of spirorhodamines. *Photochem Photobiol Sci* 11(6):1081–1086. doi: 10.1039/c2pp05402g.

Nakai, J., M. Ohkura, and K. Imoto. 2001. A high signal-to-noise Ca(2+) probe composed of a single green fluorescent protein. *Nat Biotechnol* 19(2): 137–141. doi: 10.1038/84397.

Neumann, E., M. Schaeferridder, Y. Wang, and P. H. Hofschneider. 1982. Gene-transfer into mouse lyoma cells by electroporation in high electric-fields. *EMBO J* 1(7):841–845.

Nieuwenhuizen, R. P., K. A. Lidke, M. Bates, D. L. Puig, D. Grunwald, S. Stallinga, and B. Rieger. 2013. Measuring image resolution in optical nanoscopy. *Nat Methods* 10(6):557–562. doi: 10.1038/nmeth.2448.

Olivier, N., D. Keller, V. S. Rajan, P. Gonczy, and S. Manley. 2013. Simple buffers for 3D STORM microscopy. *Biomed Opt Express* 4(6):885–899. doi: 10.1364/BOE.4.000885.

Paez-Segala, M. G., M. G. Sun, G. Shtengel, S. Viswanathan, M. A. Baird, J. J. Macklin, R. Patel et al. 2015. Fixation-resistant photoactivatable fluorescent proteins for CLEM. *Nat Methods* 12(3):215–218. doi: 10.1038/nmeth.3225.

Patil, P. V. and D. P. Ballou. 2000. The use of protocatechuate dioxygenase for maintaining anaerobic conditions in biochemical experiments. *Anal Biochem* 286(2):187–192. doi: 10.1006/abio.2000.4802.

Patterson, G. H. and J. Lippincott-Schwartz. 2002. A photoactivatable GFP for selective photolabeling of proteins and cells. *Science* 297(5588):1873–1877. doi: 10.1126/science.1074952.

Pavani, S. R., M. A. Thompson, J. S. Biteen, S. J. Lord, N. Liu, R. J. Twieg, R. Piestun, and W. E. Moerner. 2009. Three-dimensional, single-molecule fluorescence imaging beyond the diffraction limit by using a double-helix point spread function. *Proc Natl Acad Sci USA* 106(9):2995–2999. doi: 10.1073/pnas.0900245106.

Pengo, T., S. J. Holden, and S. Manley. 2015. PALMsiever: A tool to turn raw data into results for single-molecule localization microscopy. *Bioinformatics* 31(5):797–798. doi: 10.1093/bioinformatics/btu720.

Persson, F., M. Linden, C. Unoson, and J. Elf. 2013. Extracting intracellular diffusive states and transition rates from single-molecule tracking data. *Nat Methods* 10(3):265–269. doi: 10.1038/Nmeth.2367.

Prasher, D. C., V. K. Eckenrode, W. W. Ward, F. G. Prendergast, and M. J. Cormier. 1992. Primary structure of the *Aequorea victoria* green-fluorescent protein. *Gene* 111(2):229–233.

Puchner, E. M., J. M. Walter, R. Kasper, B. Huang, and W. A. Lim. 2013. Counting molecules in single organelles with superresolution microscopy allows tracking of the endosome maturation trajectory. *Proc Natl Acad Sci USA* 110(40):16015–16020. doi: 10.1073/pnas.1309676110.

Qu, X., D. Wu, L. Mets, and N. F. Scherer. 2004. Nanometer-localized multiple single-molecule fluorescence microscopy. *Proc Natl Acad Sci USA* 101(31):11298–11303. doi: 10.1073/pnas.0402155101.

Rasnik, I., S. A. McKinney, and T. Ha. 2006. Nonblinking and long-lasting single-molecule fluorescence imaging. *Nat Methods* 3(11):891–893. doi: 10.1038/nmeth934.

Rayleigh. 1896. XV. On the theory of optical images, with special reference to the microscope. *Philos Mag Series 5* 42(255):167–195. doi: 10.1080/14786449608620902.

Rocha, S., J. A. Hutchison, K. Peneva, A. Herrmann, K. Mullen, M. Skjot, C. I. Jorgensen et al. 2009. Linking phospholipase mobility to activity by single-molecule wide-field microscopy. *ChemPhysChem* 10(1):151–161. doi: 10.1002/cphc.200800537.

Roeffaers, M. B., G. De Cremer, J. Libeert, R. Ameloot, P. Dedecker, A. J. Bons, M. Buckins et al. 2009. Super-resolution reactivity mapping of nanostructured catalyst particles. *Angew Chem Int Ed Engl* 48(49):9285–9289. doi: 10.1002/anie.200904944.

Rollins, G. C., J. Y. Shin, C. Bustamante, and S. Presse. 2015. Stochastic approach to the molecular counting problem in superresolution microscopy. *Proc Natl Acad Sci USA* 112(2):E110–E118. doi: 10.1073/pnas.1408071112.

Rust, M. J., M. Bates, and X. Zhuang. 2006. Sub-diffraction-limit imaging by stochastic optical reconstruction microscopy (STORM). *Nat Methods* 3(10):793–795. doi: 10.1038/nmeth929.

Sakon, J. J. and K. R. Weninger. 2010. Detecting the conformation of individual proteins in live cells. *Nat Methods* 7(3):203–205. doi: 10.1038/Nmeth.1421.

Sauer, M., J. Hofkens, and J. Enderlein. 2010. *Handbook of Fluorescence Spectroscopy and Imaging: From Ensemble to Single Molecules*. Weinheim, Germany: John Wiley & Sons.

Schäfer, L. V., G. Groenhof, M. Boggio-Pasqua, M. A. Robb, and H. Grubmuller. 2008. Chromophore protonation state controls photoswitching of the fluoroprotein asFP595. *PLoS Comput Biol* 4(3):e1000034. doi: 10.1371/journal.pcbi.1000034.

Schäfer, L. V, G. Groenhof, A. R. Klingen, G. Matthias Ullmann, M. Boggio-Pasqua, M. A. Robb, and H. Grubmüller. 2007. Photoswitching of the fluorescent protein asFP595: Mechanism, proton pathways, and absorption spectra. *Angewandte Chemie* 119(4):536–542. doi: 10.1002/ange.200602315.

Schäfer, P., S. van de Linde, J. Lehmann, M. Sauer, and S. Doose. 2013. Methylene blue- and thiol-based oxygen depletion for super-resolution imaging. *Anal Chem* 85(6):3393–3400. doi: 10.1021/ac400035k.

Schoen, I., J. Ries, E. Klotzsch, H. Ewers, and V. Vogel. 2011. Binding-activated localization microscopy of DNA structures. *Nano Lett* 11(9):4008–4011. doi: 10.1021/nl2025954.

Schucker, K., T. Holm, C. Franke, M. Sauer, and R. Benavente. 2015. Elucidation of synaptonemal complex organization by super-resolution imaging with isotropic resolution. *Proc Natl Acad Sci USA* 112(7):2029–2033. doi: 10.1073/pnas.1414814112.

Seefeldt, B., R. Kasper, M. Beining, J. Mattay, J. Arden-Jacob, N. Kemnitzer, K. H. Drexhage, M. Heilemann, and M. Sauer. 2010. Spiropyrans as molecular optical switches. *Photochem Photobiol Sci* 9(2):213–220. doi: 10.1039/b9pp00118b.

Sengupta, P., T. Jovanovic-Talisman, and J. Lippincott-Schwartz. 2013. Quantifying spatial organization in point-localization superresolution images using pair correlation analysis. *Nat Protoc* 8(2):345–354. doi: 10.1038/nprot.2013.005.

Shaner, N. C., P. A. Steinbach, and R. Y. Tsien. 2005. A guide to choosing fluorescent proteins. *Nat Methods* 2(12):905–909. doi: 10.1038/nmeth819.

Shannon, C. E. 1949. Communication in the presence of noise. *Proc IRE* 37(1): 10–21. doi: 10.1109/Jrproc.1949.232969.

Sharonov, A. and R. M. Hochstrasser. 2006. Wide-field subdiffraction imaging by accumulated binding of diffusing probes. *Proc Natl Acad Sci USA* 103(50):18911–18916. doi: 10.1073/pnas.0609643104.

Shcherbakova, D. M., P. Sengupta, J. Lippincott-Schwartz, and V. V. Verkhusha. 2014. Photocontrollable fluorescent proteins for superresolution imaging. *Annu Rev Biophys* 43:303–329. doi: 10.1146/annurev-biophys-051013-022836.

Shcherbo, D., Shemiakina, II, A. V. Ryabova, K. E. Luker, B. T. Schmidt, E. A. Souslova, T. V. Gorodnicheva et al. 2010. Near-infrared fluorescent proteins. *Nat Methods* 7(10):827–829. doi: 10.1038/nmeth.1501.

Shtengel, G., J. A. Galbraith, C. G. Galbraith, J. Lippincott-Schwartz, J. M. Gillette, S. Manley, R. Sougrat et al. 2009. Interferometric fluorescent super-resolution microscopy resolves 3D cellular ultrastructure. *Proc Natl Acad Sci USA* 106(9):3125–3130. doi: 10.1073/pnas.0813131106.

Shu, X., A. Royant, M. Z. Lin, T. A. Aguilera, V. Lev-Ram, P. A. Steinbach, and R. Y. Tsien. 2009. Mammalian expression of infrared fluorescent proteins engineered from a bacterial phytochrome. *Science* 324(5928):804–807. doi: 10.1126/science.1168683.

Siedentopf, H. and R. Zsigmondy. 1902. Über Sichtbarmachung und Größenbestimmung ultramikoskopischer Teilchen, mit besonderer Anwendung auf Goldrubingläser. *Annalen der Physik* 315(1):1–39. doi: 10.1002/andp.19023150102.

Singer, C. 1914. Notes on the early history of microscopy. *Proc R Soc Med* 7 (Sect Hist Med):247–279.

Small, A. and S. Stahlheber. 2014. Fluorophore localization algorithms for super-resolution microscopy. *Nat Methods* 11(3):267–279. doi: 10.1038/nmeth.2844.

Soumpasis, D. M. 1983. Theoretical-analysis of fluorescence photobleaching recovery experiments. *Biophys J* 41(1):95–97.

Steinhauer, C., C. Forthmann, J. Vogelsang, and P. Tinnefeld. 2008. Superresolution microscopy on the basis of engineered dark states. *J Am Chem Soc* 130(50):16840–16841. doi: 10.1021/ja806590m.

Stelzer, E. H. K. and S. W. Hell. 1992. Fundamental improvement of resolution with a 4Pi-confocal fluorescence microscope using two-photon excitation. *Opt Commun* 93:277–282.

Stiel, A. C., M. Andresen, H. Bock, M. Hilbert, J. Schilde, A. Schonle, C. Eggeling, A. Egner, S. W. Hell, and S. Jakobs. 2008. Generation of monomeric reversibly switchable red fluorescent proteins for far-field fluorescence nanoscopy. *Biophys J* 95(6):2989–2997. doi: 10.1529/biophysj.108.130146.

Stokes, G. G. 1852. On the change of refrangibility of light. *Phil Trans R Soc Lond* 142:463–562.

Stryer, L. 1978. Fluorescence energy-transfer as a spectroscopic ruler. *Annu Rev Biochem* 47:819–846. doi: 10.1146/annurev.bi.47.070178.004131.

Subach, F. V., V. N. Malashkevich, W. D. Zencheck, H. Xiao, G. S. Filonov, S. C. Almo, and V. V. Verkhusha. 2009a. Photoactivation mechanism of PAmCherry based on crystal structures of the protein in the dark and fluorescent states. *Proc Natl Acad Sci USA* 106(50):21097–21102. doi: 10.1073/pnas.0909204106.

Subach, F. V., G. H. Patterson, S. Manley, J. M. Gillette, J. Lippincott-Schwartz, and V. V. Verkhusha. 2009b. Photoactivatable mCherry for high-resolution two-color fluorescence microscopy. *Nat Methods* 6(2):153–159. doi: 10.1038/nmeth.1298.

Subach, O. M., D. Entenberg, J. S. Condeelis, and V. V. Verkhusha. 2012. A FRET-facilitated photoswitching using an orange fluorescent protein with the fast photoconversion kinetics. *J Am Chem Soc* 134(36):14789–14799. doi: 10.1021/ja3034137.

Swoboda, M., J. Henig, H. M. Cheng, D. Brugger, D. Haltrich, N. Plumere, and M. Schlierf. 2012. Enzymatic oxygen scavenging for photostability without pH drop in single-molecule experiments. *ACS Nano* 6(7):6364–6369. doi: 10.1021/nn301895c.

Swulius, M. T. and G. J. Jensen. 2012. The helical MreB cytoskeleton in Escherichia coli MC1000/pLE7 is an artifact of the N-Terminal yellow fluorescent protein tag. *J Bacteriol* 194(23):6382–6386. doi: 10.1128/JB.00505-12.

Synge, E. H. 1928. A suggested method for extending microscopic resolution into the ultramicroscopic region. *Philos Mag Series 7* 6(35):356–362. doi: 10.1080/14786440808564615.

Szymborska, A., A. de Marco, N. Daigle, V. C. Cordes, J. A. Briggs, and J. Ellenberg. 2013. Nuclear pore scaffold structure analyzed by super-resolution microscopy and particle averaging. *Science* 341(6146):655–658. doi: 10.1126/science.1240672.

Taylor, D. L. and Y. L. Wang. 1978. Molecular cytochemistry—Incorporation of fluorescently labeled actin into living cells. *Proc Nat Acad Sci USA* 75(2):857–861. doi: 10.1073/pnas.75.2.857.

Toseland, C. P. 2013. Fluorescent labeling and modification of proteins. *J Chem Biol* 6(3):85–95. doi: 10.1007/s12154-013-0094-5.

Uno, K., H. Niikura, M. Morimoto, Y. Ishibashi, H. Miyasaka, and M. Irie. 2011. In situ preparation of highly fluorescent dyes upon photoirradiation. *J Am Chem Soc* 133(34):13558–13564. doi: 10.1021/ja204583e.

Uno, S. N., M. Kamiya, T. Yoshihara, K. Sugawara, K. Okabe, M. C. Tarhan, H. Fujita et al. 2014. A spontaneously blinking fluorophore based on intramolecular spirocyclization for live-cell super-resolution imaging. *Nat Chem* 6(8):681–689. doi: 10.1038/nchem.2002.

Valnes, K. and P. Brandtzaeg. 1985. Retardation of immunofluorescence fading during microscopy. *J Histochem Cytochem* 33(8):755–761.

van de Linde, S., I. Krstic, T. Prisner, S. Doose, M. Heilemann, and M. Sauer. 2011a. Photoinduced formation of reversible dye radicals and their impact on super-resolution imaging. *Photochem Photobiol Sci* 10(4):499–506. doi: 10.1039/c0pp00317d.

van de Linde, S., A. Löschberger, T. Klein, M. Heidbreder, S. Wolter, M. Heilemann, and M. Sauer. 2011b. Direct stochastic optical reconstruction microscopy with standard fluorescent probes. *Nat Protoc* 6(7):991–1009. doi: 10.1038/nprot.2011.336.

van de Linde, S. and M. Sauer. 2014. How to switch a fluorophore: From undesired blinking to controlled photoswitching. *Chem Soc Rev* 43(4):1076–1087. doi: 10.1039/C3cs60195a.

Vandenberg, W., M. Leutenegger, T. Lasser, J. Hofkens, and P. Dedecker. 2015. Diffraction-unlimited imaging: From pretty pictures to hard numbers. *Cell Tissue Res.* doi: 10.1007/s00441-014-2109-0.

Vaughan, J. C., G. T. Dempsey, E. Sun, and X. Zhuang. 2013. Phosphine quenching of cyanine dyes as a versatile tool for fluorescence microscopy. *J Am Chem Soc* 135(4):1197–1200. doi: 10.1021/ja3105279.

Vaughan, J. C., S. Jia, and X. Zhuang. 2012. Ultrabright photoactivatable fluorophores created by reductive caging. *Nat Methods* 9(12):1181–1184. doi: 10.1038/nmeth.2214.

Vogelsang, J., T. Cordes, C. Forthmann, C. Steinhauer, and P. Tinnefeld. 2009. Controlling the fluorescence of ordinary oxazine dyes for single-molecule switching and superresolution microscopy. *Proc Natl Acad Sci USA* 106(20):8107–8112. doi: 10.1073/pnas.0811875106.

Vogelsang, J., R. Kasper, C. Steinhauer, B. Person, M. Heilemann, M. Sauer, and P. Tinnefeld. 2008. A reducing and oxidizing system minimizes photobleaching and blinking of fluorescent dyes. *Angew Chem Int Ed Engl* 47(29):5465–5469. doi: 10.1002/anie.200801518.

Vogelsang, J., C. Steinhauer, C. Forthmann, I. H. Stein, B. Person-Skegro, T. Cordes, and P. Tinnefeld. 2010. Make them blink: Probes for super-resolution microscopy. *ChemPhysChem* 11(12):2475–2490. doi: 10.1002/cphc.201000189.

Wang, F., W. Beng Tan, Y. Zhang, X. Fan, and M. Wang. 2006. Luminescent nanomaterials for biological labelling. *Nanotechnology* 17(1):R1–R13. doi: 10.1088/0957-4484/17/1/r01.

Weber, G. 1952. Polarization of the fluorescence of macromolecules. I. Theory and experimental method. *Biochem J* 51(2):145–155.

Widengren, J., A. Chmyrov, C. Eggeling, P. A. Lofdahl, and C. A. Seidel. 2007. Strategies to improve photostabilities in ultrasensitive fluorescence spectroscopy. *J Phys Chem A* 111(3):429–440. doi: 10.1021/jp0646325.

Winterflood, C. M. and H. Ewers. 2014. Single-molecule localization microscopy using mCherry. *ChemPhysChem* 15(16):3447–3451. doi: 10.1002/cphc.201402423.

Wolter, S., U. Endesfelder, S. van de Linde, M. Heilemann, and M. Sauer. 2011. Measuring localization performance of super-resolution algorithms on very active samples. *Opt Express* 19(8):7020–7033. doi: 10.1364/OE.19.007020.

Wysocki, L. M., J. B. Grimm, A. N. Tkachuk, T. A. Brown, E. Betzig, and L. D. Lavis. 2011. Facile and general synthesis of photoactivatable xanthene dyes. *Angew Chem Int Ed Engl* 50(47):11206–11209. doi: 10.1002/anie.201104571.

You, L., D. Zha, and E. V. Anslyn. 2015. Recent advances in supramolecular analytical chemistry using optical sensing. *Chem Rev.* doi: 10.1021/cr5005524.

Zernike, F. 1942. Phase contrast, a new method for the microscopic observation of transparent objects Part II. *Physica* 9:974–986. doi: 10.1016/S0031-8914(42)80079-8.

Zhang, L., H. N. Patel, J. W. Lappe, and R. M. Wachter. 2006. Reaction progress of chromophore biogenesis in green fluorescent protein. *J Am Chem Soc* 128(14):4766–4772. doi: 10.1021/ja0580439.

Zondervan, R., F. Kulzer, S. B. Orlinskii, and M. Orrit. 2003. Photoblinking of rhodamine 6G in poly(vinyl alcohol): Radical dark state formed through the triplet. *J Phys Chem A* 107(35):6770–6776. doi: 10.1021/Jp034723r.

4

Image Analysis for Single-Molecule Localization Microscopy

Martin Ovesný, Pavel Křížek, Josef Borkovec, Zdeněk Švindrych, and Guy M. Hagen

4.1 Introduction
4.2 The Traditional Approach
4.3 Temporal Resolution in SMLM
4.4 Evaluating the Results of SMLM Processing Using Monte Carlo Simulations
4.5 Software for SMLM
4.6 Materials and Methods
Acknowledgments
References

4.1 Introduction

Single-molecule localization microscopy (SMLM) methods such as stochastic optical reconstruction microscopy (Rust et al. 2006) (STORM) and photoactivated localization microscopy (Betzig et al. 2006) (PALM) have quickly been adopted by many laboratories as a reliable method for achieving optical resolution beyond the diffraction limit. One reason for this rapid dissemination is that the method requires a fairly simple microscope setup, but perhaps more importantly, SMLM methods invite contributions from a variety of disciplines in chemistry and physics, as well as the ultimate applications of SMLM in cell biology. Probe design, labeling strategies, instrumentation development, and image analysis are all part of the process, making SMLM a truly interdisciplinary endeavor.

In SMLM, a super-resolution image is reconstructed from a sequence of conventional images of sparsely distributed single photoswitchable molecules. Because the image sequence is usually long and the positions of the molecules have to be estimated systematically with subdiffraction precision, specialized software is required for processing the data. The design, implementation, and evaluation of algorithms for processing SMLM data, and the combination of these algorithms into useful software to create super-resolution images which are informative, accurate, and beautiful has become a field in its own right (Small and Stahlheber 2014; Sage et al. 2015).

Here we review and discuss various aspects of image processing methods for SMLM. We first detail the traditional approach, followed by a discussion of some alternative methods that have recently been introduced. We also discuss how Monte Carlo simulations can be used to evaluate SMLM processing procedures, and discuss one of the software packages that is currently available.

4.2 The Traditional Approach

The traditional approach can be broken down into several steps, shown as a flowchart in Figure 4.1. Sections 4.2.1 through 4.2.7 describe each of the steps in detail.

4.2.1 Image Filtering for Noise Reduction and Feature Enhancement

Image filtering is one of the first steps in many image processing applications. This step helps to identify the molecules in the raw data by reducing noise and enhancing the features we are interested in. In this case, we are interested in point-like signals representing single molecules. Low-pass filters (e.g., a Gaussian filter or averaging filter) can be used to reduce noise in the image (Křížek et al. 2011). Recently, better results were acquired with band-pass filters, for example with a lowered Gaussian filter as was used in DAOSTORM (Holden et al. 2011), or with a filter based on wavelet transformation (Izeddin et al. 2012). The wavelet transform is commonly used in modern signal-processing applications and in our hands always provides good results, even with very low signal-to-noise ratios.

Figure 4.2 shows an example of the effect of denoising and feature enhancement using a wavelet transform-based band-pass filter. Figure 4.2a shows a wide-field (WF) image of an A431 cell expressing mCitrine-erbB3. Figure 4.2b shows a single frame of SMLM raw data. The image is from early in the acquired sequence and has many photoactivated molecules and a high background from cellular autofluorescence. Despite the high background, single molecules can be emphasized with a band-pass filter (Figure 4.2c), helping produce a high-quality SMLM reconstruction (Figure 4.4d). See Section 4.6 for a description of the sample and the microscopy setup used to acquire these images.

4.2.2 Thresholding and Detection of Molecules

Thresholding and detection involves finding the approximate positions of the molecules in the input images after treatment with the noise reduction and feature enhancement step described above. Detection, also known as "spot finding" or "blob detection" is part of the field of feature detection in image processing and has been quantitatively evaluated for single-molecule imaging (Ruusuvuori et al. 2010). The question we are asking is "what is a real molecule and what is only noise?" We have found that detection of local intensity maxima (in an 8-connected neighborhood) gives reliable results. The next step is to apply a threshold. Izeddin suggested that the threshold value can be calculated as the standard deviation of the background noise. An estimate of the background

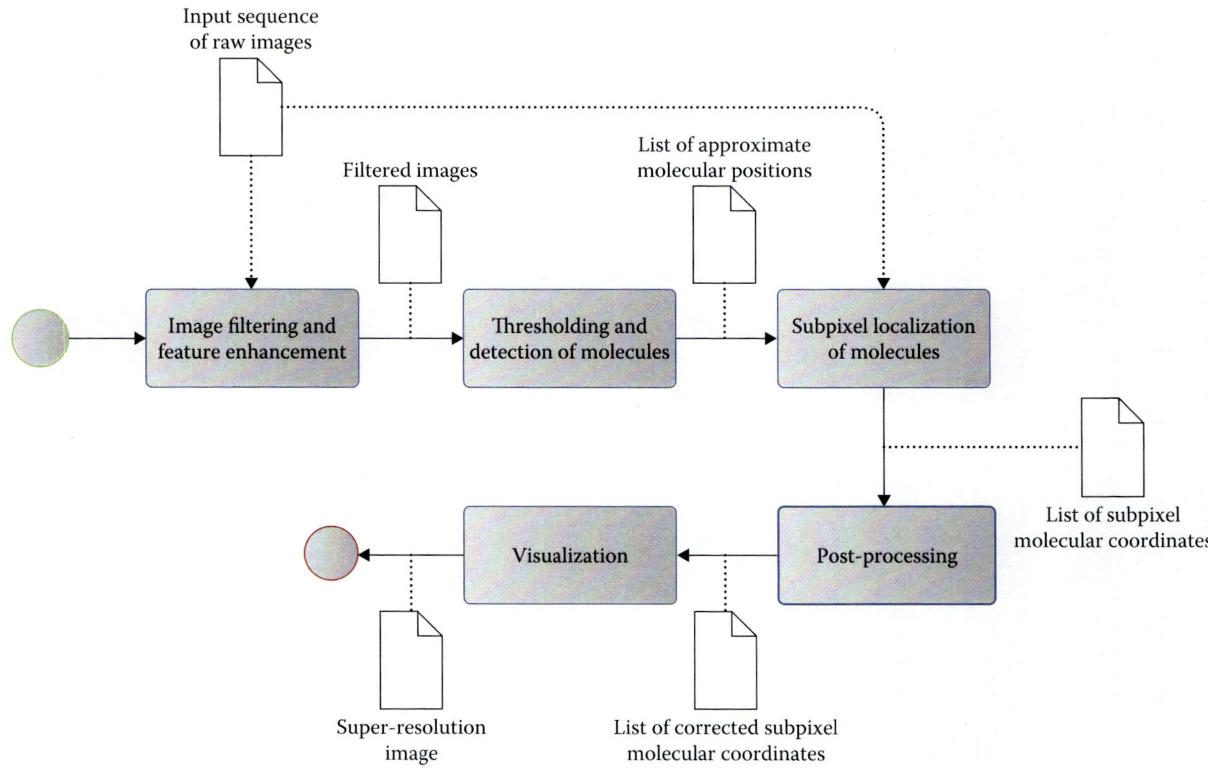

Figure 4.1

Data analysis steps for single-molecule localization super-resolution imaging. The "pages" represent data or images, while the "boxes" represent processing steps carried out by analysis software. (Reprinted with permission from Ovesný, M et al., *Bioinformatics*, 30, 2389–2390, 2014a.)

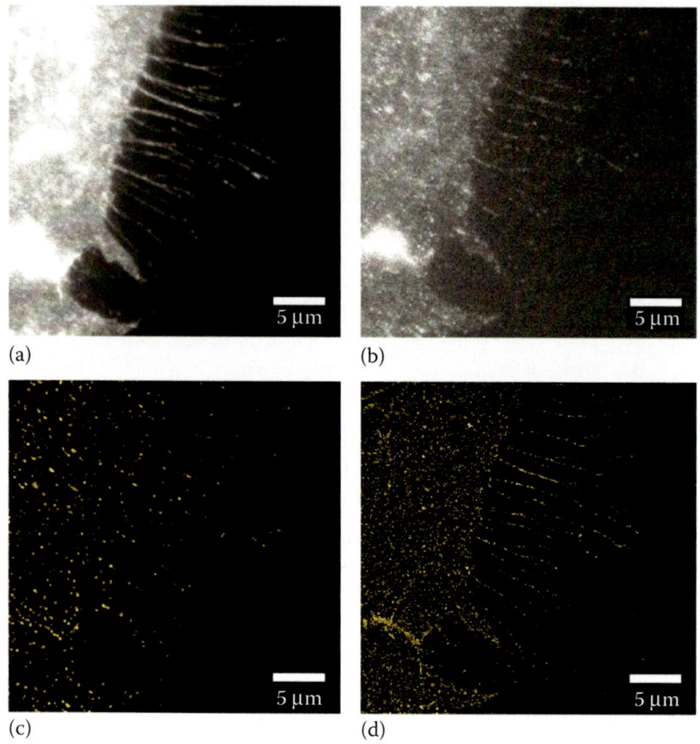

Figure 4.2

Example of feature enhancement with a band-pass filter. (a) Widefield image, (b) single frame of SMLM raw data, and (c) single frame; result of wavelet-based band-pass filter. Single molecules and groups of molecules are visible, and (d) SMLM reconstruction, 10,000 frames, ~162,500 detected molecules. The average localization accuracy according to Thompson, R E et al. (2002.) is 18 +/− 6 nm (mean +/− SD). The brightness and contrast of each image were adjusted individually to improve visibility.

noise can be obtained as the first plane of the wavelet decomposition performed in image filtering step (Izeddin et al. 2012).

We would like to emphasize that the results of detection, according to whichever algorithm is used, can be evaluated quantitatively using realistic Monte Carlo simulations based on measured parameters, see Section 4.4. This is important because it helps remove bias in the experiment.

4.2.3 Subpixel Localization

Subpixel localization of single molecules with an accuracy below the diffraction limit is the basis of SMLM methods and is the factor which produces the super-resolution effect. Prior to super-resolution microscopy, the concept of subdiffraction localization was used for many years in particle tracking algorithms (Kusumi et al. 1993; Daumas et al. 2003; Suzuki et al. 2005; Chenouard et al. 2014), and the problem of localization of a single molecule or nanoparticle has been treated theoretically rather extensively (Thompson et al. 2002; Abraham et al. 2009; Deschout et al. 2014). There were also several reports using single-molecule localization to measure size or distance relationships (Yildiz et al. 2003; Gordon et al. 2004; Qu et al. 2004; Churchman et al. 2005; Lidke et al. 2005;

Lemmer et al. 2008). Localization algorithms can be divided into two categories. Fast, nonfitting methods with analytic solutions and slower, generally more accurate methods based on fitting a point spread function (PSF) model using iterative nonlinear optimization.

To localize molecules with subpixel precision, after the detection step described above, one needs to create subimages by cutting out each molecule in a small region of interest. The size of the window should be set proportional to the width of the imaged size of the molecules in the image. This depends on the magnification of the microscope and the pixel size of the camera. For accurate single-molecule localization, it is necessary that the image of entire molecule fits into the small window and that there is only a single molecule in each window. Usually, subimage windows of 7 × 7 up to 15 × 15 pixels are used. Note that in this step the raw input images should be used.

4.2.3.1 Nonfitting Methods

Calculation of the center of mass in a local image neighborhood is a very fast method for subpixel localization of molecules and is used in QuickPALM (Henriques et al. 2010). The main idea is simply to calculate the mean pixel positions weighted by the intensity of the image data.

Another algorithm finds the subpixel position of a molecule by determining the point with maximal radial symmetry in the data as described in Parthasarathy (2012), Ma et al. (2012), and Ovesný et al. (2014a). The general idea is to find the origin of radial symmetry (i.e., the center of a molecule) as the point with the minimum distance to gradient-oriented lines passing through all data points. The calculation of each molecular position is very fast due to an analytical solution. Radial symmetry is a robust feature in SMLM data, making the algorithm resistant to noise.

4.2.3.2 Fitting Methods

The response of a microscope to a point-like source of light is described by the PSF. Because a single-molecule emitter can be treated as a point source in SMLM data, the result of fitting a PSF model to an image of a single molecule is an estimate of the subdiffraction molecular position and its intensity. The most commonly used model is a Gaussian function. It has been shown that the Gaussian function provides a very good approximation of the real PSF of a microscope (Thompson et al. 2002; Stallinga and Rieger 2010). This is partly due to pixelation effects and the presence of noise, which makes the difference between the Gaussian function and the real PSF negligible. The advantage of Gaussian PSF models are their simplicity, robustness, and computational efficiency. Gaussian models can be further modified to take into account the discrete pixel grid of a digital camera by fitting its integrated form (Smith et al. 2010; Huang et al. 2011).

To perform the fitting, one can use least squares, or weighted least squares, which assume normally (Gaussian) distributed noise. Certain maximum likelihood estimation (MLE) methods assume a Poisson noise model, which more accurately describes the photon counting process dominant in the extreme low light conditions which are present when imaging single molecules (Smith et al. 2010). When using MLE, the pixel values must be expressed in photons, requiring a calibrated camera. With higher photon counts, weighted least squares methods provide a good approximation to the MLE results.

4.2.3.3 Localization Uncertainty

To estimate limits of localization precision, one can use the concept of Fisher information and calculate the Cramér–Rao lower bound (Smith et al. 2010), which gives a lower bound on the variance of an unbiased estimator of the quantity of interest such as the position of the molecule or number of detected photons. However, other models are commonly used such as the Thomson–Larson–Webb equation (Thompson et al. 2002). Another possibility is Zhuang's method (Rust et al. 2006), which uses localization of a single molecule in many camera frames. One then computes the standard deviation of the x and y positions and takes this as the localization accuracy. This is an attractive method because it is directly based on measured quantities.

4.2.4 The "Crowded Field Problem"

High spatial densities of activated molecules can result in a "crowded field problem," in which single molecules are not adequately resolved. When there are too many photoactivated molecules in a single camera frame, their images start to overlap. The PSF models described above can fit only a single molecule, so a more sophisticated solution is needed, also see Section 4.3 below.

4.2.4.1 Multiemitter Fitting Analysis

One approach for solving the high-density problem uses multiemitter fitting (Huang et al. 2011). Here, subimages are fit with multimodal 2D Gaussian functions. First, the algorithm fits PSF1 (a single-molecule model) with an initial molecular position obtained by one of the methods described in Section 4.2. The fitted PSF is subtracted from the raw data and the position of the maximum intensity value in the residual image is taken as an approximate position of a second molecule. The fitting is now repeated on the raw data with PSF2 (a model containing two molecules) and with the initial positions estimated in the previous steps. The result of the fit is subtracted from the raw data to find an approximate position of a third molecule in the residual image. The process is repeated until the number of molecules reaches a user-defined limit. This process is rather computationally intensive and can be accelerated using parallel processing on a graphics processing unit (GPU) (Huang et al. 2011).

Because a model with more parameters will always be able to fit the data at least as well as a model with fewer parameters (Bevington 2003), statistical tests are required to determine whether the more complex model provides a significantly better fit of the underlying data. The statistical tests are usually based on pair-wise model comparison. Here a fit by PSF1 is compared with a fit by PSF2, the better of the two is compared with a fit by PSF3, and so on. Pair-wise comparisons are based on an F-test (for least squares methods) (Bevington 2003) or on a log-likelihood ratio test (for MLE methods) (Huang et al. 2011).

When using multiple-emitter fitting analysis, repeated localizations of a single molecule in one frame may occur due to overlapping fitting sub-regions. To solve this problem molecules with a mutual distance smaller than their localization uncertainty are grouped together, and in each group, only the molecule with the smallest localization uncertainty is kept.

4.2.5 Post Processing

Post processing may involve several steps depending on the nature of the acquired data. The following steps can be applied to the list of coordinates generated in the previous steps.

4.2.5.1 Drift Correction

Long time series data acquisitions usually suffer from sample drift. Applying drift correction may result in a significant improvement in resolution. There are two commonly used methods for lateral drift correction. The first is based on tracking fiducial markers (usually fluorescent beads or gold nanoparticles) present in the sample. Coverslips with embedded fiducial markers are also available (www.hestzig.com). The second method is based on cross-correlation of similar structures in reconstructed super-resolution images (Mlodzianoski et al. 2011). In this approach, the whole image sequence is broken into several (perhaps 10–20) subsequences. Super-resolution images are formed from each of these subsequences and compared using cross-correlation methods to calculate the sample drift in time. Drift for each frame is then interpolated from the results of the cross-correlation to achieve a final result in which the molecular positions are adjusted to remove sample drift. Correlation methods can correct drift in Z as well (McGorty et al. 2013).

Correlation-based drift correction has the advantage that no fiducial markers in the sample are needed. However, the method needs many localizations to form an image with enough features to allow the cross-correlation to be calculated accurately. To overcome this drawback a redundant cross-correlation-based drift method has been developed (Wang et al. 2014). This method calculates differential images between all pairs of frames in the acquired sequence and then estimates the drift for all of the frames as the solution to a single optimization problem. This method can be applied to data where the number of localizations is low.

4.2.5.2 Merging

In some SMLM experiments (particularly in *direct* STORM [*d*STORM] [Heilemann et al. 2008]), a single photoactivated molecule may appear intermittently in several sequential images before bleaching completely. If this is the case, one can identify such sequences of molecular locations and combine them into one single molecule. The new position of the resulting signal can be calculated as the mean value of the original data, while the intensity is calculated as the sum. The localization uncertainty of the merged molecule is calculated from the new values. The distance within which molecules are merged together in the subsequent frames, as well as the allowed number of frames in which the molecule is not detected, are usually user-specified values. The merging algorithm is usually based on finding nearest-neighbors within a specified radius between active molecules from the previous frames and the current frame.

4.2.5.3 Filtering

The next step is usually to remove unwanted molecules according to user-defined criteria. Histograms of particular measured parameters are useful to select thresholds to define the criteria. However, some reports of SMLM algorithms removed molecules which were, for example, too big or too small, or decisions were made based on other parameters such as photon count or asymmetry. This was performed without any quantitative guidance or measurements to help decide where to place the appropriate thresholds. This is not a systematic approach and can result in measurements containing experimenter bias, however unintentional. Every relevant parameter can be measured and evaluated according to appropriate statistical tests. We feel this is, in general, the right way to proceed because the overall algorithm can then be based as much as possible on measured quantities and will help in generating reproducible results. Such decisions should be further supported by Monte Carlo simulations (Křížek et al. 2011).

4.2.6 Visualization Methods

SMLM data contains the (x, y, z) positions of the localized molecules. This differs considerably from the intensity-based data which are acquired in conventional fluorescence microscopy. Visualization (also called rendering) of SMLM data involves creation of a new super-resolution image based on the list of molecular coordinates (Baddeley et al. 2010). In the three-dimensional case, each rendered image slice usually contains a visualization of molecules with axial positions in a desired range. The magnification ratio of the new super-resolution image to the original image is usually a user-specified parameter. Specialized software is available that performs visualization and post-processing functions (PALMSiever) (Pengo et al. 2014), while another software package is available that specializes in 3D visualization (ViSP) (El Beheiry and Dahan 2013).

There are several methods for visualization reviewed in Baddeley et al. (2010). Figure 4.3 shows a comparison of visualization methods such as Gaussian rendering, jittered histogram rendering (Křížek et al. 2011), and adaptive shifted histogram (ASH) rendering methods (Ovesný et al. 2014a) on simulated data. Figure 4.4 shows comparison of visualization methods using a publicly available dataset acquired using a real cellular sample (Tubulin AF647, contributed by Nicolas Olivier and Suliana Manley) from the single-molecule challenge website (Benchmarking of SMLM Software 2013), http://bigwww.epfl.ch/palm/. We recommend trying a few different methods before making the final choice.

4.2.7 Further Analysis

Once the initial analysis is complete, SMLM experiments can be further analyzed in several ways. Colocalization analysis is a commonly used approach to measure spatial overlap between two (or more) different fluorescent labels, each having a different emission wavelength. Traditional colocalization analysis of laser scanning confocal microscopy images has several limitations (Ronneberger et al. 2008), however, SMLM methods offer a new paradigm for colocalization studies due to the coordinate-based nature of the data (Gunkel et al. 2009; Malkusch et al. 2012; Smirnov et al. 2014).

A second important post-processing application is the analysis of molecular clustering. Several methods have been suggested for identifying and analyzing clusters of molecules, especially cell membrane proteins (Owen et al. 2010, 2012; Rossy et al. 2014; Rubin-Delanchy et al. 2015). Indeed, measurements and visualization of potential interactions of membrane proteins has been reenergized by the advent of super-resolution imaging. Several studies have reported clustering phenomena, for example in T cells (Williamson et al. 2011; Rossy et al. 2013) and in cardiac myocytes (Baddeley et al. 2009). In addition to its visualization features, the software package PALMsiever can analyze clustering (Pengo et al. 2014) using the classic density-based spatial clustering algorithm DBSCAN (Ester et al. 1996).

When evaluating clustering, overcounting the number of molecules can be a serious problem. Pair correlation analysis can be used to reduce overcounting

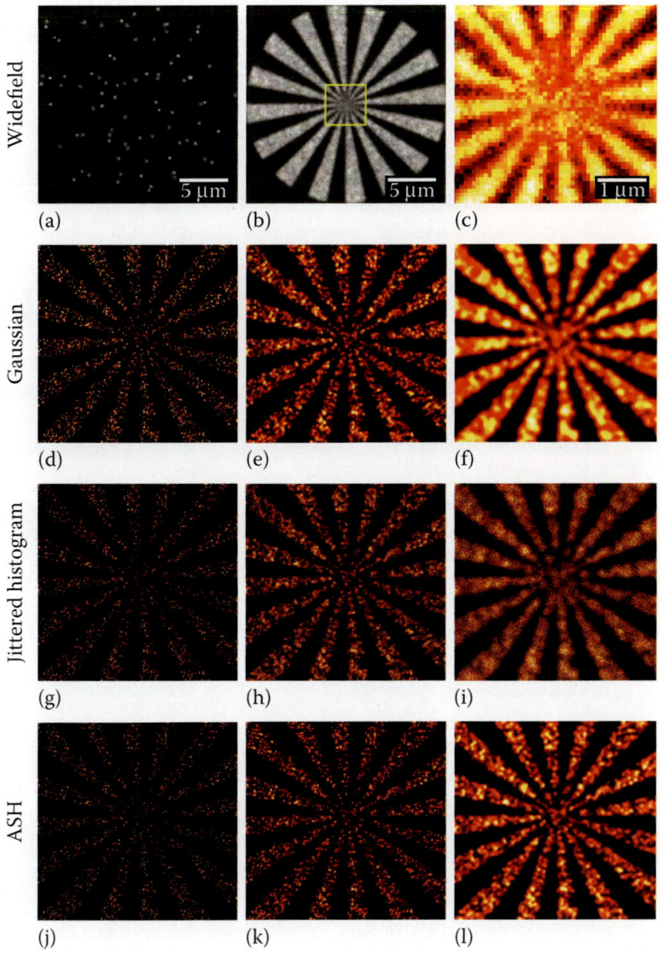

Figure 4.3

Comparison of visualization methods on simulated data. Widefield images: (a) a single frame of raw data, (b) average intensity time projection, and (c) zoomed region indicated in (b). (d–l) Super-resolution SMLM images of the region indicated in (b) magnified 10×. Gaussian rendering with lateral uncertainty set to: (d) computed localization uncertainty, (e) 20 nm, (f) 50 nm. Jittered histogram with number of averages set to 100 and with lateral uncertainty set to: (g) computed localization uncertainty, (h) 20 nm, (i) 50 nm. Average shifted histogram with the number of lateral shifts set to: (j) 2, (k) 4, and (l) 8. (Reprinted with permission from Ovesný, M et al., *Bioinformatics*, 30, 2389–2390, 2014a.)

(Veatch et al. 2012; Sengupta et al. 2013). This is important in *d*STORM experiments where the same fluorescent dye molecule might be detected several times in the image series.

Another interesting possibility is the creation of a density map based on the coordinates of the detected molecules. Figure 4.5 shows a WF image of an A431 cell expressing mCitrine-erbB3, a 3D visualization of the cell acquired using an astigmatic imaging setup (the *z*-position of each molecule is color coded depending on its calculated axial position), and a density map calculated using the

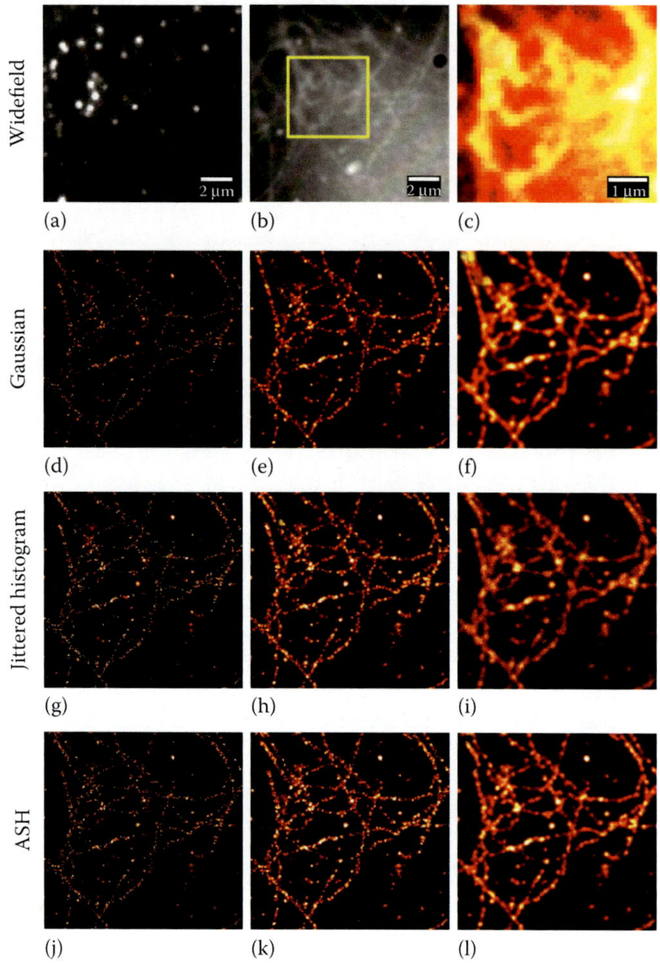

Figure 4.4

Comparison of visualization methods using the publically available dataset "Tubulin AF647." Widefield images: (a) a single frame of raw data, (b) average intensity time projection, (c) zoomed region indicated in (b). (d–l) Super-resolution SMLM images of the region indicated in (b) magnified 10×. Gaussian rendering with lateral uncertainty set to: (d) computed localization uncertainty, (e) 20 nm, (f) 50 nm. Jittered histogram with number of averages set to 100 and with lateral uncertainty set to: (g) computed localization uncertainty, (h) 20 nm, (i) 50 nm. Average shifted histogram with the number of lateral shifts set to: (j) 2, (k) 4, and (l) 8. (Reprinted with permission from Ovesný, M et al., *Bioinformatics*, 30, 2389–2390, 2014a.)

SMLM coordinate data (color coded according to the local density, and overlaid with the WF image). See Section 4.6 for a description of the sample and the microscopy setup used to acquire these images.

4.3 Temporal Resolution in SMLM

One drawback of SMLM is the fact that recording a long sequence of diffraction-limited images is required for reconstruction of a single super-resolution image. This problem has been addressed by increasing the density of photoactivated

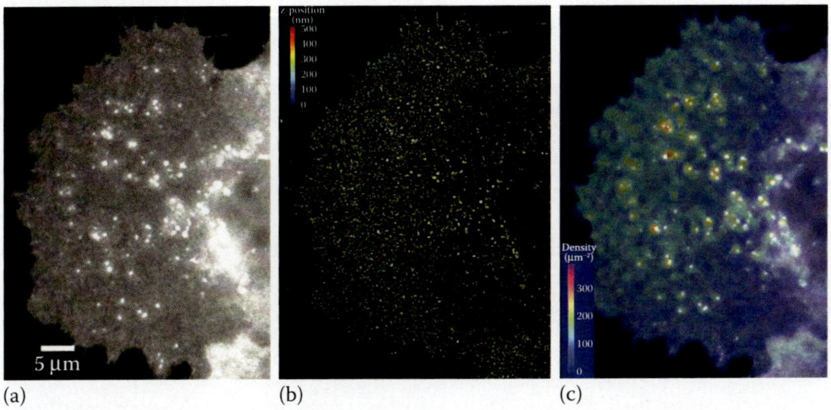

Figure 4.5

Images of an A431 cell expressing mCitrine-erbB3. (a) Widefield, (b) 3D visualization from astigmatic imaging, color coded according to estimated axial position, and (c) local density computed from localization data, color coded according to density, and overlaid on the widefield image.

fluorophores, thus the same number of molecules can be acquired within a shorter sequence, which effectively increases the temporal resolution and thus expands the possibilities for applying SMLM to the observation of dynamic processes in live cells (Subach et al. 2010).

However, when increasing the density, the images of the molecules start to overlap which poses a challenge in the analysis of such images. The goal of image analysis is then to detect the blinking events without a significant drop in spatial resolution. This renders the classical approach described in the previous section not applicable for high-density imaging. The first attempts to analyze such images were based on drawing a fitting region around isolated groups of molecules and then fitting multiple molecules as described in Section 4.2.4.1. Some of these include multiemitter fitting analysis (Huang et al. 2011) or DAOSTORM (Holden et al. 2011), which was further extended for 3D analysis (Babcock et al. 2012).

Later, new concepts were introduced which apply advanced techniques for spatial localization using ideas from the field of compressed sensing and some of them also incorporate the temporal information of intensities changing throughout the sequence. The idea of using reconstruction techniques from compressed sensing was first introduced in CSSTORM (Zhu et al. 2012). The main idea is to estimate a super-resolution image for each frame of the sequence using an optimization program with specific constraints, such as nonnegativity and sparsity. Then the molecules are detected as local maxima in the estimated super-resolution image. This method has proven to be robust to noise and to provide much higher detection rates compared to multiemitter fitting. On the other hand, the method is also very computationally intensive and has high demands on computer memory. This was addressed by applying an l_1-homotopy method (Babcock et al. 2013) which made the computation orders of magnitude faster, but this didn't solve the problem with high demands on memory. Therefore, the analysis was practically limited to 2D images and only small regions could be analyzed at one time, therefore the image had to be divided into overlapping tiles which were analyzed and the results combined together. A new fast algorithm called FALCON (Min et al. 2014b) solved the issues with memory and thus could be applied to larger images and was recently extended for 3D analysis (Min et al. 2014a). It is important to mention that all the methods based on compressed sensing used a simplified noise model based on Gaussian statistics. Since the algorithms work with photon counts and it is well established that photon-counting processes follow Poisson statistics, the simplified Gaussian noise model introduces approximation errors. This was recently addressed in our group's new algorithm, 3denseSTORM (https://github.com/zitmen/3densestorm) (Ovesný et al. 2014b), where the more precise Poisson noise model was assumed. This algorithm is capable of both 2D and 3D analysis and is both fast and memory efficient.

The temporal information, that is the changes in intensities throughout the sequence, can be used to isolate individual blinking events as was shown in gSHRImP (Simonson et al. 2011). It is also possible to analyze the sequence as whole assuming blinking statistics formulated as a Markov process as was done in deconSTORM (Mukamel et al. 2012) or 3B (Cox et al. 2012). It is worth noting that even though 3B presents an interesting mathematical model which delivers state of the art results, the algorithm is so slow that a computational cluster is required to obtain any results, even in small images.

Recently, another method based on nonnegative matrix factorization was introduced (Mandula et al. 2014). This method was designed to make super-resolution imaging with quantum dots possible and is also based on the analysis of whole sequence. Using quantum dots as a fluorescent label might help to further increase the temporal resolution since they provide greater photostability and have much higher brightness than fluorescent proteins or organic dyes.

4.4 Evaluating the Results of SMLM Processing Using Monte Carlo Simulations

Monte Carlo simulations can be used to evaluate each of the steps of the SMLM analysis process (Křížek et al. 2011). We believe the simulations should be informed by real measurements. For example, how many photons are detected from each molecule? The distribution of molecular photon rates can be measured in a real experiment and that distribution (its shape, mean value, and standard deviation) can be incorporated into the simulation. Similarly, what is the density of photoactivated molecules in each frame? How many photons of background are detected? What kind of noise is introduced by the camera? Knowledge of these factors help make the simulations realistic and quantitative.

The basic steps in creating this kind of simulation are as follows: (1) build a simulation based on measured values by introducing artificial molecules into a blank image; (2) process the simulated data with the software package of choice using (if possible) a variety of settings for the filtering and detection steps; and (3) evaluate the results with respect to detection rate and localization accuracy.

Figure 4.6 shows an example of a simulation using a Siemens star as a density-coded mask. The intensities in each spoke of the star are interpreted as molecular densities in the simulation.

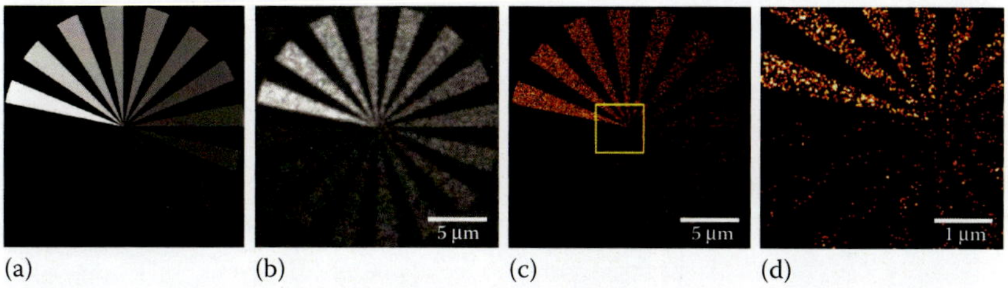

Figure 4.6

SMLM data generated with a variable spatial density of molecules. (a) Gray-scale mask, (b) average intensity time projection of raw data, (c) SMLM data reconstruction, and (d) zoomed region indicated in (c). The different intensity values in each spoke result in different densities of simulated molecules in the SMLM reconstruction. (Reprinted with permission from Ovesný, M et al., *Bioinformatics*, 30, 2389–2390, 2014a.)

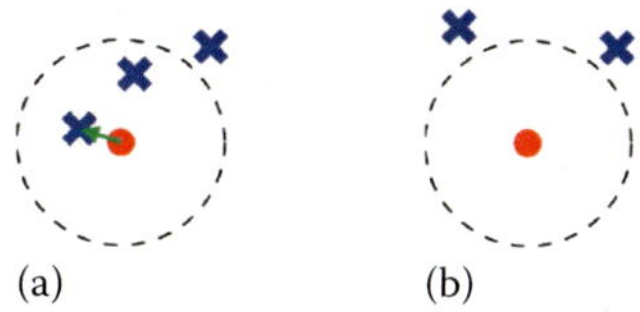

Figure 4.7

Counting localized and missed molecules. Red dot—ground-truth position of a simulated molecule, blue cross—localized molecule, green arrow—association of a localized molecule with ground-truth position, dashed circle—detection tolerance radius. (a) 1 TP + 2 FP and (b) 1 FN + 2 FP.

4.4.1 Precision and Recall

Because in a simulation the positions of the generated molecules are known, the process of performance evaluation starts by pairing the molecules which were detected and localized by the processing software with the closest molecule in the ground-truth data (Figure 4.7). The numbers of correctly and incorrectly identified molecules are counted as follows. If the distance between the paired molecules is smaller than a user-specified radius, then the localization is counted as a true positive (TP) detection and the localized molecule is associated with the ground-truth position. If the distance is greater than or equal to that radius, then the localization is counted as a false positive (FP) detection. Ground-truth molecules which were not associated with the localized molecules are counted as false negatives (FN, missed molecules).

Statistical measures related to the number of correctly or incorrectly detected molecules and missed molecules are the recall r, and the precision p (Wolter et al. 2010; Křížek et al. 2011). Their definitions are given by $r = \text{TP}/\text{TP} + \text{FN}$; $p = \text{TP}/\text{TP} + \text{FP}$. Recall measures the fraction of correctly identified molecules, and precision measures the fraction of correctly identified molecules out of all localizations. The theoretical optimum is achieved for values of recall and precision both equal to 1.0. Recall and precision can be combined into a single value, the F1 score, $\text{F1} = 2pr/p + r$. The F1 score helps quantify the results in an unbiased way.

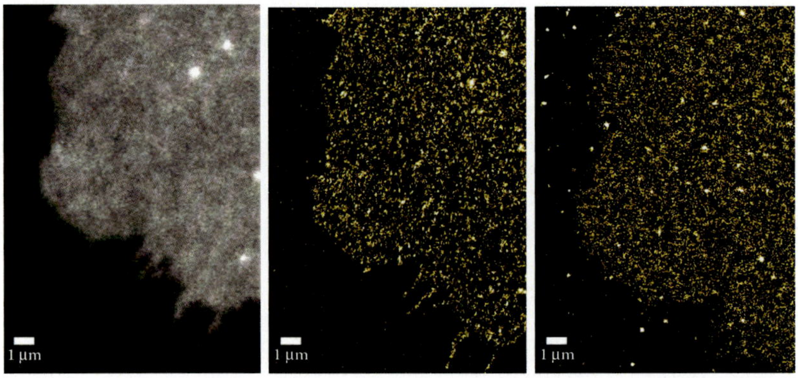

(a) (b) (c)

Figure 4.8

Images and simulation of an A431 cell expressing mCitrine-erbB3. (a) Widefield, (b) SMLM reconstruction, and (c) Monte Carlo simulation based on the widefield image.

Figure 4.8 shows a WF image of an A431 cell expressing mCitrine-erbB3, its SMLM reconstruction, and the result of Monte Carlo simulation based on the WF image. The intensities in the WF image are used to generate molecular densities in the simulation, resulting in a simulation that looks similar to the real data. Histograms of the molecular intensity, background level, fitted FWHM, and so on were measured in the SMLM data and used to create the simulation. We can then ask how different the real data is from the simulation. The exact coordinates will be different, but we can compare quantities of interest such as the molecular density.

4.5 Software for SMLM

The "single-molecule challenge" website lists over 50 different entries for analyzing SMLM data (http://bigwww.epfl.ch/smlm/) (Benchmarking of Single-Molecule Localization Microscopy Software 2013). We refer readers to this site for a comprehensive listing of available software, as well as publically available data sets. The results of the challenge were published in 2015, and quantitatively compares the various software packages (Sage et al. 2015).

4.5.1 ThunderSTORM

Our solution for SMLM data analysis is called ThunderSTORM (Ovesný et al. 2014a). ThunderSTORM is an open-source software package which runs in ImageJ. ThunderSTORM is modular, interactive, and platform-independent. The software provides a complete set of tools for processing, visualization, simulation, and quantitative analysis of data acquired by a variety of SMLM methods. ThunderSTORM was developed using a home-built SMLM system (described in Section 4.6), but the software has been tested with data acquired using commercially available Nikon N-STORM and Zeiss Elyra systems, and offers several unique capabilities compared to the analysis packages offered by these companies.

Our philosophy in developing ThunderSTORM has been to offer an extensive collection of processing and post-processing methods. We developed the software based on extensive testing with both real and simulated data. We also provide a very detailed description of the implemented methods and algorithms as well as a detailed user's guide. ThunderSTORM and the documentation is available at https://code.google.com/p/thunder-storm/.

4.6 Materials and Methods

4.6.1 Cell Lines and Reagents

A431 cells expressing mCitrine-erbB3 were a kind gift from Donna Arndt-Jovin and Thomas Jovin of the Max Planck Institute for Biophysical Chemistry (Göttingen, Germany) (Hagen et al. 2009). The cells were maintained in phenol red-free DMEM supplemented with 10% FCS, 100 U/mL penicillin, 100 U/mL streptomycin, and L-glutamate (all from Invitrogen, Carlsbad, CA) at 37°C and 100% humidity. Mowiol containing DABCO was purchased from Fluka (St. Louis, MO). Dithothreitol (DTT) was purchased from Sigma (St. Louis, MO).

4.6.2 Sample Preparation

A431 cells were grown on high precision #1.5 coverslips (Zeiss, Jena, Germany) which had been cleaned in glacial acetic acid followed by distilled water and 100% ethanol. Cells were first washed with PBS, then fixed with 4% paraformaldehyde for 15 min at room temperature. The cells were then washed with PBS, then the coverslips were mounted in mowiol containing DABCO and 50 mM DTT, then sealed on clean slides with clear nail polish. We typically waited 24 h before imaging the samples to allow the mowiol to harden.

4.6.3 Microscopy

We used the same setup as described in detail previously (Křížek et al. 2011). Briefly, we used an IX71 microscope (Olympus, Tokyo, Japan) equipped with an Olympus PlanApo TIRFM 100×/1.45 NA oil immersion objective and a front-illuminated Ixon DU885 EMCCD camera (Andor, Belfast, Northern Ireland). Laser illumination from a 473 nm, 400 mW laser (Dragon laser, ChangChun, China) was filtered using a 470/10 nm filter (Chroma, Bellows Falls, VT), then coupled into the microscope using a 0.39 NA, 600 μm diameter multimode optical fiber (M29L05, Thor Labs, Newton, NJ). The fiber output was collimated with a 2 inch, 60 mm FL lens (LA1401-A, Thor Labs) and introduced into the microscope using an Olympus L-shape fluorescence illuminator (IX2-RFAL). This configuration resulted in an evenly illuminated field. Fluorescence was isolated using a filter set appropriate for YFP (Chroma). We closed the field stop so that only a small area of the sample (~30 μm diameter) was illuminated by the full power laser. To help reduce sample drift, particularly in the axial direction, we maintained the objective temperature at 25°C using an objective heater (type OWS-1 with TC-124A controller, Warner Instruments, Hamden, CT). For 3D imaging, we placed a 500 mm focal length cylindrical lens about 30 mm in front of the camera (Huang et al. 2008).

Image sequences were acquired using Andor IQ software. We typically recorded 20,000 frames with an exposure time of 100 ms, an EM gain of 100. Image sequences were analyzed with ThunderSTORM (Ovesný et al. 2014a) software using the default settings.

Acknowledgments

This work was supported by the UCCS Center for the Biofrontiers Institute and by the Czech Science Foundation.

References

Abraham, A V, S Ram, J Chao, E S Ward, and R J Ober. 2009. Quantitative study of single molecule location estimation techniques. *Optics Express* 17(26):23352–23373.

Babcock, H P, J R Moffitt, Y Cao, and X Zhuang. 2013. Fast compressed sensing analysis for super-resolution imaging using L1-homotopy. *Optics Express* 21(23):28583.

Babcock, H, Y M Sigal, and X Zhuang. 2012. A high-density 3D localization algorithm for stochastic optical reconstruction microscopy. *Optical Nanoscopy* 1(1):1.

Baddeley, D, M B Cannell, and C Soeller. 2010. Visualization of localization microscopy data. *Microscopy and Microanalysis* 16(01):64–72.

Baddeley, D, I D Jayasinghe, L Lam, S Rossberger, M B Cannell, and C Soeller. 2009. Optical single-channel resolution imaging of the ryanodine receptor distribution in rat cardiac myocytes. *PNAS* 106(52):22275–22280.

Benchmarking of Single-Molecule Localization Microscopy Software. 2013. http://bigwww.epfl.ch/palm/.

Betzig, E, G H Patterson, R Sougrat, O W Lindwasser, S Olenych, J S Bonifacino, M W Davidson, J Lippincott-Schwartz, and H F Hess. 2006. Imaging intracellular fluorescent proteins at nanometer resolution. *Science* 313(5793):1642–1645.

Bevington, P R. 2003. *Data Reduction and Error Analysis for the Physical Sciences*. 3rd edn. New York: McGraw-Hill.

Chenouard, N, I Smal, F de Chaumont, M Maška, I F Sbalzarini, Y Gong, J Cardinale et al. 2014. Objective comparison of particle tracking methods. *Nature Methods* 11(3):281–89.

Churchman, L S, Z Okten, R S Rock, J F Dawson, and J A Spudich. 2005. Single molecule high-resolution colocalization of Cy3 and Cy5 attached to macromolecules measures intramolecular distances through time. *Proceedings of the National Academy of Sciences of the United States of America* 102(5):1419–1423.

Cox, S, E Rosten, J Monypenny, T Jovanovic-Talisman, D T Burnette, J Lippincott-Schwartz, G E Jones, and R Heintzmann. 2012. Bayesian localization microscopy reveals nanoscale podosome dynamics. *Nature Methods* 9(2):195–200.

Daumas, F, N Destainville, C Millot, A Lopez, D Dean, and L Salomé. 2003. Confined diffusion without fences of a g-protein-coupled receptor as revealed by single particle tracking. *Biophysical Journal* 84(1):356–66.

Deschout, H, F Cella Zanacchi, M Mlodzianoski, A Diaspro, J Bewersdorf, S T Hess, and K Braeckmans. 2014. Precisely and accurately localizing single emitters in fluorescence microscopy. *Nature Methods* 11(3):253–66.

El Beheiry, M and M Dahan. 2013. ViSP: Representing single-particle localizations in three dimensions. *Nature Methods* 10(8):689–90.

Ester, M, H-P Kriegel, J Sander, and X Xu. 1996. A density-based algorithm for discovering clusters in large spatial databases with noise. In *Proceedings of the 2nd International Conference on Knowledge Discovery and Data Mining*. AAAI Press, Portland, OR, pp. 226–231.

Gordon, M P, T Ha, and P R Selvin. 2004. Single-molecule high-resolution imaging with photobleaching. *Proceedings of the National Academy of Sciences of the United States of America* 101(17):6462–6465.

Gunkel, M, F Erdel, K Rippe, P Lemmer, R Kaufmann, C Hörmann, R Amberger, and C Cremer. 2009. Dual color localization microscopy of cellular nanostructures. *Biotechnology Journal* 4(6):927–938.

Hagen, G M, W Caarls, K A Lidke, A H B De Vries, C Fritsch, B George Barisas, D J Arndt-Jovin, and T M Jovin. 2009. Fluorescence recovery after photobleaching and photoconversion in multiple arbitrary regions of interest using a programmable array microscope. *Microscopy Research and Technique* 72(6):431–440.

Heilemann, M, S van de Linde, M Schüttpelz, R Kasper, B Seefeldt, A Mukherjee, P Tinnefeld, and M Sauer. 2008. Subdiffraction-resolution fluorescence imaging with conventional fluorescent probes. *Angewandte Chemie International Edition* 47:6172–6176.

Henriques, R, M Lelek, E F Fornasiero, F Valtorta, C Zimmer, and M M Mhlanga. 2010. QuickPALM: 3D real-time photoactivation nanoscopy image processing in ImageJ. *Nature Methods* 7(5):339–340.

Holden, S J, S Uphoff, and A N Kapanidis. 2011. DAOSTORM: An algorithm for high density super-resolution microscopy. *Nature Methods* 8(4):279–280.

Huang, B, S A Jones, B Brandenburg, and X Zhuang. 2008. Whole-cell 3D STORM reveals interactions between cellular structures with nanometer-scale resolution. *Nature Methods* 5(12):1047–1052.

Huang, F, S L Schwartz, J M Byars, and K A Lidke. 2011. Simultaneous multiple-emitter fitting for single molecule super-resolution imaging. *Biomedical Optics Express* 2(5):1377–1393.

Izeddin, I, J Boulanger, V Racine, C G Specht, A Kechkar, D Nair, A Triller, D Choquet, M Dahan, and J B Sibarita. 2012. Wavelet analysis for single molecule localization microscopy. *Optics Express* 20(3):2081–2095.

Křížek, P, I Raška, and G M Hagen. 2011. Minimizing detection errors in single molecule localization microscopy. *Optics Express* 19(4):3226–3235.

Kusumi, A, Y Sako, and M Yamamoto. 1993. Confined lateral diffusion of membrane receptors as studied by single particle tracking (nanovid microscopy). Effects of calcium-induced differentiation in cultured epithelial cells. *Biophysical Journal* 65(5):2021–2040.

Lemmer, P, M Gunkel, D Baddeley, R Kaufmann, A Urich, Y Weiland, J Reymann, P Müller, M Hausmann, and C Cremer. 2008. SPDM: Light microscopy with single-molecule resolution at the nanoscale. *Applied Physics B: Lasers and Optics* 93(1):1–12.

Lidke, K A, B Rieger, T M Jovin, and R Heintzmann. 2005. Superresolution by localization of quantum dots using blinking statistics. *Optics Express* 12(18):7052–7062.

Ma, H, F Long, S Zeng, and Z-L Huang. 2012. Fast and precise algorithm based on maximum radial symmetry for single molecule localization. *Optics Letters* 37(13):2481–2483.

Malkusch, S, U Endesfelder, J Mondry, M Gelléri, P J Verveer, and M Heilemann. 2012. Coordinate-based colocalization analysis of single-molecule localization microscopy data. *Histochemistry and Cell Biology* 137(1):1–10.

Mandula, O, I Š Šestak, R Heintzmann, and C K I Williams. 2014. Localisation microscopy with quantum dots using non-negative matrix factorisation. *Optics Express* 22(20):24594–24605.

McGorty, R, D Kamiyama, and B Huang. 2013. Active microscope stabilization in three dimensions using image correlation. *Optical Nanoscopy* 2(1):3.

Min, J, S J Holden, L Carlini, M Unser, S Manley, and J C Ye. 2014a. 3D high-density localization microscopy using hybrid astigmatic/biplane imaging and sparse image reconstruction. *Biomedical Optics Express* 5(11):3935–3948.

Min, J, C Vonesch, H Kirshner, L Carlini, N Olivier, S Holden, S Manley, J C Ye, and M Unser. 2014b. FALCON: Fast and unbiased reconstruction of high-density super-resolution microscopy data. *Scientific Reports* 4(January):4577.

Mlodzianoski, M J, J M Schreiner, S P Callahan, K Smolková, A Dlasková, J Santorová, P Ježek, and J Bewersdorf. 2011. Sample drift correction in 3D fluorescence photoactivation localization microscopy. *Optics Express* 19(16):15009–15019.

Mukamel, E A, H Babcock, and X Zhuang. 2012. Statistical deconvolution for super-resolution fluorescence microscopy. *Biophysical Journal* 102(10):2391–2400.

Ovesný, M, P Křížek, J Borkovec, Z Švindrych, and G M Hagen. 2014a. ThunderSTORM: A comprehensive ImageJ plug-in for PALM and STORM data analysis and super-resolution imaging. *Bioinformatics* (Oxford, England) 30(16):2389–2390.

Ovesný, M, P Křížek, Z Švindrych, and G M. Hagen. 2014b. High density 3D localization microscopy using sparse support recovery. *Optics Express* 22(25):31263–31276.

Owen, D M, C Rentero, A Magenau, D Williamson, M Rodriguez, K Gaus, and J Rossy. 2010. PALM imaging and cluster analysis of protein heterogeneity at the cell surface. *Journal of Biophotonics* 3(7):446–454.

Owen, D M, D Williamson, A Magenau, and K Gaus. 2012. Optical techniques for imaging membrane domains in live cells (live-cell palm of protein clustering). *Methods Enzymol* 504:221–235.

Parthasarathy, R. 2012. Rapid, accurate particle tracking by calculation of radial symmetry centers. *Nature Methods* 9(7):724–726.

Pengo, T, S J Holden, and S Manley. 2014. PALMsiever: A tool to turn raw data into results for single-molecule localization microscopy. *Bioinformatics* 31(5):797–798.

Qu, X, D Wu, L Mets, and N F Scherer. 2004. Nanometer-localized multiple single-molecule fluorescence microscopy. *Proceedings of the National Academy of Sciences of the United States of America* 101(31):11298–11303.

Ronneberger, O, D Baddeley, F Scheipl, P J Verveer, H Burkhardt, C Cremer, L Fahrmeir, T Cremer, and B Joffe. 2008. Spatial quantitative analysis of fluorescently labeled nuclear structures: Problems, methods, pitfalls. *Chromosome Research* 16(3):523–562.

Rossy, J, E Cohen, K Gaus, and D M Owen. 2014. Method for co-cluster analysis in multichannel single-molecule localisation data. *Histochemistry and Cell Biology* 141(6):605–612.

Rossy, J, D M Owen, D J Williamson, Z Yang, and K Gaus. 2013. Conformational states of the kinase Lck regulate clustering in early T cell signaling. *Nature Immunology* 14(1):82–89.

Rubin-Delanchy, P, G L Burn, J Griffié, D J Williamson, N A Heard, A P Cope, and D M Owen. 2015. Bayesian cluster identification in single-molecule localization microscopy data. *Nature Methods* 12(11):1072–1076.

Rust, M J, M Bates, and X Zhuang. 2006. Sub-diffraction-limit imaging by stochastic optical reconstruction microscopy (STORM). *Nature Methods* 3(10):793–795.

Ruusuvuori, P, T Aijö, S Chowdhury, C Garmendia-Torres, J Selinummi, M Birbaumer, A M Dudley, L Pelkmans, and O Yli-Harja. 2010. Evaluation of methods for detection of fluorescence labeled subcellular objects in microscope images. *BMC Bioinformatics* 11(1):248.

Sage, D, H Kirshner, T Pengo, N Stuurman, J Min, S Manley, and M Unser. 2015. Quantitative evaluation of software packages for single-molecule localization microscopy. *Nature Methods* 12(8):717–724.

Sengupta, P, T Jovanovic-Talisman, and J Lippincott-Schwartz. 2013. Quantifying spatial organization in point-localization superresolution images using pair correlation analysis. *Nature Protocols* 8(2):345–354.

Simonson, P D, E Rothenberg, and P R Selvin. 2011. Single-molecule-based superresolution images in the presence of multiple fluorophores. *Nano Letters* 11(11):5090–5096.

Small, A and S Stahlheber. 2014. Fluorophore localization algorithms for superresolution microscopy. *Nature Methods* 11(3):267–279.

Smirnov, E, J Borkovec, L Kováčik, S Svidenská, A Schröfel, M Skalníková, Z Švindrych et al. 2014. Separation of replication and transcription domains in nucleoli. *Journal of Structural Biology* 188(3):259–266.

Smith, C S, N Joseph, B Rieger, and K A Lidke. 2010. Fast, single-molecule localization that achieves theoretically minimum uncertainty. *Nature Methods* 7(5):373–375.

Stallinga, S and B Rieger. 2010. Accuracy of the gaussian point spread function model in 2D localization microscopy. *Optics Express* 18(24):24461–24476.

Subach, F V, G H Patterson, M Renz, J Lippincott-Schwartz, and V V Verkhusha. 2010. Bright monomeric photoactivatable red fluorescent protein for two-color super-resolution sptPALM of live cells. *Journal of the American Chemical Society* 132(18):6481–6491.

Suzuki, K, K Ritchie, E Kajikawa, T Fujiwara, and A Kusumi. 2005. Rapid hop diffusion of a G-protein-coupled receptor in the plasma membrane as revealed by single-molecule techniques. *Biophysical Journal* 88(5):3659–3680.

Thompson, R E, D R Larson, and W W Webb. 2002. Precise nanometer localization analysis for individual fluorescent probes. *Biophysical Journal* 82(5):2775–2783.

Veatch, S L, B B Machta, S A Shelby, E N Chiang, D A Holowka, and B A Baird. 2012. Correlation functions quantify super-resolution images and estimate apparent clustering due to over-counting. *PLoS One* 7(2):e31457.

Wang, Y, J Schnitzbauer, Z Hu, X Li, Y Cheng, Z-L Huang, and B Huang. 2014. Localization events-based sample drift correction for localization microscopy with redundant cross-correlation algorithm. *Optics Express* 22(13):15982–15991.

Williamson, D J, D M Owen, J Rossy, A Magenau, M Wehrmann, J J Gooding, and K Gaus. 2011. Pre-existing clusters of the adaptor Lat do not participate in early T cell signaling events. *Nature Immunology* 12(7):655–662.

Wolter, S, M Schuttpelz, M Tscherepanow, S Van De Linde, M Heilemann, and M Sauer. 2010. Real-time computation of subdiffraction-resolution fluorescence images. *Journal of Microscopy* 237(1):12–22.

Yildiz, A, J N Forkey, S A McKinney, T Ha, Y E Goldman, and P R Selvin. 2003. Myosin V walks hand-over-hand: Single fluorophore imaging with 1.5-nm localization. *Science* 300(5628):2061–2065.

Zhu, L, W Zhang, D Elnatan, and B Huang. 2012. Faster STORM using compressed sensing. *Nature Methods* 9(7):721–723.

5

Deconvolution of Nanoscopic Images

Hans van der Voort

5.1 Introduction
5.2 Microscopic Image Formation
5.3 Deconvolution Algorithms
5.4 Iterative Deconvolution Methods
5.5 Bayesian Deconvolution
5.6 Space Variant PSFs
5.7 Summary
Acknowledgments
Appendix
References

5.1 Introduction

Deconvolution is an image restoration technique which aims at undoing, literally "unrolling," the blurring effect any imaging system has on the "true image," the object distribution. In doing so, deconvolution is a natural companion of super-resolution imaging. If a particular image could be deconvolved perfectly one would obtain boundless resolution, and it would not even be necessary to build complex microscopes. Unfortunately, the gain which can be achieved by deconvolution is limited. One obvious source of this limit is that most optical systems entirely stop transferring information about the object beyond a certain amount of detail. It would seem obvious that deconvolution techniques cannot "roll back" information which is not present in the image, but as will be shown in favorable conditions this boundary can be overcome, albeit within limits.

Fluorescent labeling techniques play a key role in modern microscopy, not only because of the wide possibilities to label-specific biological structures, but also because the incoherency of the emitted light in most cases results in a predicable, low artifact linear imaging system. However, as resolution improves and moves into the nanoscopic regime, fewer and fewer fluorescent molecules contribute to the light recorded in a particular pixel or voxel. In a typical high-resolution image, 100 detected photons per pixel in the brightest parts can be considered excellent, but cases where no more than 10 are present do occur. Clearly, effective treatment of the resulting high noise levels must be an integral part of a successful deconvolution scheme.

The aim of this chapter is to show the reader the close relation between microscopic image formation and deconvolution techniques, and to provide insight in the basic construction and properties of a limited number of well known and successful deconvolution algorithms. The focus is on deconvolution of low signal super-resolution images, in particular data from stimulated emission depletion (STED) nanoscopes [1], but some issues related to deconvolution of conventional widefield fluorescence images will be discussed as well. Finally, we discuss solutions of practical issues which affect the application range of deconvolution methods.

5.2 Microscopic Image Formation

In an incoherent imaging system such as a fluorescence microscope objects are imaged independently. In other words, the imaging in a fluorescence microscope follows the superposition principle: when two objects A and B are imaged simultaneously, the result is equal to the sum of the independently imaged objects, $Image(A + B) = Image(A) + Image(B)$. With the superposition principle, the image of an object can be computed by substituting each point of an object by a PSF with corresponding intensity, and subsequently summing all the PSFs. This operation is called a "convolution" of the object with a PSF, see Figure 5.1.

$$g(\mathbf{x}) = h(\mathbf{x}) * f(\mathbf{x}), \tag{5.1}$$

with $\mathbf{x}$ a point in the image, $g(\mathbf{x})$ the image, $h(\mathbf{x})$ the PSF, and $f(\mathbf{x})$ the object. The PSF does not necessarily have a fixed shape, it may have a different shape in different locations of the specimen as is indeed the case in many practical applications.

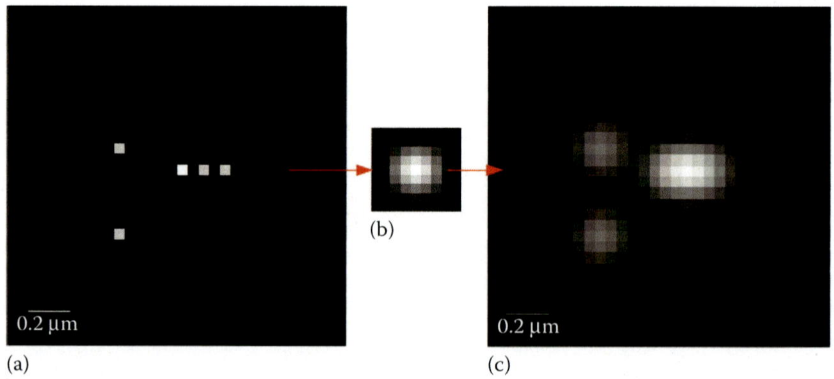

Figure 5.1

Each point in of the object (a) is replaced by a PSF (b) at the corresponding position in the image (c), and with a total intensity proportional to the intensity of the point. To carry out such a discrete convolution operation for each of the N samples in the object and M PSF samples NM multiplications and additions are needed. In many cases the PSF is as large as the object, so N^2 operations are needed.

It is important to note at this point that the imaging properties of not all microscopes can be described by a convolution operation. For example, in coherent reflection microscopes phase relations between reflecting objects in the specimen cause the contributions from different parts of the object to strongly interact with each other so that the superposition principle does not hold for the final intensity image. The same is true for microscopes with partial coherence like bright-field microscopes. Strictly speaking, such images cannot be deconvolved. Still, it has been shown that when specific conditions are met deconvolution of bright-field data is still possible [2]. Fortunately, in the case of fluorescence, the randomness of the moments at which each of the excited molecules emits destroys all phase relations between emitting parts of the object so that each is imaged independent from the others.

5.2.1 Spatial Frequencies

From the perspective of deconvolution and image acquisition (Section 5.5.4), it is important to understand to what degree a particular microscope type transfers "spatial frequencies" from the specimen to the image. Just as the superposition principle can be applied to split an image into point objects, it can be applied to split an image into spatial frequencies. Just as with the point objects, the spatial frequencies act as building blocks of the image. An important property of linear imaging systems with a constant PSF is that when the sinusoidal shaped spatial frequencies are imaged, their shape is unaffected, only their amplitude and phase will be modified. This is summarized by the "convolution theorem" which states that the Fourier transform of the convolution of two functions is equal to the product of the Fourier transforms of the functions:

$$\mathcal{F}(h(\mathbf{x}) * f(\mathbf{x})) = \mathcal{F}(h(\mathbf{x}))\mathcal{F}(f(\mathbf{x})) \tag{5.2}$$

Here, the Fourier transform of $h(\mathbf{x})$, $\mathcal{F}(h(\mathbf{x})) = H(\mathbf{u})$, $\mathbf{u}$ a spatial frequency, is also called the "optical transfer function" (OTF). The OTF determines how and if a certain constituent spatial frequency of the object is transferred to the image. Typically, high spatial frequencies, representing fine details in the image, will be transferred more weakly by the microscope than low frequencies which represent structures at a coarser scale. Since it is convention to normalize the integral of the PSF to 1, $H(0) = 1$, the OTF zero frequency representing the integral of the PSF.

An important characteristic of most microscopic OTFs is the "cutoff" frequency or "bandwidth:" a limit beyond no spatial frequency is transferred to the image (see Figure 5.2).

5.2.2 Bandwidth of Microscopes

Since we are interested in the 3D imaging properties of microscopes, we must also consider the dependency of the OTF and its bandwidth on the direction of the spatial frequencies. In the ideal case, the imaging properties and therefore the bandwidth would be the same in all directions. Then, the surface representing the bandwidth in all directions would be a sphere, and objects would be imaged independently of their orientation. However, in general the axial imaging properties of microscopes differ vastly from the lateral properties, more so for low numerical aperture (NA) systems, see Figure 5.3.

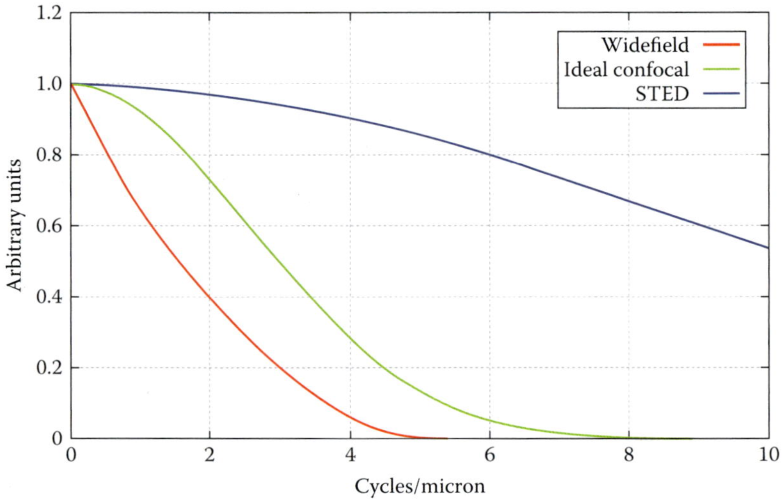

Figure 5.2

Lateral 2D optical transfer function of a widefield microscope (red), a confocal microscope with infinitely small pinhole (green), and a STED microscope with a saturation factor of 50 (blue, see text). The confocal band limit is to the right edge of the plot at 10 cycles/micron. The widefield band limit is half of the confocal band limit, here at 5 cycles/micron. The STED OTF has no band limit. It is convention to normalize the zero frequency, corresponding with the integral of the PSF, to 1.

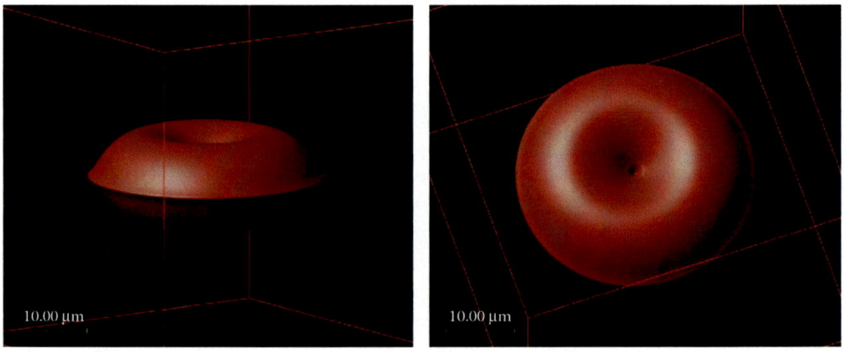

(a) (b)

Figure 5.3

Bandwidth surface of a widefield microscope with an NA of 1.3 and emission at 520 nm. (a) Side view, optical axis is vertical. The flattish shape means the bandwidth in the lateral directions is higher than the maximum bandwidth in the axial direction, corresponding with higher lateral than axial resolution. (b) View at low angle from the optical axis, clearly showing the central depression, often called "missing cone" making up a doughnut like shape of the surface. Spatial frequencies in the missing cone, for example representing horizontal structures like cell membranes, are not imaged.

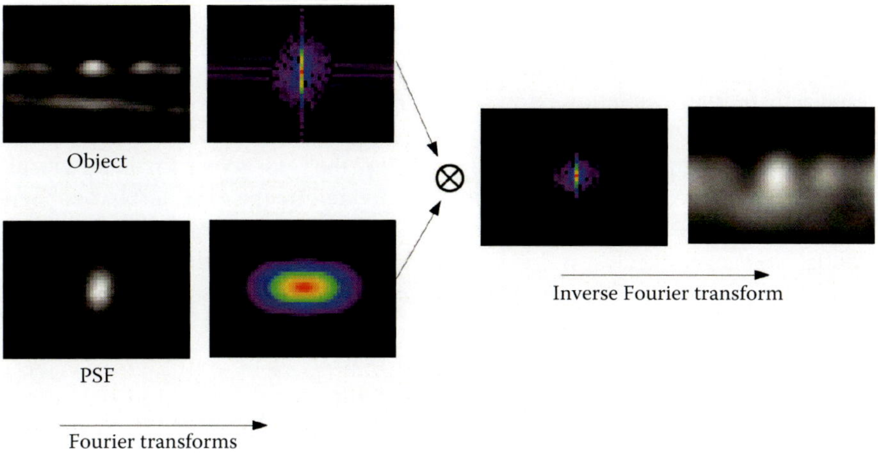

Figure 5.4

Fast computation of a convolution using the convolution theorem. The ⊗ symbol indicates point by point multiplication of the object spectrum $F(\mathbf{u})$ and $H(\mathbf{u})$ (see text) to find the image spectrum $G(\mathbf{u})$.

5.2.3 Efficient Computation of a Convolution

Since convolution operations are a key building block in any deconvolution algorithm, see Section 5.3, it is of key importance to reduce the computational effort involved in computing a convolution in the straightforward fashion outlined in Figure 5.1. The convolution theorem allows us to compute a convolution in the following steps:

- Two forward Fourier transforms $h(\mathbf{x}) \xrightarrow{\mathcal{F}} H(\mathbf{u})$ and $f(\mathbf{x}) \xrightarrow{\mathcal{F}} F(\mathbf{u})$,
- A Point-by-point multiplication in the frequency domain $H(\mathbf{u})F(\mathbf{u})$,
- And one inverse Fourier transform $G(\mathbf{u}) \xrightarrow{\mathcal{F}^{-1}} g(\mathbf{x})$ to obtain the final image (see Figure 5.4)

Computationally, the most expensive operations here are the Fourier transforms with about 2.5 $N \log_2(N)$ operations each [3,4], far less than the $2N^2$ operations needed for the straightforward method. For example, for a fair sized $1024 \times 1024 \times 64$ voxel image the difference in computational effort is about 5.10^5. Without this gain, deconvolution would be hardly feasible, but there is a price to pay: a fixed shape PSF and edge artifacts. Solutions on dealing with PSFs vary due to refractive index (RI) mismatch-induced spherical aberration as discussed in Section 5.6.

5.3 Deconvolution Algorithms

5.3.1 Linear Inversion

The scheme for computing a convolution in Figure 5.4 at first glance suggests that a convolution can be undone by reversal of the scheme: by inverting the last Fourier transform in the scheme a spectrum of the image is easily obtained. Dividing this spectrum by the OTF would then result in the object spectrum. One subsequent inverse Fourier transform would yield the object. However, the existence of a bandwidth causes large parts of the OTF to be zero, blocking the division by the OTF. Even in cases where the OTF is never truly zero like STED, it will be very small for high spatial frequencies. Since there is always noise present in the image, especially in high-resolution fluorescence images, this noise is also present in the image spectrum. Because the spectrum of photon noise is evenly distributed over spatial frequencies, it dominates the areas where the OTF is very small. As a result, when in these areas the image spectrum is divided by the OTF, all that happens is amplification of the noise components. Still, by a careful trade-off of boosting spatial frequencies against noise amplification, it is possible to restore images to moderate degree with Tikhonov–Miller regularized inversion [5,6].

$$F(\mathbf{u}) = \frac{H^*(\mathbf{u})G(\mathbf{u})}{|H(\mathbf{u})|^2 + \alpha} \tag{5.3}$$

where:
α is the regularization factor

In this equation, for values of **u** where the OTF $H(\mathbf{u})$ is zero, $F(\mathbf{u})$ will be set to zero. For values of **u** where $|H(\mathbf{u})| < \alpha$, $F(\mathbf{u}) < G(\mathbf{u})$ so that amplification of spatial frequencies for which this is the case will not take place. The property of the Tikhonov–Miller method to set all spatial frequencies beyond the band limit of the microscope to zero limits the resolution gain which can be obtained by it.

5.3.2 Super-Resolution

Applied to microscopes, the term "super-resolution" usually refers to the ability to solve "beyond the diffraction limit." For fluorescence, this diffraction limit is often associated with the Rayleigh's resolution criterion which corresponds with the radius of the first dark ring of the PSF: $d_{\text{Rayleigh}} = 0.61\ \lambda/NA$. However, in high-aperture PSFs the dark ring tends to disappear making this criterion impractical. Alternative criteria, for example the full width at half maximum (FWHM) of the PSF, are easier to measure but yield different results. Using 3 it is easy to reduce the size of the PSF, so with this definition Tikhonov–Miller inversion qualifies as super-resolution technique. In any case, the particular set of choices made in establishing a criterion remain somewhat arbitrary. In turn, this makes "super-resolution" a rather qualitative term.

Instead of evaluating the width of the PSF in some fashion, it is also possible to use the extent of the OTF as criterion. Since the OTF for most microscopes is band limited, bandwidth can serve as criterion to establish whether a particular technique qualifies as super-resolution technique. While this seems less arbitrary, a bandwidth reference must still be selected, involving an arbitrary choice. Assuming a small pinhole, in the confocal microscope the PSF is the product of an excitation distribution and a detection distribution both similar to the PSF of a widefield microscope. According to the convolution theorem (Equation 5.2) the OTF of the combined system is the convolution of two widefield OTFs, resulting in the combined OTF and its bandwidth twice as wide (see also Figure 5.2) as that of the equivalent widefield OTF. With the widefield bandwidth as reference the confocal microscope would qualify as super-resolution system. In addition to these issues, microscopes based on nonlinear effects like STED do not have a bandwidth which can be readily determined. So also with a bandwidth-based criterion it is hard to define whether a particular microscope is a super-resolution system or not.

For a deconvolution technique to qualify as "super-resolution" capable we can say it should be able to restore spatial frequencies outside the pass band of the microscope at hand. Tikhonov–Miller linear inversion does not have this capability so does not qualify as super-resolution technique. For the case of nonlinear widefield deconvolution this leads to clear results (see Figure 5.10). However, to apply the same criterion to confocal images with their poorly filled band (see Figure 5.2) seems hardly fair, while with this criterion super-resolution deconvolution on STED data is not possible since the STED nanoscopes do not have a band limit. To avoid all ambiguity, when in the remainder of this chapter the term "super-resolution" is used, it refers to an image showing spatial frequency component outside the band of an equivalent confocal image.

5.4 Iterative Deconvolution Methods

To overcome the limitations of inverse filtering it is necessary to use all available knowledge about the object and imaging process in addition to knowledge of the microscope properties in the form of the PSF:

- Knowledge about the distribution of fluorescent dye density. Even without specifying a precise distribution, it is certain that the density is nonnegative
- Knowledge about the noise model
- Knowledge about the extent or shape of the object. This is only available in rare cases but potentially powerful

Since it is difficult to find closed form solutions like Equation 5.3 which are capable of including these items, iterative schemes are used to approach the solution step by step. An example of such a method is given in Figure 5.5. This method is based upon the idea to improve an estimate of the object by adding a correction image which becomes zero when a perfect match is obtained between measured image g and, dropping the image function argument vectors $\mathbf{x}$, the imaged estimate $h^* f_{\text{perfect}}$. To reach this situation the method of fixed-point iteration is applied to improve a particular imperfect estimate f_k by a correction image c_k which becomes zero when the solution is reached:

$$\begin{aligned} c_k &= \tau(g - h * f_k), \text{ and} \\ f_{k+1} &= f_k + c_k, \end{aligned} \tag{5.4}$$

τ a step size parameter. The advantages of this scheme are that (1) the a priori information of nonnegativity can be easily included by removing the negative values from each new f_{k+1} and (2) iterations can be stopped when the result starts getting worse. Stopping the iterations then has a similar function as the regularization parameter α in Tikhonov–Miller inversion (Equation 5.3). It can be shown [8,9] that after many iterations, and provided the measured image does not contain out of band noise [9], the original van Cittert algorithm behaves like Tikhonov–Miller inversion with $\alpha = 0$, so without any regularization. Unfortunately, the algorithm is not very stable in the presence of noise due to the lack of build-in regularization, and also it only converges to a solution when, depending on the properties of the PSF, the step size parameter τ is sufficiently small.

In Figure 5.5, the algorithm is applied to data from a 2-photon microscope with a noise level which can be considered quite good for a high-resolution image,

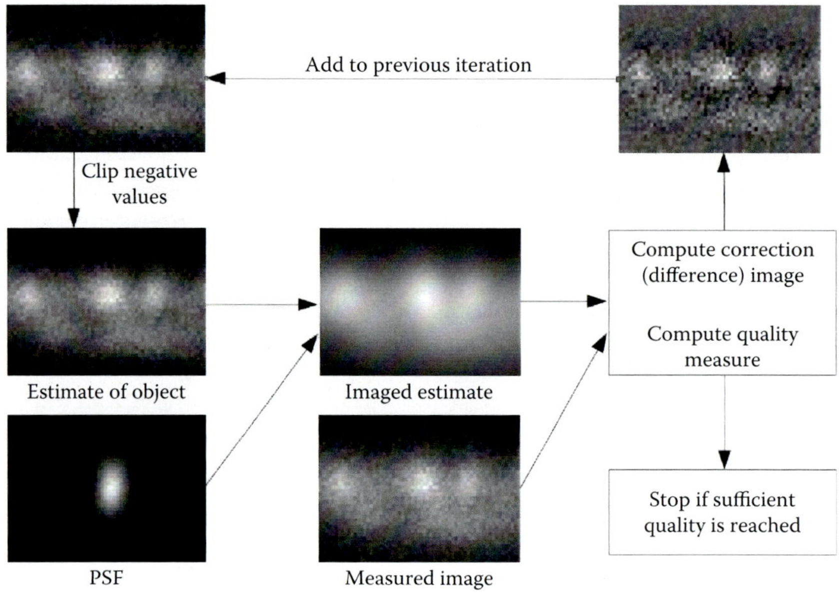

Figure 5.5

The van Cittert method (1931) is one of the most simple deconvolution algorithms. It can be started with a blurred version of the measured image. In the first step the current estimate is convolved with the PSF to compute the estimated image. In the second step the difference image (top-right in the diagram) between the estimated and measured image is computed. In this step also a quality measure can be computed. In the next step a fraction of the difference image (0.4 in this case) is added to the current estimate. The following step is a modification of the original algorithm: since the difference image is likely to contain negative pixel values, adding it to the current estimate might introduce negative pixel values. In the modification these are now clipped off to yield a new estimate. Shown here is the state of the algorithm at the fourth iteration, measured image is a two-photon excitation, and photon counted confocal image sampled at 40 and 81 nm in the lateral and axial directions, respectively [7]. Iterations can be stopped interactively by the user, or stopped after the quality measure does not improve anymore or starts deteriorating.

more than 200 detected photons per pixel in the brightest parts. When an image is sampled according to the Nyquist criterion, this means that the elliptically shaped band pass region (see Figure 5.6) tightly fits into a rectangular box containing all spatial frequencies of the image. Photon "white" noise is equally distributed among the spatial frequencies in this box, so there will be out of band noise in the corner regions of the image spectrum. Since the corner regions take up nearly half of the spectral volume, the amount of out of band noise is appreciable in any well-sampled high-resolution image.

Figure 5.7 shows how an *X-Z* plane of the estimate changes over many iterations and how quickly noise amplification sets in. To make a reproducible selection from these results an objective stopping criterion is needed. Since the algorithm aims to reduce the correction image toward zero, a reasonable criterion would be to minimize the total content of the difference image. Taking the negative pixel values present in the difference image into account, this can be done by computing the squared norm of the difference image:

$$C_k = \|g - h * f_k\|^2 \tag{5.5}$$

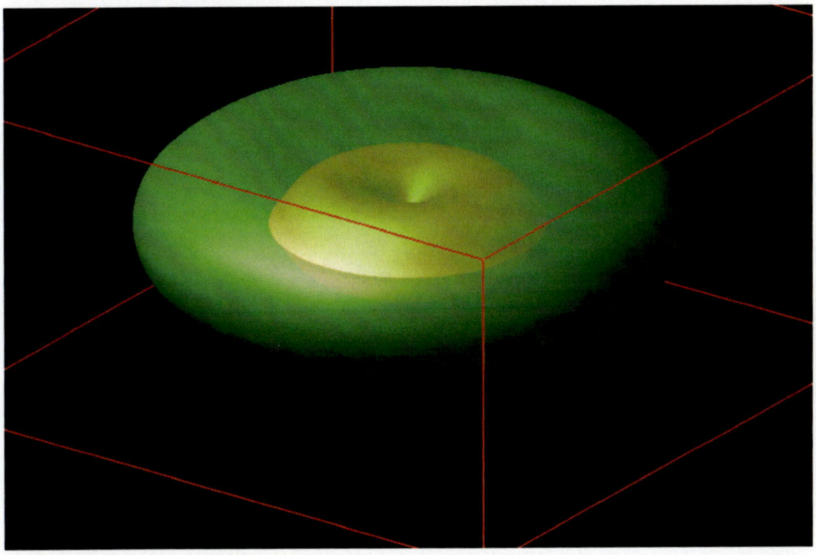

Figure 5.6

Bandwidth region of confocal (green) and equivalent widefield (red) compared.

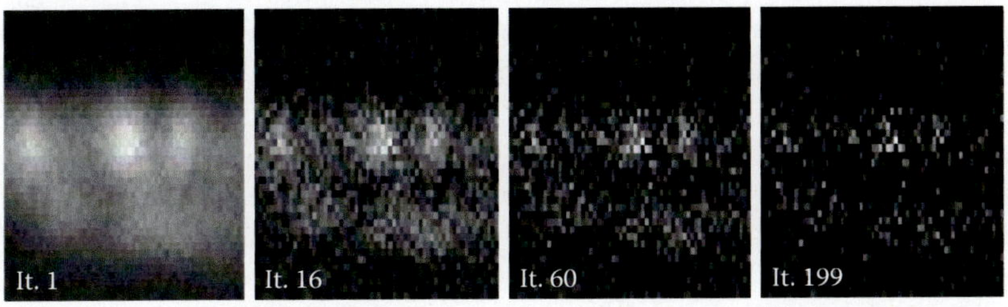

(a) (b) (c) (d)

Figure 5.7

Results of the modified (see Figure 5.5) van Cittert algorithm. XZ slices from the same data set and conditions as in Figure 5.5. (a–d) after the first iteration, the estimate with the lowest squared norm C_k of the difference image (see Equation 5.5 and shown top-right in Figure 5.5), after the 60th iteration were the result becomes already quite sparse, and after the 200th iteration where the result becomes even more sparse but stable.

Here, $||.||^2$ represents taking the squared norm of the image, that is, the sum over all pixels squared. For example, for a three pixel image this would be $C(f) = f_0^2 + f_1^2 + f_2^2$. In Figure 5.7 the best result is found by selecting the estimate resulting in the lowest value of C_k.

Clearly, there is a discrepancy between the criterion used for selecting the best result, and the final image to which the modified van Cittard algorithm converges. To devise an algorithm which converges on a result which also optimizes the criterion used, the algorithm should be derived from the criterion itself. This will also allow for inclusion of a priori information into the criterion and obviates the need to include the a priori information in an ad-hoc fashion as described in Figure 5.5. In the next section the procedure to derive a criterion-driven deconvolution algorithm will be outlined.

5.4.1 The Penalty Function

The choice made in the previous section for the selection criterion could have been expanded by including an energy constraint on the result. An energy constraint, where each pixel contributes with the square of its value, could have penalized the very bright "high-energy" pixels in the extreme results of Figure 5.7, so that a smoother result could have been obtained. Including an energy constraint is equivalent with a priori information about the object that it must have a certain smoothness. Such a penalty function could look as follows:

$$\Phi(f_k)=\left\|g-h*f_k\right\|^2+\alpha\left\|f_k\right\|^2 \tag{5.6}$$

The first term in the right-hand side of this equation is the stopping criterion as was used to select the "best" result from the van Cittert iterations in Figure 5.7 and the second term is the energy constraint on f_k. If the a priori information regarding the nonnegativity is not used, Equation 5.6 is solved by Equation 5.3. If it is used, an iterative algorithm can be constructed where nonnegativity is enforced in an ad-hoc fashion after each update of the estimate, just as outlined in Figure 5.5. From the numerical point of view, the attractiveness of this iterative constrained Tikhonov–Miller (ICTM, [8,10]) algorithm lies in the fact that it is fairly easy to compute the "gradient" of the penalty function. The penalty gradient is an image where each pixel represents the amount of change of the total penalty if the corresponding pixel in the current estimate is changed by a certain tiny fraction. Figure 5.8 shows how the gradient can be associated with a change in direction. On a small scale, changing the estimate a short way in the gradient direction will cause the most rapid decrease of the penalty compared to other directions. Since the aim is to minimize the penalty, this is called the "steepest descent" direction. As shown in Figure 5.8, the steepest descent does not necessarily lead to the penalty minimum, multiple steps may be necessary. A steepest descent algorithm takes successive steps

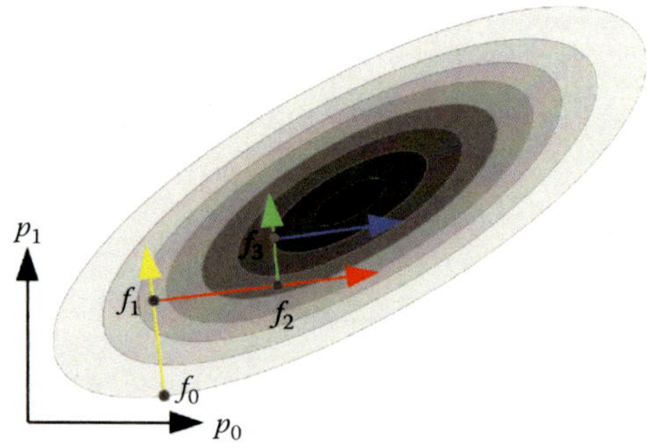

Figure 5.8

Steepest descent algorithm acting on a penalty function for two pixel values, f_0 and f_1, of a two-pixel image. In the center of the valley the penalty in minimal. From the first estimate f_0 a search is started along the direction of the steepest descent at f_0 (yellow arrow) for the lowest value of the penalty. The result is the new estimate f_1. A new search is started at f_1 along the direction of the steepest descent at f_1, yielding estimate f_2, and so on. The result is a zigzag path toward the bottom of the valley.

while maximizing the effect of each step. For most penalty functions, computing a new descent direction is the most time consuming part in a steepest descent iteration. Therefore, once a new direction is established, it is often worthwhile to employ relatively simple sub iterations to find the best step length.

5.5 Bayesian Deconvolution

The penalty functions introduced in the previous sections do not include explicitly available priori information, nor do they give a satisfying justification of the choices made. When recording a low light image with photon counts in the range of 5–100 photon in the brightest pixel, it is clear that a second recording of the same object will result in a different image due to the random nature of the arrival time of the photons at the detector. Therefore, even with perfect knowledge of the imaging process one cannot hope to restore the object from an image with such a random character. The best one can do is compute the most "likely" object to give rise to the measured image. To do so, the so-called "posterior probability" $P(f|g)$ that a certain object gave rise to the given measured image must be computed. With the aid of Bayes' rule of conditional probabilities:

$$P(f|g)P(g) = P(g|f)P(f) \tag{5.7}$$

the posterior probability can be related to the easier to compute likelihood $P(g|f)$ of finding a measure image g from a known object f. For the purpose of optimizing $P(f|g)$, $P(g)$ is independent of f so can be ignored. This means that for finding the maximum a posteriori (MAP) probability $P(f|g)$ two terms need to be computed: a likelihood term $P(g|f)$ and a priori probability term $P(f)$ for the object estimate.

To construct a deconvolution algorithm aimed at finding the MAP probability we need the following ingredients:

- The likelihood term. In the section below it will be shown how this term can be obtained from the known PSF and the noise model.
- A choice for the object probability, a summary of all what is known about the object.
- An optimization scheme, for example an iterative scheme based on the gradient of the MAP probability, so that it becomes possible to construct an algorithm similar to the steepest descent method discussed in the previous section.

In the sections below these items will be discussed one by one.

5.5.1 Computing the Likelihood

For the low photon count images we are interested in, the major source of noise by far is Poisson noise due the random arrival times of the photons at the detector. Using the well-known Poisson distribution and the known PSF we can compute $P(g|f)$ from a current estimate. For a single image pixel with expected value $\lambda > 0$ the probability of measuring n photons ($n \geq 0$) is

$$P_{\text{single}}(n|\lambda) = \frac{e^{-\lambda}\lambda^{n}}{n!} \tag{5.8}$$

The expectancy values λ_i for each pixel i of the imaged estimate are computed by convolving the current estimate with the PSF. Some distance away from the object the expectancy values are likely to drop toward the background value which in a confocal or STED microscope may be much smaller than 1. For $\lambda \ll 1$ and $n = 1$ Equation 5.8 reduces to $P(1|\lambda) \approx \lambda$.

The total probability to detect an image with pixel values n_i, given an estimate image with λ_i values each resulting in a per-pixel probability P_i, is the product over all P_i:

$$P(g|f) = \prod P_i \tag{5.9}$$

Since Equation 5.8 has the capability of computing the probability of detecting photons for noninteger valued backgrounds, it makes sense to include the background in the imaging model of Equation 5.1 instead of simply subtracting it from the measured image. Writing out the convolution operation as a summation:

$$\lambda_i = \sum_j h_{i,j} f_j + b_i, \tag{5.10}$$

Here, the f_j are the pixels of the object, all ordered in a single column vector. For a 3D image, this ordering can be realized as, starting at the bottom plane, concatenating all X-direction lines, then concatenating all lines in the next plane, and so on. The $h_{i,j}$ form a huge $N \times N$ matrix $\mathbf{H}$ of which each row i contains a complete 3D PSF for a particular pixel in the image. The benefit of this notation is that space variant PSFs are automatically included. As pointed out in 6 direct evaluation of this sum is impractical, but for the current argument that does not matter.

In contrast with the methods outlined in Section 5.3 the resulting imaged estimate Λ is not used directly but interpreted as a expectancy image from which with the help of Equations 5.8 and 5.9 the likelihood term is computed.

5.5.2 The Object Prior Probability

The most simple choice here is to assign equal probability to all objects. This simplifies the MAP estimation into a maximum likelihood estimation (MLE). To see how this works out, consider the single pixel case of Equation 5.8. To estimate the MAP, using Bayes rule (Equation 5.7) and omitting $P(g)$ since it is constant, we get: $P_{\text{single}}(\lambda|n) \sim \phi(\lambda) = \left((e^{-\lambda}\lambda^n)/n!\right)$ which needs to be maximized with respect to λ. It is more convenient to maximize:

$$\log(\phi) = -\lambda + n\log(\lambda) - \log(n!). \tag{5.11}$$

The maximum is found when $\left((\partial \log(\phi))/\partial\lambda\right) = (n/\lambda) - 1 = 0$, so $\lambda = n$, that is, the most likely expectancy value for a pixel with n detected photons is n.

This may seem a trivial result, but it is good to realize that it was obtained under the unrealistic condition that all outcomes have equal probability. With a prior probability, Equation 5.11 would have had an extra term depending on λ, resulting in a different solution for the maximum.

To expand the single pixel scheme outlined above to find the most likely object giving rise to an observed image (MLE), Equation 5.10 is combined with Equation 5.9. Using a similar optimization method to find the maximum as used in the van Cittert iterations, the Richardson–Lucy (RL) algorithm can be found. For details see Appendix A.1.

The RL algorithm has been successfully applied to many fields including fluorescence microscopy [11]. Since the RL algorithm does not contain a priori information about the object other than that it is nonnegative, it is prone to noise amplification just like the van Cittert method. Indeed, Figure 5.9 clearly shows the RL algorithm converging on an unacceptable sparse solution where all intensity is concentrated in a few pixels. Still, Figure 5.9 also shows that acceptable results can be obtained by stopping iterations at a suitable count, the breaking off of the iterations acting as a regularization method where resolution gain is traded against noise amplification. The RL method is also capable of restoring out of band spatial frequencies, see Figure 5.10. Many variants of the RL method exist which mitigate noise amplification and also accelerate its usually slow convergence.

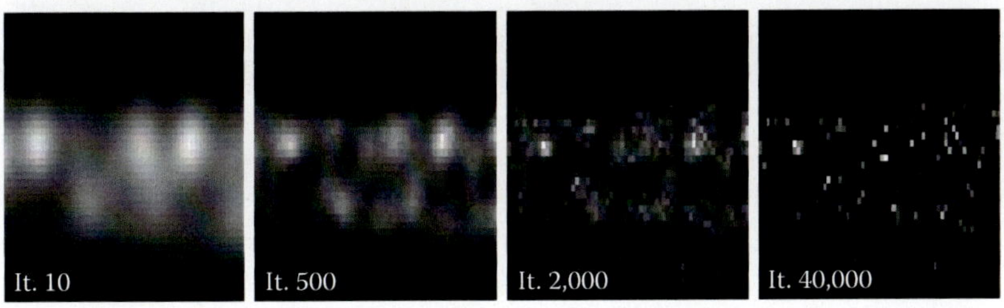

Figure 5.9

Results of the Richardson–Lucy (RL) algorithm after 10, 500, 2000, and 40,000 iterations. In this unaccelerated RL algorithm the amount of iterations needed to achieve convergence is so high that propagation of numerical errors can become a problem.

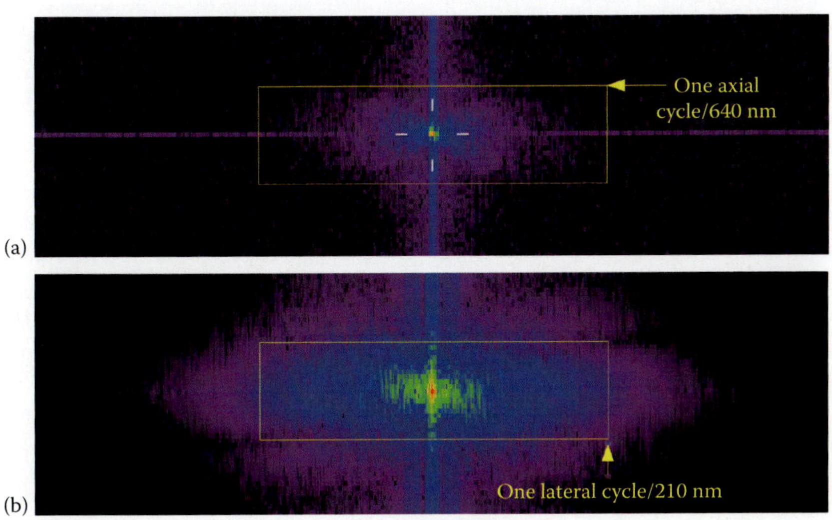

Figure 5.10

Maximum likelihood estimation (MLE) applied to widefield data. (a) Spectrum of the raw image and (b) Spectrum of the deconvolved image. The yellow box, a cross section through an axially symmetric pillbox, confines the doughnut-like pass band of the widefield microscope, see also Figure 5.3. In the top image this doughnut-like shape of the pass band can indeed be seen. The vertical and less intense horizontal bars are due to data truncation. The spectrum of the deconvolved image shows reconstructed spatial frequencies outside the pillbox, and importantly also in the missing cone.

5.5.3 Regularization

In the Bayesian framework, regularization of the solution is realized by the a priori probability term $P(f)$. This can take the shape of a Gaussian distribution, maximum entropy penalty, total variation, and Good's roughness penalty or combinations thereof [12].

Jaynes argues [13] that the maximum entropy measure offers a "fail-safe" solution in the sense that the solutions cannot show any detail for which there is no evidence in the data. The entropy of an image is easy to compute: $S(f) = -\sum f_i \log(f_i)$, so also from a computational perspective a good candidate for a penalty function. Still, if we do have extra knowledge about the data, for example that we are dealing with filaments or single-molecule fluorescence emitters, that extra knowledge will not be included when entropy is used as a penalty function since it does not take structure into account at all. Good points out [14] that "small violent ripples" will have little effect on the total entropy of a continuous object function. Instead, he proposes a measure which considers the similarity of an object function with a shifted version of itself. By defining this similarity in terms of information divergence [15] or cross entropy, Good arrives at the following penalty function:

$$\Phi(f) \propto \sum_i \frac{\Delta f_i}{f_i}, \tag{5.12}$$

where $\Delta = \left((\partial^2)/(\partial x^2)\right) + \left((\partial^2)/(\partial y^2)\right) + \left((\partial^2)/(\partial z^2)\right)$ is the Laplace operator. Since it involves second-order derivatives in all directions it triggers particularly strongly on sharp peaks, the most extreme of which are isolated pixels. Therefore, it is expected to be effective in penalizing the amplification of noise in the image background as visible in Figure 5.9. The Laplace operator is straightforward to implement, see Figure 5.11. The use of Good's roughness as a penalty function in microscopic deconvolution was pioneered by Joshi and Miller [16].

5.5.4 Sample Rate for Deconvolution

The Nyquist–Shannon theorem states that a signal can only be completely reconstructed from a sequence of samples when the sample rate is more than twice the highest spatial frequency transferred by the microscope. When the microscope is band limited (see red and green curves in Figure 5.2) no spatial frequencies from the object above the limit occur in the image, so the Nyquist rate is twice

	1	
1	–4	1
	1	

Figure 5.11

Filter window implementing a 2D Laplace operator. Centered on a pixel in the image, the output value is the sum of its four direct neighbors minus four times its own value. The sum of all components is zero, so there is no penalty for homogeneous regions. Also there is no penalty of constant slopes. An isolated point surrounded by zero is assigned a value of –4. Assuming uniformly sampled data, the filter can be extended to 3D by including the two pixels above and below the center pixel and increasing its weight to –6.

the band limit in a particular direction. For example, for the green confocal OTF curve in Figure 5.2, it is 20 samples per micron, sampling distance 50 nm. Although that curve is for an ideal confocal microscope with an infinitely small pinhole, the OTF is close to zero already at 80% of the bandwidth. Therefore, it could be argued that the effective bandwidth is less and sampling could be coarser. While there are conditions where issues like photo bleaching and time constraints make it hard to reach the Nyquist rate, it is generally not advisable to deconvolve slightly undersampled data. Regarding photobleaching constraints, it is often the total amount of photons detected from a sample which is limited, not the amount of samples taken, so taking more samples only means less photons per sample. Since the relationships between neighboring samples are taken into account during deconvolution, spreading photons over samples does not pose a problem, see Figure 5.12.

In the case of the STED microscope, there is no bandwidth limit and therefore no straightforward way to establish a sufficient sampling rate. Depending on the available signal, practical values are around one-third of the expected resolution in the lateral direction. In the axial direction, for vortex-type STED systems, equivalent confocal values can be used, or if specimen properties like bleaching allow it, 50%–80% of the confocal value. In STED systems equipped with a Z-doughnut (see below, Section 5.5.7 and Figure 5.15) where the axial size of the PSF strongly depends on the configuration and intensity of the depletion field, practical sufficient sampling intervals range from 20% to 50% of equivalent confocal values (Figures 5.13 and 5.15).

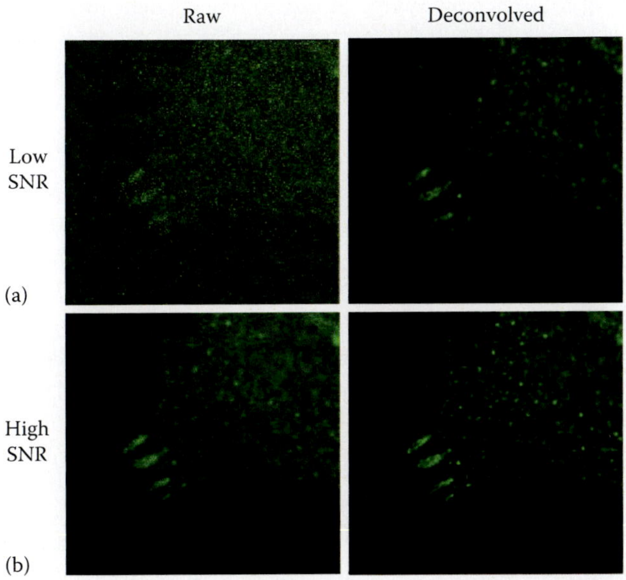

Figure 5.12

Well-sampled confocal images of the same specimen recorded at low (a) and high (b) signal strength with a Zeiss LSM510 confocal microscope. Sampled at 35.2 nm laterally and 100 nm axially the image represents a volume of 18 × 18 × 2.1 μm. (Image and deconvolved result courtesy of Pawel Pasierbek, BioOptics Facility, IMP, Vienna, Austria.)

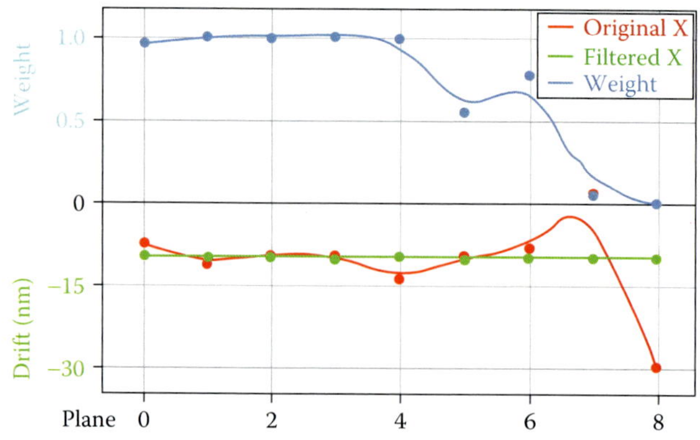

Figure 5.13

Plane-to-plane drift rate in X-direction versus plane number (horizontal). Estimated STED drift (green) from cross correlation signal (red) using confidence weights (blue, normalized to unity). The average drift rate corresponds to about −10 nm per plane. (Data used by permission from Dr. Grazvydas Lukinavicius, EPFL, Lausanne, Switzerland.)

5.5.5 Acceleration of Optimization

With the rapid increase of the volume of microscopic data in recent years, outpacing speed improvements in computer hardware, efficient deconvolution algorithms have seen renewed interest [18,19]. A computationally efficient, conjugate gradients implementation using Good's roughness penalty was first published by Verveer and Jovin [20]. Though not particularly memory efficient, when combined with recent developments in the field of conjugate gradient schemes [21,22] this type of algorithm proved to be extremely efficient in terms of computational effort and capable to deconvolve STED data sets in 5–10 iterations. This improved computational efficiency also serves to counter further increases in complexity when depth-dependent PSFs have to be considered due to the effects of spherical aberration.

5.5.6 STED Deconvolution

STED microscopes based on a vortex phase plate [23] are characterized by a very narrow but high PSF. Along the axial direction the resolution is similar to that of an equivalent confocal microscope, but laterally it is easily four times narrower. Imaging with this pencil-like PSF poses severe demands on the stability of the scanner, exacerbated by the increase in pixel dwell times needed to detect enough photons. Clearly, deconvolution will fail if one attempts to deconvolve a data set deformed by drift with an undisturbed PSF. Fortunately, the pencil-like shape of the PSF also means that adjacent XY-planes in a well-sampled data set will have a high degree of correlation. By computing the cross correlation and phase cross correlation between adjacent planes with strong correlation, and interpolating over planes without significant correlation, a drift map can be created for the complete data set. This drift map is then used to correct the measured data prior to deconvolution. Figure 5.14 shows the resolution gain in the axial but also lateral direction after drift correction.

Practical resolution gain. The resolution increase which can be obtained in typical STED data is a factor of 2 in all directions, lateral resolution numbers as low

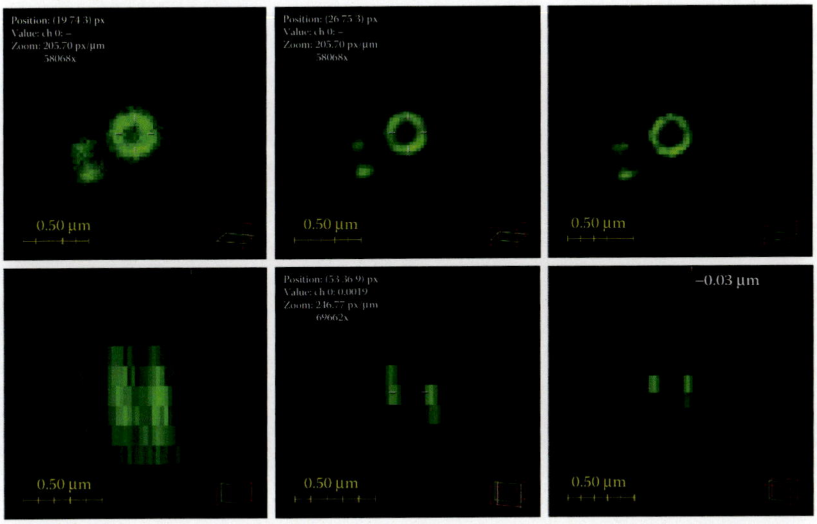

Figure 5.14

Left to right: Raw versus deconvolved versus stabilized deconvolved vortex STED image of a centriole. Top row: *xy* planes; bottom row: *xz* planes. Pixel sizes: 29 × 29 × 118 nm. After correction and deconvolution with CMLE [17] axial resolution is slightly above one axial sample distance, while lateral resolution is also improved. (Data used by permission from Dr. Grazvydas Lukinavicius, EPFL, Lausanne, Switzerland.)

as 22 nm, true nanoscopic performance, have been reported [24] for vortex phase plate-based systems. Reducing the resolution by a factor of 2 in all directions results in a reduction of the resolution volume by a factor of 8. Because energy is locally conserved during deconvolution, signal strength from small objects may become 5–10 times higher in the deconvolved image as compared to the raw data.

5.5.7 STED-3D Deconvolution

While the vortex phase plate-based STED systems offer very high lateral resolution, the high ratio between their confocal-like axial and lateral resolution does not make them suited for 3D morphological studies. An alternative phase plate configuration achieves a depletion distribution which is zero at the focus just like with a vortex doughnut-shaped depletion distribution, but shows appreciable intensity above and below the focus as well [25]. This Z-doughnut results in strongly increased axial resolution, but the gain in lateral resolution is more moderate. In the STED-3X system from Leica Microsystems [26], it is possible to superimpose both a vortex doughnut and a Z-doughnut to optimize the shape of the PSF for a specific specimen (see Figure 5.15). Though the PSFs in Figure 5.15 were computed for a small RI mismatch between a medium RI of 1.47 and immersion RI of 1.515, the 100% Z-doughnut PSF is already asymmetric in Z. This means that RI mismatches should be minimized to avoid collapse of the Z-doughnut, especially deeper in the specimen. For results obtained with low photon count STED-3X data and deconvolution with Good's roughness regularization and conjugate gradients deconvolution, see Figure 5.16.

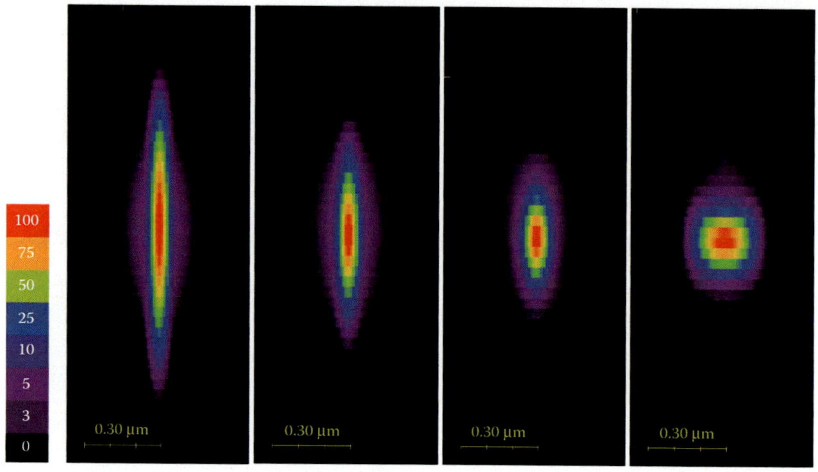

Figure 5.15

STED-3D theoretical PSFs, NA 1.4, medium RI 1.47, saturation factor 40 [27]. Pixel size 20 × 20 × 50 nm. (a–d) 0%, 15%, 50%, 100% Z-doughnut versus vortex doughnut, respectively. It is assumed that effectively a small fraction of the molecules does not get depleted, the so-called immunity fraction, here set to 5%. This leads to a confocal PSF component in the STED PSF, resulting in a "shoulder," left and right of the PSF, especially visible in the left-most 100% vortex PSF.

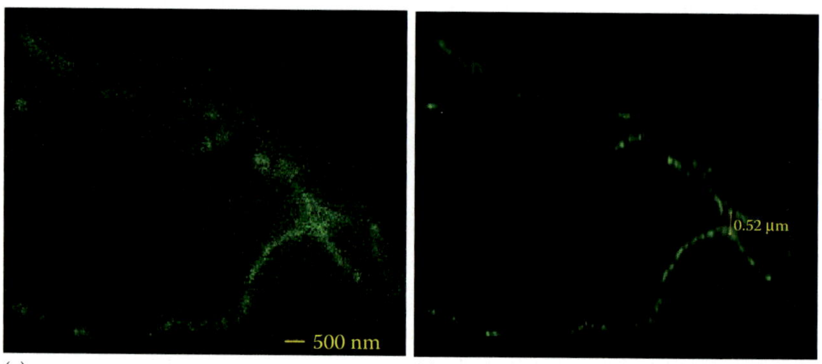

Figure 5.16

Dino flagellate image recorded with a STED-3X system from Leica Microsystems [25]. Saturation factor 20, 50% Z-doughnut. The brightest pixel represents 17 detected photons. (a) *X-Z* plane, raw data. (b) Same plane after deconvolution with 2 PSFs (see Section 5.6.1), Good's roughness penalty as implemented in Huygens Professional [17], 10 iterations, showing effective denoising of structures and background. Estimated *Z*-resolution after deconvolution <120 nm. Due to the high noise, resolution in raw data cannot be reliably measured. (Data courtesy Timo Zimmermann, CRG-Centre for Genomic Regulation, Barcelona, Spain.)

5.6 Space Variant PSFs

In many practical cases, the RI of the specimen is not matching the design RI, resulting in depth-dependent spherical aberration, and due to that in a changing PSF. For a worst-case scenario where a high NA oil immersion lens is used to image into water, see Figure 5.17. In this case, total internal reflection clamps the

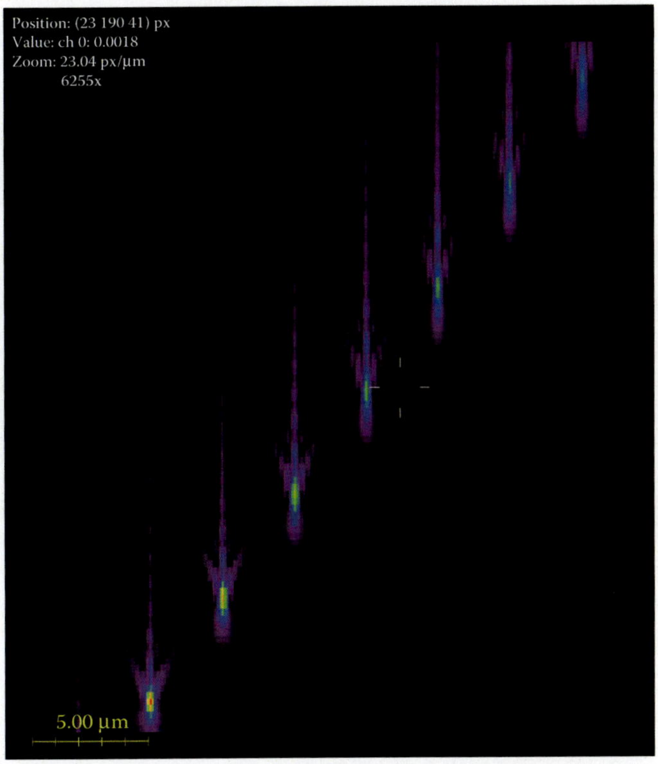

Figure 5.17

Widefield PSFs at different depth, from 0 to 25 µm from the cover slip. NA: 1.4, oil immersion into water. Though the NA is larger than the medium RI causing total internal reflection, a situation to be avoided, these condition are still sometimes encountered in practical data. For the used false color table, see Figure 5.15.

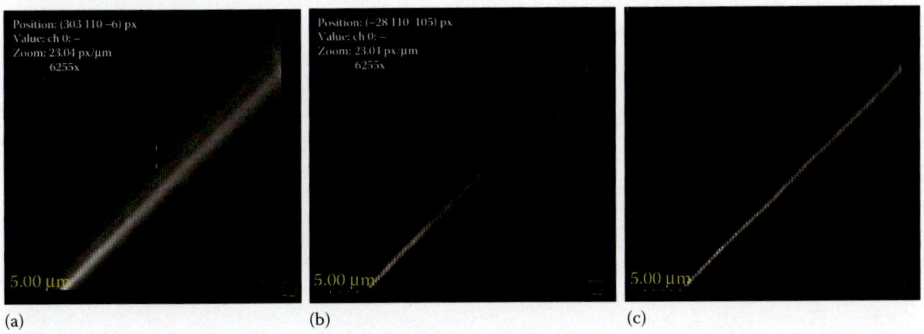

Figure 5.18

Diagonal row of points in the *X-Z* plane as imaged in widefield. (a) $NA = 1.4$ widefield image suffering from severe oil/water refractive index mismatch. (b) Deconvolved with a single PSF. (c) Deconvolved with 5 PSFs.

effective NA to the RI of water, large amounts of spherical aberration distort the PSF, more so at deeper locations. Clearly, deconvolution with a single PSF will fail to provide good results over the entire image depth range, see Figure 5.18. The most straightforward solution to solve this is to split the image into slabs thin enough for the PSF to be considered constant within a slab. Subsequently, the slabs are deconvolved individually and glued together. In cases of low spherical aberration the slabs can be fairly thick; in a typical image only a few are needed.

To avoid visible seams the slabs are chosen to overlap. In the overlap region values of neighboring slabs are summed according to smooth cosine-shaped weighting functions. The necessary overlap size is related to the axial extend of the support of the PSF, that is, the length of the PSF tail along Z. A large object located just above a particular slab will contribute to a point within the slab proportional to its overlap with the PSF tail. To be able to ignore that contribution for all locations in the slab, the transition region must be large enough to put enough distance between the slab and such objects. For confocal or vortex STED cases where the PSF has limited support, this overlap becomes sufficiently small after 4–6 times the axial resolution, unless large (>2 Airy disk) pinholes are used. However, in the case of a widefield PSF due to conservation of energy, the *X-Y* plane-wise integrated intensity is constant. This means that large far-off objects may still contribute to a particular slab, so this slab strategy is problematic for nonsparse widefield images. In sparse widefield images containing small objects the contributions of objects to another slab may be ignored.

As long as the volume of the overlap areas is not much larger than the net slab volume the computational and especially memory efficiency of this technique is quite good, in fact it is often used to reduce memory usage for large data sets. In cases where the overlap areas need to become very large compared to the slab volume a different strategy is needed.

5.6.1 Hybrid FFT/Spatial Convolution

The basic problem with space variant PSFs is not that it is difficult to compute a convolution with them (see Equation 5.10), but that FFTs are needed to reduce the computational effort as pointed out in Section 5.2.3. At the cost of some computational efficiency this can be solved by the procedure outlined in Figure 5.19.

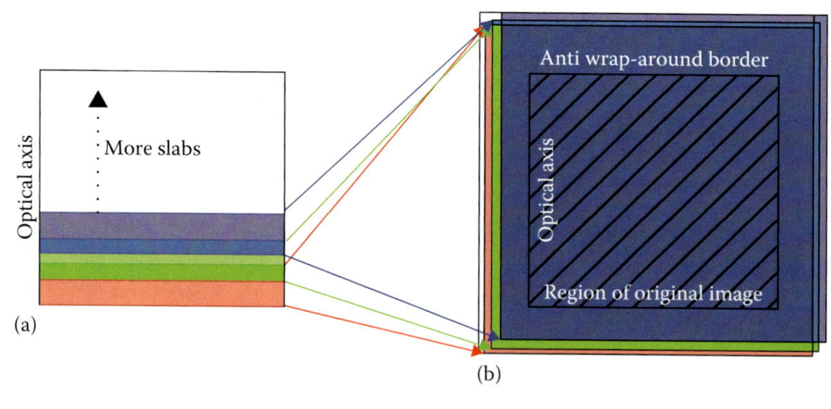

Figure 5.19

Hybrid FFT/spatial convolution. On (a), the data to be deconvolved with a space variant PSF is split into multiple deeply or completely overlapping slabs in such a way that the sum of all slabs is equal to the original data. Next, each slab is convolved with its appropriate PSF using FFTs. The slab is padded to avoid the wrap-around effect which arises from the fact that both data and PSF are taken to be periodic and of the same size. When an object at the image edge is convolved with a PSF, the tails of the convolved object which would fall over the edge appear at the opposing edge instead. Padding using criteria as mentioned in Section 5.6 solves this issue. The final result on (b) is obtained by summation of all per-slab results.

In the case of PSFs with limited axial support, the padded slabs will be smaller than the original data, increasing computational efficiency.

5.7 Summary

Intuitively it would seem that deconvolution can only function in low noise conditions, like encountered in high signal widefield and confocal images. Perhaps the most important finding in this chapter is that even in low photon count data like in Figure 5.16 deconvolution can significantly increase resolution and signal strength, pushing the signal levels at which it is still practical to acquire images to even lower levels. Also the application area of deconvolution is not limited to images recorded under perfect optical conditions: distortions due to sources like thermal drift can be corrected, the detrimental effects of RI mismatch-induced spherical aberration can be compensated. Obviously, when high-resolution images are needed one should minimize issues like RI mismatch, mechanical instability, chromatic aberration, or crosstalk between recorded channels while avoiding acquisition errors like undersampling, sampling calibration errors, or saturation of the detector. With all these measures in place, deconvolution will be able to recover details and reduce noise in any kind of microscopic fluorescence image.

Acknowledgments

The images shown in this chapter were kindly provided by Hiroshi Kano, Martin Schrader and Stefan Hell (then at University of Turku, Finland, Figures 5.4, 5.5, 5.7, 5.9), James Evans (then at the Whilehead institute, Boston, MA, Figure 5.10), Pawel Pasierbek, (IMP, Vienna, Austria, Figure 5.12), Grazvydas Lukinavicius (EPFL, Lausanne, Switzerland, Figures 5.13 and 5.14), and Timo Zimmermann (CRG-Centre for Genomic Regulation, Barcelona, Spain, Figure 5.16). Special thanks to my colleagues at Scientific Volume Imaging, Nicolaas van der Voort for helpful comments, and Alberto Diaspro for suggestions toward this contribution.

Appendix

A.1 The RL Algorithm

The single pixel example discussed in Section 5.5.2 can be expanded to a many pixel imaging process with Equation 5.9 and by noting that the expectancy value λ_i for each pixel is given by Equation 5.10. In Section 5.5.2 the concept of ordering all pixels of an image into a vector was introduced, with the convolution operation in the form of a matrix multiplication. This allows us to rewrite Equation 5.10 in a more compact form: $\Lambda = \mathbf{H}f + \mathbf{b}$, with Λ the expectancy image, $\mathbf{H}$ the matrix $h_{i,j}$ containing the PSF coefficients, and $\mathbf{b}$ the background image, in many practical cases just a single constant.

For reasons of convention and also to conform to the notion of a penalty function as introduced in Equation 5.6, the maximization scheme outlined above is turned into a minimization by computing the total negative log-likelihood over all pixels:

$$\Phi = \sum_i \lambda_i - n\log(\lambda_i) + \log(n!) \tag{A.1}$$

where the product in Equation 5.9 is transformed into a summation due to taking the logarithm, and where the sign is changed to turn the maximization problem into a more conventional minimization problem. At the minimum, at the bottom of the valley (see Figure 5.8) all derivatives of $\mathbf{\Phi}$ with respect to all object pixels will be zero. This yields

$$\frac{\partial \Phi}{\partial f_i} = \frac{\partial(\mathbf{H}f + \mathbf{b})}{\partial f_i}\left(1 - \frac{\mathbf{g}}{\mathbf{H}f + \mathbf{b}}\right) = \mathbf{H}^{\mathrm{T}}\left(\mathbf{1} - \frac{\mathbf{g}}{\mathbf{H}f + \mathbf{b}}\right) = 0 \tag{A.2}$$

With the sum of all elements of $\mathbf{H}$, for all positions of the PSF $\mathbf{H}^{\mathrm{T}}\mathbf{1} = \mathbf{1}$, at the minimum of $\mathbf{\Phi}(f)$:

$$\mathbf{H}^{\mathrm{T}}\left(\frac{\mathbf{g}}{\mathbf{H}f + \mathbf{b}}\right) = \mathbf{1} \tag{A.3}$$

To solve this, a procedure similar to the one used in the van Cittert method can be used to derive an iterative algorithm. Since Equation A.3 becomes unity at the minimum we seek, the following multiplicative fixed-point iteration might work:

$$f_{k+1} = f_k\left(\mathbf{H}^{\mathrm{T}}\left(\frac{\mathbf{g}}{\mathbf{H}f + \mathbf{b}}\right)\right) \tag{A.4}$$

This is the famous RL algorithm, which has been proven to converge [28,29]. Interestingly, when the matrix multiplication with $\mathbf{H}^{\mathrm{T}}$ is left out in Equation A.4, the sequence is still a fixed-point iteration, known as the Gold algorithm [30].

References

1. Hell, S. W., and J. Wichmann, Breaking the diffraction resolution limit by stimulated emission: Stimulated-emission-depletion fluorescence microscopy, *Optics Letters*, 19(11), 780–782, 1994.
2. Oberlaender, M., P. Broser, B. Sakmann, and S. Hippler, Shack-hartmann wave front measurements in cortical tissue for deconvolution of large three-dimensional mosaic transmitted light brightfield micrographs, *Journal of Microscopy*, 233(2), 275–289, 2009.
3. Frigo, M., and S. G. Johnson, The design and implementation of FFTW3, *Proceedings of the IEEE*, 93(2), 216–231, 2005. Special issue on "Program Generation, Optimization, and Platform Adaptation."
4. FFT Benchmark Methodology. http://www.fftw.org/speed/method.html. accessed November 21, 2014.
5. Tikhonov, A. N., and V. Y. Arsenin, Solutions of ill-posed problems, V. H. Winston and Sons, Washington DC, 1977.
6. Bertero, M., P. Boccacci, G. Brakenhoff, F. Malfanti, and H. van der Voort, Three-dimensional image restoration in fluorescence confocal scanning microscopy, In *International Congress on Optical Science and Engineering*, pp. 86–91, International Society for Optics and Photonics, 1989.

7. Kano, H., H. T. Van Der Voort, M. Schrader, G. Van Kernpen, and S. W. Hell, Avalanche photodiode detection with object scanning and image restoration provides 2–4 fold resolution increase in two-photon fluorescence microscopy, *Bioimaging*, 4(3), 187–197, 1996.
8. Lagendijk, R. L., and J. Biemond, *Iterative Identification and Restoration of Images*, vol. 118. Springer, 1990.
9. Bertero, M., and P. Boccacci, *Introduction to Inverse Problems in Imaging*. CRC Press, 2010.
10. van der Voort, H., and K. Strasters, Restoration of confocal images for quantitative image analysis, *Journal of Microscopy*, 178(2), 165–181, 1995.
11. Holmes, T. J., Maximum-likelihood image restoration adapted for noncoherent optical imaging, *JOSA A*, 5(5), 666–673, 1988.
12. Mignotte, M., Fusion of regularization terms for image restoration, *Journal of Electronic Imaging*, 19(3), 033004–033004, 2010.
13. Jaynes, E. T., On the rationale of maximum-entropy methods, *Proceedings of the IEEE*, 70(9), 939–952, 1982.
14. Goodd, I., and R. Gaskins, Nonparametric roughness penalties for probability densities, *Biometrika*, 58(2), 255–277, 1971.
15. Csiszar, I., Why least squares and maximum entropy? An axiomatic approach to inference for linear inverse problems, *The Annals of Statistics*, 8(4), 2032–2066, 1991.
16. Joshi, S., and M. I. Miller, Maximum α posteriori estimation with goods roughness for three-dimensional optical-sectioning microscopy, *JOSA A*, 10(5), 1078–1085, 1993.
17. Huygens deconvolution software by Scientific Volume Imaging, The Netherlands. http://www.svi.nl. accessed January 4, 2015.
18. Zanella, R., P. Boccacci, L. Zanni, and M. Bertero, Efficient gradient projection methods for edge-preserving removal of poisson noise, *Inverse Problems*, 25(4), 045010, 2009.
19. Bonettini, S., R. Zanella, and L. Zanni, A scaled gradient projection method for constrained image deblurring, *Inverse Problems*, 25(1), 015002, 2009.
20. Verveer, P. J., and T. M. Jovin, Image restoration based on goods roughness penalty with application to fluorescence microscopy, *JOSA A*, 15(5), 1077–1083, 1998.
21. Hager, W. W., and H. Zhang, A survey of nonlinear conjugate gradient methods, *Pacific Journal of Optimization*, 2(1), 35–58, 2006.
22. Nakamura, W., Y. Narushima, and H. Yabe, Nonlinear conjugate gradient methods with sufficient descent properties for unconstrained optimization, *Journal of Industrial and Management Optimization*, 9(3), 595–619, 2013.
23. Török, P., and P. Munro, The use of gauss-laguerre vector beams in sted microscopy, *Optics Express*, 12(15), 3605–3617, 2004.
24. Schoonderwoert, V., R. Dijkstra, G. Luckinavicius, O. Kobler, and H. van der Voort, Huygens sted deconvolution increases signal-to-noise and image resolution towards 22 nm, *Microscopy Today*, 21(6), 38–44, 2013.
25. Keller, J., A. Sch, S. W. Hell, Efficient fluorescence inhibition patterns for resolft microscopy, *Optics Express*, 15(6), 3361–3371, 2007.
26. Leica TCS SP8 STED-3X microscope by Leica Microsystems, Germany. http://www.leica-microsystems.com/products/confocal-microscopes/leica-tcs-sp8-configurable-confocal/details/product/leica-tcs-sp8-sted-3x/. accessed January 4, 2015.

27. Harke, B., J. Keller, C. K. Ullal, V. Westphal, A. Schönle, and S. W. Hell, Resolution scaling in STED microscopy, *Optics Express*, 16(6), 4154–4162, 2008.
28. Shepp, L. A., and Y. Vardi, Maximum likelihood reconstruction for emission tomography, *Medical Imaging, IEEE Transactions on*, 1(2), 113–122, 1982.
29. Snyder, D. L., T. J. Schulz, and A. O'Sullivan, Deblurring subject to nonnegativity constraints, *Signal Processing, IEEE Transactions on*, 40(5), 1143–1150, 1992.
30. Bandžuch, P., M. Morháč, and J. Krištiak, Study of the van cittert and gold iterative methods of deconvolution and their application in the deconvolution of experimental spectra of positron annihilation, *Nuclear Instruments and Methods in Physics Research Section A: Accelerators, Spectrometers, Detectors and Associated Equipment*, 384(2), 506–515, 1997.

SECTION II
The Methods

6

Re-Scan Confocal Microscopy

Giulia M.R. De Luca, Ronald M.P. Breedlijk, and Erik M.M. Manders

6.1 Introduction
6.2 How the Confocal Works
6.3 Image Scanning Microscopy
6.4 RCM: The "Optics-Only" Alternative for ISM
6.5 Biological Applications
6.6 RCM Compared with Confocal Microscopy
6.7 Conclusion and Discussion
Bibliography

6.1 Introduction

As described in the previous chapters, the lateral resolution of an optical microscope is limited as characterized by Abbe's law. During the past decades, a broad range of techniques has been developed to circumvent this resolution barrier in both axial and lateral direction. If we define the term "super-resolution" as any optical microscopy technique that gives a resolution better than the diffraction limit as described by Abbe's law, the confocal microscope can be regarded as the first super-resolution technique in microscopy. In this chapter we will explain why confocal microscopy does not only provide better resolution in the axial direction (optical sectioning; the main reason why confocal microscopy became so popular), but also provides super-resolution in the lateral direction. Further, we will explain why the lateral super-resolution properties of the confocal microscope are in general not exploited in biomedical research. Here, we will describe a new technology, re-scan confocal microscopy (RCM) that provides both lateral and axial super-resolution. In addition (and maybe more importantly than the super-resolution properties), RCM allows imaging with an improved signal-to-noise ratio (SNR) leading to smoother and crisper images.

6.2 How the Confocal Works

RCM is based on the standard confocal microscope as it was developed in the early 1980s (Brakenhoff et al. 1979). The improved axial resolution (=optical sectioning) is the most important property of confocal microscopy and is often the reason why biomedical researchers decide to use this technique. However, most researchers do not realize and do not understand under which conditions the confocal microscope has improved lateral resolution (Mc Cutchen 1967, Sheppard and Wilson 1978, Cox et al. 1982, Cox and Sheppard 2003). Therefore, we will here resume the basics of improved axial and lateral resolution in confocal microscopy compared to widefield.

6.2.1 Optical Sectioning in Confocal Microscopy

A confocal microscope is a point scanning technique in which a focused laser beam is directed into the sample in a scanning fashion by two scanning mirrors (Figure 6.1i). The focused beam excites fluorescent molecules in the sample. The in-focus molecules are excited strongly (and therefore fluoresce brightly) and the out-of-focus molecules are excited less efficiently (and therefore show dim fluorescence). The light emitted by the fluorescent molecules in the sample is collected by the objective and directed, via the same scanning mirrors and lenses, toward a pinhole in such a way that the in-focus light will pass the pinhole and the out-of-focus light will mainly be blocked by the pinhole (Brakenhoff et al. 1979). As a result of this simple set-up, the in-focus molecules in the sample are both excited efficiently and detected efficiently. On the other hand, the out-of-focus signal from the sample is strongly suppressed by two effects: low excitation efficiency and low detection efficiency. This out-of-focus suppression is called optical sectioning and is the most important property of the confocal microscope.

6.2.2 Lateral Resolution in the Confocal Microscopy

In addition to the advantage of optical sectioning, there is another important advantageous property of the confocal microscope: super-resolution in the lateral direction. As explained above, axial resolution improvement (sectioning) is caused by less effective excitation and detection of out-of-focus structures. Improvement of lateral resolution is established in a similar way: less effective excitation of off-axis fluorescent molecules and less effective detection of off-axis fluorescence signal. So, on-axis molecules are both excited and detected more efficiently than off-axis molecules. Figure 6.1a and e show a molecule excited by the laser beam at two time points during the scanning movement of the laser spot. In Figure 6.1a, the fluorescent molecule is positioned on the optical axis of the laser beam and therefore excited with maximum efficiency and will therefore emit the maximum fluorescence intensity (100% in Figure 6.1b). A little later the axis of the optical system does not coincide anymore with the position of the molecule due to the scanning movement of the excitation spot (Figure 6.1e) and therefore the molecule will be excited with a lower efficiency and, consequently, the emitted fluorescence will be dimmer (20% in the example of Figure 6.1f).

Now, let's regard the situation where the microscope user decided to adjust the pinhole diameter to 1.0 airy unit (AU). With this relatively open pinhole (Figure 6.1c and g) most of the fluorescence signal emitted by the molecule will pass the pinhole regardless of the position (on- or off-axis) of the molecule. In the off-axis situation (Figure 6.1g), not all the emitted light will pass the pinhole

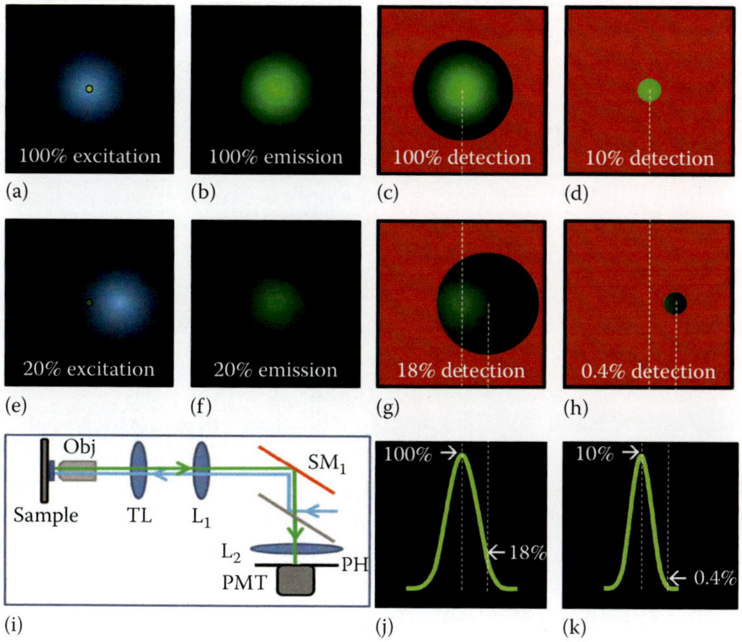

Figure 6.1

Illustration of improved resolution in confocal microscopy. In a standard confocal microscopy setup (j) the scanning mirrors SM_1 direct the laser light to the sample where they excite fluorescent molecules. The emitted light from the molecules is de-scanned via the same mirror unit SM_1, directed through the detection pinhole and detected by a photomultiplier tube (PMT) or other detector. During the scanning of the laser in the sample, the illumination (blue blurry spot in a and e) excites a fluorescent molecule (green point in a and e) with different efficiencies. When the center of the illumination coincides with the position of the fluorescent molecule (a), the excitation efficiency is maximum (100%; b) and when the molecule is off-axis of the laser beam (e), the excitation efficiency is lower (20% in this example; f). The emitted light is projected on the pinhole, which is co-aligned with the excitation laser. The detection efficiency is influenced by two factors: the position of the emission spot with respect to the center of the pinhole and the diameter of the pinhole. The detection efficiency reaches its maximum if the excitation is in the center of the molecule and if the pinhole is sufficiently open (c). If the pinhole is closed down to 0.2 AU (d) the detection efficiency will diminish to 10% (in this example). When the molecule is off-axis (g), even with an open pinhole (1 AU), the detection efficiency reduces the detected light (let's say to 18%). When the emission spot is off-axis and the pinhole is closed down to 0.2 AU (h), the two effects are multiplicative and therefore the detected light is very little (in this example 0.4%). The lateral PSF of the confocal microscope is described by the spatial integral of the light distribution that passes the pinhole, as a function of the position of the laser beam (and pinhole). The lateral PSF acquired by a confocal microscope with the pinhole at 1 AU and at 0.2 AU are shown in (k) and (l). It is clear that the diameter of the pinhole influences the lateral resolution of confocal microscope: The smaller the pinhole, the better the resolution, but on the cost of detection efficiency.

and therefore approximately 18% (in this example) of the emitted light will be detected. Figure 6.1j shows the light that passes the pinhole as a function of the position on the laser spot. This distribution of detected light (2J) is as broad as the diffraction-limited distribution of light (Figure 6.1a and b). This means that, in case of pinhole diameter of 1 AU, the confocal microscope does not have improved lateral resolution. This is an important disadvantage of this large pinhole, but it compensates the great advantage of the very high detection efficiency and, consequently, a relatively high SNR of the final image.

Lateral resolution can only be improved by further reduction of the pinhole diameter; in this example down to 0.2 AU (Figure 6.1c and g). Only a small part of the emitted light (10% in this example) will pass the relatively small pinhole. At the moment, the molecule gets an off-axis position (Figure 6.1g) and the detected light will be strongly reduced not only because off-axis molecules show dimmer emission, but also because the small pinhole catches only a dim part of the distribution of emitted light from that molecule. In this example, the off-axis suppression of fluorescence signal is 25-fold (from 10% to 0.4%). This implies that the effective distribution of detection (Figure 6.1k) is narrower than the diffraction-limited distribution of light (2A and 2B) which leads to a sharper image and an image recorded with pinhole of 1 AU (Figure 6.1j). The drawback of this small pinhole diameter is that the total detection efficiency is low, which leads to noisy images. In conclusion, confocal images recorded with a very small pinhole are sharper but noisier.

This is why most biologists are of the opinion "Better less sharp than more noisy" and therefore decide to open the pinhole to 1.0 AU. In this way they sacrifice resolution for SNR. In the next paragraphs we will explain how simple modifications of the confocal microscope, image scanning microscopy (ISM) and RCM, can provide sharper images with improved SNR.

6.2.3 Loss of Super-Resolution Information

The standard confocal microscope setup as described above with a pinhole diameter of 1.0 AU has high detection efficiency but the resolution is poor. This poor resolution is related to loss of information during the detection process, where all the light passing the pinhole is integrated in a single pixel value. This phenomenon becomes clear if we have a closer look at Figure 6.1g. This figure depicts the distribution of light, emitted by a fluorescent molecule, projected on the pinhole. It is clear that the image of the molecule is positioned off-axis, at the left-hand side of the optical axis (which is aligned with the center of the pinhole). So, this distribution contains spatial information about the position of molecules. The detector behind the pinhole, however, sums-up all the light that passes the pinhole and stores this spatially integrated signal, represented as a pixel value. Due to this spatial integration of light, the spatial information on the position of the molecule is lost. This lost information is exactly the information that is needed to obtain super-resolution. We will see in the next paragraph how this information is maintained in ISM and RCM, leading to super-resolution images.

6.3 Image Scanning Microscopy

In 1980, Professor Sheppard, pioneer in confocal microscopy, was the first to realize the potential of using a camera at the position of the pinhole (Sheppard 1988). This method would provide improved lateral resolution due to the preservation of the spatial information on the distribution of light. ISM (Muller and Enderlein 2010) was the first technology where a sensitive, fast camera (EMCCD- or a sCMOS camera) was placed at the position of the pinhole. At every position of the laser beam in the sample, the camera snaps an image of the light distribution at the position of the pinhole. It is clear that, in this way, the precious spatial information is maintained and no light is lost by squeezing any pinhole (to be clear, there isn't any pinhole). With some post-processing steps of the acquired series of "pinhole images" the final image can be reconstructed. Depending on the post-processing algorithm, a final image can be reconstructed that is identical to (1) a standard widefield fluorescence image, (2) a standard confocal image, or (3) an image with improved lateral resolution.

6.3.1 Reconstruction of ISM "Pinhole-Image" Series

As explained above, in ISM a "pinhole image" is recorded for every scanning position (x, y) of the laser spot. The "pinhole images" are in general relatively small (e.g., 16 × 16 pixels). If we add up all the "pinhole images," but not before shifting them each to a coordinate (x', y') that compensates for the position of the laser beam (x, y), the light from molecules in the sample will be placed at the "right" position in the final image and therefore, the final image will be identical to a widefield image. So in this case the relationship between the position (x, y) of the laser beam in the sample and the shifted position (x', y') in the image is $x' = M_{shift}{}^*M_{opt}{}^*x$ and $y' = M_{shift}{}^*M_{opt}{}^*y$ where $M_{shift} = 1$ and M_{opt} equals the optical magnification of the microscope (from sample to camera).

It is also possible to transform the ISM microscope to a confocal microscope. The way to do this is simple: after shifting the "pinhole images" to their right positions, but before adding them up, the "pinhole images" will be masked in such a way that the pixels within a certain distance from the center of the image keep their value and all the other pixels are set to zero. In this way as artificial pinhole with a specific diameter is created and, consequently, the signal from the out-of-focus planes will be set to zero and therefore, after adding-up the images at the right position, the final image will be an optical section of the sample.

Another reconstruction procedure, suggested by Sheppard in 1988 and re-described by Muller and Enderlein in 2010 as ISM, suggests to overcompensate the movement of the laser beam by a factor of 2. This means that a "pinhole image" recorded at laser-beam position (x, y) will be shifted to the position (x', y') with the translation $x' = M_{shift}{}^*M_{opt}{}^*x$ and $y' = M_{shift}{}^*M_{opt}{}^*y$ where $M_{shift} = 2$. This will result in a final image where the objects seem to be a little smeared out compared to the situation where $M_{shift} = 1$. And in fact it is the image of a single point source in the sample that will be smeared out by a factor of $\sqrt{2}$ due to re-positioning of the light to the "wrong" position. However, it has to be realized that the image is now magnified by a factor of 2 (since $M_{shift} = 2$). If we now resize the image by a factor of 1/2 in order to compensate for the extra 2× magnification, the $\sqrt{2}$ "smear" will also be resized and therefore the net result of this reconstruction is that the image is $\sqrt{2}$ sharper than the $M_{shift} = 1$ reconstruction. It is clear that this reconstruction procedure for improved lateral resolution can easily be combined with the artificial pinhole as described above which will lead to an image with both optical sectioning properties and lateral super-resolution properties.

6.3.2 Possibilities and Limitations of ISM

Compared to confocal microscopy this new ISM technology has three important advantages. The most important advantage is that the super-resolution properties that previously only could be achieved with a confocal with an almost closed pinhole can now be achieved with any pinhole size. That means that there is no loss of light in order to gain resolution. The second advantage of ISM is that the image can be acquired by more sensitive detector (EMCCD: QE ≈ 95%, sCMOS: QE ≈ 75%) than the detectors used in the standard confocal (GaASP: QE ≈ 45%,

PMT: QE ≈ 15%). Moreover, the third advantage is as follows: due to improved resolution, the light is focused on a smaller area of the camera chip and therefore there is more light per pixel and consequently a better SNR. These three reasons for improved SNR are important advantages of ISM and later in this chapter we will show that RSM has the same sensitivity advantages.

A drawback of ISM technology is that a "pinhole image" has to be taken for every single pixel of the final image. Although the "pinhole images" can be relatively small (between 32 and 256 pixels is enough; Muller and Enderlein 2010), the total memory capacity needed for storage of a single 2D image is large (16–128 Mb for a single 512*512 pixels image). Moreover, the total acquisition time can be a limiting factor if the camera is not fast enough. Another limiting factor is the speed of data transfer and the time that is needed for image reconstruction. As far as we know, there is only one commercial system that is based on ISM (Airyscan, Zeiss, Jena, Germany). In this system ISM is combined with deconvolution techniques that allows extra improvement of the images. Note that such deconvolution step is time consuming and sensitive for adjustment of settings in the software.

6.4 RCM: The "Optics-Only" Alternative for ISM

Inspired by the publications about the ISM technology (Sheppard 1988, Muller and Enderlein 2010), three alternative methods have been published where the reconstruction procedure is performed on-the-fly by the optics of the set-up itself. Although there are some essential differences, the methods described are all based on simultaneous scanning of the sample and re-scanning of the emitted light from the sample on a camera. York et al. (2013) (Shroff and York 2013) designed and tested an optical set-up called instant SIM (ISIM) where multiple laser spots are scanning the sample in parallel. With multilens arrays and multipinhole arrays they managed to obtain imaging with improved resolution at very high frame-rate (100 fps). In the optical photon re-assignment (OPRA) set-up of Roth et al. (2013) (Sheppard et al. 2013), a single scanning mirror was used for both scanning and re-scanning. The photon re-assignment is established by extra demagnification optics in the optical emission path. In RCM (De Luca et al. 2013) separate scanning mirrors (SM_1 and SM_2 in Figure 6.2) were used for scanning and re-scanning which makes the set-up relatively simple to build and flexible to use. Here, we will describe the principles of RCM and how RCM can be used in biological applications as a normal confocal microscope, but with improved lateral resolution and improved sensitivity properties.

6.4.1 The RCM Set-Up and Principle

RCM is a new fully optical method which overcomes the limitations on acquisition and processing time of ISM as suggested by Sheppard and Enderlein (Sheppard 1988, Muller and Enderlein 2010), while maintaining the advantages in lateral resolution improvement. The RCM consists of a standard confocal microscope equipped with a re-scanning unit for the detection of light. The light emitted by the sample is repositioned on a camera chip by an extra set of scanning mirrors (SM_2 in Figure 6.2). This extra set of mirrors and a simple optical relay system forming the re-scanning unit directs and projects the light on the camera in a scanning way. Since the trigger of the camera is synchronized with

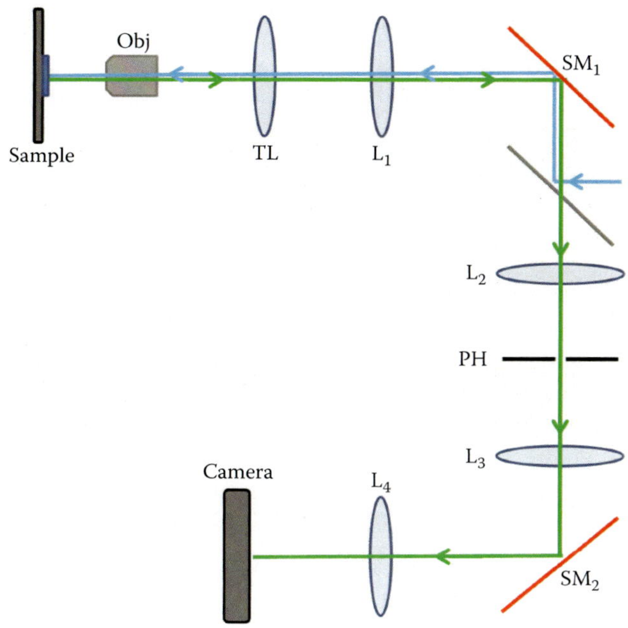

Figure 6.2

The RCM set-up consists of two parts: a standard confocal microscope and a re-scan unit. In the confocal microscope a focused laser is directed to the sample by scanner SM_1 and the emission light from the sample is projected on the pinhole via the same scanner (SM_1). Behind the pinhole of the standard confocal microscope, the RCM has the additional re-scanning unit (SM_2), which directs the light that passes the pinhole on a camera chip. By tuning the sweep factor (related to the angular amplitudes of SM_1 and SM_2) to $M = 2$, the resolution of this microscope is improved by a factor of $\sqrt{2}$.

the start of the scanning process of a single optical section, the camera integrates all the emitted light from the sample and therefore acts as an optical integrator of the "pinhole images" in the ISM technique.

6.4.2 The Sweep Factor *M*

To understand the principle of the improved lateral resolution in RCM, we have to realize that there are two different effects that contribute to the final magnification of the RCM microscope: (1) optical magnification and (2) mechanical magnification. Optical magnification (M_{optical}) is determined by the properties of all lenses along the optical path of the RCM. A first step in magnification is performed by the objective lens (OL in Figure 6.2) and the tube lens (TL). Scan lens L_1 and lens L_2 form a second magnification step and lens L_3 and re-scan-lens L_4 form a final magnification between the pinhole plane and the camera plane. These lenses determine the optical magnification (M_{optical}) of the system from sample to camera.

Another source of magnification, which is very specific for the RCM technology, is the mechanical magnification formed by the ratio of angular amplitudes (sweep) of the scanning mirrors and de-scanning mirrors. The scanning mirrors SM_1 are responsible for the scanning movement of the laser spot through the sample. The angular amplitude (or sweep) of SM_1 is directly coupled to a specific field-of-view in the sample. The same scanning unit SM_1 also has the function of de-scanner that directs the emitted light from sample toward the pinhole. So far, everything is the same as a standard confocal microscope. The second set of scanning mirrors, re-scan unit SM_2, is synchronized with SM_1 and directs the emission light that passed the pinhole on the camera chip in a scanning manner. So, the sweep of SM_2 is directly responsible for the size of the image of the field-of-view projected on the camera chip. The ratio between the field-of-view inside the sample and the image of the field-of-view projected on the camera determines the total magnification of the RCM system (M_{total}) and is the product of optical and mechanical magnification. In conclusion, the mechanical magnification of RCM is defined by the sweep factor M, where $M = M_{total}/M_{optical}$. This sweep factor M in RCM has the same function as the shift factor M_{shift} in the reconstruction procedure in ISM.

6.4.3 Resolution Improvement by RCM

Similar to ISM, where the best resolution improvement is reached when $M_{shift} = 2$, also in RCM the best resolution improvement is established for a sweep factor $M = 2$. This can be understood as follows. When the re-scanning unit "writes" the image on the camera chip with no extra sweep (so $M = 1$), the emission light of objects in the sample is projected at exactly the same place as widefield microscope with the same optical magnification. However, when the sweep factor is set to $M = 2$, the image of the objects will be smeared out over the camera chip. Due to this smearing effect, a point spread function (PSF) would increase by a factor of $\sqrt{2}$. However, since the total magnification is also increased, but increased by a factor of 2 due to the extra sweep, the net effect of this double sweep is that the width of the PSF (W) becomes $\sqrt{2}$ narrower.

The full width at half maximum (FWHM; W) of the final PSF is a function of M as described by

$$W^2 = \left(\frac{(M-1)W_{exc}}{M}\right)^2 + \left(\frac{W_{em}}{M}\right)^2$$

where W_{exc} and W_{em} are the FWHM of the excitation and emission PFS, respectively. This equation confirms that for $M = 1$, the width of the final PSF is only determined by the emission PSF (W_{em}). In this case the image is indistinguishable from an image made by widefield fluorescence microscopy. On the other hand, for $M \rightarrow \infty$, the width of the final PSF is only determined by the excitation PSF (W_{exc}). The width of the RCM PSF (W) reaches a minimum of $1/\sqrt{2}$ at $M \approx 2$. The exact optimal value of M depends on the ratio W_{exc}/W_{em} which depends of the Stokes shift of the fluorophore (Muller and Enderlein 2010). In the approximation of $W_{exc} \approx W_{em}$, (negligible Stokes shift) the best resolution is reached at $M = 2$. In conclusion, the lateral resolution in RCM with $M = 2$ is improved by a factor of $\sqrt{2}$ as compared to confocal microscopy with a pinhole diameter of 1 AU or larger.

In Figure 6.3a two microspheres with a diameter of 100 nm and separated approximately 230 nm from each other are imaged with the RCM with $M = 1$ and $M = 2$. The difference in resolution and resolving power is clearly visible. The diameter of the image of the beads decreases from 240 nm for $M = 1$ to 170 nm for $M = 2$. This experimentally confirms the theoretical $W_{RCM} = (1/\sqrt{2}) * W_{em}$.

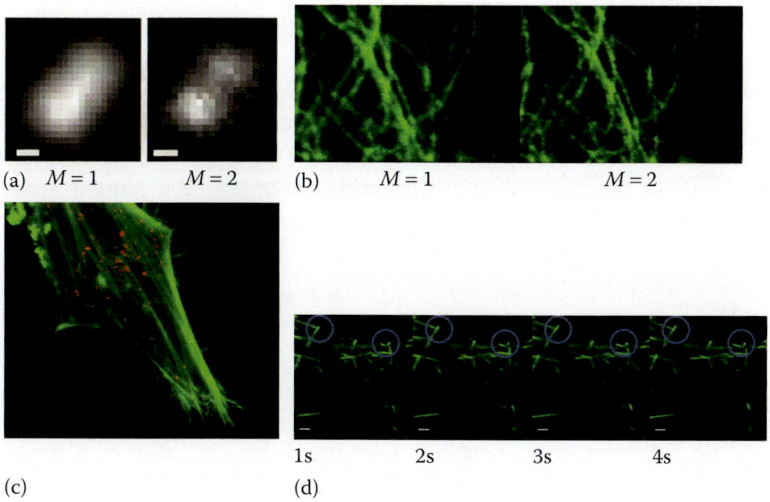

Figure 6.3

The increased resolution by RCM. The image of two 100 nm fluorescent beads (a) cannot be resolved with widefield resolution (RCM with $M = 1$) but are clearly distinguishable with RCM resolution ($M = 2$; scale bar is 100 nm). The measured diameter of the image of the beats decreased from 240 nm ($M = 1$) down to 170 nm ($M = 2$). The image of Huvec cells with Atto 488-labeled microtubules, (b) looks crisper and shows more contrast with RCM resolution ($M = 2$) compared to widefield resolution (RCM with $M = 1$). RCM is also suitable for multicolor applications (c). HeLa cells are labeled for actin (GFP) and the mitochondria (RFP). The combination of extra resolution and high sensitivity offered by RCM is suitable for live-cell imaging. As an example, we show in (d) four consecutive time points from an acquisition of growing ends of microtubules expressing EB3-GFP images at 1 fps. (Scale bar 1 μm.)

6.5 Biological Applications

For some super-resolution techniques the sample has to be labeled with specific fluorescent molecules and embedded in specific buffers (Rust et al. 2006, Henriques and Mhlanga 2009). The RCM does not require any specific sample preparation steps. Any sample that is suitable for standard confocal microscopy is suitable for RCM. The improvement in resolution in a biological sample is shown in Figure 6.3b for Huvec cells with fluorescently labeled microtubules (Atto 488). The image on the right-hand side ($M = 2$) looks crisper and has a better contrast.

Apart from single color imaging, it is also possible to detect two or more colors by alternating the excitation light and/or emission filters in consecutive frames. Figure 6.3c shows a two-color image acquired with RCM with 488 and 561 laser excitation. Actin in HeLa cells was labeled with GFP (green) and mitochondria were labeled with RFP (red).

To show that RCM is also suitable for live cell imaging, an example of four consecutive frames of time series of a cell expressing EB3-GFP which labels the growing end of microtubules is depicted in Figure 6.3d. Here the frame rate is 1 fps but it is possible to increase the frame rate by using resonant scanners or by decreasing the field of view.

6.6 RCM Compared with Confocal Microscopy

It is important to realize that RCM is, first of all, a confocal microscopy technology. The only difference with standard confocal microscopy is that RCM uses a re-scan unit in combination with a sensitive camera behind the pinhole, instead of a "one-pixel" detector such as a PMT or GaAsP detector in the confocal microscope. Although the modification to the confocal microscope is not dramatic, the effect of this relative simple technology is significant.

6.6.1 Improvement of Lateral Resolution

Measurements on fluorescent microbeads have shown that the resolution of RCM is improved by a factor of $\sqrt{2}$ as compared to standard confocal microscopy. The FWHM of the image of 100 nm yellow-green fluorescent beads measured with a 100×, 1.49 NA TIRF oil objective (Nikon) was measured on a confocal microscope and an RCM microscope. For the confocal microscope, the FWHM was 170 nm only when the pinhole diameter was reduced to 0.2 AU. For 1 AU and larger the FWHM was more than 235 nm (see Figure 6.4). Here, we see clearly a difference by a factor of $\sqrt{2}$ as promised by the optical theory of confocal microscopy (Brakenhoff et al. 1979). With the re-scan microscope the FWHM was always 170 nm independent from the diameter of the pinhole.

Since the setting of pinhole diameter does not influence the lateral resolution anymore by applying the RCM technology, it does not make sense to squeeze the pinhole to a diameter smaller than 1 AU (indicated by red crosses in Figure 6.4). This means that in RCM the pinhole diameter is exclusively regulating the sectioning properties. Measurements on fluorescent beads have shown that the optical sections are around 10% thinner in RCM as compared to standard confocal microscopy (De Luca et al. data not shown).

6.6.2 Improvement of SNR

Measurements on both biological samples and test samples (fluorescent plastic or beads) have shown that the SNR in RCM is 2–3 times better than standard confocal microscopy (De Luca et al. data not shown). There are three reasons for this improvement. First of all, it is not necessary in RCM to squeeze the pinhole to a very small diameter in order to obtain the resolution that RCM provides (see red crosses in Figure 6.4). Therefore, precious light is not lost in RCM and consequently the SNR is improved. The second reason is that the cameras used in RCM have a higher quantum efficiency (EMCCD: QE $\approx$ 95%, sCMOS: QE $\approx$ 75%) than the less sensitive "one-pixel" detectors used in confocal microscopy (GaASP: QE $\approx$ 45%, PMT: QE $\approx$ 15%). A third reason for higher SNR in RCM is related to the improved resolution of RCM. Objects that are smaller than the PSF of the microscope will be imaged as a blurry spot. But, since RCM images are $\sqrt{2}$ sharper than confocal images, the area of the blurry spot in RCM will be 2× smaller ($(\sqrt{2})^2$) than the spot in the confocal microscope. Since the light is 2× more concentrated, the SNR will be $\sqrt{2}$ times higher in this spot.

Higher SNR also means that there is the possibility to reduce the light dose for a single image (by reducing laser power or exposure time) which implies reduced phototoxicity. For some biological applications this may even be more important than the improved resolution in RCM.

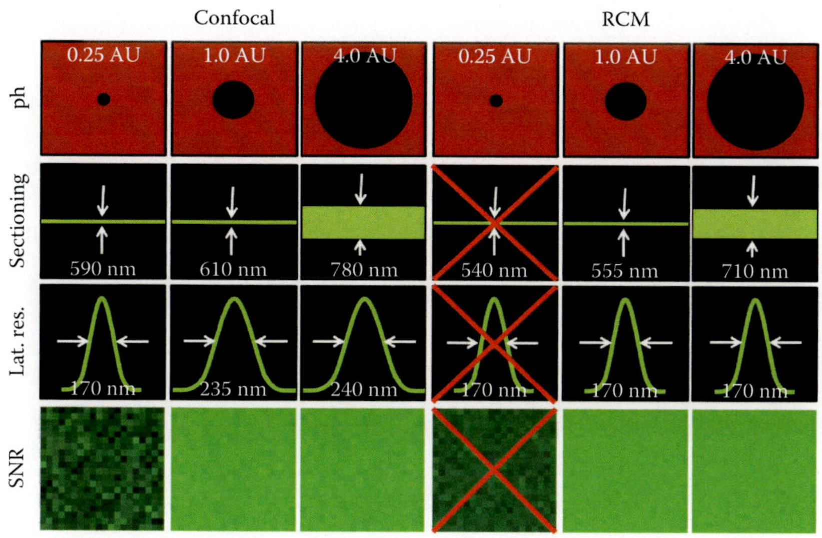

Figure 6.4

Summary of the properties of RCM compared to standard confocal microscopy. In confocal microscopy, closing the pinhole (0.25 AU) helps obtaining good optical sectioning and gives extra lateral resolution (170 nm), but on costs of a good SNR. Many users choose to set the pinhole to 1.0 AU since they prefer a good SNR and are willing to give in lateral resolution (235 nm). A very open pinhole does not significantly change the lateral resolution or the SNR (of the in-focus signal), but gives a thicker optical section. In conclusion, in confocal microscopy the pinhole size has an influence on (1) optical sectioning, (2) lateral resolution, and (3) SNR of the in-focus information. While in confocal microscopy the pinhole causes a compromise between lateral resolution and SNR, in RCM this trade-off is not valid anymore. The lateral resolution is always improved (170 nm) and thus independent from the pinhole diameter. The red crosses through the images at pinhole diameter of 0.25 AU indicating that a very closed pinhole is not advantageous anymore in RCM; the very closed pinholes only decrease SNR and do not improve lateral or axial resolution. Note that, due to a small side effect, the sectioning properties of RCM is around 10% better than in confocal microscopy (De Luca, data not shown). In conclusion, in RCM the pinhole has only an effect on the thickness of the optical section. The lateral resolution of RCM is always good (170 nm) and the SNR is a little better (2–3 times) than standard confocal microscopy. (Numbers given in this figure are based on measurements on fluorescent 100 nm beads.)

6.7 Conclusion and Discussion

The RCM technology is very well comparable to the ISM technique that has become both a technological and a commercial success (Airyscan, Zeiss) due to the combination with deconvolution techniques. Note that also RCM images can be deconvolved with sophisticated software (e.g., Huygens, SVI). By deconvolution the images will be improved; an extra step in both resolution and SNR (data not shown). It is good to realize that deconvolution is an additional step applicable to "optics-only" RCM image.

At the moment of the publication of this chapter, the (re-)inventors of this technology in Amsterdam are making the RCM technology commonly available (www.re-scan.nl) and they are working on further developments. It should be possible to increase the frame rate by using resonance scanners (up to 30 fps). Multicolor imaging for ratio imaging or FRET imaging should be developed and also combination with techniques like fluorescence recovery after photobleaching and life time imaging, two photon microscopy and stimulated emission depletion microscopy are new steps to be taken.

As a conclusion, we underline the improved lateral resolution of the RCM as compared to the conventional confocal microscopy while the sectioning capabilities are maintained. Moreover, since RCM is an "optics-only" method, post-processing of the image (which may introduce artifacts) is not required. Also, the improved SNR and possible reduction of phototoxicity by using more sensitive cameras is another important factor for the quality of the RCM image. Because of these improvements, we expect that RCM is applicable for biological applications where extra resolution and photon economy are an essential part of the imaging strategy.

Bibliography

Brakenhoff, G. J., P. Blom, and P. Barends, Confocal scanning light microscopy with high aperture immersion lenses, *Journal of Microscopy* 117:219–232 (1979).

Cox, G., and C. J. R. Sheppard, Practical limits of resolution in confocal and non-linear microscopy, *Microscopy Research and Technique* 63(1):18–22 (2003).

Cox, J., C. J. R. Sheppard, and T. Wilson, Super-resolution by confocal fluorescent microscopy, *Optik* 60:391–396 (1982).

De Luca, G. M. R., R. M. P. Breedijk, E. M. M. Manders et al., Re-scan confocal microscopy: Scanning twice for better resolution, *Biomedical Optics Express* 4(11):2644–2656 (2013).

De Luca, G. M. R., R. M. P. Breedijk, E. M. M. Manders et al., Re-scan Confocal Microscopy (RCM) improves the resolution of confocal microscopy and increases the sensitivity, *submitted.*

Heintzmann, R., S. Roth, C. Sheppard et al., Verfahren zum optisch hochaufgelösten Rasterscanning eines Objekts, Patent Publication Number DE 102013005927 A1 (2013).

Henriques, R., and M. M. Mhlanga, PALM and STORM: What hides beyond the Rayleigh limit?, *Biotechnology Journal* 4(6):846–857 (2009).

Mc Cutchen, C. W., Superresolution in microscopy and the Abbe resolution limit, *Journal of Optical Society of America* 57(10):1190–1192 (1967).

Muller, C. B., and J. Enderlein, Image scanning microscopy, *Physical Review Letters* 104:198101 (2010).

Roth, S., C. J. R. Sheppard, K. Wicker, and R. Heintzmann, Optical photon reassignment microscopy (OPRA), *Optical Nanoscopy* 2(1):5 (2013).

Rust, M. J., M. Bates, and X. Zhuang, Sub-diffraction-limit imaging by stochastic optical reconstruction microscopy (STORM), *Nature Methods* 3(10):793–795 (2006).

Sheppard, C. J. R., Super-resolution in confocal imaging, *Optik* 80:53–54 (1988).

Sheppard, C. J. R., S. B. Mehta, and R. Heintzmann, Superresolution by image scanning microscopy using pixel reassignment, *Optics Letters* 38(15):2889–2892 (2013).

Sheppard, C. J. R., G. Vicidomini, and A. Diaspro, Image scanning microscopy with small detector arrays, *Focus on Microscopy Proceedings*, 2014.

Sheppard, C. J. R., and T. Wilson, Image formation in scanning microscopes with partially coherent source and detector, *Optica Acta* 25(4):315–325 (1978).

Shroff, H., and G. York, Multi-focal structured illumination microscopy and methods, Patent Publication Number WO2013/126762 A1 (2013).

York, A. G., P. Chandris, H. Shroff et al. Instant super-resolution imaging in live cells and embryos via analog image processing, *Nature Methods* 10(11):1122–1126 (2013).

7

Structured Illumination Microscopy

Sara Abrahamsson, Graeme Ball, Kai Wicker, Rainer Heintzmann, and Lothar Schermelleh

7.1 Introduction
7.2 Super-Resolution Imaging Using SIM
7.3 Different Flavors of SIM
7.4 SIM Imaging in Practice
7.5 Summary
References

7.1 Introduction

Structured illumination microscopy (SIM) is a super-resolution fluorescence microscopy technique that can provide both increased resolution, beyond the diffraction limit of light, and excellent optical sectioning. Several commercial super-resolution SIM systems are available, and are accessible to users at many academic imaging facilities: DeltaVision OMX (by GE/Applied Precision), ELYRA (by Zeiss), and N-SIM (by Nikon). These systems provide many of the advantages of the traditional, widefield, light microscope. Fluorescently labeled fixed or living biological specimens can be imaged with visible light, mounted on a microscope slide under a coverslip or in a glass-bottom cell culture dish. In terms of light dose, SIM is mild on the sample compared to other super-resolution techniques. Standard, linear SIM achieves a factor of two resolution improvement, producing images with around ~100 nm lateral and ~300 nm axial resolution. Even finer details can be resolved when linear SIM is combined with TIRF. Using a nonlinear SIM approach, true super-resolution imaging with theoretically unlimited resolution is possible. There is no hard limit to what resolution can be obtained, but the signal-to-noise ratio (SNR) of the data in practice determines what resolution can be reached (Heintzmann et al. 2002, Gustafsson 2005). The highest resolution nonlinear SIM data reported in biological imaging to this date is ~50 nm (Rego et al. 2012).

7.2 Super-Resolution Imaging Using SIM

7.2.1 The Abbe Diffraction Limit of Resolution

In an optical microscope, resolution is limited by diffraction to around 200 nm when imaging using visible light. This diffraction limit of resolution was originally described by Ernst Abbe (1873). Abbe formulated how the wavelength of the light and the numerical aperture (NA) of the objective fundamentally determine the smallest peak-to-peak dimension ($d_{\min}$) of sample structures that can be optically resolved (Equation 7.1). The NA is defined by the aperture half-angle α over which the objective collects light, and the refractive index n of the immersion medium (Equation 7.2). Since there is a practical limit to the angle over which the microscope objective can collect light from the specimen, there is also a limit to how fine details we can resolve optically.

$$d_{\min} = \frac{\lambda}{(2\,NA)} \tag{7.1}$$

$$NA = n \sin(\alpha) \tag{7.2}$$

As an example, let us calculate the Abbe resolution limit for imaging a sample labeled with green fluorescent protein (GFP) with emission around 530 nm, using an oil immersion objective with $NA = 1.4$. We should here, if all the optics are perfect and aberration-free, be able to resolve details only as small as $d_{\min} = \lambda/(2\ NA) = 530/(2{*}1.4) = 189$ nm. Super-resolution imaging techniques can take us beyond this limit, and obtain images with even higher resolution than what is suggested by the Abbe law of resolution (Figure 7.1). This is possible by circumventing the assumed conditions under which the resolution "law" is valid. Some of these assumptions are uniform illumination, linear processes, and limited NA.

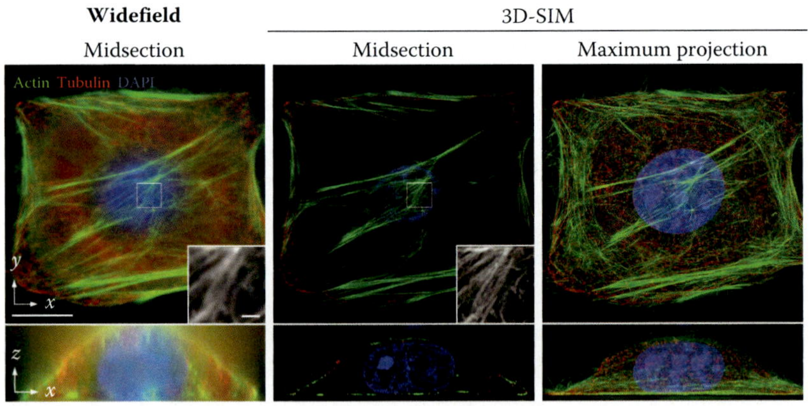

Figure 7.1

Linear three-dimensional structured illumination microscopy (3D-SIM) provides excellent optical sectioning and a factor two increased lateral and axial resolution compared to a regular widefield image. The data set was imaged using the DeltaVision OMX microscope and shows a formaldehyde-fixed mouse epithelial cell labeled with fluorescent Phalloidin (actin, green) antibodies against tubulin (red) and 4′6-diamidino-2-phenylindole (DAPI, blue). The widefield image stack was generated from averaging five phase steps of the first angle. Scale bar = 10 μm (inset 1 μm).

Super-resolution microscopy techniques can be divided into two groups: statistical localization methods, such as PALM/STORM (Shroff et al. 2008), and true imaging methods, such as STED microscopy (Hell and Wichmann 1994); dual objective imaging (Gustafsson et al. 1995); and SIM (Heintzmann and Cremer 1999, Gustafsson 2000, 2005).

7.2.2 How SIM Works

SIM uses nonuniform illumination light to circumvent the diffraction limit of resolution. By merely structuring the fluorescence excitation light that illuminates the specimen through the objective, it is possible to obtain a factor two improvement of resolution compared to regular, widefield imaging. In its true super-resolution form, SIM relies on nonuniform illumination combined with nonlinear processes, such as fluorescence saturation or photoswitching. Theoretically, nonlinear SIM can produce images of arbitrarily high resolution. In practice, the number of photons available from the sample determines the SNR of the raw data and thereby what resolution is achievable (Heintzmann et al. 2002, Gustafsson 2005).

In a typical SIM imaging experiment, the fluorescence excitation light is carefully controlled to form a periodic pattern of fine stripes that illuminate the specimen. During imaging, the pattern of the stripes mixes with the structure of the specimen to produce patterns analogous to Moiré fringes (Figure 7.2). The patterns that arise in the mixing are large enough to be resolvable by the microscope, even though the underlying structure itself is not. In this manner, details in the sample that are too fine to be imaged by the microscope optics (due to the limit of diffraction) are made visible in the form of low-resolution information. Since the stripe pattern is known, the underlying sample structure can be computed from the SIM data using a reconstruction algorithm of choice. In a typical SIM acquisition sequence, multiple images are recorded while the illumination stripes are translated and rotated across the sample. The super-resolution image is then computed from the resulting data set, for example, using the commercial software provided by the manufacturer of the system that was used to record the data or a free, open source software such as the ImageJ plugin fairSIM (Müller et al. 2016).

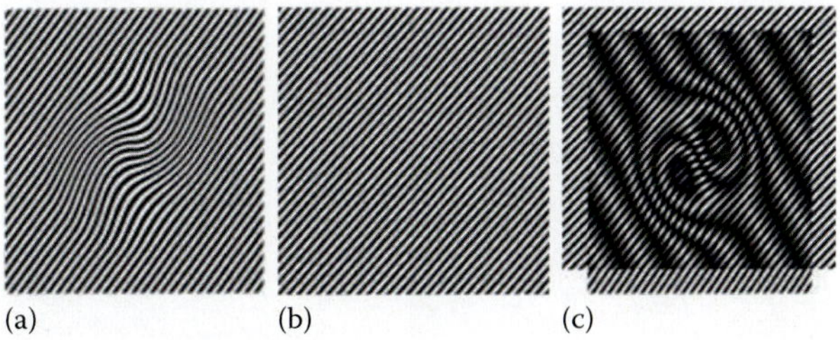

Figure 7.2

Resolution extension through the Moiré effect. If an unknown sample structure (a) is multiplied by a known regular illumination pattern (b), a beat pattern (Moiré fringes) will appear (c). The Moiré fringes occur at the spatial difference frequencies between the pattern frequency and each spatial frequency component of the sample structure and can be coarse enough to be observed through the microscope even if the original unknown pattern is unresolvable. Otherwise-unobservable sample information can be deduced from the fringes and computationally restored. (Figure and text reprinted from Figure 7.1 Gustafsson, M.G.L., *Proc. Natl. Acad. Sci.*, 102, 13081–13086, 2005. with permission. Copyright National Academy of Sciences.)

7.2.3 The SIM Data Acquisition Sequence

To capture and decode the specimen structure using SIM, we need to record a data set where the stripes of the structured illumination pattern are translated across the specimen multiple times. When the stripes are translated in tiny steps across the specimen in the direction orthogonal to the line pattern (this is referred to as "phase" translation), super-resolution information is obtained about the underlying structure in the direction of the translation. To obtain close to isotropic resolution in all directions, the stripe pattern is translated across the sample in several different orientations. Figure 7.3 illustrates the acquisition of a three-dimensional (3D)-SIM data set with five "phase" translations in three directions separated by 60° exemplified on a "lawn" of fluorescent beads.

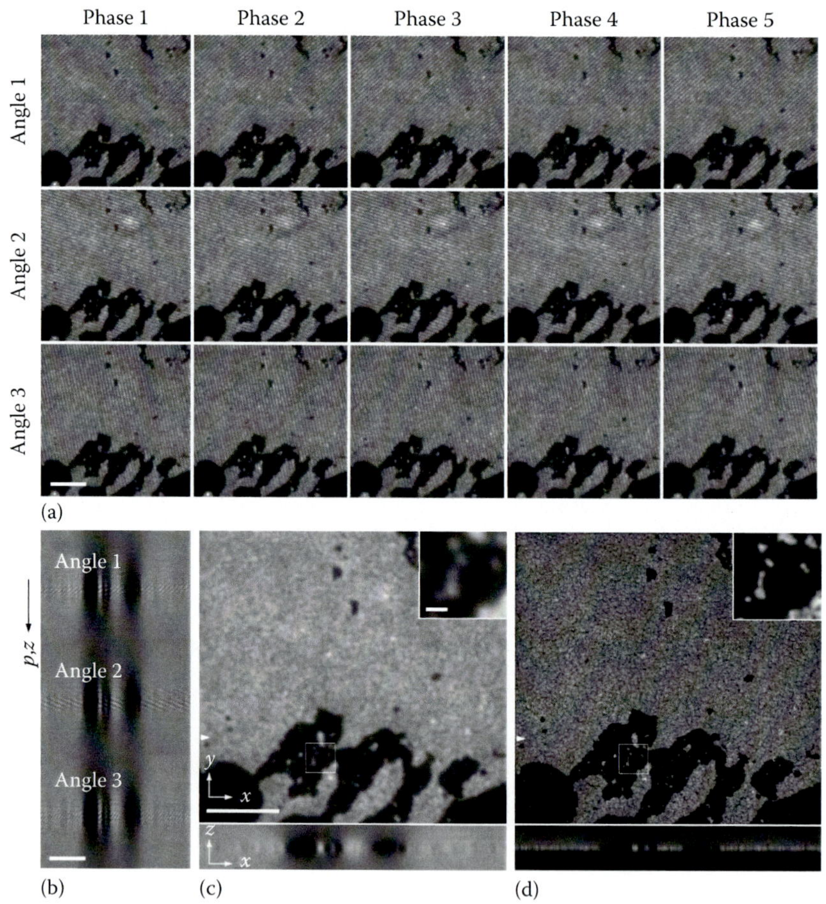

Figure 7.3

SIM data acquisition sequence illustrated with 3D-SIM data set of a lawn of green, 100 nm diameter fluorescent beads. (a) Five phase steps and three orientations of the structured illumination pattern from a single focal plane of the 3D-SIM data acquisition sequence. (b) Orthogonal/axial view of the raw data set in the order of acquisition. (c) Lateral (*xy*) and orthogonal/axial (*xz*) view of a widefield image generated from average projecting the phase-shifted raw images for each plane. (d) Corresponding reconstructed 3D-SIM image visualizing the optical sectioning with resolution increase in all three spatial dimensions. Scale bars = 5 μm (inset 0.5 μm).

7.2.4 SIM Data Reconstruction

7.2.4.1 Fourier Space

Understanding SIM image formation and reconstruction is often more intuitive in Fourier (frequency) space. By frequency, we here mean "spatial frequency," which denotes how small or large the detail in a pattern is. Fine details have high spatial frequency, and coarse details have low spatial frequency. Natural images consist of a distribution of many different spatial frequencies. In the computational world, a representation of Fourier space can be obtained by applying the 2D fast Fourier transform (FFT) to the image. Figure 7.4 shows a photograph (Figure 7.4a) and the strength of its Fourier transform (Figure 7.4b–c) to illustrate this concept. In Fourier space, frequency information is distributed with the lower frequencies toward the center, and higher frequencies toward the periphery. This is apparent when we intentionally blur the photograph (Figure 7.4d) and inspect the Fourier transform of the blurred image (Figure 7.4e–f). Here, the higher spatial frequency information is missing. (All image processing in Figure 7.4 was done using the free, open-source software Fiji/ImageJ, Schindelin et al. 2012.) The mathematical Fourier transform operation in reality describes what a lens does to light. In the microscope, the Fourier plane is the back focal plane of the objective (usually located inside the objective itself). This plane is often referred to as the pupil plane. The extent of the pupil plane (which is determined by the NA) defines the highest spatial frequencies (i.e., the finest details) that can be captured and visualized by the imaging system.

7.2.4.2 SIM Reconstruction in Fourier Space

Figure 7.5 illustrates SIM data reconstruction in Fourier space. SIM data contains higher-frequency information from the specimen than what can be resolved optically. This information has been down-modulated into lower frequencies, that are resolvable by the microscope. In Fourier space, this means that information that belongs further out in Fourier space has been relocated to inside the

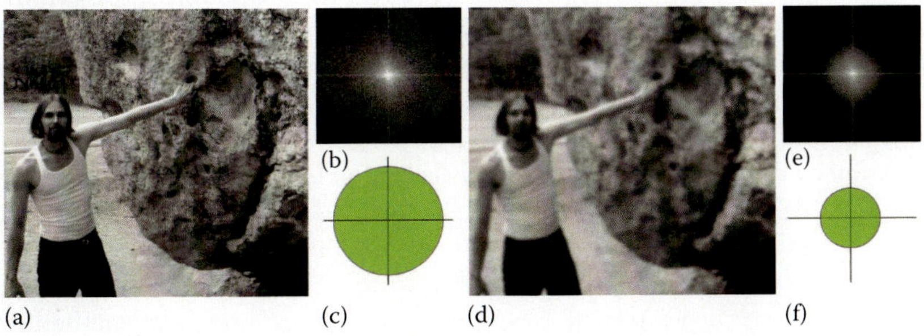

Figure 7.4

Image data and its 2D Fourier transform, illustrating image spatial frequency content and its distribution in Fourier space. (a) Photograph of a rock climber by a boulder at Turtle Lake, Durango, Colorado. (b) 2D fast Fourier transform (FFT) of the photograph in (a) shows the spatial frequency distribution of the image. (c) Conceptual illustration of the extent of spatial frequencies present in Fourier space of the data in (b). (d) Photograph in (a) after applying a Gaussian blur to destroy fine details (i.e., the high spatial frequencies), rendering an image of significantly worse resolution than the original. (e) Fourier transform of (d) shows how the higher frequency components in Fourier space are missing. (f) Conceptual illustration of the extent of spatial frequencies present in Fourier space of the data in (e).

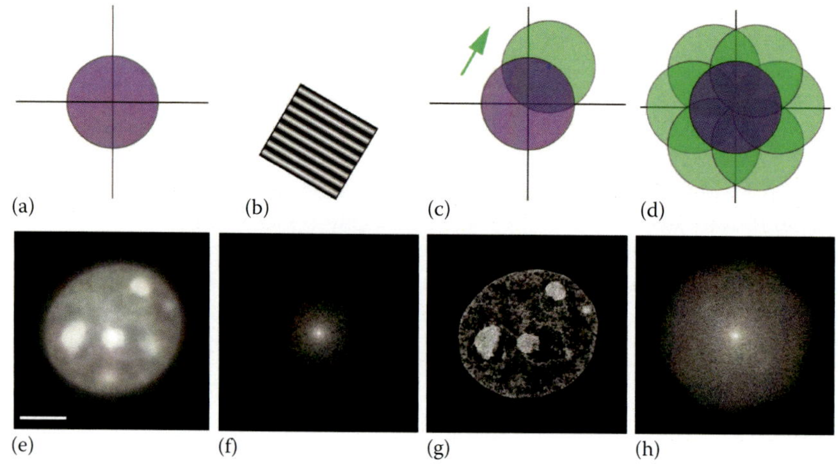

Figure 7.5

Illustration of SIM data reconstruction in Fourier space, conceptually and on a live HeLa cell expressing Histone H2B-GFP. (a) Spatial support region in Fourier space of the microscope, limited by the Abbe law of resolution. In the case of structured illumination, this region contains down-modulated, higher-order spatial frequency information that belongs further out in Fourier space. (b) The illumination pattern in a certain orientation provides information about underlying structures in the sample. (c) The SIM reconstruction algorithm separates the higher-frequency information from the unmodulated and the sum frequency term and repositions it to its proper position in Fourier space in one direction at a time. (d) Reconstructed SIM data in Fourier space with three line pattern orientations. To obtain the final image, all extracted and repositioned components are joined together and the result is Fourier transformed back to real (image) space. (e) Widefield microscopy of the HeLa cell. (f) Fourier transform of the image in (e) shows the spatial frequency content of the data within the support of the optical microscope (corresponding to the purple circle in a). (g) SIM reconstructed image from the reconstructed Fourier space data in (h).

purple circle in Figure 7.5a. The higher-frequency information is present in the data, but located at the "wrong" position. The goal of the SIM reconstruction is to extract this down-modulated information and transfer it to its proper position further out in Fourier space. The individual components of different orders of the illumination line pattern (Figure 7.5b) are computationally separated from the acquired data. They are then shifted (laterally) to their true origin in Fourier space (Figure 7.5c). The final SIM image is obtained by Fourier transforming the reconstructed data from all illumination orientations (Figure 7.5d) back to image space. An illustration of regular widefield and reconstructed SIM image data and their respective Fourier transforms is shown in Figure 7.5e–h. A mathematical treatment of SIM image formation and reconstruction is beyond the scope of this chapter, but can be found in several other works. For those interested, we recommend Gustafsson et al. (2008) and Wicker et al. (2013).

7.3 Different Flavors of SIM

7.3.1 Linear 2D-SIM

In this simplest form of SIM, the illumination pattern is generated by imaging a grating onto the sample with the help of laser illumination, while blocking the zero order. Two laser beams then interfere to form the illumination pattern, namely the

+1 and −1 order diffracted beams from the grating. The beams enter the objective close to the edge of its back aperture. This ensures close to 100% contrast of the illumination structure at the highest possible spatial frequency inside the sample. Linear 2D-SIM data sets typically contain a series of nine images, with three phase translations of the line pattern for three orientations. Linear SIM obtains approximately a factor two improvement of resolution, rendering the finest feature in the sample that can be resolved around 100 nm. The first 2D-SIM experiments were made by Heintzmann and Cremer (1999) and Gustafsson (2000).

7.3.2 Linear 3D-SIM

When SIM is applied to 3D imaging, the resolution along the optical axis is also improved two-fold (to ~300 nm). This is achieved by introducing an axial structure to the illumination light. The 3D-SIM illumination pattern is here generated by 3-beam interference of the +1, 0, and −1 order diffracted laser beams. When the zeroth order from the illumination grating is not suppressed, it causes an axial beating of frequencies, sometimes referred to as the Talbot effect. 3D-SIM was demonstrated by Gustafsson et al. (2008). The technique provides excellent optical sectioning and strong contrast enhancement by removing the out-of-focus "blur" light from surrounding structures in the specimen that can otherwise deteriorate the image (Figure 7.1). This feature is often crucial in biological imaging, where limited contrast can be a bigger problem than limited resolution (Schermelleh et al. 2008).

3D-SIM data sets typically contain a series of 15 images per focal plane, with 5 phases and 3 orientations (Figure 7.3). Implementing fast pattern generation using a liquid crystal spatial light modulator or inferometric galvo-scanning devices, and using fast cameras, these data sets can be recorded on a timescale of seconds. This allows 3D-SIM imaging of slowly moving, living samples with excellent resolution and contrast, as implemented by Shao et al. (2011), Fiolka et al. (2012), and Lesterlin et al. (2014).

7.3.3 Nonlinear SIM with Theoretically Unlimited Resolution

There is a limit to how fine lines we can produce using a linear SIM illumination pattern—the same diffraction barrier that limits how fine details we can resolve in the sample also limits the finest pattern that can be imaged via the objective into the sample. Therefore, the resolution we can obtain using linear SIM is limited to a factor two. By taking advantage of a nonlinear sample response, we can go beyond this limit and produce true super-resolution images. When a nonlinearity (e.g., the saturation effects of state transitions) is present in the interaction between the lines illuminating the sample and the fluorescent signal, the illumination lines (or the gaps between them) in the effective illumination pattern can be made arbitrarily small (Figure 7.6). Mathematically, this generates additional high-frequency harmonics in the Fourier series of sine waves that describes the illumination pattern.

Passing the super fine lines of the nonlinear SIM illumination pattern across the sample in smaller "phase" steps than in the linear SIM case, many overlapping Moiré components can be separated computationally and reassembled into an image of arbitrarily high resolution (Heintzmann et al. 2002).

In Figure 7.8, we show a reproduction of nonlinear SIM data from the original paper by Gustafsson (2005). This paper was first to experimentally show that, given enough photons, even the aforementioned resolution improvement of a factor two can be surpassed, and there is no hard limit of resolution when using

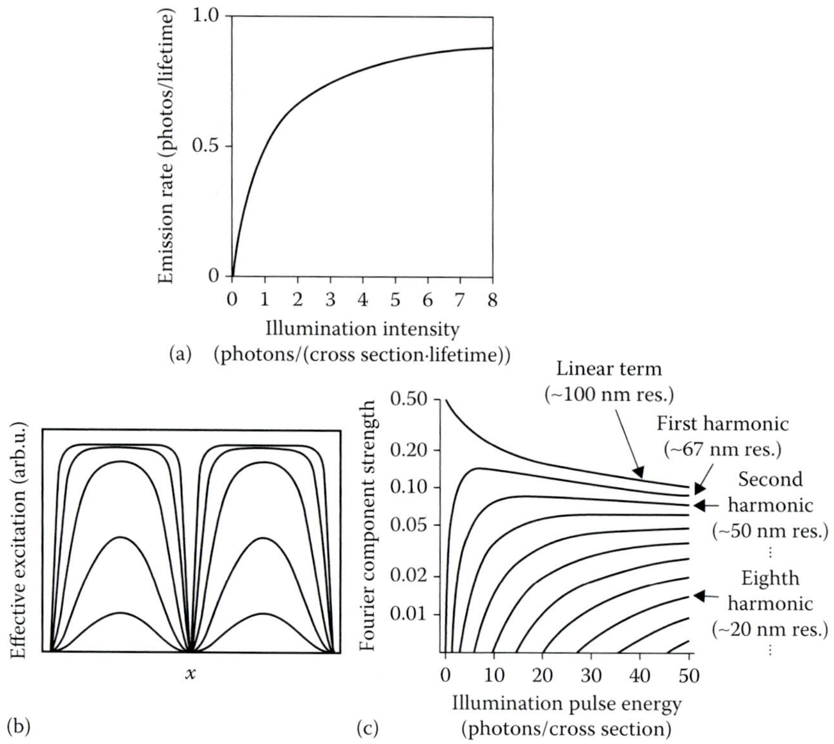

Figure 7.6

Generation of harmonics by nonlinear fluorescence. (a) The nonlinear dependence of the fluorescent emission rate on the illumination intensity in the saturation regime. (b) The emission pattern resulting from sinusoidally patterned illumination with peak pulse energy densities of (from the bottom to top curve) 0.25, 1, 4, 16, and 64 times the saturation threshold. (c) Amplitude of each Fourier component of such emission patterns as a function of the illumination pulse energy (log scale). Equivalently, the curves indicate the fraction of the image signal that stems from each harmonic of the illumination pattern. As the illumination energy increases, more and more harmonics come into play (lower curves). The calculations used the scalar model of light and were for steady-state illumination (a) and for the pulse length of 0.64 ns used in the experiments and a realistic fluorescent lifetime of 3.5 ns. (b and c). (Original figure and figure text reprinted from Figure 3, Gustafsson, M.G.L., *Proc. Natl. Acad. Sci.*, 102, 13081–13086, 2005. with permission. Copyright National Academy of Sciences.)

visible light imaging in widefield fluorescence microscopy. In Gustafsson's experiment, simple fluorescence saturation was used to effectively render finer spatial frequencies, corresponding to higher harmonics, in the illumination pattern. Using three higher harmonics (Figures 7.6 and 7.7), Gustafsson obtained images with <50 nm resolution (Figure 7.8). By using even higher order nonlinearities, the resolution of nonlinear SIM could theoretically be extended arbitrarily. For example, using the eighth harmonic of the sinusoidal stripe pattern should yield 20 nm resolution. Since nonlinear SIM requires additional images in each data set, more light must be collected from the sample to form the image. In practice, the number of photons available from the sample determines the actually obtainable resolution. Nonlinear SIM of biological samples was pioneered by Rego et al. (2012) at <50 nm resolution using photoswitchable fluorophores.

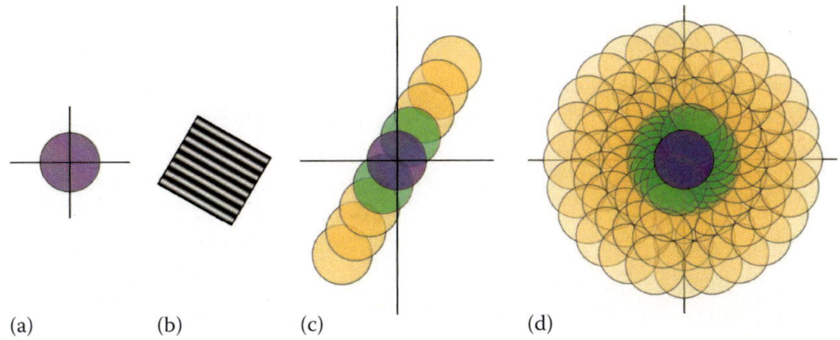

Figure 7.7

When SIM is used in combination with nonlinear processes, resolution can theoretically be unlimited. Nonlinearities arising when the structured illumination pattern interacts with the sample give rise to higher-frequency harmonics, which are used to reveal successively finer details in the sample. SIM image reconstruction for nonlinear SIM using three higher harmonics, as performed by Gustafsson (2005). (a) Spatial information available in microscope by diffraction limit. (b) Structured illumination stripe pattern at 60° orientation. (c) Resolution extension using three higher order harmonics in 60° direction. (d) Complete resolution extension using 24 orientations.

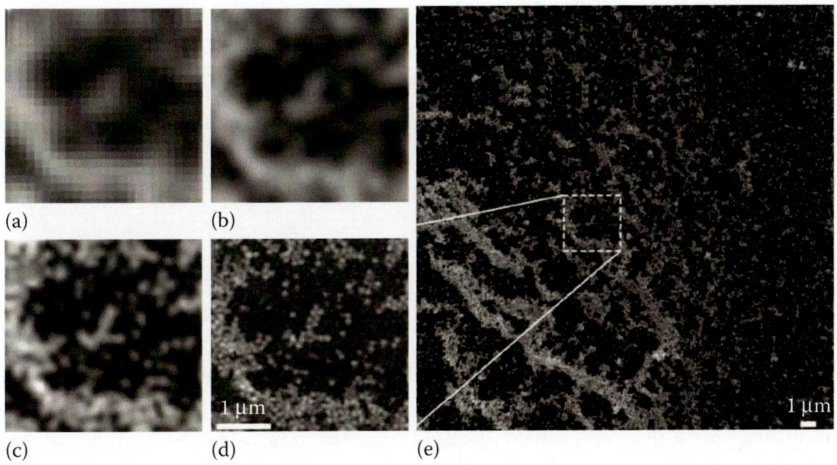

Figure 7.8

Nonlinear SIM is a true super-resolution imaging technique, capable of improving resolution beyond the factor two achieved by linear SIM. This figure is a reproduction from the original paper (Gustafsson, 2005) that demonstrated saturated SIM (SSIM) imaging 50 nm fluorescent beads. (a) Widefield fluorescence image of the bead "lawn." Due to the diffraction limit of resolution, the widefield microscope can only resolve features as fine as around 200 nm. These 50 nm beads can therefore not be individually resolved. (b) Deconvolution of the data in (a) still fails to resolve the beads. (c) Linear SIM imaging of the beads provides a resolution improvement to around 100 nm, which still fails to resolve individual beads. (d) Nonlinear saturated structured illumination microscopy (SSIM) of the beads renders them individually resolvable. (Figure reprinted from Gustafsson, M.G.L., *Proc. Natl. Acad. Sci.*, 102, 13081–13086, 2005. with permission. Copyright National Academy of Sciences.)

7.4 SIM Imaging in Practice

7.4.1 Sample Preparation

When imaging using SIM, special care must be taken with sample preparation to obtain the best possible data for image reconstruction. The sample must be mounted in a medium with the proper refractive index for the objective used and be imaged with the appropriate refractive index immersion oil and the correct thickness coverslip. To target several different molecules of interest, multiple colors can be imaged using SIM. This type of imaging poses an extra challenge when the refractive index does not match identically for different wavelengths.

7.4.2 Data Reconstruction

When analyzing SIM data it is crucial to keep in mind that the image reconstruction can result in artifacts (false images) if the raw data is not good enough, or if the user inputs bad parameters to the software. For SIM reconstruction to be possible, the data must have good modulation in the structured illumination pattern. Photobleaching and sample movement during acquisition must be minimized, since these factors can otherwise deteriorate the data and make reconstruction difficult or impossible. Finally, if improper parameters are given to the reconstruction software, noise can be amplified and create artifacts—false features in the data.

The commercial SIM systems come with software suites for data reconstruction, but even these demand a certain level of knowledge from the user to produce proper results. To assure that the reconstructed image is true, and not corrupted by artifacts, the user must carefully examine the data and the reconstructed image. Fourier transforming the reconstructed image and examining it in Fourier (frequency) space is a first step. This can be easily done using a computer with an image analysis program of choice, such as MATLAB® or one of the free, open-source, platform-independent software suites ImageJ (http://imagej.nih.gov/ij/) and Fiji (http://fiji.sc/Fiji) (Schindelin et al. 2012). A first inspection of the SIM data can simply be done by opening the reconstructed image, applying a 2D-FFT, and inspecting that the result looks reasonable (see the Fourier transform examples in Figures 7.4 and 7.5). A user-friendly and more thorough quality assessment of SIM data can be performed with the ImageJ plugin "SIMcheck" (Ball et al. 2015). The toolset comprises a collection of checks, calibration tools, and image processing utilities, and delivers objective quality control parameter to judge both input and output data quality. Figure 7.9 shows an example of how "SIMcheck" was used to examine 3D-SIM raw and reconstructed data. The "SIMcheck" software and an extensive manual explaining SIM image reconstruction can be downloaded from the Micron Oxford website (http://www.micron.ox.ac.uk).

7.5 Summary

SIM is a super-resolution microscopy technique featuring excellent resolution improvement and optical sectioning with much increased contrast while using a relatively mild light dose. It can be applied to image both living and fixed samples. Different flavors of SIM exist, including true super-resolution methods, which use switchable fluorophores or other nonlinear processes to improve resolution beyond the factor two (~100 nm) of linear SIM. To decide which SIM technique is

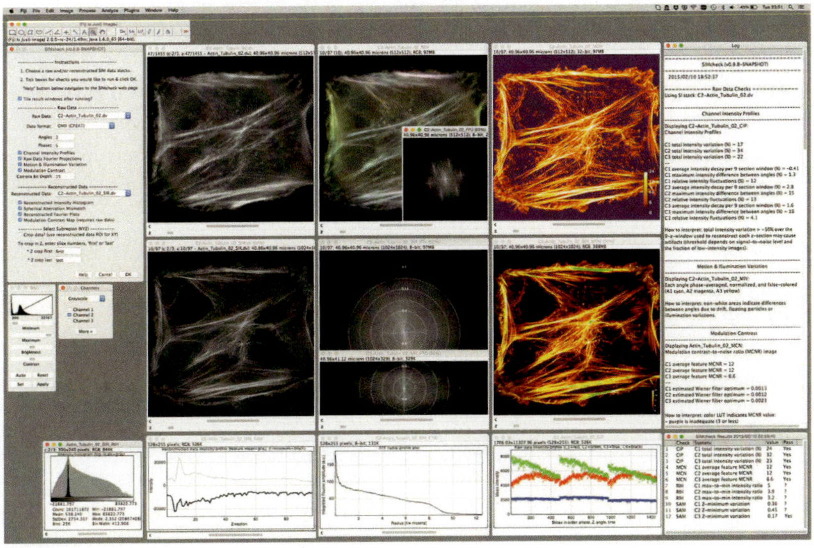

Figure 7.9

Screenshot of the "SIMcheck" plugin suite for ImageJ/Fiji while used to inspect the 3D-SIM data set shown in Figure 7.1.

right for a certain imaging experiment, parameters such as data acquisition time, light dose, and desired resolution extension should be considered. To obtain good data, particular care must be taken with proper sample mounting and refractive index matching. Care must also be taken during data reconstruction, to not produce artifacts and corrupt images. Good protocols for SIM imaging are usually available from the imaging facilities that host commercial SIM microscopes, and the free "SIMcheck" software can aid in verifying both the quality of the raw data and the integrity of the reconstructed images.

References

Abbe, E., Beitrage zur Theorie des Mikroskops und der mikroskopischen Wahrnehmung, *Arch. Mikroskop Anat.*, 9 (1873): 413–420.

Ball, G., Demmerle, J., Kaufmann, R., Davis, I., Dobbie, I. M., and Schermelleh, L. SIMcheck: a Toolbox for Successful Super-resolution Structured Illumination Microscopy, *Sci. Rep.*, 5 (2015): 15915.

Fiolka, R., Shao, L., Rego, E.H., Davidson, M.W., and Gustafsson, M.G.L., Time-lapse two-color 3D imaging of live cells with doubled resolution using structured illumination, *PNAS*, 109 (2012): 5311–5315.

Gustafsson, M.G.L. Surpassing the lateral resolution limit by a factor of two using structured illumination microscopy. *J. Microsc.*, 198 (2000): 82–87.

Gustafsson, M.G.L., Agard, D.A., and Sedat, J.W. Sevenfold improvement of axial resolution in 3D wide-field microscopy using two objective lenses, *Proceedings of SPIE*, 2412 (1995): 147–155.

Gustafsson, M.G.L., Nonlinear structured-illumination microscopy: wide-field fluorescence imaging with theoretically unlimited resolution, *Proc. Natl. Acad. Sci.*, 102(37) (2005): 13081–13086.

Gustafsson, M.G.L., Shao, L., Carlton, P.M., Wang, C.J.R., Golubovskaya, I.N., Cande, W.Z., Agard, D.A., and Sedat, J.W. Three-dimensional resolution doubling in wide-field fluorescence microscopy by structured illumination, *Biophys. J.*, 94 (2008): 4957–4970.

Heintzmann, R. and Cremer, C. Lateral modulated excitation microscopy: improvement of resolution by using a diffraction grating, *Proc. SPIE*, 3568 (1999): 185–196.

Heintzmann, R., Jovin, T. M. and Cremer, C. Saturated patterned excitation microscopy (SPEM)—A novel concept for optical resolution improvement, *J. Opt. Soc. Am. A*, 19 (2002): 1599–1609.

Hell, S.W. and Wichman, J. Breaking the diffraction resolution limit by stimulated emission: Stimulated-emission-depletion fluorescence microscopy, *Opt. Lett.*, 19 (1994): 780–782.

Lesterlin, C., Ball, G., Schermelleh, L., and Sherratt, D.J. RecA bundles mediate homology pairing between distant sisters during DNA break repair, *Nature*, 506 (2014): 249–253.

Müller, M., Mönkemöller, V., Hennig, S., Hübner W., and Huser, T. Open-source image reconstruction of super-resolution structured illumination microscopy data in ImageJ, *Nat. Commun.*, 7 (2016): 10980.

Rego, E.H., Shao, L., Macklin, J.J., Winoto, L., Johansson, G., Kamps-Hughes, N., Davidson, M.W., and Gustafsson, M.G.L. Nonlinear structured-illumination microscopy with a photoswitchable protein reveals cellular structures at 50-nm resolution, *PNAS*, 109 (2012): E135–E143.

Schermelleh, L., Carlton, P.M., Haase, S., Shao, L., Winoto, L., Kner, P., Burke, B., Cardoso, M.C., Agard, D.A., and Gustafsson, M.G.L. Subdiffraction multicolor imaging of the nuclear periphery with 3D structured illumination microscopy, *Science*, 320 (2008): 1332–1336.

Schindelin, J., Arganda-Carreras, I., and Frise, E. et al. Fiji: An open-source platform for biological-image analysis, *Nat. Methods*, 9(7) (2012): 676–682.

Shao, L., Kner, P., Rego, E.H., and Gustafsson, M.G.L. Super-resolution 3D microscopy of live whole cells using structured illumination, *Nat. Methods*, 8 (2011): 1044–1046.

Shroff, H., Galbraith, C.G., Galbraith, J.A., and Betzig, E. Live-cell photoactivated localization microscopy of nanoscale adhesion dynamics, *Nat. Methods*, 5 (2008): 417–423.

Wicker, K., Mandula, O., Best, G., Fiolka, R., and Heintzmann, R. Phase optimization for structured illumination microscopy, *Opt. Express*, 21(2) (2013): 2032–2049.

8

Correlative Nanoscopy

AFM Super-Resolution (STED/STORM)

Jenu Chacko, Francesca Cella Zanacchi, Benjamin Harke, Alberto Diaspro, and Claudio Canale

8.1 Introduction
8.2 STED AFM and STORM AFM
8.3 Nanomanipulation
8.4 Conclusions
References

Correlative nanoscopy is a new term coined from microscopy methods which correlate two or more different techniques that can visualize and study nanometer-scale details of the sample. A super-resolution fluorescence microscope combined with an atomic force microscope (AFM) is an example. These different techniques give different data sets from a sample, defining new functionalities of working and opening new windows for studying the sample. Stimulated emission depletion (STED) microscopy AFM and stochastic optical reconstruction microscopy (STORM) AFM are two additions in to this group and they have proven to be very useful in its unique way of identifying species from an AFM image. These combinations open up new functional modalities like nanomanipulation, fast targeting, and so on. These configurations are discussed and detailed in this chapter.

8.1 Introduction

AFM, introduced in 1986, is the most successful member of the scanning probe microscopy (SPM) family. SPM techniques were born in 1981 with the development of the first scanning tunneling microscope (STM) at IBM in Zürich by Binnig and Rohrer (1982), who won the Nobel Prize in Physics in 1986.

Since its development, the AFM has been extensively applied to study the structural view of biological systems. Because of its ability to work in liquid and in controlled conditions (temperature, pH, ionic strength, etc.), the AFM was the first microscope that provided details at the nanometer scale of biological material in wet/physiological environment. AFM allows the investigation of DNA, proteins, molecular assemblies, and cells through different types of imaging modes (Butt et al., 1990; Canale et al., 2013; Leuba et al., 1994; Messa et al., 2009; Pyne et al., 2014). Butt and coworkers resolved the surface topography of bacteriorhodopsin with a lateral resolution of ≈1.1 nm. Few years later, the same sample was imaged at subnanometer resolution.

The application of the AFM is not limited only to the morphological analysis. The working principle of this instrument is based on the control of the mechanical interaction between the sample and a sharp probe that is scanning on the sample itself. This suggests the use of the AFM as a force sensor or as a nanomanipulator with subnanometer-scale controlled movement capability. Exploiting the ability to discriminate forces in the order of tens of picoNewtons, the unfolding process of proteins and the dynamic force spectrum of molecular bonds was characterized. The unfolding process of the immunoglobulin domain of the giant protein titin was studied and characterized by AFM stretching measurements (Tskhovrebova et al., 1997). The characterization of bacteriorhodopsin molecules was completed by coupling the morphological analysis with a detailed description of the unfolding pathway of the single domain that compose the molecule (Müller et al., 1995).

The mechanical properties of cells, organelles, and membranes have also been widely investigated using AFM (Ferrera et al., 2014; Mescola et al., 2012; Radmacher, 1997; Radmacher et al., 1996). In particular, changes of cell stiffness were demonstrated as related to different diseases, from cancer to neurodegenerative disorders and these new insights were exploited both in the fields of basic investigation of pathological mechanisms as well as in diagnostics (Cross et al., 2007; Ferrera et al., 2014; Mescola et al., 2012). The adhesion of cells on functional substrates or on other cells was quantified at the single cell level, by using single cell force spectroscopy (SCFS).

In spite of this wide range of applications and results obtained, the AFM is inherently limited by the lack of chemical specificity: the AFM is not able to recognize the different components of a heterogeneous sample. This limitation could be severe in crowded molecular environments, such as biomembranes or cells. Several attempts to overcome this limitation were made in the past decades. The first way to confer a recognition capability to the AFM was based on the use of functionalized probes, particularly designed to recognize molecular species of interest (Kamruzzahan et al., 2006). The recognition of a molecule is made possible by looking at the interaction between the tip and the sample mediated by the ligand–receptor binding (Bellani et al., 2014). The limitation of this technique is linked to the extremely time-consuming analysis (thousands of points must be analyzed to obtain a map with a relatively high lateral resolution) and to the presence of a specific tip-sample adhesion. Other few sophisticated applications are based on the coupling of AFM to spectroscopic techniques like Raman spectroscopy, which are able to provide a chemical analysis of the sample. One of the most promising development, tip-enhanced Raman spectroscopy (TERS) demonstrated the capability to obtain chemical map on the sample with 10 nm lateral resolution (Schmid et al., 2013). Although the basic principles of TERS are simple, several practical constraints make this technique challenging in its application, especially working in biological environments.

The coupling between AFM and optical microscopy, in particular with fluorescence microscopy, was one of the first attempts to provide a chemical recognition to AFM by developing an integrated setup. This integration was constrained by the low optical resolution (lateral), limited by diffraction at ≈250 nm; and hence the two instruments worked at different spatial resolutions. The past decade saw the development and diffusion of new classes of optical microscopies that pave the way toward what has been defined as optical nanoscopy. Techniques such as STED, PALM (photoactivatable localization microscopy)/STORM, SIM (structured illumination microscopy), and so on demonstrated the possibility to extend the optical investigation of biological processes down to the nanometer scale. These new super-resolution approaches based on the use of fluorescence or fluorescent molecules have been awarded with the Nobel Prize in Chemistry in 2014.

The coupling between AFM and super-resolution microscopy was intended to solve the resolution mismatch; the lateral resolution of STED and PALM/STORM systems is close to the one that can be achieved with an AFM on a soft biological sample, making the coupling between AFM and super-resolution an ideal platform for correlative microscopy. In this chapter results obtained by coupling the AFM with super-resolution techniques, both STED and STORM, are described and the most important advances offered by the integration are pointed out.

8.1.1 AFM working principles

The working principle of all the SPM family members is based on the reconstruction of the surface topography from the analysis of the point-by-point interaction between a probe and the sample surface. The probe must sense a particular physical quantity. Different probes are used, hence different physical quantity are taken into account in the different types of SPMs: the tunneling current in STM, the evanescent field in scanning near-field optical microscopy (SNOM), the contact forces in AFM, and so on. The probe is generally a sharp stylus; this is a necessary requirement in order to obtain a localized interaction, hence high resolution capability. In the AFM the scanning probe is a sharp tip mounted at the free end of a miniaturized cantilever. The cantilever deflections, induced by the interaction with the sample surface, are measured with an exceptional sensitivity by using a method that is based on the principle of the optical lever (Alexander et al., 1989); a laser pointed at the free side of the cantilever is reflected on a photodetector (Figure 8.1), the bending or torsion of the cantilever induces a change in the position of the laser spot at the photodetector providing an indirect but precise measure of the tip position. The tip movement is measured by a piezoactuator and different solutions are available also depending on the application.

Different kinds of measurements can be done by AFM, but they can be grouped into two main groups: imaging and spectroscopy. In the imaging, the AFM can work in diverse modalities that can be classified in static or dynamic modes and characterized by the working force range: contact (static mode), intermittent contact, and noncontact mode (dynamic modes). In the investigation of soft and not strongly adherent materials, such as most of the biological samples, the use of dynamic modes is dominant since it minimizes the interaction between tip and sample that can result in sample damage or displacement. All these modes have been demonstrated with a combination of super-resolution techniques (Chacko et al., 2013a, 2013b, 2014; Harke et al., 2012).

In spectroscopy mode, the AFM could be employed to test the mechanical properties of a sample measuring the interaction between tip and sample in

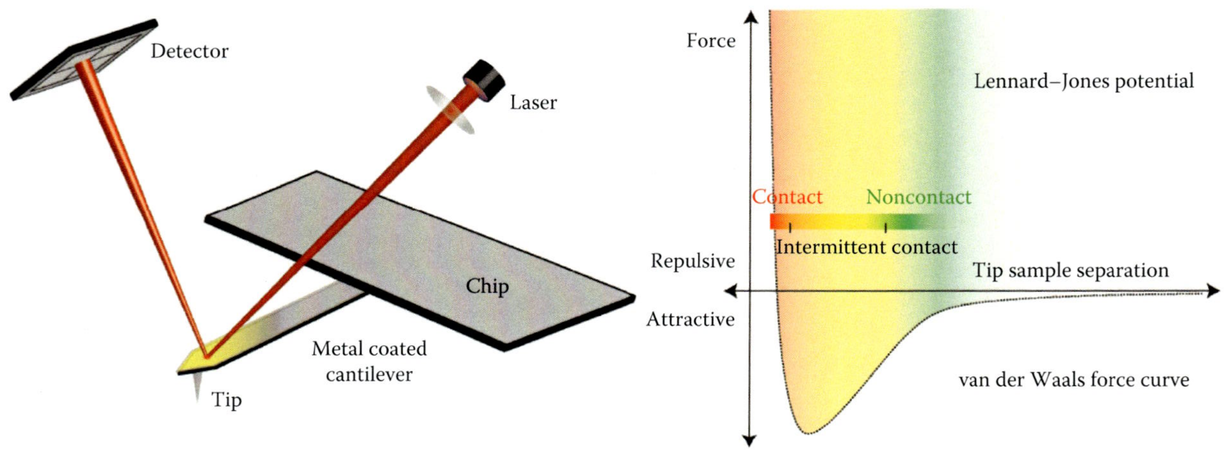

Figure 8.1

Working of AFM: cantilever and tip, representing the core of the AFM, are shown at left. The precise detection of small cantilever deflection is obtained by using a method based on the optical lever principle. A laser is reflected on the backside of the cantilever itself, hitting a quadrant photodiode (QPD). Deflections induce a change of trajectory of the reflected beam that is clearly revealed by the photodetector. The AFM tip is experiencing different force regimes at different distance from the sample. The different AFM modes are working at different regimes of forces.

single points; acquiring force distance (FD) curves. FD curves analysis provides a detailed description of the nature of the interaction between probe and sample: electrostatic interaction, capillary forces, and van der Waals forces can be accurately measured. Quantitative information on sample properties can be obtained by AFM force spectroscopy mode. In particular, the acquisition of a dense matrix of FD curves allows both the topographical and mechanical characterization of the sample with a lateral resolution in the order of tens of nm (Harke et al., 2012). More details on the AFM working principles and modes can be found in literature (Canale et al., 2011; Torre et al., 2011).

8.1.2 Optical

Among the super-resolution techniques used for imaging in the past decade, STED and STORM are widely implemented (Figure 8.2). Both these techniques have been applied in combination with AFM for high-resolution identification of the sample structures which is correlated with the AFM topology.

The key aspect of STED resolution relies on the selective switching off of the fluorescence emission from a conventional diffraction-limited confocal focal volume (Hell and Wichmann, 1994; Westphal et al., 2003). The fluorescence depletion is carried out by a second light source apart from the excitation light, thus inducing stimulated emission from the periphery of focal volume using a special phase mask able to confine light only on the periphery of focus. Increasing the laser power in this phase-engineered "donut-shaped" depletion will provide a higher resolution fluorescence focal spot (Harke et al., 2008). STED microscopy possesses a unique set of advantages compared to its competitors, such as fast image acquisition that enables to study fast dynamics. The last decade provided multitude of live imaging possibilities with STED, for example, vesicle movement *in vivo*, dendritic spine activity, actin dynamics within living brain cultures, and so on. (Nägerl et al., 2008; Urban et al., 2011; Westphal et al., 2008). The versatility provided by fluorescent-based techniques helped STED to be combined with other methods like fluorescence correlation spectroscopy (FCS) (Eggeling et al., 2009), fluorescence lifetime image microscopy (FLIM) (Lenz et al., 2011), fluorescence recovery after photobleaching (FRAP) (Sieber et al., 2007), multicolor imaging (Pellett et al., 2011), time resolved and steady state fluorescence probing of membranes (Mueller et al., 2011) and so on.

On the other hand, localization-based techniques, such as PALM and STORM are based on the sequential acquisition of single molecules. STORM/PALM relies on a single molecule readout routine which can be analyzed to reconstruct a map of all the positions of the localized molecules (Betzig et al., 2006; Hess et al., 2006; Rust et al., 2006). The recording experimental procedure, exploiting the photoactivation process or transitions to the triplet dark state allows to reach the sparse molecular regime and to maximize the photons collected from each single molecule, thus allowing an accurate localization by the Gaussian fit algorithm (Gould et al., 2009). Many methods and routines were developed and are freely available in the scientific community to maximize the performances of the localization process in a STORM/PALM experiment. Multicolor imaging, 3D super-resolution imaging, clustering analysis, hybrid super-resolution techniques based on selective plane illumination were implemented in order to match different needs and applications, separation, and in-depth imaging (Annibale et al., 2011; Diaspro et al., 2014; Xu et al., 2012; Zanacchi et al., 2011).

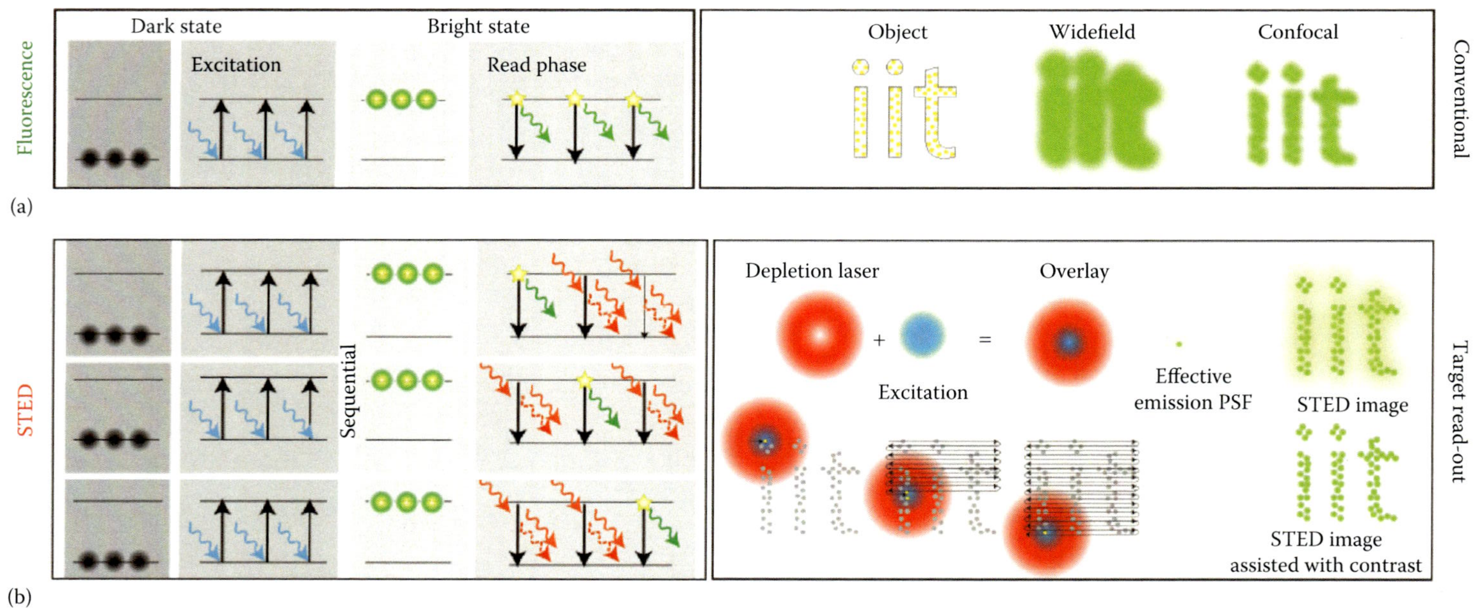

Figure 8.2

Working principle of super-resolution techniques: conventional, STED and localization techniques. The left panel shows molecular level functioning of imaging method. Three molecules are examined for (a) normal fluorescence measurement, (b) STED. (*Continued*)

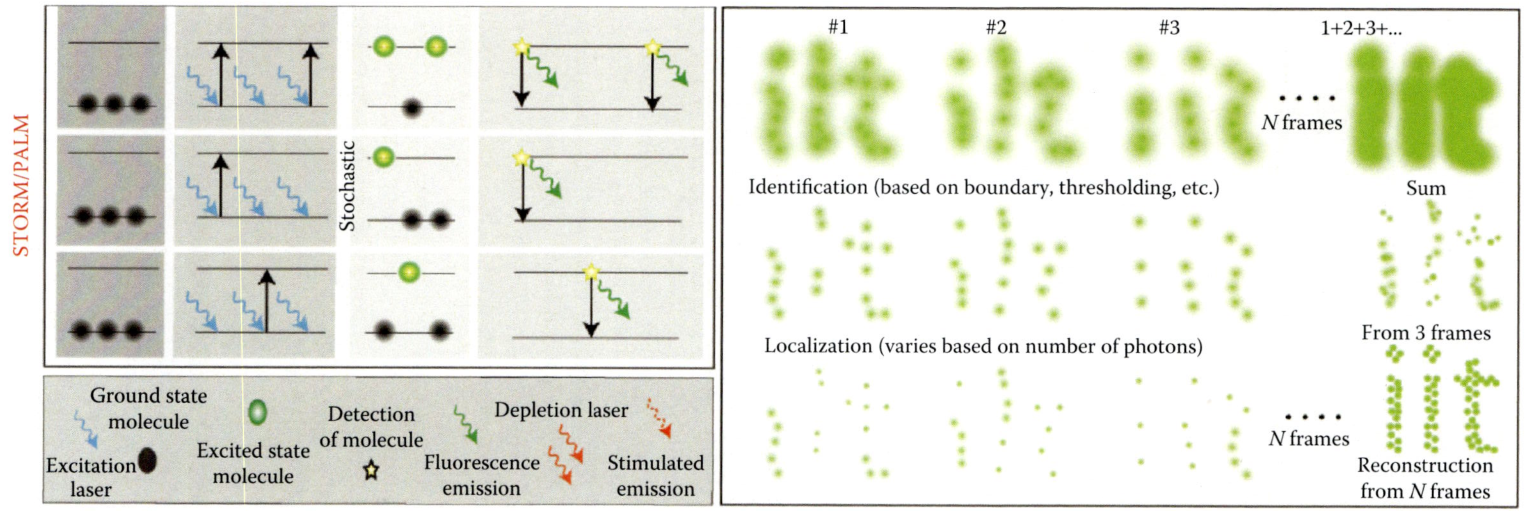

(c)

Figure 8.2

(Continued) Working principle of super-resolution techniques: conventional, STED and localization techniques. The left panel shows molecular level functioning of imaging method. (c) STORM measurements, respectively. Fluorescence is measured when a molecule comes into bright field from the dark field. STED and STORM uses higher number of readouts compared to normal fluorescence, because higher order of sampling is necessary for detailing more. The concept of switching between "on" and "off" states of a molecule can be visualized here. Both these techniques use the principle of switching off a population and create a contrast from a smaller focal volume to make the image.

These two super-resolution microscopy regimes: STED and STORM are shown in the figure above. The left panel shows molecular-level functioning behind the imaging procedure and the right panel shows structural images seen at the user end as the final output of the method. STED uses scanning while STORM uses a widefield or TIRF scheme and depends on stochastic switching off from the molecules. STED imaging can gain a higher contrast with help of several different implementations such as gated STED, mod STED, deconvolution, and so on (Ronzitti et al., 2013; Vicidomini et al., 2013; Zanella et al., 2013). On the other hand, STORM images exhibit an image quality improvement based on the molecular density of labels, better single-molecule recognition algorithms, and increasing the number of photons collected per molecule.

8.2 STED AFM and STORM AFM

One of the ideas that drive the development of correlative techniques in the past few years is that the combination of the results derived from different instruments can define a more accurate view of the sample properties. In particular, the field of microscopy has several correlative methods developed by integrating different imaging techniques or spectroscopic techniques. Trends can be seen in diffusion of correlative light and electron microscopies (Cortese et al., 2009; Sartori et al., 2007), and many other works that combine light microscopy to access more details. Among the microscopy techniques employed with electron microscope or cryoelectron tomography, fluorescence light microscopy provides noninvasive penetrating imaging with analytical and quantitative capability, allowing a three-dimensional investigation also on living cells and organisms. The preparation of the sample for fluorescence is easy and quick, providing the opportunity to look at specific molecular species by choosing appropriate tags. In the case of electron–AFM combination, electron microscopy (EM) provides detailed information on nanometric structures but it is almost impossible to work with live samples. EM must work in an UHV environment and on conductive materials for its best results and these conditions are not compatible with the work on biosystems. Even though the last decades saw the introduction of environmental EMs, these setups are limited in terms of resolution and sensitivity and the limitations to EM application are significant in biological samples. In these lists of techniques that perform imaging at high spatial resolutions and techniques imaging targets in live and physiologically relevant samples, many new correlative approaches have emerged where biomolecules are studied under a fluid environment. The possibility to work on living and dynamic systems will offer a fundamental advantage, which was sought as an important goal. Correlative approaches with AFM and fluorescence microscopy was seen as one of the solutions to this problem because of the techniques' ability to work simultaneously and address questions in live samples (Adams and Czymmek, 2007; Kassies et al., 2005).

STED AFM was one in this series of fluorescence-aided AFM and the first that demonstrated as a potential correlative microscopy/nanoscopy tool using AFM with same lateral resolution (Eifert and Kranz, 2014; Harke et al., 2012). STED microscopes provide fast super-resolution technique because of its inherent direct resolution enhancement which allows the AFM to work in a targeted mode and a high precision. The following schematic in Figure 8.3 explains the working of STED AFM.

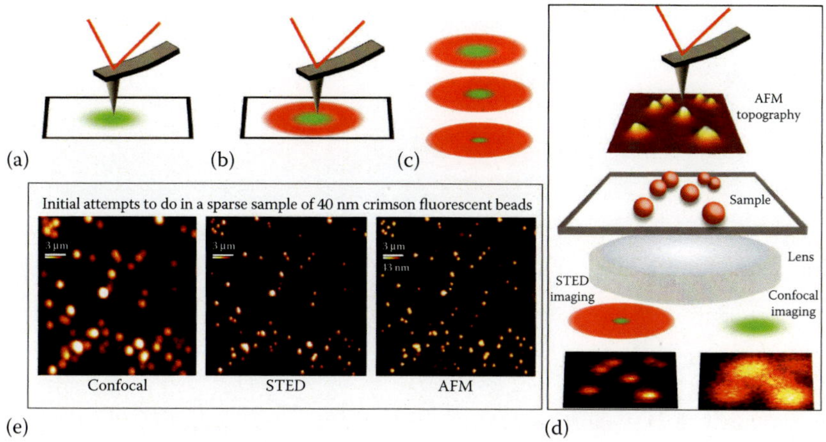

Figure 8.3

Confocal, STED, and AFM images are acquired by scanning the laser/tip on the same area. (a) and (b) shows the AFM tip working with confocal and STED mode, respectively. (c) shows how STED increases the resolution in the focal plane and the AFM tip is unaffected because it is remaining at a different height as seen in (b). 40-nm sized fluorescent bead sample is visualized in (e) by three techniques. STED and AFM images look very similar in their acquisition (e), while confocal is blurred and features appear larger due to limited resolution. (d) shows a detailed schematic and the data collection with the help of a cartoon. The low spatial resolution of confocal microscopy does not allow to distinguish between the different beads that compose the sample while STED can separate the beads and AFM can plot a 3D topological information from the sample (d).

The schematic in Figure 8.3a and b compares confocal AFM against STED AFM in terms of practical realization. Figure 8.3c shows the tunable STED resolution gaining high resolution in 2D on the focal plane. This is shown to demonstrate that the STED effect is limited to 2D and AFM tip at a different height cannot interfere with the STED imaging. Figure 8.3d shows the working of the system where a sample of 40 nm fluorescent beads is imaged by STED and confocal while AFM images from the top give a topology map visualized as a 3D surface. The final results of imaging can be compared for a wide area image of confocal, STED, and AFM in Figure 8.3e. This sample is made of fluorescent beads (crimson, Invitrogen, USA) of 40 nm diameter drop casted on poly-L-lysine coated glass. It must be noted that the lateral size of the beads appear very similar in both the STED and AFM image, which allows for a perfect correlation between the images acquired with the two techniques. The optimized ability to quantify the amount of fluorescence to the physical sizes and structures were investigated further to understand the direct correlation of techniques (Chacko et al., 2013a). The figure below characterizes the ability of STED to resolve and demonstrate the sizes as a quantity like AFM and the results are compared with confocal imaging.

Quantification ability of STED AFM can be elucidated by examining Figure 8.4. The panels show: (a) Confocal, (b) STED, and (c) AFM images of the sample consisting of a mixture of 20 and 40 nm crimson fluorescent beads. The sample shown was prepared by depositing a solution containing crimson fluorescent beads with 20 and 40 nm diameter in a ratio 4:1. The AFM image visualizes the height measured in a similar color scale plotted for fluorescence intensity in STED and confocal images. The images were analyzed with ImageJ for counting the number of

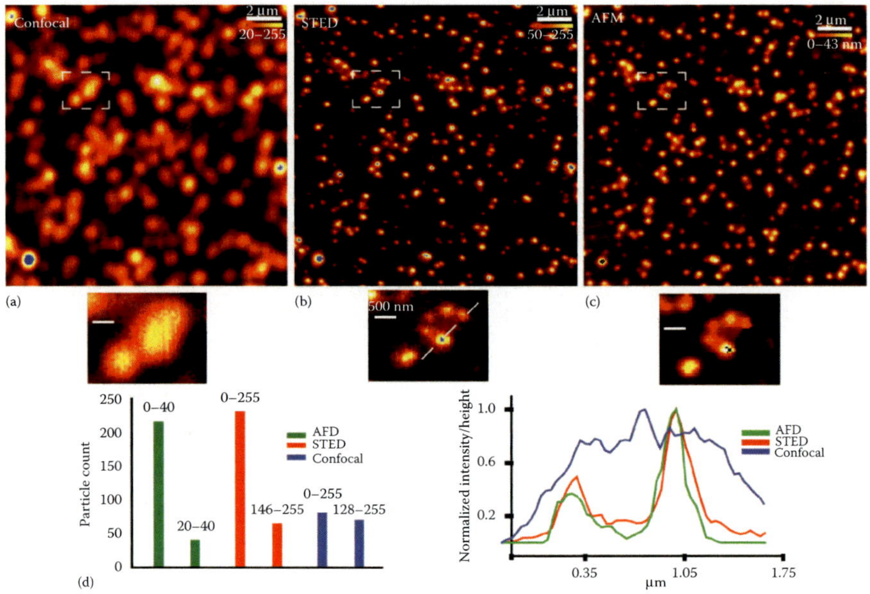

Figure 8.4

STED AFM quantification of nanospheres. Confocal, STED, and AFM images (a–c) acquired on a sample of mixed population of bead sizes 20 and 40 nm. The AFM is able to distinguish beads of different sizes and to count the number of beads within the two populations (d, green graph). Note that AFM uses height as the color map and by applying brightness-based threshold the particles can be counted. Similar result is obtained by thresholding the STED image (d, red) with its brightness as a function of its size. The confocal image is crowded and cannot be interpreted easily (d, Blue). (Modified from Chacko, J.V. et al., *PLoS One*, 8, e66608, 2013a. With permission.)

particles (shown in Figure 8.4d). The resulting graph shows the distinctive population ratio seen in AFM (green) and STED (red) both separated by a brightness size-based counting routine, while confocal is limited due to the high density of beads. Confocal can also achieve the same separation of species if the sample was not crowded because the brightness of the bead mixture helps in separation of species. Hence, it is implied that the resolution plays an important role in crowded samples and the cross-sectional line in the insets are plotted to show the resolution visualized in height image of AFM and brightness of STED while confocal brightness is blurred by conventional laws of diffraction. The counting of two species illustrates the strength of the method to portray the similarity of structure from AFM against the fluorescence emitted by the molecules of the structure.

AFM and STED show similar details in the structure, but in order to get a quantitative idea of the overlay, a side-by-side comparison of confocal or STED image overlapped with an AFM image is shown to get a graphical idea of precision achieved. We convolve the fluorescence image with the AFM image and the resultant distribution of overlap is plotted. The STED image overlap narrows the PSF five times thinner than the confocal correlative image (Figure 8.5).

STORM AFM is another correlative nanoscopy method introduced in 2013 (Chacko et al., 2013b) as a potential method to look into fixed samples that require a molecular-level specificity. It has been applied in the study of cytoskeletal structures, DNA studies, nuclear membrane studies, and so on. STORM AFM has been

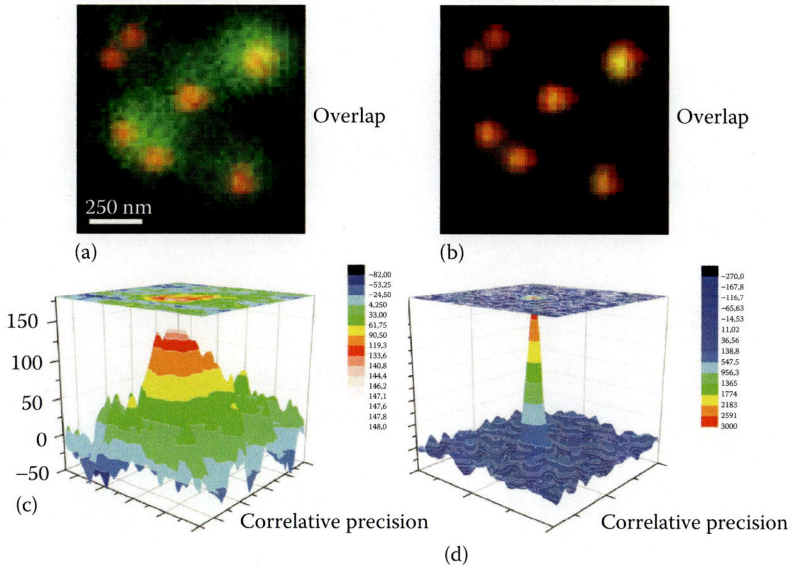

Figure 8.5

Quantification of STED AFM overlay. The image shows (a) confocal AFM overlap plotted with (b) STED AFM image. The AFM image is used to deconvolve the structure of the fluorescence image and the resultant convolution integral is plotted in the graphs in (c) and (d).

applied to the study of many macromolecules because of its ability to visualize the molecule and its structure. Another aspect of a correlative technique is its ability to aid and improve the data from the parallel technique used. Fluorescence validates and confirms the AFM topography and provides specificity to AFM imaging, while AFM topography answers the possibility of other unlabeled features in the structure and avoids errors in labeling and image reconstruction artifacts. An article from Monserrate et al. (2013) which demonstrated that most of the λ-DNA imaged with AFM image correlates with the SR image and ruled out negligible effects of incomplete labeling, rapid photobleaching, poor signal-to-noise ratio leading to exclusion during image analysis, and so forth. This emphasizes the benefits given by the coupling of different techniques which can accomplish different information and reinforce each other. In the line of commercial STORM/PALM microscopes, integration of AFM is a very easy task.

In the context of cellular imaging, the identification of a structure becomes effortful due to its natural heterogeneity. For example, images of cytoskeletal details of cell-like microtubules, actin, and so on with AFM or fluorescence. In order to reduce the ambiguity of the presence of other filaments in a region a multicolor fluorescence staining is necessary. Coloring microtubules and actin in a cell and imaging them both with fluorescence can separate these two major cytoskeletal features out. The other filaments like vimentin, keratin, and similar interfilaments are significantly smaller in size and contributes lesser for an AFM study with low probing forces. An illustration of cellular imaging is shown in Figure 8.6.

Fluorescently labeled F-actin and alpha-tubulin proteins are used to label the cytoskeleton in Figure 8.6. The cytoskeletal structures are more prominent on an AFM image when the image is far from the nucleus and the contrast of cellular features can be observed against the contrast of glass stiffness. A dual

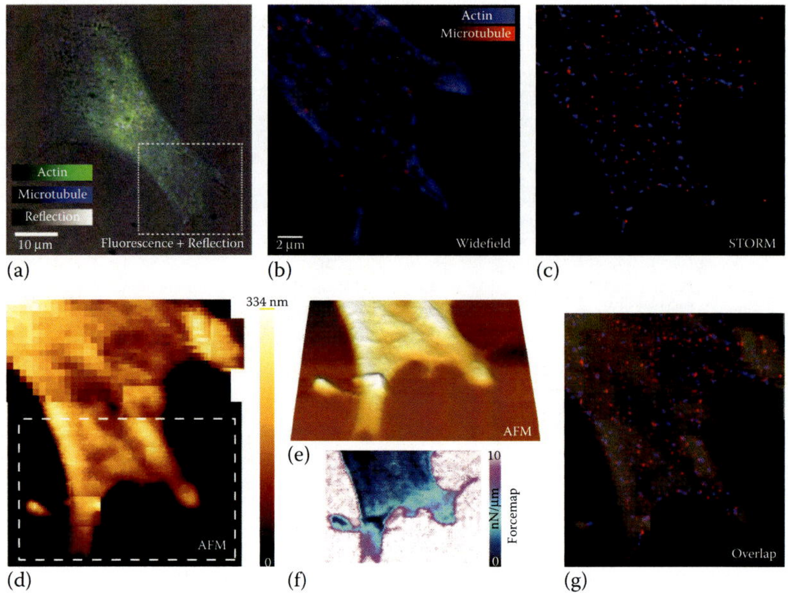

Figure 8.6

STORM AFM cytoskeleton. Actin and microtubules are labeled with two fluorescent dyes and imaged with STORM. (a) shows the widefield image of a cell and the inset shows the selected area of interest. (b) shows actin and microtubules in blue and red as seen by widefield and (c) shows the STORM image of the same area, where molecules are separated. The bottom panels show the (d) AFM modes, (e) the topology map, (f) the force map, (g) a 3D reconstruction, and an overlap of STORM image on the AFM structure to see the topology correlating with the underlying cytoskeletal details.

color STORM image requires careful staining and arduous collection of images. As shown in Figure 8.6a, the widefield image of the whole cell lets one choose an area of interest where widefield and storm images are taken. An AFM image is measured and overlaid to see actin and microtubules as shown in the figure above. Figure 8.6a shows the widefield image overlaid with the reflection channel. Figure 8.6b and c show the widefield and STORM images, respectively, where the red channel shows the microtubules labeled with Alexa 488, while the blue channel shows F-actin labeled with Alexa 647. The actin structure is well organized and forms the cell boundaries while microtubule filaments form a meshwork inside the cell. The AFM image shown in Figure 8.6d was acquired in tapping mode working in PBS, and it is presented coupled with a map of stiffness obtained from the analysis of FD curves acquired on the same area. Figure 8.6g shows the overlaid between AFM and STORM images to appreciate the filaments' structural details. It must be noted that the labeling efficiency is maintained low in these images, but the experiment is carried out as a proof of principle and it serves the purpose of distinguishing the different filaments. The staining protocol is carried out with care so that the membrane is not overcrowded with BSA molecules from the washing buffer used in immune-staining. In particular, the relative labeling efficiency in microtubules seems lesser than in actin. In this experiment, the aim is to prove the possibility to do dual color STORM with an AFM topology to study the underlying cytoskeleton. This is achieved because the localized actin and microtubules can give the structural details that are seen in AFM regardless of their sparse visualization in the fluorescence image.

In the past decades, EM served as one of the most important tools for cytoskeleton study because of its high resolution. Immunogold labeling for specific structures like actin and microtubules were used in EM from 1970s. In case of thick samples, electron-cryo-tomography was helpful in which sample sectioning was possible. Super resolution method aided with AFM offers an improvement to these techniques. AFM can give excellent Z-resolution on the surface with options to study the elasticity, adhesion, and other mechanical characteristics and can provide these results in a controlled liquid environment, offering the possibility to work on living cells or dynamic molecular systems; which is totally lost in EM. In addition, 3D imaging with STORM/STED can help in understanding underlying molecular functioning (living/dynamic systems). Furthermore, correlating single filaments can give information about the morphological effect arising from the underlying individual fibers (Figure 8.6).

From the above experiments, we know that a correlative approach based on coupling far-field fluorescence super resolution techniques with AFM, allows us to perform experiments, which are able to combine topological information, nanomechanical measurements, and specific fluorescence imaging. As a standalone technique, both STED and STORM approaches can be exploited to gain a direct and specific determination of the species with a lateral resolution in the order of tens of nanometers and both these techniques can be coupled with the AFM in order to obtain a correlative imaging method.

AFM imaging is proven difficult when applied to thick and soft samples, such as the central part of a cell. In the thicker part of the cell, AFM can provide only a smoothed cell profile and elasticity, while fluorescence microscopy can browse through the sample by acquiring multiplanar images that can be used for 3D reconstruction.

In a cell probing measurement with AFM, we can notice different contributions to topography and stiffness due to the presence of the rigid filaments that compose the cytoskeleton. These filaments are clearly detectable and we can find a perfect correspondence in the optical image in the peripheral (and thin) part of the cell (Harke et al., 2012).

Similarly, STORM can access to different planes in z with a very good 3D resolution in nanometers with use of astigmatic separation, biplanar separation, and so on (Chacko et al., 2013b). Another crude way to work in a larger axial focal range is by changing the TIRF illumination into epi-illumination or inclined illumination, which will have the drawback of a longer experimental time for localization of the molecules due to its reduced signal-to-noise ratio. STORM can adapt to perform multicolor and 3D imaging without big changes in the microscope setup, while in STED this is a complicated and toilsome optical realignment.

3D and multicolor modalities will play a key role in improving performances of correlative approaches in future, helping to overcome the eventual limitations that can occur when the structures occur in crowded environments. Many tip–probing-based approaches give access to the mechanical properties of cellular structures. The combined approach of AFM with the most advanced super resolution methods provides a golden opportunity for simultaneous topographical and mechanical analysis, with chemical recognition capability with tens of nanometer resolution. Techniques based on correlative approaches will smoothen the way to a new generation of experiments, able to couple topological information, local stiffness measurements, and specific fluorescence imaging close to the molecular level (Figure 8.7).

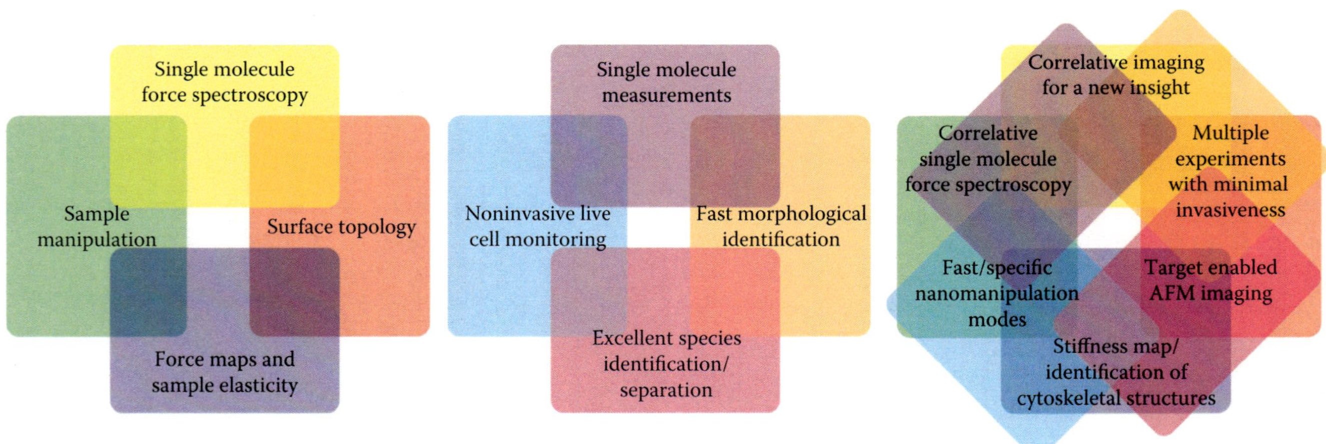

Figure 8.7

Correlative microscopy: An overview. (a) AFM, (b) Super resolution methods, and (c) AFM + SRM.

8.3 Nanomanipulation

AFM has been employed as a manipulator for many decades. The movements of the AFM tip, obtained via a piezoelectric actuator, are controlled at the subnanometric level. Furthermore, the pressure exerted by the tip on the sample can be controlled with a picoNewton precision and maintained constant in time by using a closed-loop system that constantly monitors and corrects the cantilever deflection.

These features make AFM useful as a tool for nanomanipulation. The AFM tip can indent, scratch, move, or push on a sample, inducing modification or stimulation, even to extremely localized areas. Taking advantage of these capabilities, the AFM was employed for lithographic purposes, for example, to pattern an InAs–AlSb surface quantum well by mechanical punching (Rosa et al., 1998). However, more sophisticated approaches were presented later after this. Dip-pen nanolithography (DNP) was used as positive printing methods; molecules that are present in liquid solution at the AFM tip surface are transported on the substrates via capillary forces (Jang et al., 2003). Different molecules, from thiols to proteins, were patterned via DNP (Lee et al., 2002; Piner et al., 1999). Other patterning techniques are based on exploiting of the electric mode of the AFM, local oxidation or organic molecules dissociation, and deposition under the action of an electric field created around the tip apex allow the creation of interesting features that can be exploited, for example, in the fast prototyping of Si masters for nanoimprinting (Lorenzoni et al., 2013). Another technique that allows for the very localized modification of surfaces with the attachment of organic or biomolecules directly *in situ* is nanografting (Liu et al., 2008; Xu and Liu, 1997), that has been used to study the size-dependent Young's modulus of organic thin films or to characterize the orientation and local order of biomolecules (Liu et al., 2002, 2008). Once demonstrated the potential of the AFM in manipulating micro- and nano-sized objects for different purposes, we see some limitations inherently related to the technique. The trajectory of the AFM tip movements must be defined with great accuracy, to do this two ways can be followed: first, the manipulation pathway can be chosen on the base of a previously acquired AFM image, using it as system of coordinate in which a trajectory line is drawn. This approach ensures a high accuracy especially working at the molecular level, since it takes advantage of the extremely high lateral resolution (also subnanometric) and of the exact definition of the tip position. However, there are potential sources of mismatching like thermal drift of the sample, that cannot be neglected, but however it is in the range of few nm/min. Greater limits are present like the time consumed and inability to act on dynamic processes. Another drawback is the lack of chemical specificity of the AFM, which does not allow the selective manipulation of single component that are present in a heterogeneous sample. The AFM is not an ideal technique to image large and extremely soft materials such as cells; the lateral resolution is reduced, while the AFM tip can deform or damage the sample itself, Finally, it must be noted that the effect of the nanomanipulation cannot be monitored in real-time, but just after the acquisition of a successive AFM image (e.g., for the specific AFM used in the experiment, at least 256 s are necessary to acquire an AFM image with 256 scan lines), hence, just plastic deformation or, in general, permanent effects can be distinguished.

For all these reasons, an alternative method able to drive nanomanipulation and to monitor the effect of tip-sample interaction in real-time is necessary. The natural solution has been, from the first attempts, the integration of the AFM with an optical setup. As pointed out in the previous sections, the main drawbacks of this approach are related to the aberrations that affect the optical images and to the resolution limited by diffraction (≈250 nm). Super-resolution optical imaging appeared as a good candidate to overcome these constraints. In particular, within the available super-resolution techniques, STED microscopy is the one that enable fast image acquisition, hence providing the advantage to reduce the experimental time, to work on dynamic systems and, very importantly to monitor in real-time the effect of nanomanipulation. These reasons make STED microscopy to be considered as an ultimate choice to assist nanomanipulation experiment by AFM. Figure 8.8 illustrates a few different targeted movements possible with STED AFM combination.

High precision from STED images can be applied in the manipulation for "multiple movements in a go." Figure 8.8 shows five distinct displacements that can be visualized in the STED, while confocal fails to visualize the movements. For visualization, we paint red channel showing pre-movement images and green shows post-movement images. The individual movements are color coded by their boxes and shown in Figure 8.8c–g. Note the different lengths achieved

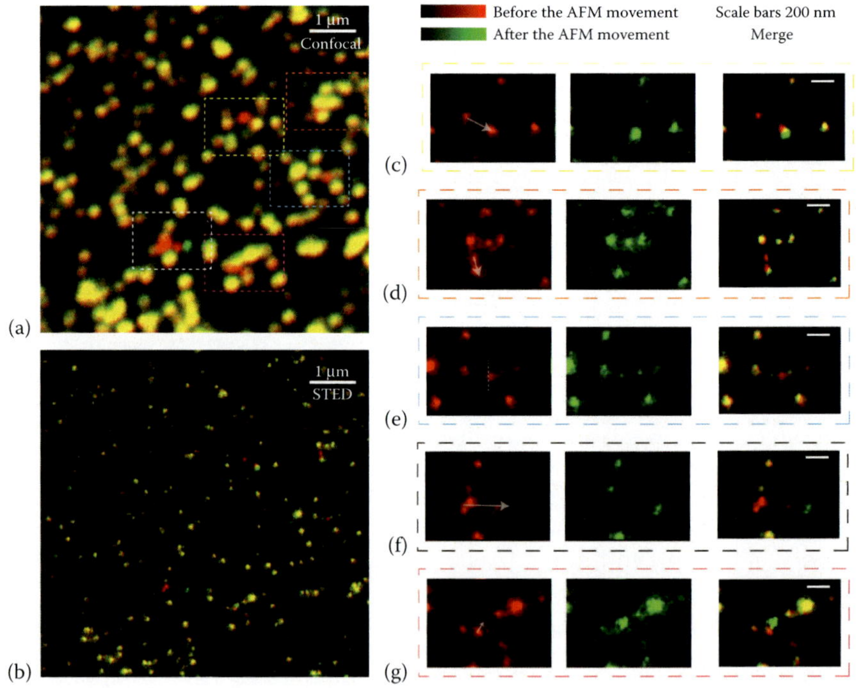

Figure 8.8

Single go measurement: multiple movements in one go. The figure shows confocal and STED images of 40 nm fluorescent beads in (a) and (b). The color channels plotted as red and green is the time axis, so that any movement will be colored as red particle that changed green. (c)–(g) show different types of movements that were assigned to the AFM with help of the STED image as a map for targeting. These panels show different motions like separation, joining, dissecting, and erasing/kicking.

in the movements and in Figure 8.8e, we are able to dissect a cluster of bead into two, while in Figure 8.8f a cluster of beads is displaced as a whole. This experiment demonstrated for the first time that it is possible to drive the AFM manipulation from an optical and fast imaging system having a precision that is comparable with the one obtained choosing the manipulation pathway from the AFM image itself, but speeding up the operation significantly by using light scan instead of an AFM probe scan for preview and monitoring.

In spite of this, the most challenging step for future development was related to the capability, not only to drive the nanomanipulation, but also to follow the effect induced by the AFM tip on the sample in real-time. Chacko et al. tested the capability of their system working on the same fluorescent beads sample. Fluorescent beads are extremely bright and stable, this means that they are scarcely affected by photobleaching; on this kind of sample the maximum resolution is achieved by STED microscopy simply working at the maximum laser power. It is ~200 mW for the system used in the experiment described above. On the other hand, during nanomanipulation the AFM tip is in contact with the sample and in the exact focus position of STED. Laser light is scattered by the AFM tip and at a regime of high laser powers, this can dramatically increase negative effects such as photobleaching or photothermal damage. STED images could be however acquired while the AFM probe is in contact with the sample (hence in the STED focal plane), but only at lower laser power, in particular, Chacko et al. found that the maximum laser power that could be employed during real-time observation of the manipulation was in the order of ~70 mW. The image shown in Figure 8.9 compares the working resolutions for video rate STED AFM imaging (50 mW) and for full resolution available on the setup by retracting the AFM tip (200 mW). The video rate resolution is a little lower, but the system is still able to provide a subdiffraction resolution live recording while the AFM tip is displacing single fluorescent beads. It is clear, looking at Figure 8.9 that the improvement that is achieved by using super-resolution can trace the trajectory followed by the nanospheres is not evident in confocal microscopy were the fluorescence emitted from different beads is often merged to form a unique and undefined spot of emission (Chacko et al., 2013a).

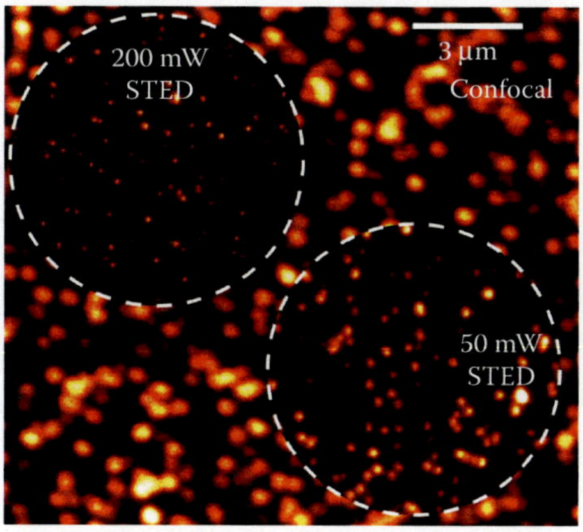

Figure 8.9

STED AFM resolution: comparing the video rate STED AFM resolution with targeted mode full STED resolution.

The improvement of STED with respect to standard confocal microscopy is particularly evident on crowded samples. Biological material, as the interior of a cell, are often characterized by crowded molecular environment, for this reason all the new capability offered by STED assistance to AFM manipulation found their natural application in biological field. AFM was used to induce mechanical or chemical stimuli to cells in its history of usage. The calcium response induced in neurons by a localized mechanical stimulus was studied (Charras and Horton, 2002). Another pioneeristic application of the AFM is in the field of nanosurgery, the idea to influence the faith of a single cell adding on it with a nanometer size chemomechanical probe. Obataya et al. (2005) were able to penetrate a cell with a thin nanofabricated needle and to monitor the penetration by using confocal microscopy, maintaining the cell alive. This method offers the great advantages with respect to standard microcapillary procedures to constantly control the displacement and the force sense by the needle during the interaction with the cell. The AFM was also used for the fine dissection of chromosome, demonstrating how it is possible to create genetic probes for a single specific cell dissecting by AFM a small and specific part of a chromosome (Di Bucchianico et al., 2011). A series of very intriguing applications have been proposed from 2009 with the development of the fluid-FM system by the group of Tomaso Zambelli at the ETH in Zurich, Switzerland. Fluid-FM is basically an AFM system that uses particular cantilever integrated with a nanofluidic control at the end with a hole at the AFM and that represents a versatile tool for nanoinjection, but also to manipulate objects, cells in particular, capturing them with gentle aspiration and releasing by applying an overpressure (Meister et al., 2009).

These experiments inspired Chacko et al. to move toward precise manipulation in a wet environment and on a biological sample. Figure 8.10 shows a submicrometric displacement of a single microtubule fiber induced by the AFM tip. Fibroblast cells were maintained in PBS and mounted on a coverslip holder

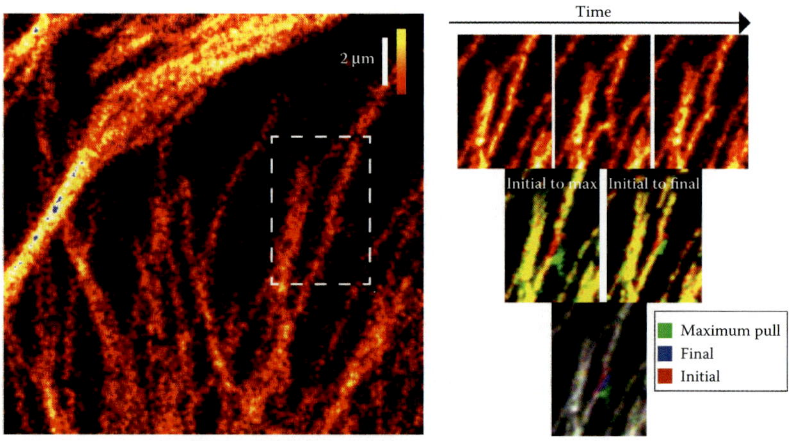

Figure 8.10

AFM precise dissection: The AFM tip is used for the precise dissection of single tubulin filament. Driving the AFM movements on the STED image acquired on the left is possible to pull laterally a single tubulin filament. The continuous acquisition of STED images confirms that the pulled filament is cut as a result of the AFM manipulation.

that ensures the perfect sealing of the sample and that reduces significantly the evaporation of the solvent (provided by JPK Instruments, Berlin, Germany). Microtubulin was labeled with Abberior Star 635P via immunostaining. In Figure 8.10, a STED image is shown where a microtubule was targeted with AFM. The movement was recorded with STED video rate imaging. AFM was able to cut a single microtubule filament and this dissection is shown in the panels. Different merge visualization is given where three frames are merged with the color codes and shown in the last row on the right panel. The white embodies the immobile tubules while red color shows the initial state, green the maximum bent position where it is broken, and blue is the final position where the tubule cannot go back to its initial state. This shows that the filament has broken.

Speed in the optical images acquisition is crucial to monitor in real time the effect of manipulation experiments. The image shows the STED image of cellular region. The smaller panels on the right show the marked area given in the STED image, where a single microtubule is broken by the sheer force applied by the AFM. These frames show merging of three time frames into one for a different visualization. First row shows three time frames. Second row shows the merger of the first and second compared to the second and third. And the last row shows the merger of all three frames. In this last frame, the combination of colors shows no movement and the single colors show the respective movement in that frame.

8.4 Conclusions

The demonstration of the effectiveness of the coupling between AFM and SRM in the study of biological samples opens the way to a new generation of integrated setups able to provide valuable correlative analysis on biosamples. The coupling between AFM and SRM offers unique advantage with respect to old technique, first of all, from the point of view of the information that the technique can provide: the use of the AFM allows the coupling between morphological cues with local mechanics, the SMR supported AFM with chemical recognition capability, and 3D investigation power on thick samples.

However, we think that the most important and pioneeristic applications should arise from the new applications offered by STED-assisted nanomanipulation, we demonstrated how easy it should be to drive precisely the interaction between AFM tip and nanometric size object and we show that this approach is not limited to working in a liquid environment and on soft biological materials. This new capabilities can be readily applied in the field of stimulus–response experiments working in the field of mechanotransduction, also related to the pathological affections of the function of mechanoreceptors. Very exciting is the idea to act on single cells and cell compartments with a nanoscalpel, being able to drive modifications at the single molecule level on living materials.

References

Adams, E.L. and Czymmek, K.J., 2007. The combined application of AFM and LSCM: Changing the way we look at innate immunity. *Imaging*, 68–76.

Alexander, S., Hellemans, L., Marti, O., Schneir, J., Elings, V., Hansma, P.K., Longmire, M., and Gurley, J., 1989. An atomic-resolution atomic-force microscope implemented using an optical lever. *Journal of Applied Physics*, 65 (1), 164–167.

Annibale, P., Vanni, S., Scarselli, M., Rothlisberger, U., and Radenovic, A., 2011. Identification of clustering artifacts in photoactivated localization microscopy. *Nature Methods*, 8 (7), 527–528.

Bellani, S., Mescola, A., Ronzitti, G., Tsushima, H., Tilve, S., Canale, C., Valtorta, F., and Chieregatti, E., 2014. GRP78 clustering at the cell surface of neurons transduces the action of exogenous alpha-synuclein. *Cell Death and Differentiation*, 21, 1971–1983.

Betzig, E., Patterson, G.H., Sougrat, R., Lindwasser, O.W., Olenych, S., Bonifacino, J.S., Davidson, M.W., Lippincott-Schwartz, J., and Hess, H.F., 2006. Imaging intracellular fluorescent proteins at nanometer resolution. *Science*, 313 (5793), 1642–1645.

Binnig, G. and Rohrer, H., 1982. Scanning tunneling microscopy. *Helvetica Physica Acta*, 55, 726–735.

Butt, H.-J., Wolff, E.K., Gould, S.A.C., Dixon Northern, B., Peterson, C.M., and Hansma, P.K., 1990. Imaging cells with the atomic force microscope. *Journal of Structural Biology*, 105 (1), 54–61.

Canale, C., Seghezza, S., Vilasi, S., Carrotta, R., Bulone, D., Diaspro, A., San Biagio, P.L., and Dante, S., 2013. Different effects of Alzheimer's peptide Aβ(1–40) oligomers and fibrils on supported lipid membranes. *Biophysical Chemistry*, 182, 23–29.

Canale, C., Torre, B., Ricci, D., and Braga, P., 2011. Recognizing and avoiding artifacts in atomic force microscopy imaging. In: P.C. Braga, and D. Ricci, eds. *Atomic Force Microscopy in Biomedical Research*. Humana Press, pp. 31–43.

Chacko, J.V., Canale, C., Harke, B., and Diaspro, A., 2013a. Sub-diffraction nano manipulation using STED AFM. *PLoS One*, 8 (6), e66608.

Chacko, J.V., Harke, B., Canale, C., and Diaspro, A., 2014. Cellular level nanomanipulation using atomic force microscope aided with superresolution imaging. *Journal of Biomedical Optics*, 19 (10), 105003–105003.

Chacko, J.V., Zanacchi, F.C., and Diaspro, A., 2013b. Probing cytoskeletal structures by coupling optical superresolution and AFM techniques for a correlative approach. *Cytoskeleton*, 70 (11), 729–740.

Charras, G.T. and Horton, M.A., 2002. Single cell mechanotransduction and its modulation analyzed by atomic force microscope indentation. *Biophysical Journal*, 82 (6), 2970–2981.

Cortese, K., Diaspro, A., and Tacchetti, C., 2009. Advanced correlative light/electron microscopy: Current methods and new developments using Tokuyasu cryosections. *The Journal of Histochemistry and Cytochemistry*, 57 (12), 1103–1112.

Cross, S.E., Jin, Y.-S., Rao, J., and Gimzewski, J.K., 2007. Nanomechanical analysis of cells from cancer patients. *Nature Nanotechnology*, 2 (12), 780–783.

Diaspro, A., Zanacchi, F.C., Bianchini, P., and Vicidomini, G., 2014. Super-resolution fluorescence optical microscopy: Targeted and stochastic read-out approaches. In: F. Benfenati, E.D. Fabrizio, and V. Torre, eds. *Novel Approaches for Single Molecule Activation and Detection*. Springer, Berlin, Germany, pp. 27–43.

Di Bucchianico, S., Poma, A.M., Giardi, M.F., Di Leandro, L., Valle, F., Biscarini, F., and Botti, D., 2011. Atomic force microscope nanolithography on chromosomes to generate single-cell genetic probes. *Journal of Nanobiotechnology*, 9 (27), 511.

Eggeling, C., Ringemann, C., Medda, R., Schwarzmann, G., Sandhoff, K., Polyakova, S., Belov, V.N. et al., 2009. Direct observation of the nanoscale dynamics of membrane lipids in a living cell. *Nature*, 457 (7233), 1159–1162.

Eifert, A. and Kranz, C., 2014. Hyphenating atomic force microscopy. *Analytical Chemistry*, 86 (11), 5190–5200.

Ferrera, D., Canale, C., Marotta, R., Mazzaro, N., Gritti, M., Mazzanti, M., Capellari, S., Cortelli, P., and Gasparini, L., 2014. Lamin B1 overexpression increases nuclear rigidity in autosomal dominant leukodystrophy fibroblasts. *The FASEB Journal*, 28 (9), 3906–3918.

Gould, T.J., Verkhusha, V.V., and Hess, S.T., 2009. Imaging biological structures with fluorescence photoactivation localization microscopy. *Nature Protocols*, 4 (3), 291–308.

Harke, B., Chacko, J.V., Haschke, H., Canale, C., and Diaspro, A., 2012. A novel nanoscopic tool by combining AFM with STED microscopy. *Optical Nanoscopy*, 1 (3).

Harke, B., Keller, J., Ullal, C.K., Westphal, V., Schönle, A., and Hell, S.W., 2008. Resolution scaling in STED microscopy. *Optics Express*, 16 (6), 4154–4162.

Hell, S.W. and Wichmann, J., 1994. Breaking the diffraction resolution limit by stimulated emission: Stimulated-emission-depletion fluorescence microscopy. *Optics Letters*, 19 (11), 780–782.

Hess, S.T., Girirajan, T.P.K., and Mason, M.D., 2006. Ultra-high resolution imaging by fluorescence photoactivation localization microscopy. *Biophysical Journal*, 91 (11), 4258–4272.

Jang, J., Schatz, G.C., and Ratner, M.A., 2003. Capillary force on a nanoscale tip in dip-pen nanolithography. *Physical Review Letters*, 90 (15), 156104.

Kamruzzahan, A.S.M., Ebner, A., Wildling, L., Kienberger, F., Riener, C.K., Hahn, C.D., Pollheimer, P.D. et al., 2006. Antibody linking to atomic force microscope tips via disulfide bond formation. *Bioconjugate Chemistry*, 17 (6), 1473–1481.

Kassies, R., Van der Werf, K.O., Lenferink, A., Hunter, C.N., Olsen, J.D., Subramaniam, V., and Otto, C., 2005. Combined AFM and confocal fluorescence microscope for applications in bio-nanotechnology. *Journal of Microscopy*, 217 (1), 109–116.

Lee, K.-B., Park, S.-J., Mirkin, C.A., Smith, J.C., and Mrksich, M., 2002. Protein nanoarrays generated by dip-pen nanolithography. *Science (New York, NY)*, 295 (5560), 1702–1705.

Lenz, M.O., Brown, A.C.N., Auksorius, E., Davis, D.M., Dunsby, C., Neil, M.A.A., and French, P.M.W., 2011. A STED-FLIM microscope applied to imaging the natural killer cell immune synapse. *Proc. SPIE* 7903, Multiphoton Microscopy in the Biomedical Sciences XI, 79032D, doi:10.1117/12.875018.

Leuba, S.H., Yang, G., Robert, C., Samori, B., van Holde, K., Zlatanova, J., and Bustamante, C., 1994. Three-dimensional structure of extended chromatin fibers as revealed by tapping-mode scanning force microscopy. *Proceedings of the National Academy of Sciences of the United States of America*, 91 (24), 11621–11625.

Liu, M., Amro, N.A., Chow, C.S., and Liu, G., 2002. Production of nanostructures of DNA on surfaces. *Nano Letters*, 2 (8), 863–867.

Liu, M., Amro, N.A., and Liu, G., 2008. Nanografting for surface physical chemistry. *Annual Review of Physical Chemistry*, 59, 367–386.

Lorenzoni, M., Giugni, A., and Torre, B., 2013. Oxidative and carbonaceous patterning of Si surface in an organic media by scanning probe lithography. *Nanoscale Research Letters*, 8 (1), 75.

Meister, A., Gabi, M., Behr, P., Studer, P., Vörös, J., Niedermann, P., Bitterli, et al., 2009. FluidFM: combining atomic force microscopy and nanofluidics in a universal liquid delivery system for single cell applications and beyond. *Nano Letters*, 9(6), 2501–2507.

Mescola, A., Vella, S., Scotto, M., Gavazzo, P., Canale, C., Diaspro, A., Pagano, A., and Vassalli, M., 2012. Probing cytoskeleton organisation of neuroblastoma cells with single-cell force spectroscopy. *Journal of Molecular Recognition*, 25 (5), 270–277.

Messa, M., Canale, C., Marconi, E., Cingolani, R., Salerno, M., and Benfenati, F., 2009. Growth cone 3-D morphology is modified by distinct micropatterned adhesion substrates. *IEEE Transactions on NanoBioscience*, 8 (2), 161–168.

Monserrate, A., Casado, S., and Flors, C., 2014. Correlative atomic force microscopy and localization-based super-resolution microscopy: Revealing labelling and image reconstruction artefacts. *ChemPhysChem*, 15 (4), 647–650.

Mueller, V., Ringemann, C., Honigmann, A., Schwarzmann, G., Medda, R., Leutenegger, M., Polyakova, S., Belov, V.N., Hell, S.W., and Eggeling, C., 2011. STED nanoscopy reveals molecular details of cholesterol- and cytoskeleton-modulated lipid interactions in living cells. *Biophysical Journal*, 101 (7), 1651–1660.

Müller, D.J., Büldt, G., and Engel, A., 1995. Force-induced conformational change of bacteriorhodopsin. *Journal of Molecular Biology*, 249 (2), 239–243.

Nägerl, U.V., Willig, K.I., Hein, B., Hell, S.W., and Bonhoeffer, T., 2008. Live-cell imaging of dendritic spines by STED microscopy. *Proceedings of the National Academy of Sciences of the United States of America*, 105 (48), 18982–18987.

Obataya, I., Nakamura, C., Han, S., Nakamura, N., and Miyake, J., 2005. Mechanical sensing of the penetration of various nanoneedles into a living cell using atomic force microscopy. *Biosensors and Bioelectronics*, 20 (8), 1652–1655.

Pellett, P.A., Sun, X., Gould, T.J., Rothman, J.E., Xu, M.-Q., Corrêa, I.R., and Bewersdorf, J., 2011. Two-color STED microscopy in living cells. *Biomedical Optics Express*, 2 (8), 2364–2371.

Piner, R.D., Zhu, J., Xu, F., Hong, S., and Mirkin, C.A., 1999. 'Dip-Pen' nanolithography. *Science*, 283 (5402), 661–663.

Pyne, A., Thompson, R., Leung, C., Roy, D., and Hoogenboom, B.W., 2014. Single-molecule reconstruction of oligonucleotide secondary structure by atomic force microscopy. *Small (Weinheim an der Bergstrasse, Germany)*, 10 (16), 3257–3261.

Radmacher, M., 1997. Measuring the elastic properties of biological samples with the AFM. *IEEE Engineering in Medicine and Biology Magazine*, 16 (2), 47–57.

Radmacher, M., Fritz, M., Kacher, C.M., Cleveland, J.P., and Hansma, P.K., 1996. Measuring the viscoelastic properties of human platelets with the atomic force microscope. *Biophysical Journal*, 70 (1), 556–567.

Ronzitti, E., Harke, B., and Diaspro, A., 2013. Frequency dependent detection in a STED microscope using modulated excitation light. *Optics Express*, 21 (1), 210–219.

Rosa, J.C., Wendel, M., Lorenz, H., Kotthaus, J.P., Thomas, M., and Kroemer, H., 1998. Direct patterning of surface quantum wells with an atomic force microscope. *Applied Physics Letters*, 73 (18), 2684–2686.

Rust, M.J., Bates, M., and Zhuang, X., 2006. Sub-diffraction-limit imaging by stochastic optical reconstruction microscopy (STORM). *Nature Methods*, 3 (10), 793–796.

Sartori, A., Gatz, R., Beck, F., Rigort, A., Baumeister, W., and Plitzko, J.M., 2007. Correlative microscopy: Bridging the gap between fluorescence light microscopy and cryo-electron tomography. *Journal of Structural Biology*, 160 (2), 135–145.

Schmid, T., Opilik, L., Blum, C., and Zenobi, R., 2013. Nanoscale chemical imaging using tip-enhanced raman spectroscopy: a critical review. *Angewandte Chemie International Edition*, 52(23), 5940–5954.

Sieber, J.J., Willig, K.I., Kutzner, C., Gerding-Reimers, C., Harke, B., Donnert, G., Rammner, B. et al., 2007. Anatomy and dynamics of a supramolecular membrane protein cluster. *Science*, 317 (5841), 1072–1076.

Torre, B., Canale, C., Ricci, D., and Braga, P., 2011. Measurement methods in atomic force microscopy. In: P.C. Braga, and D. Ricci, eds. *Atomic Force Microscopy in Biomedical Research*. Humana Press, pp. 19–29.

Tskhovrebova, L., Trinick, J., Sleep, J.A., and Simmons, R.M., 1997. Elasticity and unfolding of single molecules of the giant muscle protein titin. *Nature*, 387 (6630), 308–312.

Urban, N.T., Willig, K.I., Hell, S.W., and Nägerl, U.V., 2011. STED nanoscopy of actin dynamics in synapses deep inside living brain slices. *Biophysical Journal*, 101 (5), 1277–1284.

Vicidomini, G., Hernández, I.C., d' Amora, M., Zanacchi, F.C., Bianchini, P., and Diaspro, A., 2013. Gated cw-sted microscopy: A versatile tool for biological nanometer scale investigation. *Methods*, 66 (2), 124–130.

Westphal, V., Kastrup, L., and Hell, S.W., 2003. Lateral resolution of 28 nm (λ/25) in far-field fluorescence microscopy. *Applied Physics B*, 77 (4), 377–380.

Westphal, V., Rizzoli, S.O., Lauterbach, M.A., Kamin, D., Jahn, R., and Hell, S.W., 2008. Video-rate far-field optical nanoscopy dissects synaptic vesicle movement. *Science*, 320 (5873), 246–249.

Xu, K., Babcock, H.P., and Zhuang, X., 2012. Dual-objective STORM reveals three-dimensional filament organization in the actin cytoskeleton. *Nature Methods*, 9 (2), 185–188.

Xu, S. and Liu, G., 1997. Nanometer-scale fabrication by simultaneous nanoshaving and molecular self-assembly. *Langmuir*, 13 (2), 127–129.

Zanacchi, F.C., Lavagnino, Z., Donnorso, M.P., Del Bue, A., Furia, L., Faretta, M., and Diaspro, A., 2011. Live-cell 3D super-resolution imaging in thick biological samples. *Nature Methods*, 8 (12), 1047–1049.

Zanella, R., Zanghirati, G., Cavicchioli, R., Zanni, L., Boccacci, P., Bertero, M., and Vicidomini, G., 2013. Towards real-time image deconvolution: Application to confocal and STED microscopy. *Scientific Reports*, 3, Article number: 2523.

9

Stochastic Optical Fluctuation Imaging

Simon Christoph Stein, Anja Huss, Ingo Gregor, and Jörg Enderlein

9.1 Introduction
9.2 Theoretical Basis
9.3 SOFI Computation Algorithm
9.4 Detector Pixel Size and Fourier Interpolation
9.5 Fourier Reweighing
9.6 Fluctuation-Based PSF Estimation
9.7 Conclusion and Outlook
Acknowledgments
References

9.1 Introduction

Since the invention of stimulated emission depletion (STED) microscopy by Stefan Hell at the beginning of the 1990s [1,2], the field of super-resolution fluorescence microscopy (microscopy beyond Abbe's classical resolution limit) has seen a dramatic development with the invention and refinement of a plethora of powerful techniques and methods, see for example, [3]. From a physics point of view, these methods can be roughly divided into three categories: the first category comprises methods such as STED, ground-state depletion or GSD microscopy [4,5], or nonlinear structured-illumination microscopy (nlSIM) [6], which use the photoswitching or nonlinear response between excitation and fluorescence emission of a large number of molecules within a focal region or structured excitation pattern. The second category comprises methods which are based on the identification and localization of individual molecules within a fluorescence image. The two most prominent examples here are photoactivated localization microscopy (PALM) [7] and stochastic optical reconstruction microscopy (STORM) [8], with its most widely used variant *direct* STORM (*d*STORM) [9–11]. The third and most recent category comprises methods which directly convert the temporal information of a fluorescence signal into an enhanced spatial resolution of imaging. The first method promoting this idea was named dynamic saturation optical microscopy or DSOM [12,13], which proposed to use the temporal dynamics of a fluorescence signal upon triplet state shelving (or similar photophysical shelving into a dark state) for achieving enhanced spatial resolution with a laser scanning confocal microscope. However, similar to nlSIM, DSOM suffers from excessive photobleaching, which limits its practicability. Nonetheless, this general idea of converting temporal dynamics information into an enhanced spatial resolution has been successfully and ingenuously implemented into a fluorescence lifetime-imaging STED microscope to enhance spatial resolution and contrast [14]. A more successful, as compared to DSOM, realization of this idea was the invention of stochastic optical fluctuation imaging or SOFI [15]. In this method, one employs the stochastic temporal intensity fluctuations of emitters for enhancing the spatial resolution of an image. SOFI uses a conventional widefield microscope and does not require any change in hardware or setting. The only requirement is that the microscope has to be able to rapidly record images with high sensitivity. Only after a stack of images is recorded, SOFI evaluates the temporal fluctuations in each pixel of the images and computes a super-resolved final image. SOFI will work with any labeling where the labels exhibit stochastic, statistically independent intensity fluctuations. Typically, this is intrinsically the case for labels such as fluorescence quantum dots (QDs). However, it has been shown that all conventional fluorescence dyes which are suitable for *d*STORM can be used for SOFI, after intensity fluctuations are induced by suitable buffers [16]. Finally, also photoswitchable fluorescent proteins have been successfully used for obtaining high-quality SOFI images in biological samples [17]. Besides enhancing the spatial resolution of imaging, SOFI also provides a very efficient suppression of scattering background and autofluorescence [15], and it endows a widefield microscope with optical sectioning capability, thus allowing to use a widefield microscope for obtaining true three-dimensional images of a sample [18,19]. This chapter gives a comprehensive description of how SOFI works, and provides all the algorithmic details. Moreover, it summarizes recent progress in improving the efficiency and applicability of SOFI.

9.2 Theoretical Basis

Let us start by describing the core idea of SOFI. When imaging a sample with a widefield microscope, each emitter in the sample contributes to the final image on the detector (typically a CCD) with some intensity distribution $U(\mathbf{r}-\mathbf{r}')$ which is called the point spread function (PSF) of the microscope, where $\mathbf{r}$ denotes the position on the camera and $\mathbf{r}'$ the position of an emitter in the sample. The PSF is fully determined by the optical properties of the microscope (in particular by the numerical aperture or N.A. of the used objective), and its size defines the resolving power of the microscope: the smaller the diameter of the PSF, the smaller details can be resolved, because contributions from closely spaced emitters start to overlap, so that any information of the fine details of the emitter distribution is lost. In PALM and STORM, one circumvents this problem by having, in each recorded image, only a very small subpopulation of the labeling molecules in a fluorescent state, such that each emitter can be individually identified in the image, and its center position determined with an accuracy which is much superior to the width of the PSF. SOFI does not require this sparsity of fluorescent emitters and works even at large densities of simultaneously fluorescing emitters. To understand how it works, consider a sample which is labeled with N fluorescing molecules at positions $\mathbf{r}'_j$ and time-dependent brightness $s_j(t)$, $1 \leq j \leq N$. The image which is recorded at any moment in time t is then given by the sum

$$F(\mathbf{r},t) = \sum_{j=1}^{N} U(\mathbf{r}-\mathbf{r}'_j)\epsilon_j s_j(t) \tag{9.1}$$

where:

ϵ_j is the maximum brightness of the jth molecule
$s_j(t)$ describes its (normalized) temporal intensity fluctuations

Let us now assume that the emitting molecules are blinking, for example stochastically switching between a fluorescent and nonfluorescent state, and that this blinking of all molecules is statistically independent from each other. In that case, the temporal second-order cumulant $C_2[s_j(t), s_k(t+\tau)]$ of the fluorescence signals from two molecules has to be zero, or

$$\begin{aligned} C_2\left[s_j(t), s_k(t+\tau)\right] &\equiv \langle \delta s_j(t) \cdot \delta s_k(t+\tau) \rangle \\ &= \delta_{jk}\epsilon_j^2 f_2(\tau) \end{aligned} \tag{9.2}$$

where the angular brackets denote averaging over time t, $\delta s_j(t) = \langle s_j(t) - \langle s_j \rangle \rangle$ is the brightness variance of the jth molecule, δ_{jk} is Kronecker's symbol being one for equal indices and zero otherwise, and $f_2(\tau)$ is a function describing the second-order temporal correlation of the brightness of one molecule. For the sake of simplicity, it is assumed that this correlation function is the same for all molecules, that is, that all molecules behave statistically in the same way. Thus, when now applying this second-order cumulant operation to each pixel of the recorded movie $F(\mathbf{r}, t)$, one obtains the so-called second-order cumulant image,

$$\begin{aligned} C_2\left[F(\mathbf{r},t), F(\mathbf{r},t+\tau)\right] &= \sum_{j,k=1}^{N} U(\mathbf{r}-\mathbf{r}'_j)U(\mathbf{r}-\mathbf{r}'_k)\epsilon_j\epsilon_k \langle \delta s_j(t)\delta s_k(t+\tau)\rangle \\ &= \sum_{j=1}^{N} U^2(\mathbf{r}-\mathbf{r}'_j)\epsilon_j^2 f_2(\tau) \end{aligned} \tag{9.3}$$

where we have used, in the second line, the property of the statistical independence of the intensity fluctuations of different emitters as embodied in Equation 9.2. When inspecting Equation 9.3 one finds that each emitter of the sample contributes to the second-order cumulant image $C_2[F(\mathbf{r}, t), F(\mathbf{r}, t+\tau)]$ proportional to the second power of the PSF, which directly corresponds to an enhanced resolution! For example, assuming that the PSF can be well approximated by a Gaussian distribution with the width σ, the square of this PSF will have a width of $\sigma/\sqrt{2}$, so that the second-order SOFI image directly leads to a resolution enhancement of $\sqrt{2}$. As we will see later, the real gain in resolution is even larger and equal to 2, but only after applying a suitable deconvolution technique. However, Equation 9.3 also shows already one of the challenges of SOFI: each emitter contributes with the square of its brightness to the second-order SOFI image.

The just described simple concept of second-order SOFI can be easily generalized to higher orders. One has to realize that the second-order temporal cumulant, Equation 9.2, of two input signals $s_j(t)$ and $s_k(t)$ is defined in precisely such a way so that it is identically zero if the fluctuations of the two signals are statistically independent. One can generalize this concept to more than two input sources. For example, the third-order cumulant is defined as

$$C_3\left[s_j(t), s_k(t+\tau_1), s_l(t+\tau_2)\right] = \langle \delta s_j(t)\cdot \delta s_k(t+\tau_1)\cdot \delta s_l(t+\tau_2)\rangle \tag{9.4}$$

and the fourth-order cumulant as

$$\begin{aligned} &C_4\left[s_j(t), s_k(t+\tau_1), s_l(t+\tau_2), s_r(t+\tau_3)\right] \\ &= \langle \delta s_j(t)\cdot \delta s_k(t+\tau_1)\cdot \delta s_l(t+\tau_2)\cdot \delta s_r(t+\tau_3)\rangle \\ &\quad - \langle \delta s_j(t)\cdot \delta s_k(t+\tau_1)\rangle\langle \delta s_l(t+\tau_2)\cdot \delta s_r(t+\tau_3)\rangle \\ &\quad - \langle \delta s_j(t)\cdot \delta s_l(t+\tau_2)\rangle\langle \delta s_k(t+\tau_1)\cdot \delta s_r(t+\tau_3)\rangle \\ &\quad - \langle \delta s_j(t)\cdot \delta s_r(t+\tau_3)\rangle\langle \delta s_k(t+\tau_1)\cdot \delta s_l(t+\tau_2)\rangle \end{aligned} \tag{9.5}$$

so that, if the brightness fluctuations of all molecules are statistically independent, one has

$$\begin{aligned} C_3\left[s_j(t), s_k(t+\tau_1), s_l(t+\tau_1+\tau_2)\right] &= \delta_{jkl}\epsilon_j^3 f_3(\tau_1,\tau_2) \\ C_4\left[s_j(t), s_k(t+\tau_1), s_l(t+\tau_1+\tau_2), s_r(t+\tau_1+\tau_2+\tau_3)\right] &= \delta_{jklr}\epsilon_j^4 f_4(\tau_1,\tau_2,\tau_3) \end{aligned} \tag{9.6}$$

where the δ_{jkl} and δ_{jklr} denote Kronecker symbols which are equal to one if all indices are equal and zero otherwise, and the functions $f_3(\tau_1, \tau_2)$ and $f_4(\tau_1, \tau_2, \tau_3)$ are determined by the temporal blinking statistics of the emitters. Thus, when applying the third- or fourth-order cumulant operation to the recorded fluorescence images, one obtains, analogously to the second-order cumulant, the result

$$\begin{aligned} &C_3\left[F(\mathbf{r},t), F(\mathbf{r},t+\tau_1), F(\mathbf{r},t+\tau_1+\tau_2)\right] = \sum_{j=1}^{N} U^3(\mathbf{r}-\mathbf{r}_j')\epsilon_j^3 f_3(\tau_1,\tau_2) \\ &C_4\left[F(\mathbf{r},t), F(\mathbf{r},t+\tau_1), F(\mathbf{r},t+\tau_1+\tau_2), F(\mathbf{r},t+\tau_1+\tau_2+\tau_3)\right] = \\ &\sum_{j=1}^{N} U^4(\mathbf{r}-\mathbf{r}_j')\epsilon_j^4 f_4(\tau_1,\tau_2,\tau_3) \end{aligned} \tag{9.7}$$

The last equations directly demonstrate that each emitter contributes to an nth-order cumulant image with the nth power of its PSF, thus directly enhancing the image resolution by at least a factor of $\sqrt{n}$. For extracting all the information contained in the recorded images, it remains to integrate the cumulant images over the correlation times τ_k which then yields the definition of the final SOFI images S_n:

$$S_2(\mathbf{r}) = \left(\int d\tau_1 f_2(\tau_1)\right) \sum_{j=1}^{N} U^2(\mathbf{r}-\mathbf{r}'_j)\epsilon_j^2$$

$$S_3(\mathbf{r}) = \left(\int\int d\tau_1 d\tau_2 f_3(\tau_1,\tau_2)\right) \sum_{j=1}^{N} U^3(\mathbf{r}-\mathbf{r}'_j)\epsilon_j^3 \tag{9.8}$$

$$S_4(\mathbf{r}) = \left(\int\int\int d\tau_1 d\tau_2 d\tau_3 f_4(\tau_1,\tau_2,\tau_3)\right) \sum_{j=1}^{N} U^4(\mathbf{r}-\mathbf{r}'_j)\epsilon_j^4$$

In principle, this idea can be extended to arbitrary orders, so that one finds in general

$$S_n(\mathbf{r}) \propto \sum_{j=1}^{N} U^n(\mathbf{r}-\mathbf{r}'_j)\epsilon_j^n \tag{9.9}$$

and in Reference [15], SOFI on QDs was pushed up to the 25th order. Thus, by calculating higher and higher orders of SOFI, one can, in principle, obtain images with unlimited spatial resolution. In practice, however, the achievable order of SOFI will be limited by measurement time. All presented equations for SOFI are exact only in the limit of an infinite measurement time (infinitely long averages). The higher the order of a SOFI image, the longer one has to measure so that the temporal averages do sufficiently well approach their infinity time values. Another challenge is the already mentioned brightness skewing: each emitter contributes with the nth power of its brightness to an nth order SOFI image. Thus, the higher the order n of a SOFI image, the more it will be dominated by the brightest emitters, and contributions from dimmer emitters will be disproportionately suppressed.

9.3 SOFI Computation Algorithm

In principle, SOFI is straightforward to implement and use. First, one records a temporal stack of images $F(\mathbf{r}, t)$ (movie), where the only requirement is that the image acquisition speed is faster than the typical timescale of the intensity fluctuations of the emitters in the studied sample. For calculating an nth order SOFI image, one would then have to calculate the cumulant images using Equation 9.3 or 9.6 or similar higher-order cumulant equations. One would have to do this for *all* possible correlation times $\{\tau_j\}$, $1 \leq j \leq n$, and then to sum the results over all values of $\{\tau_j\}$ with $1 \leq \tau_j \leq \tau_{max}$ for all j, as shown in Equation 9.8, where τ_{max} is some arbitrary maximum correlation time which has to be chosen large enough so that all nth-order intensity fluctuation correlations of an emitter have decayed to zero. Values with $\tau_j = 0$ have to be excluded because they only contribute shot noise but no real information about the temporal intensity fluctuations of the emitters.

However, the just described approach is computationally prohibitive. For example, for a fourth-order SOFI image with a max correlation time τ_j of, let's say, 16 frames, one would have to calculate 560 cumulant images, which have then to be summed up for obtaining the final SOFI image. However, one can adopt a simplified computational scheme which is taken from fluorescence correlation spectroscopy (FCS) where one often encounters the task to compute a correlation curve from asynchronous single-photon counting data [20]. In this SOFI implementation, one starts with calculating the nth-order cumulant image by setting all τ-values to one, $\tau_1 = \tau_2 = ... = \tau_{n-1} = 1$,

$$C_n^{(1)}(\mathbf{r}) = C_n[F(\mathbf{r},t), F(\mathbf{r},t+1),\ldots, F(\mathbf{r},t+n-1)]. \quad (9.10)$$

In a next step, one generates a new stack of images $F^{(2)}(\mathbf{r}, t)$ by binning every two subsequent frames into a single frame, that is,

$$F^{(2)}(\mathbf{r},t/2) = F(\mathbf{r},t) + F(\mathbf{r},t+1). \quad (9.11)$$

With this new stack of frames with coarse-grained temporal resolution, one again calculates a cumulant image with all lag times τ_j equal to one,

$$C_n^{(2)}(\mathbf{r}) = C_n[F^{(2)}(\mathbf{r},t), F^{(2)}(\mathbf{r},t+1),\ldots, F^{(2)}(\mathbf{r},t+n-1)]. \quad (9.12)$$

This procedure is repeated k times until the coarse-grained time, 2^k times the frame time, is by approximately an order of magnitude larger than the typical correlation time of the intensity fluctuations, so that all the information contained in these fluctuations is captured. The final SOFI image is then calculated as the simple sum

$$S_n(\mathbf{r}) = \sum_k C_n^{(k)}(\mathbf{r}). \quad (9.13)$$

The whole procedure is visualized in Figure 9.1. As an example, for calculating a fourth-order SOFI image with a maximum correlation time of 16 frames using the just described cascaded-time approach, one has to calculate and sum only 4 cumulant images, in contrast to 560 images with the primary algorithm. Of course, the coarse-graining algorithm will not capture all possible correlation information. However, comparisons between both approaches on real data have shown that the difference in SOFI image quality between both approaches is in most cases imperceptible.

Last but not least, let us finish this section by briefly addressing photobleaching, which is a common problem in fluorescence microscopy. Because SOFI relies

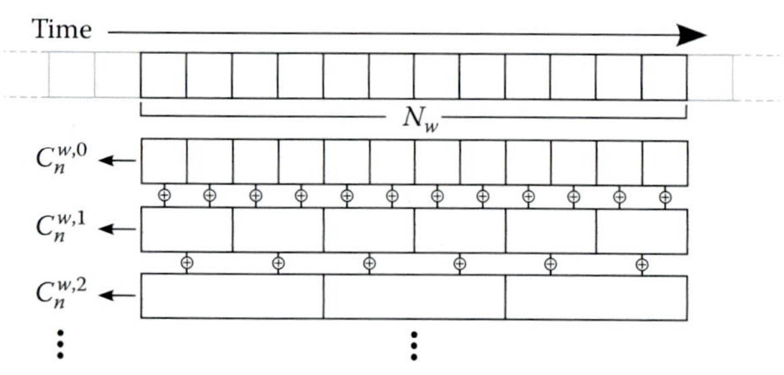

Figure 9.1

Hierarchical SOFI algorithm: For each window w with N_w frames multiple cumulant images are computed in a hierarchical way, where the time resolution is coarsened by factor 2 in each level. The sum of the cumulants of all levels approximates the integral over τ of C_n^w with exponentially growing bin size.

on the *stationarity* of the fluorescence fluctuations of the sample, it is sensitive to photobleaching artifacts. To prevent photobleaching from affecting a final SOFI image, it is recommended to divide the full stack of recorded images into substacks of N_w frames, see Figure 9.1, which has to cover a time span longer than the typical correlation time of the emitters' intensity fluctuations, but has to be much shorter than the characteristic photobleaching time. The SOFI algorithm should then be applied to each substack of the N_w frames separately, so that photobleaching will not skew the SOFI analysis. Finally, the SOFI images of all substacks can be added together to obtain the final SOFI image. This assures that the cumulant analysis of each substack is basically unaffected by photobleaching.

9.4 Detector Pixel Size and Fourier Interpolation

An important issue of SOFI is the impact of the finite pixel size of the widefield detector (e.g., EMCCD) recording the images. As was explained in section 9.2, SOFI allows, in principle, achieving arbitrarily high spatial resolution, which is only limited by the order of cumulants used and by the measurement time required for obtaining sufficiently well cumulant results. However, the raw material on which the SOFI algorithm is applied to are frames with pixels of finite size. How can one thus achieve a spatial resolution of, for example, 50 nm in a final SOFI image if the raw images with which one starts have a pixelation of, for example, 100 nm? (Figures 9.2 through 9.4).

A solution to this problem was first proposed by Dertinger et al. [21], and later perfected by Geissbuehler et al. [22]. The core idea is to generate virtual pixels interspersed between the physical pixels of the recording camera by cross-correlating signals from the original pixel grid. This approach allows indeed to generate SOFI images with arbitrarily small pixel size, which is not limited by the pixel size of the recorded raw images. However, there are several disadvantages connected with this idea. First, the SOFI brightness values of the generated virtual pixels in the final SOFI image are different from the SOFI brightness values of the primary pixels, and even between different *types* of virtual pixels, the brightness values are different. Thus, for obtaining an unbiased final SOFI image, one has to apply sophisticated brightness re-calibration procedures, which heavily

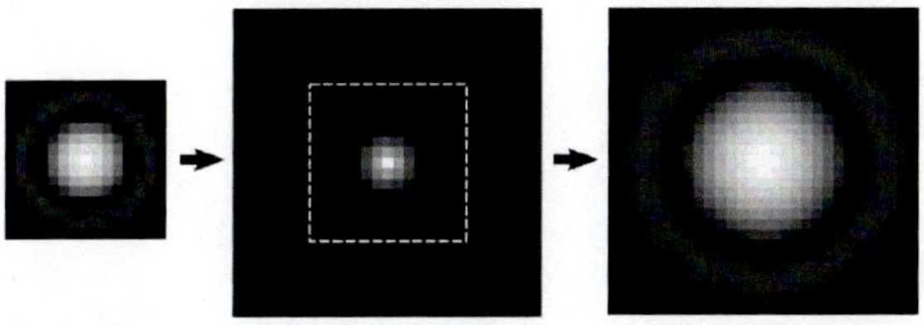

Figure 9.2

Algorithm of Fourier interpolation: Each movie frame (e.g., single emitter represented by an airy disc) is first Fourier transformed. For sufficiently small detector pixel sizes, the Fourier transform is zero on the borders due to finite support of the OTF. The Fourier transform can thus be padded with zeros without changing the frequency information, here separated by a dashed white line from the original Fourier transform. Transforming back into real space gives an artifact-free image with more pixels, where each "virtual" pixel corresponds to a smaller area than that of the original detector pixels.

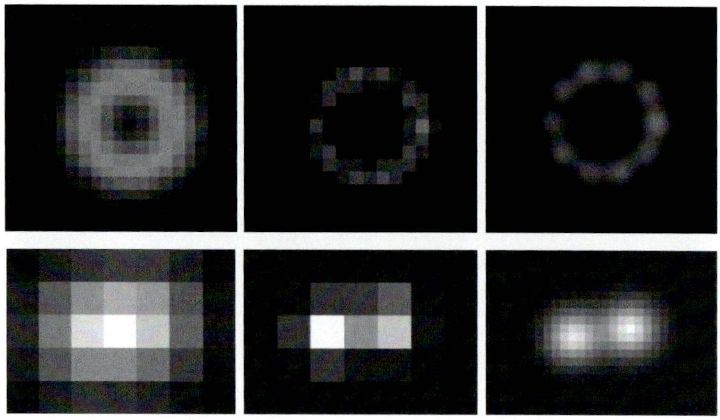

(a) (b) (c)

Figure 9.3

Two examples from simulated movies illustrating the presented Fourier interpolation in combination with SOFI. Top panels: 10 emitters in a ring. Bottom panels: Two emitters with subpixel shift (1.75 px, 0.25 px). (a) Time average of original movie. (b) Fourth-order SOFI from original movie. (c) Fourth-order SOFI from 4x-supersampled movie. It is easy to see that the Fourier interpolation improves image quality and exposes the subpixel positioning of the emitters.

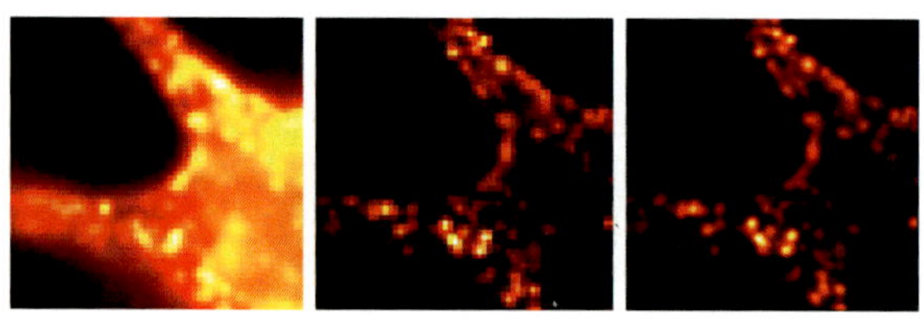

(a) (b) (c)

Figure 9.4

Fourier interpolation on a movie of blinking QDs. Rat hippocampal neuron with neurotransmitter receptor subunit $GABA_B$ R1 labeled with $QDot_525$ (raw movie contains 3000 frames recorded at 20 Hz). (a) Time average of original movie. (b) second-order SOFI without Fourier interpolation, and (c) second-order SOFI with 3x Fourier interpolation.

depend on the exact knowledge of the PSF of the imaging microscope. Second, when calculating higher-order SOFI images, the number of possible high-order cross-correlations between several pixels for obtaining one desired virtual pixel at a given position is increasing exponentially, and there is no unambiguous way to choose the optimal cross-correlation geometry. For more details, see Ref. [22].

There exists a much simpler and exact solution to this problem. The core idea is to recalculate the recorded images $F(\mathbf{r}, t)$ on a finer grid with reduced pixel size *before* calculating the SOFI images. This can be done exactly by employing the fact that Fourier transform of the PSF, which is called the optical transfer function (OTF), has only a finite support. In other words, for sufficiently large absolute values of the Fourier vectors, the OTF drops to zero. This can be used to devise an interpolation scheme for re-calculating the recorded images on a finer pixel grid without introducing any artifacts. Let us assume that the images $F(\mathbf{r}, t)$ are recorded with a camera having $N \times N$ pixels of size $d \times d$ and coordinates $\mathbf{r}_{ij}$, where $1 \leq i, j \leq N$. Thus, when directly applying the SOFI algorithm, one obtains SOFI images with the same number of pixels and pixel size as the original images. Also, the discrete two-dimensional Fourier-transformed images $\tilde{F}(\mathbf{k},t)$ which can be calculated by a fast discrete fourier transform (DFT) have the same $N \times N$ number of sampling points $\mathbf{k}_{ij}$, with a maximum length of the spatial Fourier vector of $2\pi\lceil (N-1)/2 \rceil / d$ along both x- and y-directions, where $\lceil \ \rceil$ is the ceiling function. If the camera pixel size is smaller than the resolution limit of the microscope, then the Fourier-transformed images $\tilde{F}(\mathbf{k},t)$ will be zero for large absolute values $|\mathbf{k}_{ij}|$ of the Fourier vector. Thus, padding the Fourier-transformed images with zeros will not change at all the information content of the images. However, when back-transforming the zero-padded Fourier images to real space, this results in images with smaller pixel size and larger pixel number, without changing, in any way, the real image content. By using this general idea, one can recalculate the original images on a finer grid with arbitrarily small pixel size without introducing any artifacts. In particular, when padding the Fourier images with Δ sampling points on all sides, the pixel size of the back-transformed images will be

$$d_{\text{new}} = \frac{\lceil (N-1)/2 \rceil}{\lceil (N-1)/2 \rceil + \Delta} d \tag{9.14}$$

The combination of Fourier interpolation with SOFI is called Fourier-SOFI or fSOFI, and it allows for obtaining artifact-free SOFI images on an arbitrarily refined pixel grid. In practice, the new virtual pixel size d_{new} should be less than half the resolution limit of the highest order SOFI one wants to calculate (Nyquist sampling theorem). Because the nth-order SOFI achieves a theoretical resolution that is n times better than that of the original images (see next section), d_{new} should be at least $2n$ times smaller than the resolution of the used microscope.

9.5 Fourier Reweighing

As already stated, the actual resolution enhancement of an nth SOFI image is n times, not just $\sqrt{n}$ times as one would guess from the description of the basic SOFI algorithm in Section 9.2. To see this let us recall that an nth-order SOFI image is proportional to the convolution of the nth power of the original PSF with the sample function which describes the distribution and brightness of emitters in the sample,

$$S_n(\mathbf{r}) \propto U^n(\mathbf{r}) * \sum_{j=1}^{N} \delta(\mathbf{r} - \mathbf{r}'_j)\epsilon_j^n, \tag{9.15}$$

where the star (*) denotes the convolution operation. The corresponding OTF is given by the n-fold convolution of the original OTF $\tilde{U}(\mathbf{k})$, or

$$\mathcal{F}\left[U^n(\mathbf{r})\right] = \underbrace{\tilde{U}(\mathbf{k}) * \tilde{U}(\mathbf{k}) * \ldots * \tilde{U}(\mathbf{k})}_{n\text{-times}} \tag{9.16}$$

where the symbol $\mathcal{F}$ denotes Fourier transformation. The OTFs for a conventional widefield microscope and for the second- to fourth-order SOFI are shown in Figure 9.5. The resolving power of a microscopy technique along a given

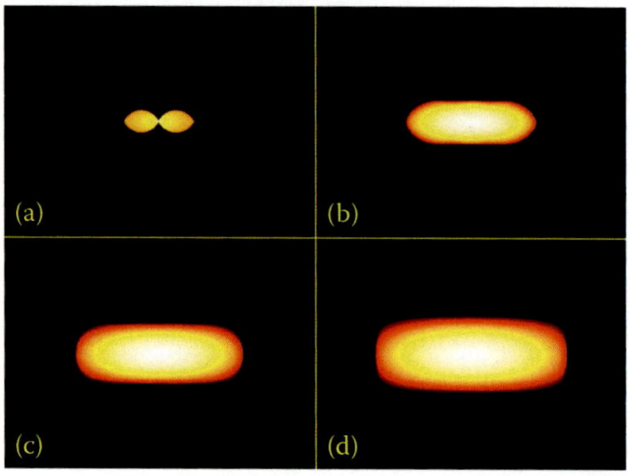

Figure 9.5

Comparison of the OTF for widefield microscopy (a) with that of second-order (b), third-order (c), and fourth-order (d) SOFI. Horizontal axis is along the lateral Fourier coordinate q and vertical axis along the axial Fourier coordinate w. The lateral extent from the optical axis of the widefield OTF in panel (a) is $2nk_0\sin\Theta$, where k_0 is the vacuum wave-vector length, n the sample's refractive index, and Θ the objective's maximum angle of light collection. The axial extension of this OTF is $2nk_0$ $(1 - \cos\Theta)$. The scale of all panels is the same.

spatial direction is proportional to the maximal extent of the OTF along the corresponding direction in Fourier space. First of all, the figure shows that all OTFs have only a finite support: along all directions, the OTF drops to zero beyond a maximal spatial frequency. For the conventional widefield microscope, an additional peculiarity is visible: the "missing cone." The missing cone refers to the fact that there are no spatial frequencies within the support of the OTF for directions with an angle toward the optical axis smaller than $\pi - \arcsin(\text{N.A.}/n)$. This results in the complete lack of any z-sectioning capability of a widefield microscope, which means that it is impossible to reconstruct a three-dimensional image of a sample by simply taking several images with the focal plane shifted along the optical axis. However, as can be seen from the panels (b) to (d) in Figure 9.5, SOFI completely removes the missing cone, which results in a superb z-sectioning capability. This was impressively demonstrated in Reference [18], where the authors were able to take high-quality three-dimensional images of cells by just using a widefield microscope and applying SOFI. In a recent work, Geissbuehler et al. [19] equipped a widefield microscope with a multiplane-imaging optics which allowed them to record, in a quasi-simultaneous manner, three-dimensional images of a sample using the intrinsic z-sectioning capability of SOFI.

Another interesting property of the SOFI-OTF which can be learned from inspecting Equation 9.16 is that its support along the lateral direction is larger than just $\sqrt{n}$ times that of the OTF of the widefield microscope. Ideally, it is n times larger, which theoretically corresponds to an n-fold increase in spatial resolution. However, the amplitudes of the spatial frequencies of the OTF are not optimal: Ideally, for a microscope with n-times improved lateral resolution, one would like to have an OTF which is the n-times stretched version of the original OTF, which would exactly correspond to an n-times reduced PSF as exemplified by the following identity:

$$\mathcal{F}\left[U(n\mathbf{r})\right] = \tilde{U}\left(\mathbf{k}/n\right). \tag{9.17}$$

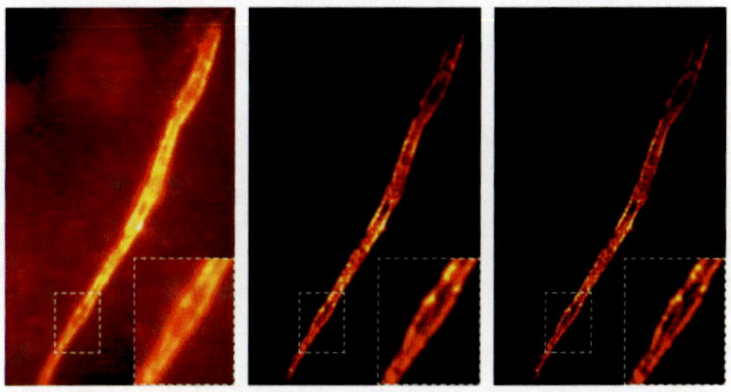

Figure 9.6

Fourier reweighing demonstrated on the movie of blinking organic dyes from Figure 9.7 (rotated). The PSF estimate from Figure 9.7c was fitted with an airy disc model and fed into the deconvolution algorithm. (a) Time average of movie, (b) second-order SOFI, and (c) second-order with Fourier reweighing. The resolution improvement is clearly visible, especially in the inset.

Because both $\tilde{U}(\mathbf{k}/n)$ and $\tilde{U}(\mathbf{k})*\tilde{U}(\mathbf{k})*\ldots*\tilde{U}(\mathbf{k})$ with $n-1$ convolutions have the same support, it is theoretically possible to transform the latter OTF into the first. The problem here is that both OTFs tend to zero when approaching the edge of their support. In principle, various deconvolution algorithms can be used to approximate this conversion, see for example, Reference [23], which uses a Richardson–Lucy approach. One of most straightforward methods is a Fourier reweighing procedure as proposed in Reference [16]. Fourier reweighing first transforms a SOFI image into Fourier space, then reweighs the Fourier amplitudes in an appropriate way, and then transforms the result back into real space, as shown in the following scheme:

$$
\begin{array}{l}
S_n(\mathbf{r}) \xrightarrow{\mathcal{F}} \tilde{S}_n(\mathbf{q}) \\
\qquad\qquad \downarrow \cdot \dfrac{\tilde{U}(\mathbf{q}/n)}{\underbrace{\tilde{U}(\mathbf{q})*\tilde{U}(\mathbf{q})*\ldots*\tilde{U}(\mathbf{q})}_{n\text{-times}}+\alpha} \\
S'_n(\mathbf{r}) \xleftarrow{\mathcal{F}^{-1}} \tilde{S}'_n(\mathbf{q})
\end{array}
\tag{9.18}
$$

Here $\tilde{U}(\mathbf{q})$ denotes the *axial projection* of the full OTF, $\tilde{U}(\mathbf{q})=\int_{-\infty}^{\infty} dw\,\tilde{U}(\mathbf{q},w)$, where $\mathbf{q}$ is the lateral and w the axial components of the Fourier vector. The scheme in Equation 9.18 is strictly correct only for emitters within the focal plane, for which, however, one typically desires to obtain the best possible lateral resolution. The factor $\alpha > 0$ is a small positive number which prevents division by zero. An example of a reweighed image is given in Figure 9.6. In praxis, the achievable resolution enhancement depends on the signal-to-noise ratio of the SOFI image, which mostly determines how small the value of α can be chosen. Too low values for α result in noise amplification and introduce artifacts within the de-convolved image.

9.6 Fluctuation-Based PSF Estimation

For successfully performing the Fourier reweighing procedure as discussed in the preceding section, an *a priori* knowledge of the PSF/OTF is required. In the original publication of SOFI, Reference [15], the PSF and OTF were calculated theoretically assuming a well-adjusted, aberration-free optics complying to the specifications of the used microscope optics. A more precise method would be to measure the PSF explicitly by using so-called PSF beads, small fluorescent polymer beads of a size which is much smaller than the PSF. However, such measurements are tedious and time-consuming. When doing SOFI, the stochastic blinking of the emitters offers an alternative way of explicitly estimating the PSF without any additional measurement. Recall that the fluorescence image $F(\mathbf{r}, t)$ is given by a linear superposition of single-emitter contributions as shown in Equation 9.1. The (lateral) Fourier transform $\tilde{F}(\mathbf{k},t)$ of such an image is given by

$$\tilde{F}(\mathbf{q},t)=\sum_{j=1}^{N}\int_{-\infty}^{\infty} dw\,\tilde{U}(\mathbf{q},w)\cdot e^{-iwz'_j-i\mathbf{q}\mathbf{r}'_j}\cdot\epsilon_j\cdot s_j(t), \tag{9.19}$$

where $\tilde{U}(\mathbf{k})=\tilde{U}(\mathbf{q},w)$ is the OTF and, as before, $\mathbf{q}$ and w are the lateral and axial Fourier coordinates, respectively. Note that the emitters' positions simply transfer into a phase factor in Fourier space. Computation of the second-order cumulant of this Fourier transform with its time-shifted complex conjugate yields

$$\begin{aligned}
&\langle\delta\tilde{F}(\mathbf{q},t)\cdot\delta\tilde{F}^*(\mathbf{q},t+\tau)\rangle=\\
&=\sum_{j,l=1}^{N}\int\int_{-\infty}^{\infty} dw\,dw'\,\tilde{U}(\mathbf{q},w)\tilde{U}^*(\mathbf{q},w')\cdot e^{i\mathbf{q}(\mathbf{r}'_l-\mathbf{r}'_j)}\\
&\quad\cdot e^{iw'z'_l-iwz'_j}\cdot\epsilon_j\epsilon_l\cdot\langle\delta s_j(t)\cdot\delta s_l(t+\tau)\rangle\\
&\sum_{j=1}^{N}\int\int_{-\infty}^{\infty} dw\,dw'\,\tilde{U}(\mathbf{q},w)\tilde{U}^*(\mathbf{q},w')\cdot e^{i(w'-w)z'_j}\cdot\epsilon_j^2 f_2(\tau)
\end{aligned} \tag{9.20}$$

where the star superscript denotes complex conjugation, and where we again used the property that $\langle s_j(t)s_l(t+\tau)\rangle = \delta_{jl}f_2(\tau)$, for example, that the cumulant is nonzero only for the same emitter which leads to the cancellation of the $\mathbf{q}$-related phase factors. If the number of blinking emitters in the sample is large enough, and if they are distributed rather randomly along the optical axis, then the sum $\sum_{j=1}^{N}\epsilon_j^2 e^{i(w'-w)z'_j}$ approaches a δ-function, $\delta(w' - w)$, times some constant pre-factor. In that case, the above expression simplifies to

$$\langle\delta\tilde{F}(\mathbf{q},t)\cdot\delta\tilde{F}^*(\mathbf{q},t+\tau)\rangle\approx f_2(\tau)\int_{-\infty}^{\infty} dw\left|\tilde{U}(\mathbf{q},w)\right|^2 \tag{9.21}$$

that is, the correlation function $\langle\delta\tilde{F}(\mathbf{q},t)\cdot\delta\tilde{F}^*(\mathbf{q},t+\tau)\rangle$ is proportional to the axial projection of the absolute square of the OTF. One can use Equation 9.21 for estimating the real OTF by fitting a model OTF against the measured correlation function $\langle\delta\tilde{F}(\mathbf{q},t)\cdot\delta\tilde{F}^*(\mathbf{q},t+\tau)\rangle$. Although any phase information is lost, this approach still yields a decent estimate of the absolute values of the OTF, which can then be used in the scheme of Equation 9.18 for performing the Fourier reweighing of SOFI images. An example is presented in Figure 9.7, where the conventional widefield image together with a second-order SOFI image is shown together with the cross section of the estimated PSF which was then used for the

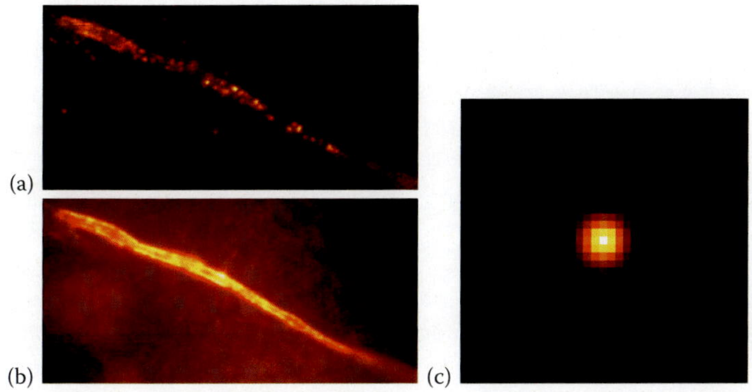

Figure 9.7

Estimation of the point spread function from a movie of a mouse hippocampal neuron stained with the blinking organic dye Alexa 647. (a) One frame of the movie, (b) time average over the movie, and (c) calculated cross section of the PSF. A deconvolution result based on this estimate is shown in Figure 9.6.

Fourier reweighing in Figure 9.6. It should be noted that the approach of OTF/PSF estimation presented here becomes even more attractive when working with the multiplane imaging system as described in Reference [19] because then one can perform the correlation of Equation 9.20 in three dimensions, which then not only gives the axial projection of the absolute square of the OTF, but the full absolute square of the OTF in all dimensions.

9.7 Conclusion and Outlook

SOFI is a relatively recent addition to the family of super-resolution fluorescence microscopy techniques. In this chapter, we could give just a first introduction into the method and its principles. One central issue of SOFI, similar to PALM and STORM, is the availability of appropriately blinking/photoswitching fluorescent labels, see for example, [24,25]. The question of what is the best photoswitching behavior for SOFI, together with a thorough comparison of SOFI with STORM, was presented in Reference [26]. Whereas the first publications on SOFI used blinking QDs for imaging, there are meanwhile several publications which successfully realized SOFI using photoswitchable proteins [17,27] or organic dyes [16]. A particularly clever approach to SOFI was presented in Reference [28], where the authors used donor molecules for labeling structures of interest and then used the FRET-induced blinking of by-diffusing acceptor molecules for performing SOFI. This approach is reminiscent of point accumulation for imaging in nanoscale topography (PAINT) in the context of PALM/STORM. Another particular variant of labeling was recently presented in Reference [29] where the authors used multicolor QD labeling for significantly boosting the label density and acquisition speed of SOFI.

One of the open challenges of SOFI is the so-called brightness skewing: As has been shown in Section 9.2, an emitter with fluorescence brightness ϵ_j contributes to an nth-order SOFI image with an apparent brightness of ϵ_j^n. Thus, for increasing orders of SOFI, dimmer emitters become increasingly discriminated with respect to brighter emitters, which significantly skews the brightness distribution in a SOFI image. Geissbuehler et al. [22] presented a first attempt to solve this problem by applying a dedicated deconvolution procedure to SOFI images, which they called balanced or bSOFI. Even more interestingly, they used their approach for extracting additional information from the temporal fluctuations of the emitters from which a SOFI image is built. For example, the fluctuation correlation time can deliver information about local oxygen concentration or pH values in a sample. Such an approach offers the fascinating prospect of joining super-resolution with quantitative information about the local environment in a sample.

In the future, two major advances can be expected in the further development of SOFI. First, the development of hardware-based approaches for the rapid calculation of the image cumulants. At the moment, SOFI cumulant calculations from raw movie files are performed on standard computers using programming environments such as MATLAB® or C. With such an approach, the calculation of, for example, a fourth-order SOFI image from a stack of 1000 raw frames takes typically several seconds to minutes, depending also on subpixel oversampling as described in Section 9.4. However, it will be relatively straightforward to implement the SOFI algorithm on dedicated hardware such as an FPGA or a graphics card. This will speed up the calculations by orders of magnitude which would enable real-time SOFI calculations. Second, there is still large room for advances in the development of better deconvolution algorithms for SOFI, which will allow to extract the full resolution intrinsically contained in a SOFI image, and to perfectly rectify the brightness skewing.

Acknowledgments

This work was supported by the German Research Foundation via the Cluster of Excellence "Center for Nanoscale Microscopy and Molecular Physiology of the Brain" (CNMPB) and via the Collaborative Research Center SFB 937 "Collective behavior of soft and biological matter," project A11, and by the German Ministry of Education and Research via the Bernstein Center for Computational Neuroscience Göttingen (BCCN).

References

1. Hell, S. W. and Jan Wichmann. Breaking the diffraction resolution limit by stimulated emission: Stimulated-emission-depletion fluorescence microscopy. *Opt. Lett.*, 19:780–782, 1994.
2. Klar, T. A., and Stefan W. Hell. Subdiffraction resolution in far-field fluorescence microscopy. *Opt. Lett.*, 24:954–956, 1999.
3. Jörg Enderlein. Advanced fluorescence microscopy. In Anders Brahme, editor, *Comprehensive Biomedical Physics*, vol. 4, pp. 111–151. Elsevier, 2014.
4. Hell, S. W. and Matthias Kroug. Ground-state-depletion fluorescence microscopy: A concept for breaking the diffraction resolution limit. *Appl. Phys. B*, 60:495–497, 1995.
5. Bretschneider, S., Christian Eggeling, and Stefan Hell. Breaking the diffraction barrier in fluorescence microscopy by optical shelving. *Phys. Rev. Lett.*, 98:218103, 2007.
6. Gustafsson, M. G. L. Nonlinear structured-illumination microscopy: Wide-field fluorescence imaging with theoretically unlimited resolution. *Proc. Nat. Acad. Sci. USA*, 102:13081–13086, 2005.

7. Betzig, E., George H. Patterson, Rachid Sougrat, O. Wolf Lindwasser, Scott Olenych, Juan S. Bonifacino, Michael W. Davidson, Jennifer Lippincott-Schwartz, and Harald F. Hess. Imaging intracellular fluorescent proteins at nanometer resolution. *Science*, 313:1642–1645, 2006.
8. Rust, M. J., Mark Bates, and Xiaowei Zhuang. Sub-diffraction-limit imaging by stochastic optical reconstruction microscopy (STORM). *Nat. Methods*, 3:793–795, 2006.
9. Van De Linde, S., R. Kasper, M. Heilemann, and M. Sauer. Photoswitching microscopy with standard fluorophores. *Appl. Phys. B*, 93:725–731, 2008.
10. Van De Linde, S., Ulrike Endesfelder, Anindita Mukherjee, Mark Schüttpelz, Gerd Wiebusch, Steve Wolter, Mike Heilemann, Markus Sauer, and M Schuttpelz. Multicolor photoswitching microscopy for sub-diffraction-resolution fluorescence imaging. *Photochem. Photobiol. Sci.*, 8:465–469, 2009.
11. Van De Linde, S., Mike Heilemann, and Markus Sauer. Live-cell super-resolution imaging with synthetic fluorophores. *Annu. Rev. Phys. Chem.*, 63:519–540, 2012.
12. Enderlein, J. and Thomas Ruckstuhl. The efficiency of surface-plasmon coupled emission for sensitive fluorescence detection. *Opt. Expr.*, 13:8855–8865, 2005.
13. Humpolícková, J., Aleš Benda, Radek Macháň, Jörg Enderlein, and Martin Hof. Dynamic saturation optical microscopy: Employing dark-state formation kinetics for resolution enhancement. *Phys. Chem. Chem. Phys.*, 12:12457–65, 2010.
14. Lanzano, L., Ivan C Hernandez, Marco Castello, Enrico Gratton, Alberto Diaspro, and Guiseppe Vicidomini. Encoding and decoding spatio-temporal information for background-free super-resolution microscopy. *Nature Commun.*, 6:6701, 2015.
15. Dertinger, T., Ryan Colyer, Gopal Iyer, Shimon Weiss, and Jörg Enderlein. Fast, background-free, 3D super-resolution optical fluctuation imaging (SOFI). *Proc. Nat. Acad. Sci. USA*, 106:22287–22292, 2009.
16. Dertinger, T., Mike Heilemann, Robert Vogel, Markus Sauer, and Shimon Weiss. Superresolution optical fluctuation imaging with organic dyes. *Ang. Chem. Int. Ed.*, 49:9441–9443, 2010.
17. Dedecker, P., Gary C. H. Mo, Thomas Dertinger, and Jin Zhang. Widely accessible method for superresolution fluorescence imaging of living systems. *Proc. Nat. Acad. Sci. USA*, 109:10909–10914, 2012.
18. Dertinger, T., Jianmin Xu, Omeed Foroutan Naini, Robert Vogel, and Shimon Weiss. SOFI-based 3D superresolution sectioning with a widefield microscope. *Opt. Nanoscopy*, 1:2, 2012.
19. Geissbuehler, S., Azat Sharipov, Aurélien Godinat, Noelia L Bocchio, Patrick A Sandoz, Anja Huss, Nickels A Jensen et al. Live-cell multiplane three-dimensional super-resolution optical fluctuation imaging. *Nat. Commun.*, 5:5830, 2014.
20. Wahl, M., Ingo Gregor, Mattias Patting, and Jörg Enderlein. Fast calculation of fluorescence correlation data with asynchronous time-correlated single-photon counting. *Opt. Expr.*, 11:3583–3591, 2003.
21. Dertinger, T., Ryan Colyer, Robert Vogel, Jörg Enderlein, and Shimon Weiss. Achieving increased resolution and more pixels with Superresolution Optical Fluctuation Imaging (SOFI). *Opt. Expr.*, 18:18875–18885, 2010.

22. Geissbuehler, S., Noelia L Bocchio, Claudio Dellagiacoma, Corinne Berclaz, Marcel Leutenegger, and Theo Lasser. Mapping molecular statistics with balanced super-resolution optical fluctuation imaging (bSOFI). *Opt. Nanoscopy*, 1:4, 2012.
23. Ingaramo, M., Andrew G York, Eelco Hoogendoorn, Marten Postma, Hari Shroff, and George H Patterson. Richardson-Lucy deconvolution as a general tool for combining images with complementary strengths. *ChemPhysChem*, 15:794–800, 2014.
24. Hedde, P. N. and G. Ulrich Nienhaus. Super-resolution localization microscopy with photoactivatable fluorescent marker proteins. *Protoplasma*, 251:349–362, 2014.
25. Chozinski, T. J., Lauren A. Gagnon, and Joshua C Vaughan. Twinkle, twinkle little star: Photoswitchable fluorophores for super-resolution imaging. *FEBS Lett.*, 588:3603–3612, 2014.
26. Geissbuehler, S., Claudio Dellagiacoma, and Theo Lasser. Comparison between SOFI and STORM. *Biomed. Opt. Expr.*, 2:810–813, 2011.
27. Zhang, X., Xuanze Chen, Zhiping Zeng, Mingshu Zhang, Yujie Sun, Peng Xi, Jianxin Peng, and Pingyong Xu. Development of a reversibly switchable fluorescent protein for super-resolution optical fluctuation imaging (SOFI). *ACS Nano*, 9:2659–2667, 2015.
28. Cho, S., Jaeduck Jang, Chaeyeon Song, Heeyoung Lee, Prabhakar Ganesan, Tae-Young Yoon, Mahn Won Kim, Myung Chul Choi, Hyotcherl Ihee, Won Do Heo, and Yongkeun Park. Simple super-resolution live-cell imaging based on diffusion-assisted Förster resonance energy transfer. *Sci. Rep.*, 3:1208, 2013.
29. Zeng, Z., Xuanze Chen, Hening Wang, Ning Huang, Chunyan Shan, Hao Zhang, Junlin Teng, and Peng Xi. Fast super-resolution imaging with ultrahigh labeling density achieved by joint tagging super-resolution optical fluctuation imaging. *Sci. Rep.*, 5:8359, 2015.

10

Super-Resolution Two-Photon Excitation Microscopy Utilizing Transmissive Liquid Crystal Devices

Kohei Otomo, Terumasa Hibi, Yuichi Kozawa, Sari Ipponjima, Shunichi Sato, and Tomomi Nemoto

10.1 Introduction
10.2 Focusing Laser Beams Using a High Numerical Aperture Lens
10.3 Beam Modulation by LCDs
10.4 TPLSM Utilizing an HRP Beam
10.5 Two-Photon Excitation STED Microscopy
10.6 Summary and Outlook
Acknowledgments
References

10.1 Introduction

Fluorescence microscopy is widely used in medical and biological research. A targeted fluorophore generally absorbs one photon with an energy equal to the energy difference between the electronic ground state and the first excited state of the molecule. The phenomenon in which a molecule simultaneously absorbs two photons, both having approximately twice the wavelength of the abovementioned single photon, can also occur, but it is extremely rare (Figure 10.1); this phenomenon is called the two-photon excitation process. It is also possible for a fluorophore to simultaneously absorb three photons, each having approximately one-third of the total excitation energy; this is called a three-photon excitation. Multiphoton excitation was first predicted in theory by Göppert-Mayer in 1931, at the dawn of the quantum mechanical era (Göppert-Mayer 1931).

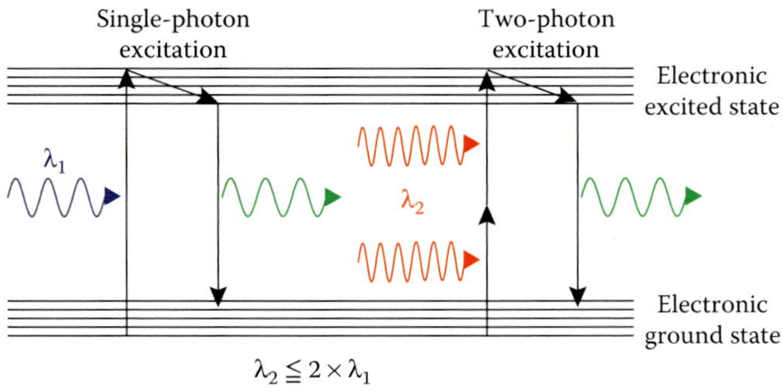

Figure 10.1

Jablonski diagrams for single-photon excitation and two-photon excitation.

About 60 years later, the first multiphoton excitation images (using two-photon excitation) of biological specimens were demonstrated (Denk et al. 1990).

According to the development of laser techniques, laser scanning microscopy (LSM), which uses a laser as an excitation light source for fluorescence microscopy, has also been developed. In an LSM system, focused excitation laser light scans specimens sequentially point-by-point, line-by-line, or taking multiple points at once. Light emitted by fluorescence passes through some optical filters and is detected by photomultiplier tubes (PMTs) or avalanche photodiodes. Finally, pixel information is assembled into one image by a computer. In particular, confocal LSM (CLSM) is the most popular methodology; this idea was first described in a patent application by Minsky in 1957 (Minsky 1988). CLSM utilizes the pinhole in front of the detector not to detect fluorescent signals out of focus of the objective lens (dashed line in Figure 10.2a), but to allow optically sectioned images to be acquired (Conchello and Lichtman 2005). Conversely, the development of mode-locked titanium:sapphire (Ti:Sa) laser techniques (Spence et al. 1991) largely contributed to the establishment of a two-photon excitation LSM (TPLSM) technique.

TPLSM for biological specimens has several advantages (Denk et al. 2006; Nemoto 2008; Hoover and Squier 2013). First, excitation is spatially localized

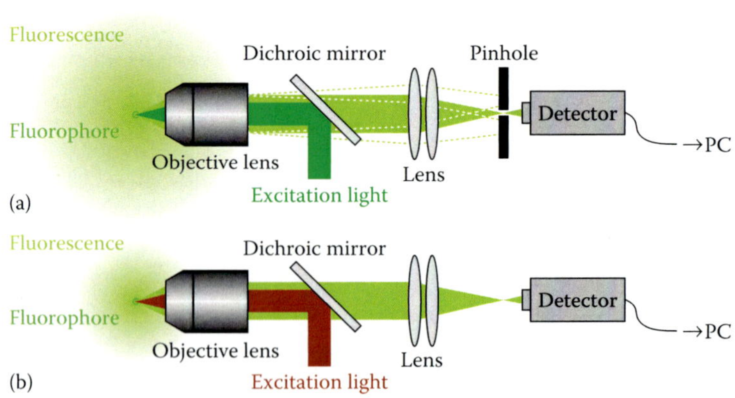

Figure 10.2

CLSM (a) and TPLSM (b).

at the focus of the excitation laser light used in TPLSM due to a nonlinear dependence on photon density, which allows for the acquisition of optically sectioned fluorescence images. Compared with single-photon excitation, two-photon excitation fluorescence appears only in a tiny volume at the focus, as shown in Figure 10.3. Because three-dimensional resolution depends on the localization of excitation to the focal volume in contrast with the CLSM case, out-of-focus photobleaching, photodamage, and the attenuation of the excitation beam by out-of-focus absorption rarely occur. Furthermore, since a confocal pinhole is not required (Figure 10.2b), its alignment and adjustment are not needed. Second, the excitation induced by near-infrared (NIR) laser lights offers superior penetration depth and is less invasive for living specimens. Therefore, TPLSM has become an essential intravital analytical method in life science and medical research.

Most conventional TPLSM systems detect fluorescent signals with a nondescanned detector (NDD) which collects all the emitted light (Denk et al. 2006; Nemoto 2008). Therefore, its spatial resolution corresponds to the square values of the intensity distribution around the focal pattern of excitation light. Conversely, the spatial resolution of CLSM is also related to the size of confocal pinhole. As shown in Figure 10.4, the intensity distribution around the focal pattern of visible light (450 nm) is sharper than the square values of the intensity distribution of NIR light with twice the wavelength (900 nm). Moreover, the effective spatial resolution of CLSM becomes smaller according to the size of the confocal pinhole in exchange for its fluorescent signals decreasing. Since fluorophore TPLSM targets have spectral properties that are visible in nature, an NIR laser is generally used as the excitation light source. Therefore, in comparison with conventional CLSM, the spatial resolution with TPLSM tends to be inferior, mainly due to the wavelength of the excitation laser light.

Recently, the spatial resolution of TPLSM was improved by applying the stimulated emission depletion (STED) microscopy proposed by Hell (Hell and Wichmann 1994; Moneron and Hell 2009). The STED process is induced by a donut-shaped, high-power laser light (STED light) that is superimposed with excitation laser light. The STED light wavelength is usually set to be in the longer wavelength range of the emission spectra of fluorophores. By using optical filters to separate the stimulated emission light, a diffraction unlimited central spot can be selected (Hell 2003). In 2009, Moneron and Hell used TPLSM with the

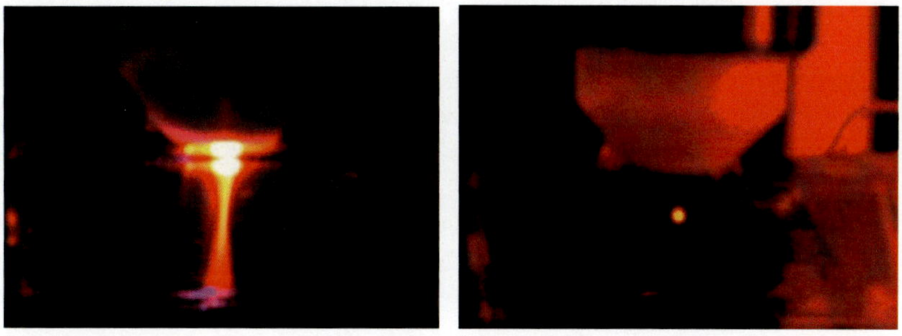

(a) (b)

Figure 10.3

Fluorescence emitting areas under the objective lens during a (a) single- and (b) two-photon excitation processes.

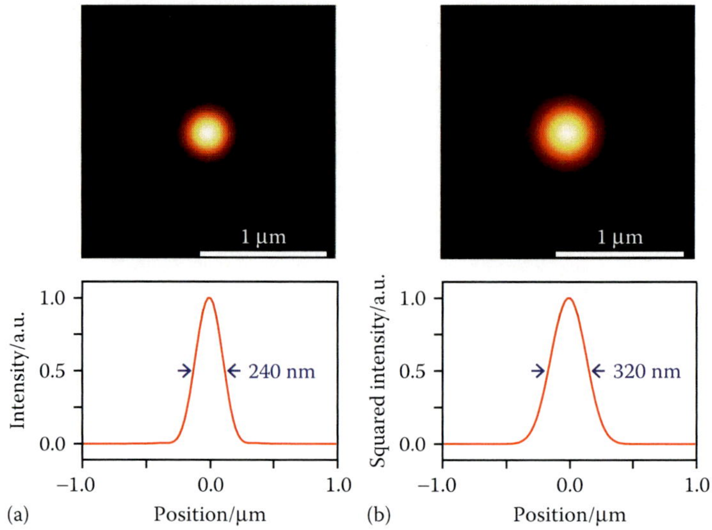

Figure 10.4

(a) Numerically calculated intensity distribution of circularly polarized visible light ($\lambda = 450$ nm). (b) Numerically calculated squared intensity distribution of circularly polarized near-infrared light ($\lambda = 900$ nm). Both are calculated as if for an objective lens with NA. 1.20. The upper and lower panels are images and vertical intensity profiles, respectively.

STED process (TP-STED) to visualize nanostructures of <100 nm marked with fluorescent dyes (Moneron and Hell 2009). Shortly after this study, Li et al. (2009) revealed that TP-STED is also applicable for green fluorescent protein. Moreover in 2013, TP-STED was used to reveal nanostructures of acute biological tissues at depths of 50–100 μm from the surface (Bethge et al. 2013; Takasaki et al. 2013). Unlike most super-resolution microscopy, which is only suitable for thin specimens, TP-STED methodology is expected to visualize biological nanostructures even in living animals. Hence, super-resolution two-photon excitation microscopy has the potential to visualize deeper regions of thick specimens than can conventionally be observed. To achieve super-resolution two-photon excitation microscopy, a smaller effective focal spot is needed. Recently, a variety of laser beam modulation methodologies were proposed for biological LSM, as described in later sections.

In this chapter, we first describe the focusing properties of various laser beams. Next, we describe our beam modulation method utilizing transmissive liquid crystal (LC) devices (tLCDs) and show their applications to super-resolution two-photon excitation microscopy. Finally, areas of future research are discussed based on our results and those of other works.

10.2 Focusing Laser Beams Using a High Numerical Aperture Lens

Linearly polarized (LP) Gaussian beams are practical and commonly used beams produced by commercially available laser devices. Circularly polarized beams, which are generally converted from LP ones by passage through a quarter-wave plate, are also used for many laser applications including LSM. However, in nature, a laser beam is characterized by a great variety of spatial "modes" which exhibit the spatial distributions of intensity, polarization, and phase on the beam cross section (Figure 10.5). The intensity distribution of a laser beam is also referred to as its "transverse mode."

Hermite–Gaussian (HG) and Laguerre–Gaussian (LG) modes are well-known transverse modes derived as solutions to the scalar paraxial wave equation for different coordinate systems (Siegman 1986). These mode beams are specified by two mode indices and are expressed as HG_{mn} or LG_{pl}. In both cases, the fundamental mode (HG_{00} or LG_{00}) corresponds to a Gaussian beam. Meanwhile, higher-order mode beams exhibit unique intensity distributions with dark nodes according to the mode indices. In particular, higher-order LG_{pl} beams with $l > 0$ present donut-shaped intensity distributions with a dark core in the center. The dark core of LG_{pl} beams arises from a spiral phase variation along the azimuthal direction, ϕ, of the beam cross section, as shown in Figure 10.5c, since the phase on the beam axis is undefined. This spiral phase of LG_{pl} beams is expressed by $\exp(il\phi)$, where l is referred to as a topological charge in the field of singular optics (Yao and Padgett 2011). On account of this spiral phase, a donut-shaped beam with a nonzero topological charge is also called a vortex beam.

Besides the intensity and phase distributions, polarization is another essential property that characterizes a laser beam. The transverse modes described above are based on scalar theory, which means that the polarization distribution of light beams is assumed to be linear or homogeneous on the beam cross section. Conversely, cylindrical vector beams are laser beams that have spatially inhomogeneous and axially symmetric polarization distributions. Radial and azimuthal polarizations shown in Figure 10.6b are the most fundamental polarization states

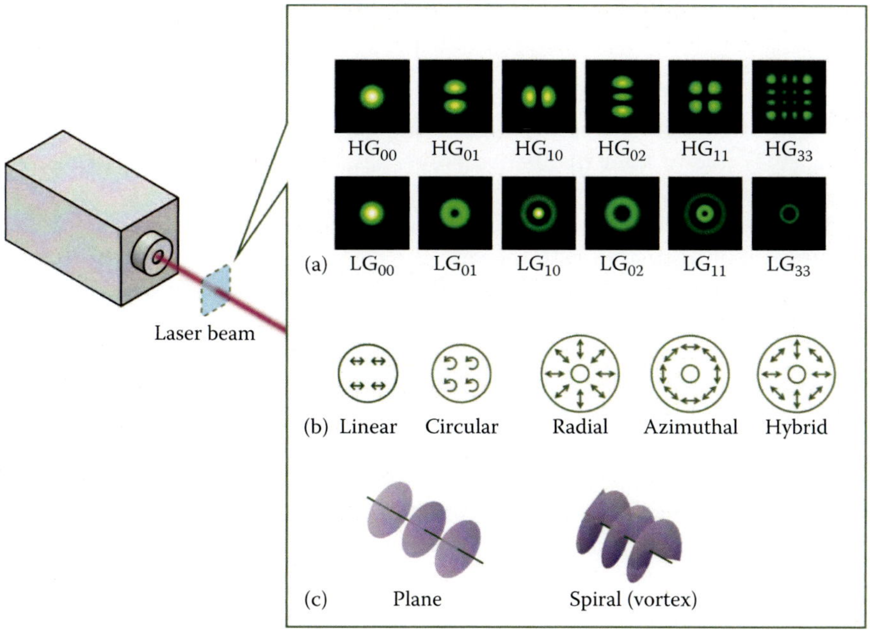

Figure 10.5

(a) Various intensity, (b) polarization, and (c) phase distributions on the cross section of a laser beam.

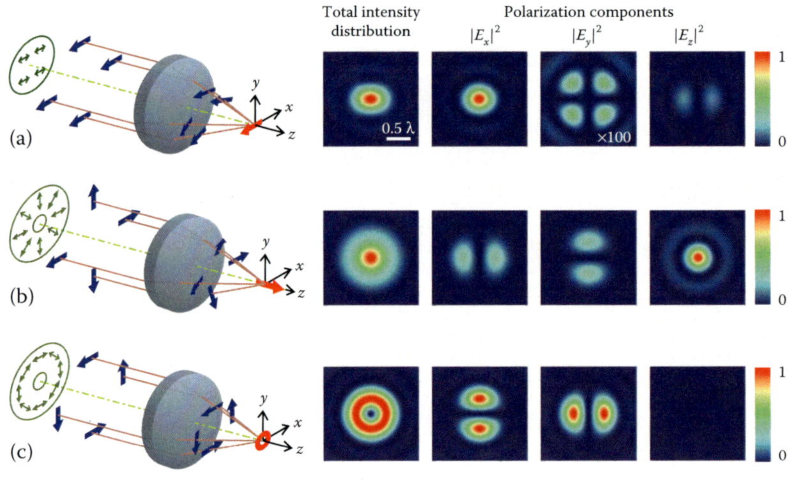

Figure 10.6

Tight focusing of linearly (a), radially (b), and azimuthally (c) polarized beams. The calculated total intensity distribution, and the $|E_x|^2$, $|E_y|^2$, and $|E_z|^2$ components at the focus (*xy*-plane) are shown in each column. The focusing condition is $NA = 1.2$ with $n = 1.33$.

for cylindrical vector beams. Hybrid polarization is a combined state of radial and azimuthal polarizations. For cylindrical vector beams, higher-order transverse modes in which the electric field is represented as a LG mode, namely vector LG beams, can be derived as solutions of the paraxial vector wave equation (Tovar 1998). While the intensity distribution of a vector LG beam is identical to that of an LP LG beam, the vector LG beam itself contains no spiral phase variation on the beam cross section. Instead, the axially symmetric polarization distribution leads to a donut-shaped pattern for a vector LG beam because the polarization direction on the beam axis is undefined. These spatially inhomogeneous polarizations have recently attracted much attention (Zhan 2009; Züchner et al. 2011) due to their distinctive properties, especially in the tight focusing condition as we will discuss below.

For LSM, the spot sizes and the shapes of excitation laser beams are important features that directly relate to the spatial resolution and imaging performance. In scalar diffraction theory, the intensity distribution at the focus of a lens for a plane wave is represented by a familiar "Airy pattern" (Born and Wolf 1999), which is mathematically expressed as $[2J_1(k\mathrm{NA}r)/k\mathrm{NA}r]^2$, where J_1 is Bessel function of the first kind of order one, k is the wavenumber of the focused light, NA is the numerical aperture of the lens, and r is the lateral distance from the focus. The Airy pattern produced by a plane wave with a wavelength of λ gives the focal spot with a full width at half maximum (FWHM) size of $0.51\lambda/NA$.

Strictly speaking, however, the formation of an Airy pattern at the focus is no longer valid under the tight focusing condition using a high NA objective lens, which is the condition usually used in LSM systems. Under tight focusing conditions, the dominant difference between a rigorous focal spot and an Airy pattern is the generation of a nonnegligible, longitudinal component of the electric field at the focus (Richards and Wolf 1959), which is not considered in scalar diffraction theory. Figure 10.6 displays the calculated intensity distributions and the polarization components at the foci of laser beams having linear, radial, and azimuthal polarizations, which are assumed to have been focused with a water-immersion objective lens ($NA = 1.2$ and a refractive index $n = 1.33$). These numerical calculations were performed on the basis of vector diffraction theory (Richards and Wolf 1959; Youngworth and Brown 2000), which precisely deals with the effect of polarization under the tight focusing conditions.

As shown in Figure 10.6a, for the focusing of a Gaussian beam polarized linearly along the x-axis, not only the $|E_x|^2$ polarization component but also $|E_y|^2$ and $|E_z|^2$ appear at the focus. While the y-component $|E_y|^2$ is almost negligible, the z-component $|E_z|^2$ with a two-lobe pattern along the polarization axis is considerable, which accounts for the elongation of the focal spot along the x-axis. Consequently, compared to an Airy pattern, the spot size of a tightly focused, LP Gaussian beam substantially increases along the polarization direction. In the example shown in Figure 10.6a, the FWHM values of the focal spot along the x- and y-axes are $0.76\lambda_m$ and $0.66\lambda_m$, respectively, whereas the corresponding FWHM value of an Airy pattern in this condition is $0.57\lambda_m$, where λ_m represents the beam wavelength in the medium, that is, $\lambda_m = \lambda/n$. This elongation that causes an elliptical focal spot can be overcome by focusing a circularly polarized beam; the beam produces a perfectly circular focal spot even under the tight focusing condition, because the polarization-dependent elongation is averaged over the lateral directions. However, the unavoidable longitudinal component still results in an increase in the size of focal spot of a circularly polarized beam compared to an Airy pattern.

By contrast, the longitudinal component generated by focusing a radially polarized beam forms a strong and circular spot at the focus (Youngworth and Brown 2000). Owing to the cylindrically symmetric polarization distribution of a radially polarized beam, the longitudinal components of all the focused rays are superposed at the focus to produce a bright spot, as shown in Figure 10.6b. Meanwhile, no longitudinal components appear in the focusing of an azimuthally polarized beam, meaning that it can maintain a donut-shaped intensity distribution at the focus, even under high NA focusing.

One of the most promising features of cylindrical vector beams is that the longitudinal component of a radially polarized beam is capable of creating a smaller focal spot than a conventional LP beam under the tight focusing condition (Quabis et al. 2000). In the past, many research groups have proposed various methods to creating and enhancing the longitudinal component by applying, for example, an annular mask (Quabis et al. 2000; Sheppard and Choudhury 2004) or a binary phase shifter (Helseth 2001; Wang et al. 2008) for the focusing of a radially polarized beam.

Numerical studies have recently revealed that a higher-order radially polarized LG (HRP-LG_{p1}) mode beam can also produce a smaller focal spot as its radial mode order p increases (Kozawa and Sato 2007, 2012). An HRP-LG_{p1} beam possesses $p + 1$ concentric rings that present a specific node spacing and a phase shift of π between the adjacent rings on the beam cross section. Typical examples of the calculated intensity distributions at the focus ($NA = 1.2$, $n = 1.33$) of HRP beams, as well as LP Gaussian beams, are summarized in Figure 10.7. As the radial mode index p increases, the sharp focal spot at the center on the focal plane (xy-plane) becomes pronounced, while the many side lobes around the center are also apparent. Furthermore, the focal spots of HRP beams are extended along the z-axis, suggesting that the depth of focus of an HRP beam can be significantly enhanced

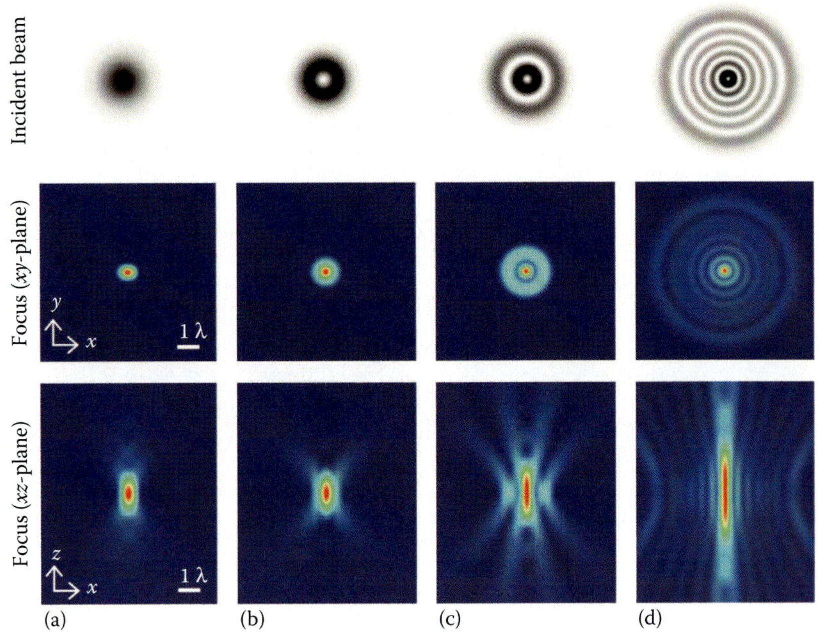

Figure 10.7

Calculated intensity distributions in the *xy*- and *xz*-planes near the focus of *x*-polarized Gaussian (a), radially polarized LG_{01} (b), LG_{11} (c), and LG_{51} (d) beams under a focusing condition of $NA = 1.2$ with $n = 1.33$.

compared to the focusing of an LP beam. The lateral FWHM size of the focal spot for an HRP-LG_{51} beam is $0.51\lambda_m$, which is reduced by approximately 33% compared to that for an *x*-polarized Gaussian beam ($0.76\lambda_m$) estimated along the polarization axis (*x*-axis). Such a noticeably small focal spot is preferable for improving the spatial resolution in CLSM (Kozawa et al. 2011). Moreover, as we will see later, this smaller focal spot property can be further enhanced when we consider two-photon excitation LSM (Ipponjima et al. 2014), in which the two-photon point spread function (PSF) corresponds to the square of the intensity distribution of the focused excitation beam.

The longitudinal component generated under the tight focusing condition also has an influence on the formation of the donut-shaped pattern used as a depletion beam in STED microscopy. As previously mentioned, vortex beams such as those with LG_{01} modes are generally employed for STED. The intensity distribution at the focus of a tightly focused vortex beam strongly depends on its polarization state (Iketaki et al. 2007; Hao et al. 2010), as shown in Figure 10.8. For the vortex beam having a spiral phase variation of $\exp(i\phi)$, the longitudinal component at the focus covers the center of the donut-shaped pattern, except for left-handed circular polarization. If the direction of the spiral phase variation is flipped to $\exp(-i\phi)$, right-handed circular polarization is required to obtain a completely dark spot at the focus. These behaviors of circularly polarized vortex beams imply that a small amount of phase shift caused by, for example, optical components will readily change its donut-shaped pattern into a bright spot at the focus, which is unfavorable in STED microscopy. Therefore, the polarization

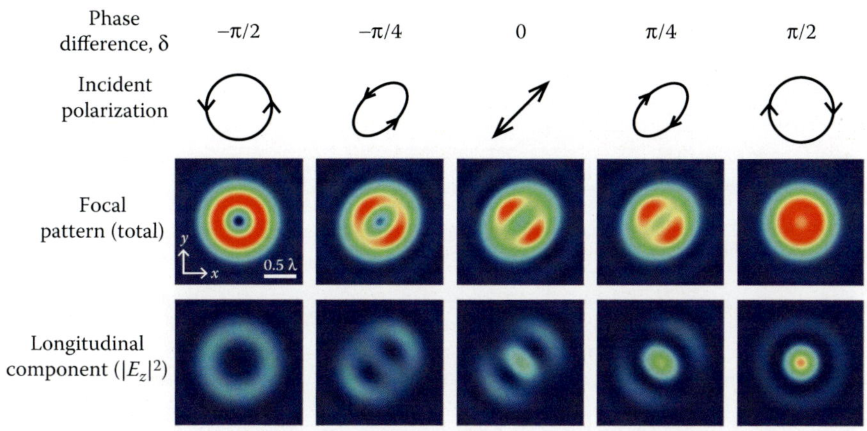

Figure 10.8

The calculated total intensity distributions (middle row) and the longitudinal components (bottom row) at the focus of a vortex beam having an azimuthal phase variation of exp($i\phi$) with different incident polarization states varying from left-handed (left panel) to right-handed (right panel) through a linear polarization at 45° (center panel). The focusing condition is $NA = 1.2$ with $n = 1.33$.

state of a STED light beam should be precisely controlled in order to achieve an ultimately enhanced spatial resolution in STED microscopy. It should be noted that an azimuthally polarized beam is also able to produce a perfect donut shape, even under the tight focusing condition, as already shown in Figure 10.6c. Owing to the cylindrical symmetry of the polarization distribution, the donut-shaped pattern of an azimuthally polarized beam can be utilized in STED microscopy to realize orientation imaging of a single molecule with enhanced spatial resolution (Reuss et al. 2010).

10.3 Beam Modulation by LCDs

One method for generating the various beams mentioned in the previous section, spatial light modulators (SLMs) have been used in recent years (Maurer et al. 2011). SLMs are objects that impose some form of spatially varying modulation on a light beam. SLMs are applicable for optical microscopy systems when integrated onto the optical path; therefore, conventional laser devices that generate Gaussian beams are usable as light sources. SLMs have recently been used for optical microscopy, not only to modulate a variety of beams but also to remove optical aberrations. SLMs are classified into two groups: variable-shaped mirrors, such as deformable mirrors or digital micromirror devices, in which actuators or membranes are moved by electric or magnetic fields (Booth 2014); and LC devices (Maurer et al. 2011). Both of them modulate the beam; the former uses its inhomogeneous reflection pattern, the latter enables optical retardation by transmissions to LC molecules.

Since our experimental setup is based on LC devices, we will focus on them. As shown in Figure 10.9a, the LC molecule acts as the refractive index ellipsoid that possesses two refractive index components, n_o and n_e, along its perpendicular axes. Changing the applied voltage across the layer holding LC molecules

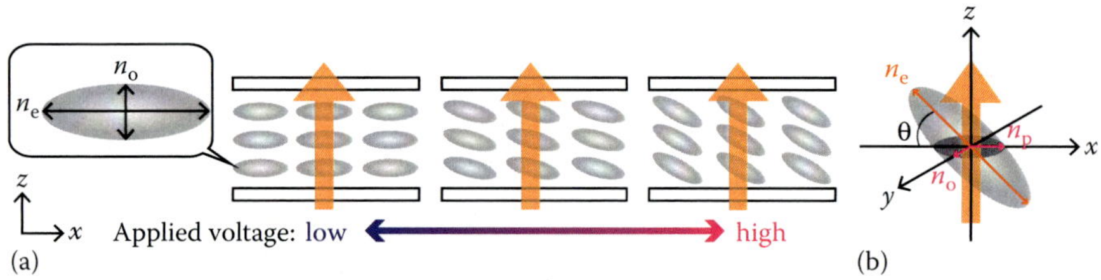

Figure 10.9

(a) Schematic of an LC device. Changing the voltage across the layer causes the LC molecules to rotate along an axis perpendicular to both the light propagation direction and the polarization vector. (b) The electric field of light passing through LC molecules inclined at an angle θ relative to the *x*-axis direction is affected by two refractive index components along the *x*- and *y*-axes. One refractive index component along the *x*-axis, n_p, depends on θ, the other refractive index component along the *y*-axis, n_o, does not.

causes the angle of the LC molecules to tilt. The electric field of light passing through the area where the LC molecules incline at an angle θ relative to the *x*-axis in Figure 10.9b is affected by two refractive index components, n_p and n_o, along the *x*- and *y*-axes, respectively. As shown in Figure 10.9b, n_p depends on the angle θ and is mathematically expressed as $[n_o n_e/(n_o^2\cos^2\theta + n_e^2\sin^2\theta)^{0.5}]$. Conversely, n_o does not depend on the angle θ. Therefore, phase modulation of the beam whose polarization corresponds to the *x*-axis is possible by tuning the applied voltage. When the LP light is polarized at a certain angle between the *x*- and *y*-axes, the *x*- and *y*-components are affected by n_p and n_o, respectively. If a some phase shift between the *x*- and *y*- components is generated, LP light will be converted to elliptical or circularly polarized light. In other words, LC devices are able to modulate the polarization properties of light. Moreover, LC devices are compatible with lights over a wide wavelength region by tuning the applied voltage.

LC on silicon (LCOS) devices are the representative products of SLMs using LC molecules. LCOS technology has been developed for image and video display applications by combining the light-modulating properties of LC materials with the advantages of high-performance silicon complementary metal oxide semiconductor (CMOS) technology (Zhang et al. 2014). Currently, several types of LCOS devices are commercially available and used in a variety of applications, including laser processing, laser manipulations, optical pulse shaping, adaptive optics, and some super-resolution microscopies (Gould et al. 2012). Most commercially available LCOS devices adopt the reflection method, in which an incident ray of light is reflected by a mirror after modulations by LC molecules. Hence, the surface reflection of devices and the distortion of mirrors cause a decline in the conversion efficiency. Conversely, we have utilized tLCDs to develop super-resolution microscopy systems (Kozawa et al. 2011; Ipponjima et al. 2014; Otomo et al. 2014). tLCDs enable the modulation of laser beam properties, such as phase, polarization, and intensity at a high conversion efficiency, since reflected components are removed from the optical path. In addition, compared with reflective LC devices wherein installation makes the optical system larger, tLCDs can be installed compactly just by placing them in the optical path.

10.4 TPLSM Utilizing an HRP Beam

In this section, we show our introduction of tLCDs into TPLSM to generate an HRP beam and demonstrate the enhancement of the lateral spatial resolution with an extension of the depth of field (DOF) (Ipponjima et al. 2014).

First, we numerically calculated the PSFs for two-photon excitation, in the cases of both an LP and an HRP beam, by squaring the intensity distribution around the focal point (Figure 10.10). The PSF calculated for the LP beam in the focal plane showed that the distribution was slightly elongated along the polarized direction (x-axis) compared with the perpendicular direction (y-axis) (Figure 10.10a). The FWHM values of the LP beam along the x- and y-axes were estimated to be 0.55 and 0.42 λ_m, respectively. In contrast, the PSF calculated for the HRP beam in the focal plane was cylindrically symmetric (Figure 10.10c). The FWHM value in the HRP beam was estimated to be 0.36 λ_m. From these FWHM values in the LP and the HRP beams, it was demonstrated that the HRP beam improved the lateral spatial resolution. Along the z-axis, the PSF in the HRP beam showed longitudinal elongation (Figure 10.10d). The FWHM values along the z-axis in the LP and HRP beams were estimated to be 1.19 and 3.79 λ_m, respectively (Figure 10.10f). Thus, we predicted that such an HRP beam would enable the visualization of objects at a higher lateral resolution and with a deeper DOF in TPLSM.

To experimentally confirm this improvement of the lateral spatial resolution in TPLSM utilizing an HRP beam, we installed our developed tLCDs, which could convert an LP beam into an HRP beam, in front of an objective lens.

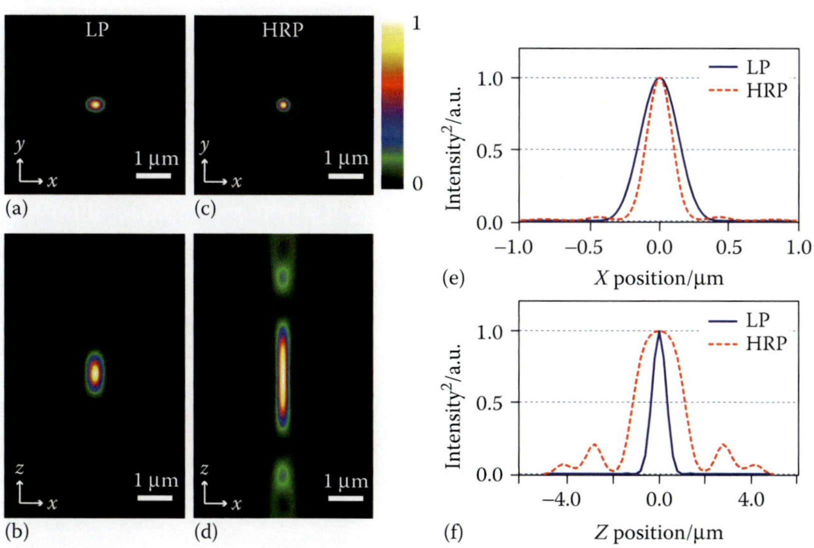

Figure 10.10

Lateral (x–y) and axial (x–z) calculated intensity distributions for TPLSM and the intensity profiles with 800 nm excitation. (a) and (c): Intensity patterns in the focal plane ($z = 0$) for the focusing of the LP beam (polarized along the x-axis) and the HRP beam, respectively. (b) and (d): Calculated intensity distributions in the x–z plane ($y = 0$) for the focusing of the LP beam and the HRP beam, respectively. (e) and (f): Intensity profiles across the center of the intensity distribution along the x- and z-axes, respectively. The solid (blue) and dashed (red) lines are the intensity profiles for the LP beam and the HRP beam, respectively.

Figure 10.11a shows our system, in which a mode-locked Ti:Sa laser (Tsunami, Spectra Physics) generated femtosecond NIR pulses at wavelengths of 800, 860, and 910 nm. An outgoing beam passed through a negative chirper, to compensate for the pulse shape at the position of the specimen, and a polarizing filter. After an LP beam entered the galvanomirror scanner (FV1000 MPE, Olympus) and an upright microscope (BX61WI, Olympus), it was converted into an HRP beam by two tLCDs; the 6-rings-divided phase mask (tLCD-6; Figure 10.11b) and the 12-pieces-divided polarization converter (tLCD-12; Figure 10.11c) (Kozawa et al. 2011), which were inserted between a revolver and a water-immersion objective lens (UPlanSApo 60XW, NA 1.20, Olympus). Fluorescent signals emitted from the specimens were passed through emission filters and detected by a PMT.

An image in LSM is given by the convolution of the PSF and sample structures. Since the measured width of an object with a sufficiently small diameter is close to the actual resolution in LSM, a tiny fluorescent bead is often used for estimation of the PSF. Accordingly, we evaluated spatial resolution by measuring the fluorescent intensity distributions of an isolated fluorescent yellow-green bead with 0.17-μm diameter (0.17-μm YG bead) embedded in agarose gel (Figure 10.12). The FWHM values in the LP beam along the x- and y-axes were measured as 294 and 260 nm, respectively; a slight elongation along the polarization direction was confirmed (Figure 10.12a). In the HRP beam, the intensity pattern in the focal plane was almost cylindrically symmetric (Figure 10.12c). The FWHM values in the HRP beam were 230 and 231 nm on the x- and y-axes, respectively. Furthermore, the x–z intensity pattern of the HRP beam showed that the DOF was significantly extended along the z-axis relative to that of the LP beam (Figure 10.12b and d). The FWHM values along the z-axis were determined to be 916 nm (LP) and 3793 nm (HRP), respectively (Figure 10.12f). These results demonstrated that the lateral spatial resolution was improved, and the DOF became deeper in TPLSM by using an HRP beam.

At the glass–water interface, the intensity pattern of a 0.17 μm fluorescent bead illuminated with an 800 nm LP beam was significantly elongated along the x-axis in comparison with that in agarose gel (Figures 10.12a and 10.13a). The lateral intensity pattern in the HRP beam at the focus also remained cylindrically symmetric at the interface (Figures 10.12c and 10.13c). The FWHM values along the x-axis in the LP and HRP beams were determined to be 510 and 188 nm,

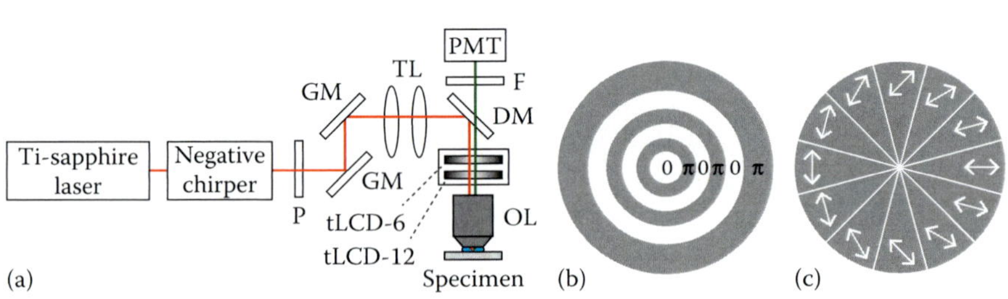

Figure 10.11

(a) Schematic of our TPLSM system utilizing an HRP beam. F: filter; GM: galvano mirrors; tLCD: transmissive liquid crystal device; 6: 6-ring divided phase mask; 12: 12-piece divided polarization converter; OL: objective lens; P: polarizer; PMT: photomultiplier tube; TL: tube lenses. (b) Theoretical phase distribution generated by tLCD-6. (c) Schematic of the tLCD-12 behaving as a segmented half-wave plate. White arrows indicate the orientation of the fast axis in each divided region.

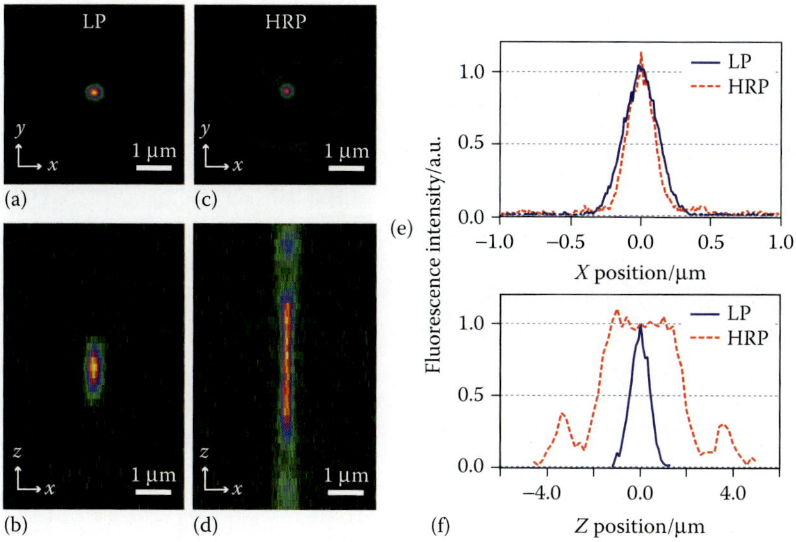

Figure 10.12

Lateral (x–y) and axial (x–z) images of fluorescent beads with a diameter of 0.17 μm in agarose gel and the intensity profiles under 800 nm excitation. (a) and (c): Intensity patterns in the focal plane ($z = 0$) for focusing the LP beam (polarized along the x-axis) and the converted HRP beam, respectively. (b) and (d): Measured intensity distributions in the x–z plane for the focusing of the LP beam and the HRP beam, respectively. (e) and (f): Intensity profiles across the center of the intensity distribution along the x- and z-axes, respectively. The solid (blue) and dashed (red) lines are the intensity profiles with the LP beam and the HRP beam, respectively.

respectively (Figure 10.13e). The FWHM value in the HRP beam at the interface became smaller than that in agarose gel, while that in the LP beam became larger. When the polarization direction of the LP beam was changed by rotating the half-wavelength plate, the elongated direction of the intensity distributions was in accord with the polarization direction. In addition, the intensity distributions of all isolated beads in the field of view were elongated along the polarization direction. Thus, this elongation could be interpreted by the enhancement of the relative contribution of the longitudinal electric field, similar to the result obtained at the glass–air interface (Dehez et al. 2009). The intensity distribution in the x–z plane around the focus of the HRP beam was also elongated along the z-axis (Figure 10.13d).

Next, we observed fluorescent yellow–green beads with 0.20 μm diameters aggregated on the coverslip. While the fluorescence image obtained by the LP beam was very blurred and individual beads could not be distinguished (Figure 10.14a), in the same field, each of the beads could be distinguished by the HRP beam (Figure 10.14b). This higher spatial resolution imaging with an HRP beam may have been realized by its smaller focal spot at the glass–water interface, as shown in Figure 10.13.

To confirm that this method of improving the lateral resolution is applicable to biological samples, we observed microtubules immunostained with Alexa Fluor 488 dye in fixed COS-7 cells on a coverslip. Fluorescence images obtained using an 800 nm HRP beam clearly revealed fine structures of microtubules that

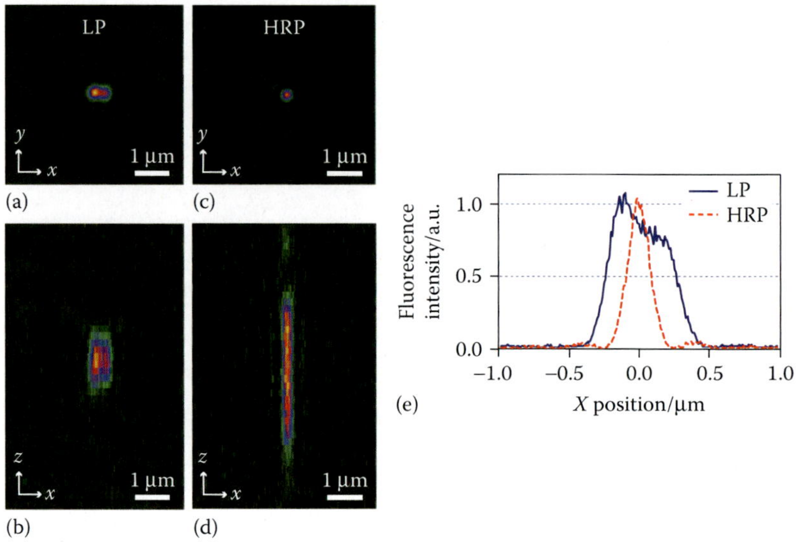

Figure 10.13

Lateral (*x*–*y*) and axial (*x*–*z*) images of fluorescent beads with a diameter of 0.17-μm at the glass–water interface (see the inset in [e]) and the intensity profiles under 800 nm excitation. (a) and (c): Images with the LP beam (polarized along the *x*-axis) and the HRP beam in the focal plane ($z = 0$), respectively. The beads are attached to the cover glass and surrounded by water. (b) and (d): Measured intensity distributions in the *x*–*z* plane for the focusing of the LP beam and the HRP beam, respectively. (e) Intensity profiles along the *x*-axis. The blue and red lines indicate the intensity profiles with the LP beam and the HRP beam, respectively.

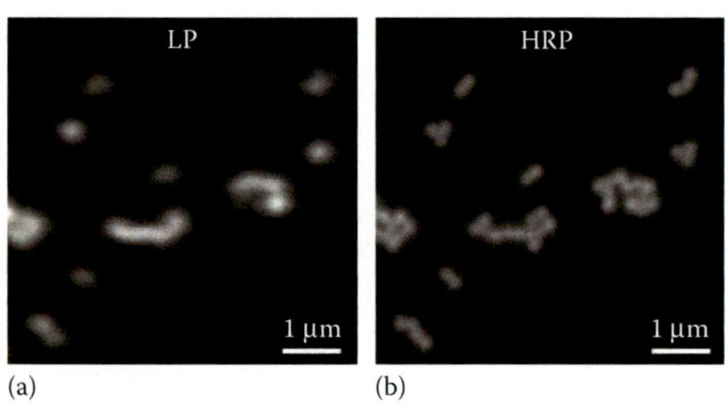

Figure 10.14

Two-photon images of aggregated fluorescent beads with a diameter of 0.20-μm at the glass–water interface by the LP beam (a) and the HRP beam (b) in the focal plane.

the LP beam could not identify at the same wavelength in the same field of view (Figure 10.15a and b). Slight contrast degradation was observed in the case of the HRP beam focusing, probably resulting from slightly remaining side lobes (Figures 10.12 and 10.13). Moreover, similar results were obtained in the samples stained with other fluorescent dyes, including Alexa Fluor 405 or Alexa Fluor 546 (Figure 10.15c–f). Because our tLCDs are applicable to various excitation wavelengths (Kozawa et al. 2011), fluorescent images obtained by other HRP beams at 910 nm excitation also demonstrated similarly improved resolution

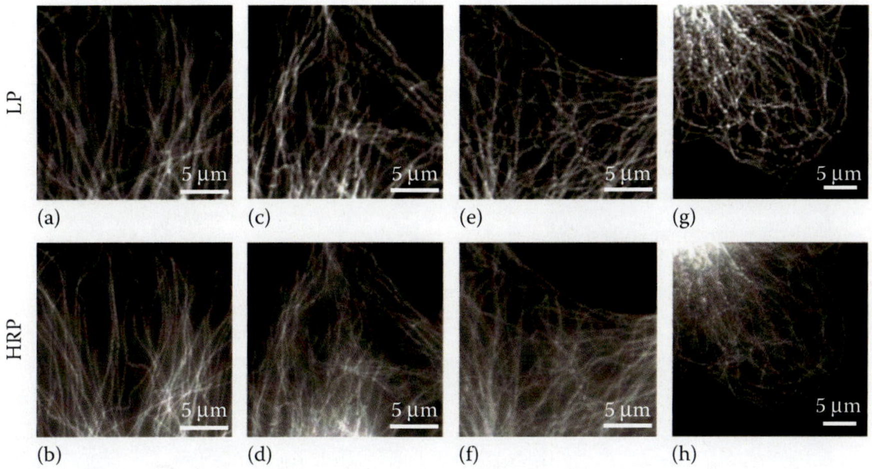

Figure 10.15

Images of microtubules in fixed COS-7 cells that were visualized by various fluorescent dyes with the LP beam (a, c, e, and g) and the HRP beam (b, d, f, and h). These microtubules in fixed COS-7 cells were stained by Alexa Fluor 488 (a and b), Alexa Fluor 405 (c and d), and Alexa Fluor 546 (e–h). Excitation wavelengths of 800 nm (a–f) and 910 nm (g and h) were used.

(Figure 10.15g and h). These results suggest that HRP beams are superior for the visualizing fine intracellular structures using various fluorescent dyes.

Finally, we observed structures moving in a living cell by labeling them with a fluorescent protein. F-actin in living COS-7 cells was visualized by Lifeact (Riedl et al. 2008) fused with mTFP1 (Ai et al. 2006). The time series of images taken by the HRP beam revealed filopodia movement at a higher spatial resolution compared with the LP beam (Figure 10.16). While some filopodia in images were out of focus during cell motility in the case of the LP beam, they were kept in focus in the HRP beam (Figure 10.16b). This was probably because the DOF in the HRP beam was deeper than that in the LP beam. Next, we measured the resolution with the LP and HRP beams in time-lapse images. Because simultaneous observation of living cells with LP and HRP beams was not possible, the resolution was estimated using the intensity profile of the line perpendicular to the tip by assuming that the width of the tip of the filopodia was smaller than our achieved lateral spatial resolution. The widths measured from these images with the LP and HRP beams were 294 and 229 nm, respectively (Figure 10.16c). These values were consistent with the FWHM values from the fluorescent beads in agarose gel (Figure 10.12). These results indicated that the HRP beam could also be utilized to improve the spatial resolution in living cells and for in-focus observation of living cells for 15 min, even if they are moving along the longitudinal axis.

10.5 Two-Photon Excitation STED Microscopy

In this section, we show a newly constructed system for TP-STED microscopy consisting of a STED laser light source added into a conventional TPLSM system. The STED light beam was modulated by tLCDs to create a vortex beam. Here, we demonstrate this system's effectiveness for phase shift compensations, its applicability for multiple wavelengths, and its use in super-resolution microscopy (Otomo et al. 2014).

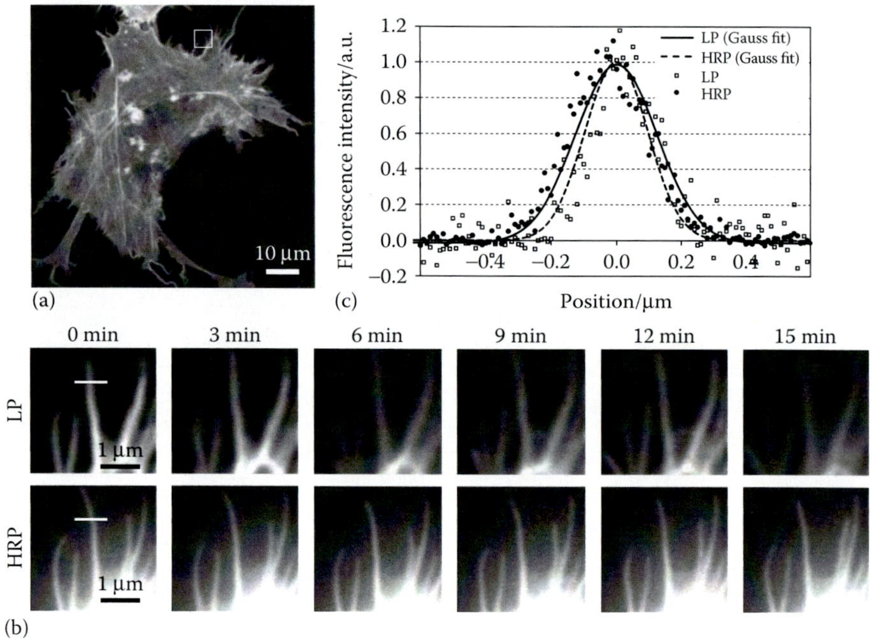

Figure 10.16

(a) COS-7 cells expressing Lifeact-tagged mTFP1. Time-lapse images in (b) were obtained in the area indicated by a rectangle in (a) for 15 min with 15 s per frame with the LP beam (upper panels) and the HRP beam (lower panels). The intensity profiles in (c) represented by filled circles (LP beam) and open squares (HRP beam) show the intensity distributions along the white lines in (b). The curves in (c) represent the Gaussian fit to the intensity profile for the LP beam (solid line) and the HRP beam (dashed line), respectively. The excitation wavelength was 860 nm.

A schematic of our optical setup is shown in Figure 10.17a. A two-photon excitation light (wavelength 900 nm) was generated using a Ti:Sa laser, and its pulse shape at the position of specimen was compensated by a negative chirper, as in the HRP experiment. The light source for STED was an optically pumped, 577 nm, semiconductor, continuous-wave (CW) laser (Genesis CX 577-3000, Coherent). Spatial mode cleaning of the STED light was accomplished using a

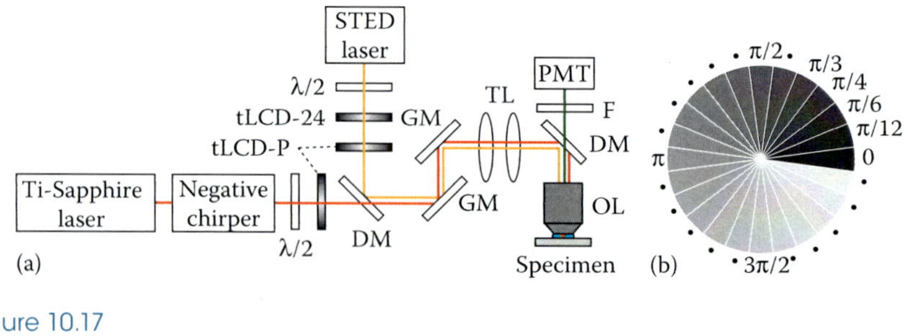

Figure 10.17

(a) Schematic of our TP-STED system using tLCDs. DM: dichroic mirror; F: filter; GM: galvano mirrors; tLCD: transmissive liquid crystal device; P: plain cell; 24: 24-piece divided phase mask; M: mirror; OL: objective lens; PMT: photomultiplier tube; TL: tube lenses; λ/2: halfwave plate. (b) Theoretical phase distribution of a vortex beam generated by tLCD-24.

spatial filter. The linear polarization directions for both beams were oriented toward LC molecules in tLCDs by passing through half-wave plates. Two types of light, which were modulated by tLCDs as noted below, were merged at the first dichroic mirror (RDM680, Olympus) and introduced into a galvanomirror scanner and an upright microscope, as was used in the HRP experiment. Both beams were reflected by the second dichroic mirror (custom-made filter combining a 577 nm notch with an 880 nm-edge short pass filter; Asahi Spectra) and were focused on a specimen using a water immersion objective lens ($NA = 1.20$). Fluorescent light was collected by the objective lens, passed through the second dichroic mirror and emission filters, and finally detected with a PMT.

We used two types of tLCDs; one was a plain cell tLCD (tLCD-P) with homogeneously aligned LC molecules functioning as an applied voltage-dependent variable wave plate. The other was a 24-piece divided phase mask (tLCD-24) that enabled production of a spiral-phase distribution with 24 steps around the center of the beam. The two-photon excitation light was modulated into a circular polarization at the position of the objective lens by one tLCD-P. As shown by the phase distribution in Figure 10.17b, by using the tLCD-24 mask, the STED light was converted into a vortex beam for which the topological charge was 1. This vortex beam was passed through a tLCD-P installed in the optical path, which tuned its polarization at the position of the objective lens.

For STED microscopy, the focus of a vortex beam should have a steep central dark spot, indicating that a higher spatial resolution can be achieved as the intensity at the center of the dark spot is reduced (Hell 2003). As shown in Figure 10.8, a vortex beam depends on the polarization of a focused laser beam. To focus a vortex beam, the chirality of the circular polarization was chosen so as to create a fine dark spot at the focus. Figure 10.18a shows the calculated focal pattern and the intensity profile across the center for the focusing of a circularly polarized

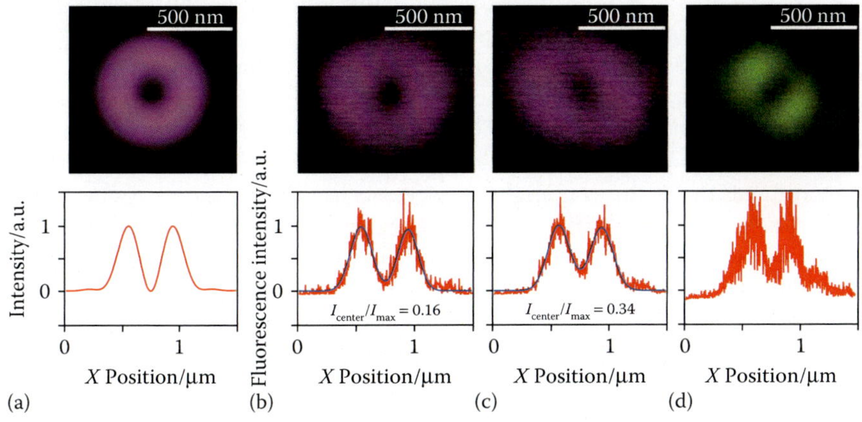

Figure 10.18

(a) Calculated intensity distribution in the focal plane for focusing of the circularly polarized 577 nm vortex beam created by tLCD-24. (b)–(d): Fluorescence images of a fluorescent bead directly excited by vortex beams. The lower panels show the fluorescence intensity profiles across the intensity center along the *x*-axis. (b) Image of an orange fluorescent bead with a diameter of 0.17 μm excited by 577 nm light with circular polarization at the objective lens. (c) Image of an orange fluorescent bead with a diameter of 0.17 μm excited by 577 nm light with circular polarization at tLCDs. (d) Image of a yellow-green fluorescent bead with a diameter of 0.17 μm excited by 473 nm light with circular polarization at the objective lens.

vortex beam produced by the tLCD-24. As shown in Figure 10.18a, the donut-shaped pattern was obtained at the focus, indicating that the effect of the 24-step discrete phase variation in the tLCD-24 was negligible in forming an ideal donut-shaped pattern as the STED light. Therefore, we first estimated the PSFs of various vortex beams created using our setup by acquiring fluorescent images of a tiny fluorescent bead directly excited by STED lights.

For the case of a circularly polarized vortex beam at the position of the tLCDs, a difference in phase shifts between the s- and p-polarized beams generated in the optical path resulted in elliptical polarization (aspect ratio of approximately 8) at the position of the objective lens. By tuning the voltage applied to a tLCD-P (see Figure 10.17a), we generated circularly polarized light at the position of the objective lens and confirmed PSF with a tiny fluorescent bead image, as shown in Figure 10.18b. The intensity distribution of the generated circularly polarized vortex beam at the focus was almost identical to the calculated ideal focal pattern shown in Figure 10.18a. In contrast, a slightly distorted donut shape was obtained when the applied voltage was tuned so as to produce a circularly polarized vortex beam at the position of the tLCDs (Figure 10.18b). The ratios of central intensity to maximum intensity (I_{center}/I_{max}) were estimated for the images in Figure 10.18a and b. The I_{center}/I_{max} values were 0.16 and 0.34 for the images in Figure 10.18a and b, respectively. Thus, the phase shift caused by the microscope's optics was compensated for by adjusting the voltage applied to the tLCD-P.

tLCDs can modify optical retardation by tuning applied voltages and are applicable for beams over a wide wavelength region from visible to NIR (Kozawa et al. 2011; Ipponjima et al. 2014). Thus, to demonstrate the vortex beam of another wavelength using the same tLCDs, a diode-pumped solid-state laser of 473 nm (SDL-473-005T, Shanghai Dream Laser Technology) was used. A 473 nm light was introduced into the same optical path as the 577 nm STED light, except for a spatial filter and a second dichroic mirror; in this case, the second dichroic mirror was replaced with a 505 nm-long pass filter (DM505, Olympus). We confirmed that vortex beams of different wavelengths were provided by the same set of tLCDs. Figure 10.18c shows an image of a 0.17 μm YG bead that was directly excited by a 473 nm vortex beam. Although its PSF was rather distorted, the same set of tLCDs produced a vortex beam through the application of a different suitable voltage. This distortion might have been caused by some optical aberrations, including astigmatism of the 473 nm laser beam. As will be discussed in the next section, astigmatism is known to perturb the hollow focal pattern of a vortex beam (Fernández et al. 2009; Deng et al. 2010).

The focal planes of the two-photon excitation light and STED light used in our system were verified. Figure 10.19 shows an orthogonal view of the *xyz*-fluorescent images of two different types of fluorescent beads that were placed on the same cover slip. Because several previous studies on STED employed circularly polarized beams for excitation, we adopted another tLCD-P to generate a laser beam that was circularly polarized at the position of the objective lens. Each of their focal planes became nearly identical by simply introducing these laser beams in parallel into the corresponding laser inputs of the microscope.

Next, to evaluate the spatial resolution of our TP-STED system, fluorescent images of green beads with diameters of 0.10 μm were acquired with both a conventional TPLSM system and our TP-STED microscope. As shown in Figure 10.20a, after applying the STED light, the fluorescent bead image which had exhibited an isotropic shape in the focal plane became elliptical as the

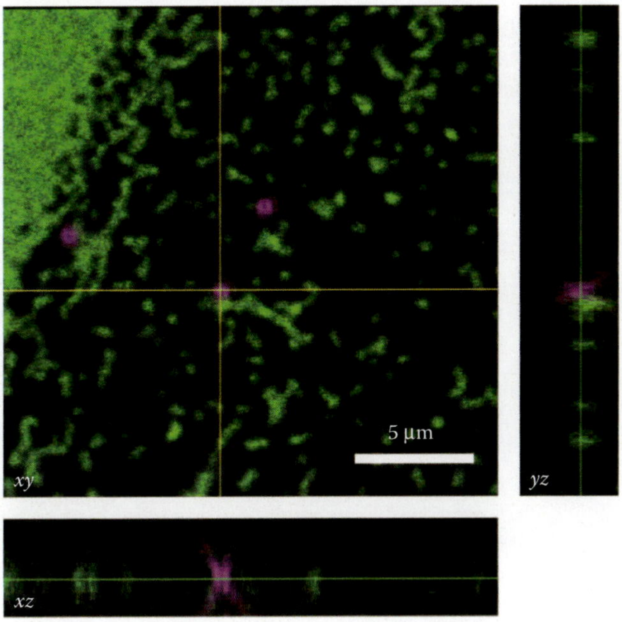

Figure 10.19

Merged fluorescence image of green fluorescent beads with diameters of 0.10 μm (green) and orange fluorescent beads with diameters of 0.17 μm (magenta) placed on the same cover slip.

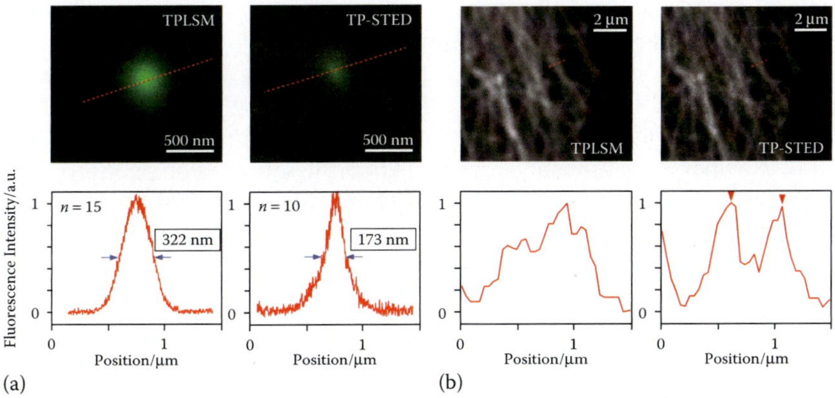

Figure 10.20

Comparisons of TPLSM and TP-STED images. (a) A green fluorescent bead with a diameter of 0.10 μm. The lower panels show the averaged fluorescence intensity profiles of *n* beads across the intensity center along the red dashed lines in the fluorescent images. Inlet length value indicates the full width at half maximum. (b) Microtubule networks in fixed COS-7 cells after immunostaining with antibodies conjugated with the fluorescent dye ATTO 425. The lower panels show the fluorescence intensity profiles across the red dashed lines in the fluorescent images.

size itself became smaller. Using the averaged fluorescence intensity profiles acquired with conventional TPLSM and our TP-STED, we estimated that the FWHM values were 322 and 173 nm, respectively. To estimate more accurately the PSFs and spatial resolution of the microscopy, measurements of fluorescent images of smaller-sized beads would be required. However, the sensitivity of our conventional TPLSM system was insufficient to clearly visualize such a small fluorescent bead due to weak fluorescence.

Finally, we demonstrated that our method was applicable for biological specimens by observing microtubules in fixed COS-7 cells that were immunostained with fluorescent dye-marked antibodies. ATTO 425 was chosen as the fluorescent dye because the wavelength of STED light did not overlap with its absorption band. As shown in Figure 10.20b, TP-STED could more clearly visualize the fine network structures of microtubules.

10.6 Summary and Outlook

As described in the two previous sections, the lateral spatial resolution of TPLSM for biological specimens was improved by utilizing tLCDs. Because tLCDs are electrically controllable, they can be used at a wide range of wavelengths. Our techniques are applicable to fixed and living cells using various fluorescent dyes or proteins and have the potential to visualize nanostructures of the deep regions of various biological specimens.

At first, we demonstrated that an HRP beam in the NIR region was generated from an LP beam by placing only a pair of tLCDs in front of the objective lens. The lateral spatial resolution in TPLSM was improved and the DOF was expanded using an HRP beam similar to a "light needle" (Wang et al. 2008; Kozawa et al. 2011). In our previous work with CLSM (Kozawa et al. 2011), the DOF was suppressed because of the confocal pinhole. In contrast, the absence of the confocal pinhole in TPLSM resulted in a deeper DOF without degradation at a higher lateral resolution. This feature is attractive for biological researchers in several situations. As shown in Figure 10.16, in-focus time-lapse images of longitudinally moving objects were obtained. In addition, a deeper DOF can significantly minimize the acquisition time for 3D image by reducing the number of 2D images that must be captured from the same region. Conversely, this methodology has the merit that almost all fluorescent molecules used in conventional TPLSM are applicable. Since some super-resolution microscopies require selecting suitable fluorescent dyes or proteins (Hell 2007), this feature is important for biologists who do not want to change fluorescent dyes that are reliable for their observations.

Next, by adding tLCDs and a CW yellow laser to our STED, we developed a new TP-STED system with the following useful properties. First, by adjusting the voltage applied to a tLCD-P, we compensated for various phase shifts in the optical path to acquire a fine donut-shaped PSF and confirmed that this vortex beam, which had a 24-part helical phase distribution of 2π created by a tLCD-24 (see Figure 10.17b), was sufficient for STED microscopy. Second, our TP-STED microscopy system was employable at different wavelengths of the STED laser by only tuning the voltages applied to tLCDs. Finally, by modifying a conventional TPLSM system just by adding tLCDs and a depletion laser light source, we developed a TP-STED system. This methodology achieved super-resolution microscopy more readily. Moreover, one feature of TPLSM is that various fluorescent dyes can be excited simultaneously using a single wavelength light (Nemoto 2008; Ogata et al. 2012). Thus, combined with a wavelength-tunable light source for STED, this methodology might provide for super-resolution microscopy to visualize multiple components in living specimens using a variety of fluorophores.

The spatial resolution of our present TP-STED microscopy system was estimated to be >100 nm. In contrast to the TPLSM system utilizing an HRP beam, the spatial resolution of the TP-STED system in previous studies was reported as <100 nm (Moneron and Hell 2009; Li et al. 2009; Bethge et al. 2013; Takasaki et al. 2013). One reason might be that we used a CW laser for STED; the STED efficiency was reported to be superior under pulse-type STED microscopy (Bethge et al. 2013; Takasaki et al. 2013). Another was that the vortex beam for STED was slightly distorted, which resulted in a finite intensity at the center of the focus (Figure 10.18a) due to optical aberrations. The optical aberration arising from mirrors and the focusing lens in the microscope setup may have had an influence on the donut-shaped STED beam at the focus produced by a focused vortex beam with circular polarization. In particular, astigmatism in the focusing of a vortex beam is known to cause a significant deterioration of its donut shape at the focus (Singh et al. 2009; Deng et al. 2010). Figure 10.21 shows the calculated focal patterns for the focusing of a circularly polarized vortex beam with several degrees of astigmatism based on a theoretical formula (Deng et al. 2010). The wave-front aberration function for primary astigmatism at the pupil plane of a focusing objective lens can be expressed as $\Phi\ (\rho, \varphi) = \exp\ (ikA_{as}\ \rho^2\cos^2\ \varphi)$, where ρ and φ are the lateral and azimuthal positions at the pupil, respectively, and A_{as} is the aberration coefficient for astigmatism. As the degree of astigmatism increases, the donut-shaped, dark spot pattern of a vortex beam is changed into a two-lobe pattern. In addition, as shown in Figure 10.21e, the intensity at the center is no longer zero for the vortex beam affected by astigmatism. This degradation of the donut pattern may cause the depletion of fluorescence emission by the STED effect, even at the center of an excitation focal spot, leading to an inhibited resolution enhancement in STED microscopy. In the deeper regions of biological specimens which TPLSM targets, various optical aberrations are known to appear and hamper the proper focusing of lights. By introducing other types of SLMs, such as LCOS devices, deformable mirrors,

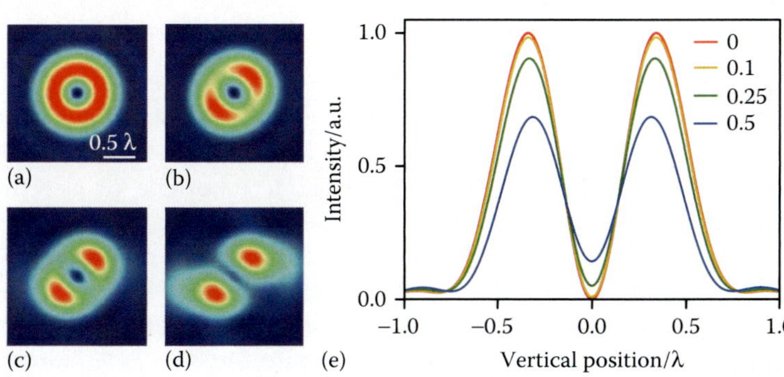

Figure 10.21

The intensity distributions at the focus of a circularly polarized vortex beam without astigmatism (a) and with astigmatism of A_{as} = 0.1 (b), 0.25 (c), and 0.5 (d). The intensity profiles along the vertical axis through the center of the beam shown in (a–d) are plotted in (e).

or the new types of tLCDs, processes for canceling the optical aberrations are required (Fernández et al. 2009; Shaw et al. 2010).

From another point of view, the temporal resolution of optical microscopy is equally significant as the spatial resolution in understanding biological events. As most TPLSM systems employ a single-point laser scanning method using moving mirrors, their temporal resolution primarily depends on the speed of the physical movement of these mirrors. Therefore, several improvements of the temporal resolution of TPLSM have been demonstrated in recent years. The most popular method is utilizing resonant mirrors which have enabled video-rate scanning TPLSM (Fan et al. 1999). If the specimen has a sufficient fluorescent signal or the detector has a sufficient sensitivity, the temporal resolution of our tLCD-based TPLSM system can be improved simply by exchanging scanning mirrors to resonant ones. Conversely, alternative scanning methods besides single-point laser scanning have also been proposed. One method is based on light-sheet microscopy (Huisken et al. 2004). By utilizing a two-photon excitation process, light-sheet microscopy has been reported to attain deeper penetration depths (Truong et al. 2011; Zanacchi et al. 2013) and reduce the influence of light scattering (Maruyama et al. 2014). Recently, Chen et al. (2014) developed a new light-sheet microscopy possessing a high spatiotemporal resolution by utilizing Bessel beams; this achievement strongly suggests the applicability of HRP beams. The other is a multipoint laser scanning method utilizing spinning disk scanners (Bewersdorf et al. 1998; Fujita et al. 1999; Shimozawa et al. 2013), beam splitting units (Fittinghoff et al. 2000; Nielsen et al. 2001), or SLMs (Matsumoto et al. 2014). In our group's recent study, the field of view of multipoint scanning TPLSM was dramatically improved by utilizing a high-peak power Yb-based laser while maintaining a high temporal resolution (Otomo et al. 2015). As the next step, we must consider the spatial resolution of this system by combining with our tLCD methodologies.

TPLSM has extended the range of LSM, especially where dynamic imaging in living specimens is needed. In the near future, we hope that super-resolution two-photon excitation microscopy including our tLCD methodologies flourishes, not only for visualizing a variety of biological nanostructures, but also for assisting diagnosis and surgery.

Acknowledgments

We thank Dr. N. Hashimoto, Mr M. Kurihara, and Mr K. Matsumoto of Citizen Holdings Co., Ltd. Tokorozawa, Japan for kindly providing tLCDs. We are also grateful for the helpful advice from Professor H. Yokoyama of New Industry Creation Hatchery Center, Tohoku University, Sendai, Japan; Dr. K. Kobayashi and Dr. Y. Matsuo of the Nikon Imaging Center at Hokkaido University, Sapporo, Japan; and Dr. K. Iijima and Dr. R. Kawakami of the Laboratory of Molecular and Cellular Biophysics in the Research Institute for Electronic Science, Hokkaido University, Sapporo, Japan. This work was supported by Core Research for Evolutional Science and Technology (CREST), Japan Science and Technology Agency (JST), Japan; JSPS KAKENHI Grants Numbers 25840044, 25560411, 22300131, 22113005, 26242082, and 25-1699 of the Ministry of Education, Culture, Sports, Science and Technology (MEXT) Japan; Nano-Macro Materials, Devices and System Research Alliance (MEXT); and Network Joint Research Center for Materials and Devices (MEXT).

References

Ai, H.-W., Henderson, J. N., Remington, S. J., and Campbell, R. E., Directed evolution of a monomeric, bright and photostable version of Clavularia cyan fluorescent protein: Structural characterization and applications in fluorescence imaging, *Biochemical Journal* 400 (2006): 531–540.

Bethge, P., Chéreau, R., Avignone, E. et al., Two-photon excitation STED microscopy in two colors in acute brain slices, *Biophysical Journal* 104 (2013): 778–785.

Bewersdorf, J., Pick, R., and Hell, S. W., Two-photon excitation STED microscopy in two colors in acute brain slices, *Optics Letter* 23 (1998), 655.

Booth, M. J., Adaptive optical microscopy: The ongoing quest for a perfect image, *Light: Science & Applications* 3 (2014): e165.

Born, M. and Wolf, E., *Principles of Optics: Electromagnetic Theory of Propagation, Interference and Diffraction of Light* (7th ed.) (Cambridge: Cambridge University Press, 1999).

Chen, B.-C., Legant, W. R., Wang, K. et al., Lattice light-sheet microscopy: Imaging molecules to embryos at high spatiotemporal resolution, *Science* 346 (2014): 1257998.

Conchello, J.-A. and Lichtman, J. W., Optical sectioning microscopy, *Nature Methods* 2 (2005): 920–931.

Dehez, H., Piché, M., and De Koninck, Y., Enhanced resolution in two-photon imaging using a TM01 laser beam at a dielectric interface, *Optics Letter* 34 (2009): 3601–3603.

Deng, S., Liu, L., Cheng, Y. et al., Effects of primary aberrations on the fluorescence depletion patterns of STED microscopy, *Optics Express* 18 (2010): 1657–1666.

Denk, W., Piston, D. W., and Webb, W. W., Multi-photon molecular excitation in laser-scanning microscopy, in *Handbook of Biological Confocal Microscopy* (3rd ed.), J. B. Pawley (ed.) (New York: Springer Science + Business Media, 2006), pp. 535–549.

Denk, W., Strickler, J. H., and Webb, W. W., Two-photon laser scanning fluorescence microscopy, *Science* 248 (1990): 73–76.

Fan, G. Y., Fujisaki, H., Miyawaki, A. et al., Video-rate scanning two-photon excitation fluorescence microscopy and ratio imaging with cameleons, *Biophysical Journal* 76 (1999): 2412–2420.

Fernández, E. J., Prieto, P. M., and Artal, P., Wave-aberration control with a liquid crystal on silicon (LCOS) spatial phase modulator, *Optics Express* 17 (2009): 11013–11025.

Fittinghoff, D. N., Wiseman, P. W., and Squier, J. A., Widefield multiphoton and temporally decorrelated multifocal multi-photon microscopy, *Optics Express* 7 (2000): 273–279.

Fujita, K., Nakamura, O., Kaneko, T. et al., Real-time imaging of two-photon-induced fluorescence with a microlens-array scanner and a regenerative amplifier, *Journal of Microscopy* 194 (1999): 528–531.

Göppert-Mayer, V. M., Uber Elementarrakte mit zwei Quantensprungen, *Annalen der Physik* 9 (1931): 273–294.

Gould, T. J., Burke, D., Bewersdorf, J., and Booth, M. J., Adaptive optics enables 3D STED microscopy in aberrating specimens, *Optics Express* 20 (2012): 20998–21009.

Hao, X., Kuang, C., Wang, T., and Liu, X., Effects of polarization on the de-excitation dark focal spot in STED microscopy, *Journal of Optics* 12 (2010): 115707.

Hell, S. W., Toward fluorescence nanoscopy, *Nature Biotechnology* 21 (2003): 1347–1355.

Hell, S. W., Far-field optical nanoscopy, *Science* 316 (2007): 1353–1358.

Hell, S. W. and Wichmann, J., Breaking the diffraction resolution limit by stimulated emission: stimulated-emission-depletion fluorescence microscopy, *Optics Letter* 19 (1994): 780–782.

Helseth, L. E., Roles of polarization, phase and amplitude in solid immersion lens systems, *Optics Communications* 191 (2001): 161–172.

Hoover, E. E. and Squier, J. A., Advances in multiphoton microscopy technology, *Nature Photonics* 7 (2013): 93–101.

Huisken, J., Swoger, J., Del Bene, F. et al., Optical sectioning deep inside live embryos by selective plane illumination microscopy, *Science* 32 (2004): 1007–1009.

Iketaki, Y., Watanabe, T., Bokor, N., and Fujii, M., Investigation of the center intensity of first- and second-order Laguerre—Gaussian beams with linear and circular polarization, *Optics Letters* 32 (2007): 2357–2359.

Ipponjima, S., Hibi, T., Kozawa, Y. et al., Improvement of lateral resolution and extension of depth of field in two-photon microscopy by a higher-order radially polarized beam, *Microscopy (Oxford)* 63 (2014): 23–32.

Kozawa, Y., Hibi, T., Sato, A. et al., Lateral resolution enhancement of laser scanning microscopy by a higher-order radially polarized mode beam, *Optics Express* 19 (2011): 15947–15954.

Kozawa, Y. and Sato, S., Sharper focal spot formed by higher-order radially polarized laser beams, *Journal of the Optical Society of America A* 24 (2007): 1793–1798.

Kozawa, Y. and Sato, S., Focusing of higher-order radially polarized Laguerre—Gaussian beam, *Journal of the Optical Society of America A* 29 (2012): 2439–2443.

Li, Q., Wu, S. S., and Chou, K. C., Subdiffraction-limit two-photon fluorescence microscopy for GFP-tagged cell imaging, *Biophysical Journal* 97 (2009): 3224–3228.

Maruyama, M., Oshima, M., Kajiura-Kobayashi, H. et al., Wide field intravital imaging by two-photon-excitation digital-scanned light-sheet microscopy (2p-DSLM) with a high-pulse energy laser, *Biomedical Optics Express* 5 (2014): 3311–3325.

Matsumoto, N., Okazaki, S., Fukushi, Y. et al., An adaptive approach for uniform scanning in multifocal multiphoton microscopy with a spatial light modulator, *Optics Express* 22 (2014): 633–645.

Maurer, C., Jasacher, A., Bernet, S., and Ritsch-Marte, M., What spatial light modulators can do for optical microscopy, *Lasers & Photonics Review* 5 (2011): 81–101.

Minsky, M., Memoir on inventing the confocal scanning microscope, *Scanning* 10 (1988): 128–139.

Moneron, G. and Hell, S. W., Two-photon excitation STED microscopy, *Optics Express* 17 (2009): 14567–14573.

Nemoto, T., Living cell functions and morphology revealed by two-photon microscopy in intact neural and secretory organs, *Molecules and Cells* 26 (2008), 113–120.

Nielsen, T., Fricke, M., Hellwig, D., and Andersen, P., High efficiency beam splitter for multifocal multi-photon microscopy, *Journal of Microscopy* 201 (2001): 368–376.

Ogata, S., Miki, T., Seino, S. et al., A novel function of Noc2 in agonist-induced intracellular Ca^{2+} increase during zymogen-granule exocytosis in pancreatic acinar cells, *PLoS One* 7 (2012), e37048.

Otomo, K., Hibi, T., Kozawa, Y. et al., Two-photon excitation STED microscopy by utilizing transmissive liquid crystal devices, *Optics Express* 22 (2014): 28215–28221.

Otomo, K., Hibi, T., Murata, T. et al., Multi-point scanning two-photon excitation microscopy by utilising a high-peak-power 1042-nm laser, *Analytical Sciences* 31 (2015): 307–313.

Quabis, S., Dorn, R., Eberler, M. et al., Focusing light to a tighter spot, *Optics Communications* 179 (2000): 1–7.

Reuss, M., Engelhardt J., and Hell, S. W., Birefringent device converts a standard scanning microscope into a STED microscope that also maps molecular orientation, *Optics Express* 18 (2010): 1049–1058.

Richards, B. and Wolf, E., Electromagnetic diffraction in optical systems II. Structure of the image field in an aplanatic system, *Proceedings of the Royal Society of London. Series A, Mathematical and Physical Sciences* 253 (1959): 358–379.

Riedl, J., Crevenna, A. H., Kessenbrock, K. et al., Lifeact: A versatile marker to visualize F-actin, *Nature Methods* 5 (2008): 605–607.

Shaw, M., Hall, S., Knox, S. et al., Characterization of deformable mirrors for spherical aberration correction in optical sectioning microscopy, *Optics Express* 18 (2010): 6900–6913.

Sheppard, C. J. R. and Choudhury, A., Annular pupils, radial polarization, and superresolution, *Applied Optics* 43 (2004); 4322–4327.

Shimozawa, T., Yamagata, K., Kondo, T. et al., Improving spinning disk confocal microscopy by preventing pinhole cross-talk for intravital imaging, *Proceeding of the National Academy of Sciences of the United States of America* 110 (2013): 3399–3404.

Siegman, A. E., *Lasers* (Mill Valley, CA: University Science Books, 1986).

Singh, R. K., Senthilkumaran P., and Singh, K., Tight focusing of vortex beams in presence of primary astigmatism, *Journal of the Optical Society of America A* 26 (2009): 576–588.

Spence, D. E., Kean, P. N., and Sibbett, W., 60-fsec pulse generation from a self-mode-locked Ti:sapphire laser, *Optics Letters* 16 (1991): 42–44.

Takasaki, K. T., Ding, J. B., and Sabatini, B. L., Live-cell superresolution imaging by pulsed STED two-photon excitation microscopy, *Biophysical Journal* 104 (2013): 770–777.

Tovar, A. A., Production and propagation of cylindrically polarized Laguerre–Gaussian laser beams, *Journal of the Optical Society of America A* 15 (1998): 2705–2711.

Truong, T. V., Supatto, W., Koos, D. S. et al., Deep and fast live imaging with two-photon scanned light-sheet microscopy, *Nature Methods* 8 (2011): 757–760.

Wang, H., Shi, L., Lukyanchuk, B. et al., Creation of a needle of longitudinally polarized light in vacuum using binary optics, *Nature Photonics* 2 (2008): 501–505.

Yao, A. M. and Padgett, M. J., Orbital angular momentum: origins, behavior and applications, *Advances in Optics and Photonics* 3 (2011): 161–204.

Youngworth, K. and Brown, T., Focusing of high numerical aperture cylindrical-vector beams, *Optics Express* 7 (2000): 77–87.

Zanacchi, F. C., Lavagnino, Z., Faretta, M. et al., Light-sheet confined super-resolution using two-photon photoactivation, *PLoS One* 8 (2013): e67667.
Zhan, Q., Cylindrical vector beams: From mathematical concepts to applications, *Advances in Optics and Photonics* 1 (2009): 1–57.
Zhang, Z., You, Z., and Chu, D., Fundamentals of phase-only liquid crystal on silicon (LCOS) devices, *Light: Science & Applications* 3 (2014): e213.
Züchner, T., Failla, A., and Meixner, A., Light microscopy with doughnut modes: A concept to detect, characterize, and manipulate individual nanoobjects, *Angewandte Chemie (International ed. in English)* 50 (2011): 5274–5293.

11

Fluorescent Nanodiamonds in Biological and Biomedical Imaging and Sensing

Metin Kayci, Flavio M. Mor, and Aleksandra Radenovic

11.1 Introduction
11.2 General Properties of FNDs as Biomarkers and Biosensors
11.3 Nanoscale Imaging of Biological Systems with NV Centers Hosted in Nanodiamonds
11.4 Super-Resolution Microscopy with FNDs
11.5 Nanoscale Sensing of Biological Systems with NV Centers Hosted in Nanodiamonds
11.6 Manipulation Techniques
11.7 Conclusion
11.8 List of Acronyms
References

11.1 Introduction

Fluorescent biomarkers such as fluorescent proteins (FPs) [1], organic dyes [2], and quantum dots (QDs) [3] have been widely used for biological and biomedical imaging due to their fair brightness and biocompatibility [4,5]. However, these common fluorescent biomarkers will photobleach, blink, or unfortunately exhibit both. Therefore, recent efforts have been directed toward development of a biocompatible luminescent/fluorescent labels that neither photobleach nor blink. In addition, higher molecular brightness in such ideal fluorescent probes would allow deeper and more sensitive fluorescence tomography. Fluorescent nanodiamonds (FNDs) present a good candidate for such ideal fluorescent probe since they are 5–100 nm large [6], biocompatible particles with excellent photostability and have surface that can be easily functionalized [7]. These features have directed their use toward numerous demanding fluorescent imagining modalities such as

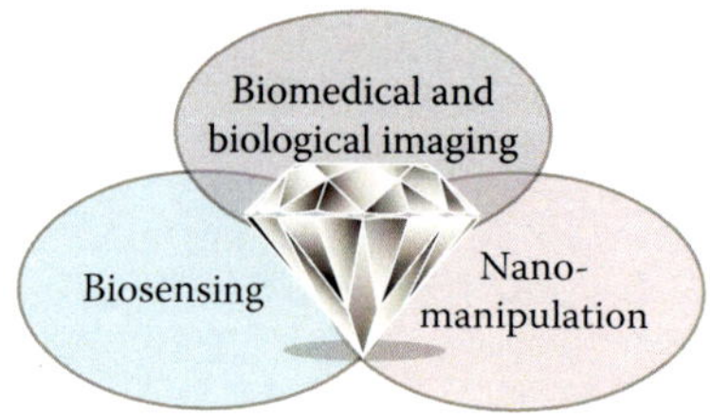

Figure 11.1

Fluorescent nanodiamonds are applied in numerous research fields. We have selected three areas that could be of potential interest to life scientists.

fluorescence lifetime imaging microscopy (FLIM) [8], FNDs are used as donors in fluorescence resonance energy transfer experiments (FRETs) [9], as probes in long-term *in vivo* tracking [10] together with *in vivo* single particle tracking [11], as well as in both super-resolution microscopy modalities such as localization-based [12], and by using stimulated emission depletion (STED) microscopy [13] which in turn is scanning-based. Moreover, the nitrogen vacancy (NV) centers in FNDs can act as optically readable sensors that could measure with very high precision variety of physical quantities including ultrasensitive magnetometry [14], sensing of the electrical fields [15], ion concentrations [16], and thermometry [17]. From a point of view of biological applications, it is extremely exciting to employ these nanoscale sensors in living cells as demonstrated by Kucsko et al. [17]. On the other hand, recent demonstration of the electron spin resonance (ESR) signal detection from a single protein using as a sensor single NV center in bulk diamond paved the way for the future experiments in the cells where the bulk diamond should be replaced by a nanodiamond with similar NV magnetic sensitivity [18].

In this chapter, from the vast body of work, we will focus first on the physical properties and optical characterization of FNDs that could be of interest to life scientists. Next, we outline the applications of FNDs in bioimaging and biosensing (Figure 11.1). Finally, we show interesting applications using optical tweezers and microfluidics chips that allow three-dimensional (3D) nanomanipulation of these multipurpose probes. Three-dimensional nanomanipulation is an essential prerequisite for their use in closed microfluidic and intracellular environments.

11.2 General Properties of FNDs as Biomarkers and Biosensors

In the ideal case, diamond is transparent, due to its large bandgap (5.5 eV), however presence of a defect can induce a level structure shortening of its bandgap that is sufficient to permit the excitation in the visible spectrum (750–400 nm corresponding to 0.65–3.1 eV). If these defects are stable inside the crystal lattice they perturb the energy level structure by introducing narrow levels within the bandgap.

If the defect transition energy is smaller than the bandgap of the hosting diamond then the emission from the defect will not be absorbed by the hosting material, which is in our case diamond being an indirect wide bandgap semiconductor. As the resulting sharp zero phonon line (ZPL) transition has a specific location in the emission spectrum and the diamond acquires a characteristic color the defect is so-called color center. More than 500 color centers have been investigated in the wide bandgap of diamond but only a few of them have been identified as bright and stable [19]. The two most common centers are the neutral center, NV^0 and the negatively charged center, NV^-. Among these color centers, NV^- center is identified to be the most promising one thanks to its spectral and spin properties compatible to applications ranging from quantum information processing to nanoscale metrology. NV^- center in diamond crystals is formed in C_{3v} symmetry by a nitrogen impurity adjacent to a carbon vacancy (shown in Figure 11.2a).

These defects might occur in nature under high pressure and temperature, conditions that occur in the protoplanetary disks of certain types of stars [23,24] and later isolated and identified in the meteorites [25]. Similarly in artificial nanodiamonds, NV centers are produced either under high pressure and temperature or can be generated through electron and ion irradiations [26]. The irradiation damages in the diamond lattice can be annealed to diffuse the vacancies

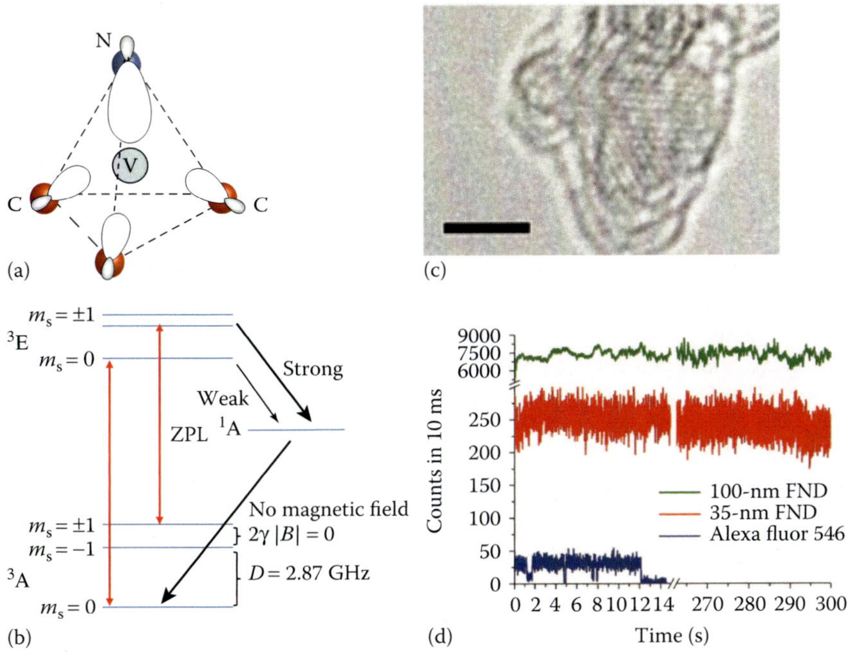

Figure 11.2

(a) A crystal model of the single NV^- center (shown in gray) hosted in the diamond crystal lattice having a substitutional nitrogen (shown in blue) adjacent to a carbon vacancy. (From Kayci, M. et al., *Nano Lett.*, 14, 5335, 2014. With permission.) (b) Energy diagram of the ground state ^3A, excited state ^3E, and metastable singlet state ^1A represented for NV. (From Kayci, M. et al., *Nano Lett.*, 14, 5335, 2014. With permission.) (c) High-resolution TEM image of 5 nm large nanodiamond revealing diamond (111) crystal planes. Scale bar: 2 nm. (Reprinted with permission from Smith, B. R. et al., *Small*, 5, 1649, 2009.) (d) Fluorescence time traces for a single 100-nm FND (green), a single 35-nm FND (red), and a single Alexa Fluor 546 dye (blue). To allow long-term observation dye molecule attached to a single dsDNA molecule, while FNDs were adhered to the glass coverslip. In the 300 s time window, fluorescence signal has been stable and no blinking or photobleaching has been observed for both sizes of FNDs. (Reprinted with permission from Fu, C. C. et al., *Proc. Natl. Acad. Sci. USA*, 104, 727, 2007.)

Table 11.1 Comparison of Fluorescent Properties for Single NV^-, Quantum Dot, Typical Organic Dye, EGFP Protein, and Cellular Proteins That Give Rise to Cellular Autofluorescence

Property	NV^- Defect in Nanodiamond	Quantum Dot [36]	Organic Dye [36]	EGFP	Autofluorescence of Protein Clusters
Size	>4 nm [21,37]	3–10 nm	<1 nm	~3 nm	>3 nm
Quantum yield	0.7–0.8	0.1–0.8	0.5–1	0.6	0.26 [38]
Lifetime	25 ns	10–100 ns	1–10 ns	1–4 ns [39]	NA
Molecular brightness	500–100 kcps	50–200 kcps	10–150 [40]	25 kcps [41]	2.1 kcps [41]
Bleaching	No	No	Yes	Yes	Yes
Blinking	No	Yes	Yes	Yes	Yes
Emission spectrum	680–800 nm	IR-UV selected by size	IR-UV selected by type	500–520 nm	250–650 nm
Toxicity	Low	From low to high	Dye dependent	None	None
Thermal stability	High	High	Low	Low	Low

to the implanted nitrogen atoms. Recently, defect hosting in nanodiamonds of size as small as 5 nm has been synthesized [21,27] (shown in Figure 11.2c). This opens a gate for nanoscale drug delivery and real-time monitoring of various physical quantities such as temperature [28], pressure [29], magnetic field, and ionic concentrations [30] in cellular environment.

A single NV^- defect exhibits two level quantum system and as there is a short lag between the excitation and subsequent decay it emits a photon at a time, enabling single-photon applications. It has nonphotobleaching nonblinking stable fluorescence characteristics with quantum yield close to unity at room temperature. In terms of bioimaging, these unique properties are extremely fitting, for example, broad fluorescence emission conveniently centered around 700 nm guarantees negligible interference with cellular autofluorescence (see Table 11.1), while the fluorescence signal of these defects, hosted in NDs larger than 5 nm, is extremely stable devoid of blinking and photobleaching [14,31]. Furthermore, it has been demonstrated that diamond nanocrystals hosting NV^- defects are biologically inert and do not induce cytotoxicity and apoptosis in the most tested cell lines [7,32]. Yet, numerous functionalization protocols took advantage of either hydrophilic oxygen-terminated diamond surface or hydrogen-terminated surface resulting in the applications where FNDs were coupled to proteins [33] and DNA molecules [34]. Recently, FNDs functionalization has been simplified exploiting silica encapsulation allowing better colloidal stability [35]. All above-mentioned properties make FNDs as promising candidate for next-generation biomarkers. In order to put in the context the performance of this newly emerged biomarker, Table 11.1 lists literature reported values of fluorescent properties for single NV^-, QD, typical organic dye, EGFP protein, and cellular proteins that give rise to autofluorescence. It is evident that single NV^- center will outperform both QDs and EPGF protein cluster while cellular autofluorescence is an order of magnitude lower.

In addition to their use in imaging, NV^- defects hosted in nanodiamonds are extensively employed in quantum computing thanks to their long spin coherence time [42], and optically addressable spin states [43].

At room temperature, NV^- center has a sharp ZPL, arising by zero–zero vibronic transitions, that is well resolved from the phonon side bands (PSBs), arising by phonon-assisted transitions. The indistinguishable photons in the ZPL make it an outstanding emitter for quantum optic experiments [44,45].

The basic motivation behind NV^- centers as building block of the hardware for quantum computing is its high level control properties on a solid-state platform. Owing to its atom-like structure, NV^- defects are promising alternative to trapped ions that are isolated from environmental couplings hence offering high-precision measurements [46]. While the isolation approach imposes special requirements (e.g., high vacuum and low temperature) on the environment and complicates integration stage and scaling up of the technology, the NV^- defects hosted in diamond are capable of operating under ambient conditions. As single NV^- center has a nuclear spin with long coherence time which can be used as quantum register and an electron spin which can be used as control/detection unit of nuclear spins it has been intensely studied for the realization of quantum processors.

NV^- centers in diamond have two unpaired electrons forming electron spin triplets, $S = 1$, in the ground and excited states. Optic excitation operates as a pump for $m_s = 0$ sublevel of ground state, ^3A. Given that $m_s = 0$ and degenerate sublevels $m_s = \pm 1$ are separated by zero field splitting parameter D, a resonant microwave signal will excite optically populated $m_s = 0$ state to the $m_s = \pm 1$ sates at this frequency (as shown in Figure 11.2b). As $m_s = \pm 1$ states have higher probability of growing intersystem crossing (ISC), which leads to a decrease in the fluorescence, this mechanism serves as basis of optically detected magnetic resonance (ODMR). The ODMR spectrum of NV center is a powerful tool used in nanoscale sensing as it reflects local physical perturbations influencing the state transitions. It has been shown that the splitting parameter D is temperature [17] and pressure [29] dependent that single NV center can be used as sensitive nanoscale sensor (see Table 11.2).

In the presence of a static magnetic field B aligned to NV center quantization axis, field $m_s = \pm 1$ degeneracy shifts by $\Delta = m_s \gamma B$ where γ is NV gyromagnetic ratio. When the field is not aligned, multiple centers in nanodiamond provide four resonance shifts in ODMR spectrum each for one crystallographic axis projection. Using this property one can resolve vector magnetic field by a reference magnetic field fixed in orientation.

Physical quantities can also be resolved via pulsed ODMR spectroscopy as single NV^- center can form an effective two-level quantum system. In Ramsey pulse sequence scheme the phase acquired in the free precession time is proportional to the external magnetic field. Therefore, the spin-dependent PL of NV center can be used for DC magnetometry [47]. Moreover, a spin-echo pulse

Table 11.2 NV-Based Sensor Typical Sensitivities

Property	Typical Sensitivity	Coupling Mechanism	Reference(s)
Pressure	6.8 bar $\sqrt{\text{Hz}}$	Via zero field splitting parameter D	[29]
Temperature	0.13 K$\sqrt{\text{Hz}}$	Via zero field splitting parameter D	[28]
Magnetic field	0.36 μT/$\sqrt{\text{Hz}}$	Zeeman effect	[14,48,55]
Electric field	($\varepsilon_{xy} = 280\ \varepsilon_z$ 5.8) V cm^{-1}/$\sqrt{\text{Hz}}$	Stark effect and spin-orbit coupling	[15,56]
Orientation	0.1°$\sqrt{\text{Hz}}$	Via zero field splitting parameter D	

Source: Schirhagl, R. et al., *Annu. Rev. Phys. Chem.*, 65, 83 2014. With permission.

sequence synchronized to an AC field behaves as phase additive on each half of the sequence so can be used for AC magnetometry [48]. In such pulsed experiments as the optic and microwave excitations are not performed simultaneously, the microwave excitation tuned down will reduce the power broadening and the optic excitation tuned up will increase the collected photons resulting in an enhancement in sensitivity compared to continuous wave excitation experiments.

High-spatial-resolution magnetic-field detection cannot be performed in cell-like environments with methodologies such as superconducting quantum interference devices (SQUIDs) [49], the Hall effect in semiconductors [50], atomic vapor-based magnetometry [51] as well as magnetic resonance force microscopy [52]. Measurement of weak magnetic fields in biological samples with nanometer spatial resolution is thus an important problem to study. The possibility to detect such weak magnetic fields has become possible through taking advantage of above-mentioned quantum properties of NV centers. As a sensor, NV centers are better suited to probe magnetic fields compared to electric fields, due to its moderate sensitivity to applied electric field compared to QDs [15]. Table 11.2 shows that NV centers embedded in single FNDs make possible the measurement of magnetic fields with hundreds of nT $Hz^{-1/2}$ sensitivity due to the fact that the applied static or oscillating magnetic field cause the relative energy shift between two Zeeman sublevels while in the case of electric filed sensing, the electric inaction caused by Stark effect and spin orbit coupling is much weaker and reflected in lower sensitivity.

Researchers showed the possibility to sensitively detect individual charges in 3D by the rotation of a magnetic bias field [15]. This methodology is designed around a quantum metrology technique based on a single NV defect center spin in diamond, which could reach a sensitivity of 202 V cm^{-1} $Hz^{-1/2}$. This is equivalent to a single elementary charge placed at a distance of ~150 nm from the nanodiamond particle probe and sensing NV spin within 1 s of averaging was required. In addition, 100-s averaging was needed to sense a single electron charge from a distance of 35 nm with a signal-to-noise ratio of more than 1000. Similarly to magnetic field detection [47], the sensitivity on electric field measurements is much better when performed with the field-induced phase accumulation method [15]. Increasing the evolution time, τ, from 8 to 80 μs allows a precise determination of minimally detected electrical fields as small as 7 V cm^{-1}.

Temperature sensing at nanoscale and with high sensitivity is also possible using single defects in diamond. The spatial resolution is related to the size of the nanodiamond, as in NSOM, whereas the sensitivity depends on the NV concentration [28]. The temperature noise floor that can be reached corresponds to 5 mK $Hz^{-1/2}$ for single defects in bulk sensors. In contrast, the temperature noise floor is 130 mK $Hz^{-1/2}$ with a precision down to 1 mK for nanocrystal sizes [53]. In consequence, temperature can be probed over length scales of a few tens of nanometers. The unique possibility to combine such accuracy and position resolution with the high photostability of FNDs should enable detection of heat produced by chemical reactions taking place among single molecules. In the absence of an external magnetic field, the temperature shifts the ground state electron spin polarization, $m_s = 0$ with respect to the sublevels $m_s = \pm 1$. Experimentally, this shift, referred to as axial ZFS, was found to vary significantly as of function of temperature [28]. To sense the temperature one optically measures either the parameter D (~2.87 GHz) with the ODMR technique [28], which corresponds to the ZFS, or the ground state spin coherence time (~1 ms) [17,54]. The principle of the latest method is to detect the NV fluorescence modulated with $\cos(2\pi\Delta D\tau)$, that is, a D-Ramsey oscillation with frequency ΔD. Fluorescence as a function of evolution time provides an increasing phase accumulation, τ, improving the temperature uncertainty [17]. Increasing τ from 50 to 250 μs makes possible to identify temperature variations as small as 1.8 mK in an ultrapure bulk diamond sample [17].

11.3 Nanoscale Imaging of Biological Systems with NV Centers Hosted in Nanodiamonds

Due to their high photostability and biocompatibility, FNDs are well-suited to far- and near-field probes for imaging purposes of intracellular environments with resolution below the diffraction limit. First reported use of FNDs as biological markers for diffraction-limited imaging dates back to 2005, when a team lead by Huan-Cheng Chang demonstrated that FNDs are spontaneously internalized in HeLa cells and display no cytotoxicity [7] (see Figure 11.3a). Besides imaging,

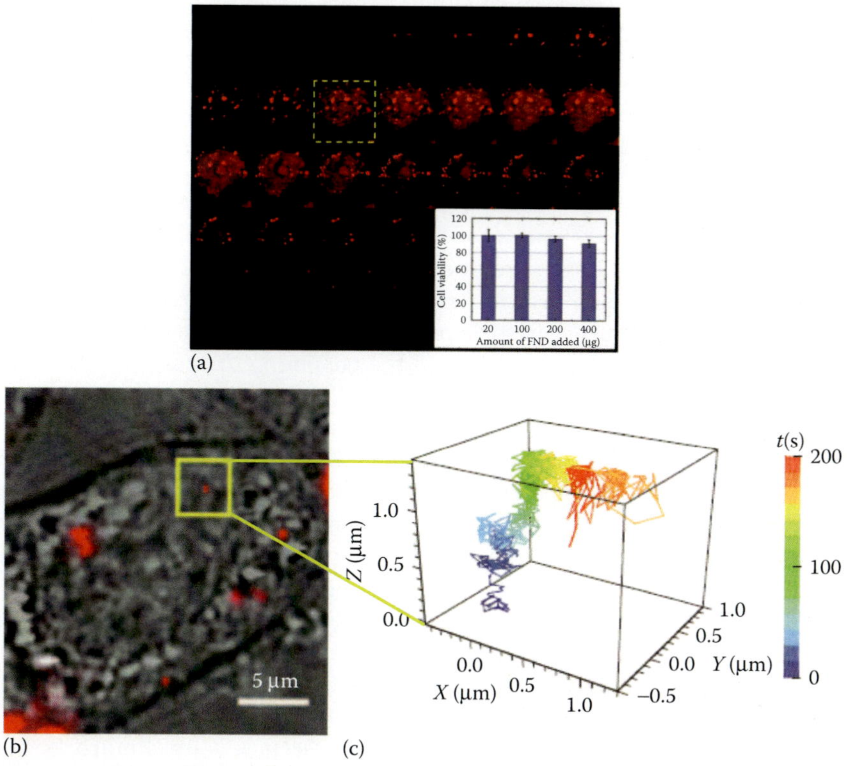

Figure 11.3

(a) The cross-sectional confocal fluorescence images of a single 293T human kidney cell after FND uptake. Inset shows cell viability test demonstrating low cytotoxicity of FNDs. (Reprinted with permission from Yu, S. J. et al., *J. Am. Chem. Soc.*, 127, 17604, 2005.) (b) Three-dimensional tracking of a single FND in the cytoplasm of a living cell. Merged bright-field image of the HeLa cell together with the epifluorescence image of the single FNDs and FND agglomerates. (c) 3D trajectory of the selected single FND particle (shown in [b]). Single FND is followed for about 200 s as it diffuses through cytoplasm. (b and c Adapted with permission from Chang, Y. R. et al., *Nat. Nanotechnol.*, 3, 284, 2008.)

single-particle tracking (SPT) techniques are instrumental when studying the biomolecular activity occurring at the intracellular level. Shortly after, their application in fluorescence imaging, excellent photostability of FNDs has been exploited for long-term 3D tracking of single 35 nm large FNDs in a living cell [6] (Figure 11.3b).

Chang et al. could track single FND for periods longer than 200 s. Since then the use of FNDs as biomarkers and as fluorescent probes for long-term tracking has exploded and now it is not only limited to the cells but includes organs and model organisms (including *C. elegans* [58], *Drosophila* [59], mice [60], and Zebrafish [61]) allowing to study a wide range of biomedical problems such as organ development [8], embryogenesis [59], stem cell proliferation [8], neuronal survival and morphogenesis [62], and cancer cell identification [63]. In addition, FNDs as biomarkers have opened the door to novel imaging and SPT modalities that are not accessible to conventional fluorescence microscopy including orientation tracking over long periods, with resolution in the millisecond timescale of intracellular events [64]. In biomedical imaging, NDs are used as contrast agents in magnetic resonance imaging (MRI) [65].

11.4 Super-Resolution Microscopy with FNDs

The past decade has been marked with the rapid progress in super-resolution microscopy. Methods based on localization and tracking of single-particle light emitters are required for applications ranging from high-resolution optical microscopy and biosensing to single-molecule detection [66]. Even though FPs and organic dyes are standard probes in the domain of super-resolution localization microscopy, nanosized photoluminescent probes with efficient linear and nonlinear optical response including FNDs [67,68], ferroelectric perovskites [69–72], and upconverters [73] start to attract researchers attention. Although all these kinds of particle probes are by its virtue highly photostable, low excitation laser power together with the possibility to obtain FNDs smaller than FPs [25] makes them the most suited labeling candidates to circumvent the diffraction resolution barrier down to few nanometers. Near-field scanning optical microscopy (NSOM) of soft biological structures and detection of molecular interactions became possible after the advent of AFM and force spectroscopy techniques. Essentially NSOM presents first nanoscopy imaging technique and it exploits the evanescent field produced by a subwavelength aperture. In 1928 Edward H. Synge (1890–1957) proposed the original idea for NSOM, however it took more than 50 years to experimentally realize this idea requiring a technological progress development of lasers, microfabrication of subwavelength apertures, precision positioners, single-photon detectors, and computers [74–76].

Interestingly, the resolution in NSOM, which allows simultaneous high-resolution measurements of topology and fluorescence, is not limited by light diffraction but by the size of the scanning probe, which is composed by a tapered optical fiber having a diameter of ~50 nm. Inspired by scanning optical microscopy, it has been reported that a single fluorescent diamond can be attached to a NSOM tip and used as a local light source [68]. Excitation of the diamond crystal at the end of tip was performed using an illumination light wave guided through the fiber holding it. The NSOM images taken with the diamond crystals as a light source of a gold mask in fluorescence mode showed, however, a resolution of only 300 nm shown in Figure 11.4a. To further increase the resolution, it was proposed to implement tips with smaller crystals with a single color center [68]. Significant

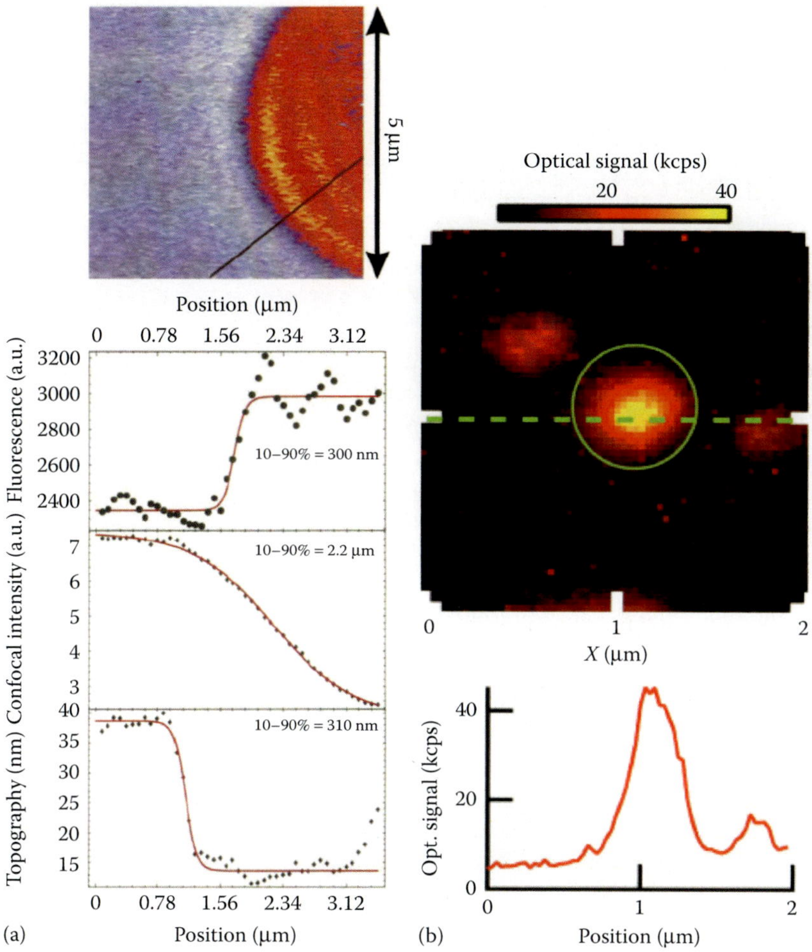

Figure 11.4

(a) First NSOM scan taken with the fluorescence nanodiamond attached to the tip. Scan is taken in a constant-gap mode across a section of a gold mask. (Adapted with permission from Kuhn, S. et al., *J. Microsc. (Oxf.)*, 202, 2, 2001.) (b) Luminescence image of a single FND. The image is recorded at an excitation optical power of 100 μW measured at the tip apex. Corresponding optical cross section. (Adapted with permission from Sonnefraud, Y. et al., *Opt. Lett.*, 33, 611, 2008.)

improvement in resolution has been achieved by using much smaller FNDs (25 nm). Figure 11.4b demonstrates that NSOM can detect nonbleaching luminescence signals that can be detected from an individual nanodiamond hosting a single NV color center [77]. Here, reported FND was smaller than 25 nm in diameter. Another possibility to overcome the diffraction limit is to exploit the effect of stimulated emission discovered by Einstein [78], and subsequently demonstrated experimentally by Stephan Hell [79,80].

In brief, STED microscopy is a far-field approach that takes advantage of the limitations of matter to be imaged in order to reconstruct a super-resolution image. Stimulated emission is a physical process where an excited atomic electron, or an excited molecular state, may drop to a lower energy level transferring its energy to an electromagnetic wave of a certain wavelength. In addition to the excitation laser, a second laser, referred to as STED laser, cancels fluorescence of the emitter located outside the center of excitation.

Technically, the doughnut-shaped de-excitation STED laser is realized with a phase modulator. STED microscopy was used to image NV centers with nanoscale resolution and using focused laser light [67]. Only the NV centers hosted in FND that are located in the subdiffraction-sized area around the doughnut, where the STED beam intensity is lower than the threshold of excited state depletion by stimulated emission, are not turned off and therefore still allowed to emit and to be identified among multiple centers present in 50–100 nm large FND (see Figure 11.5).

To achieve such performances, the intensity of the STED laser, I_{STED}, has to be in the order of GW cm^{-2} pointing that only robust markers can be used. In case of highly photostable FNDs, application of $I_{STED} = 3.7$ GW cm^{-2} was needed to reduce the focal spot, that is, full width at half maximum (FWHM), from 223 nm in diameter down to 8 nm [67] as shown in Figure 11.5a–c. Further increasing I_{STED} up to 8.6 GW cm^{-2} compresses the FWHM to values as small as 5.8 nm in 1D, which is still predicted by the theoretical inverse square-root law [81]. This value is 133 times smaller than the used wavelength for imaging the color centers in single FNDs and represents a new regime in optical-based microscopy.

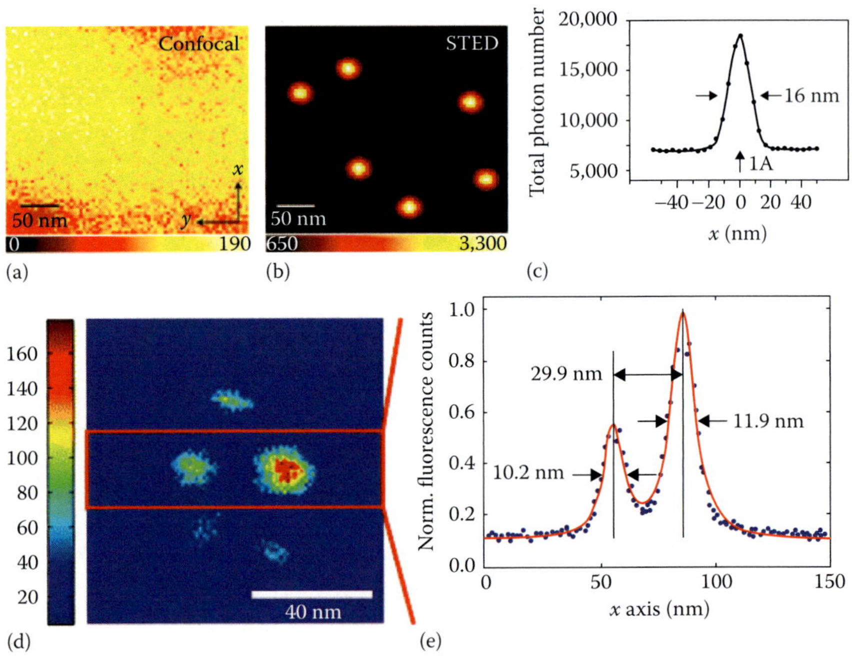

Figure 11.5

(a) Conventional confocal microscopy image which displays a featureless image, while in (b) the super-resolving STED microscopy image reveals individual luminescent NV centers inside the diamond crystal. (c) Corresponding vertically binned STED image profile showing one NV center. (a–c: Adapted with permission from Rittweger, E. et al., *Nat. Photonics*, 3, 144, 2009.) (d) Subdiffraction resolution STED image of single FND hosting 5 NV centers and (e) corresponding vertically binned STED image profile. (d and e Reprinted with permission from Arroyo-Camejo, S. et al., *Acs Nano*, 7, 10912, 2013.)

Recently, up to 5 single NV centers have been resolved from the single FND using super-resolving STED microscopy [82] (Figure 11.5d–e).

Manipulation of spin degrees of freedom using conventional far-field optical microscopy techniques is limited by diffraction. In principle, one cannot resolve spins by less than 250 nm. In contrast, the spin-reversible saturable optical linear fluorescence transitions (RESOLFT) microscopy developed specifically for far-field imaging and manipulation of individual NV electronic spins in nanodiamond revealed a nanoscale resolution [83].

The basic idea in spin-RESOLFT approach for spin detection and control with subdiffraction resolution shown in Figure 11.6a is to (i) polarize by optical pumping a NV center, (ii) coherently manipulate the NV center with resonant microwave, and (iii) detect its spin-state-dependent fluorescence [83]. To improve the spatial resolution, the sample is illuminated with a doughnut-shaped laser beam, as in STED microscopy, just before reading the spin state. In consequence, spins that are not in the center of the doughnut contribute differently to the fluorescence signal compared to spins located at few nanometers from the center of the beam. Using the spin-RESOLFT technique applied to an NV center in diamond, it is possible to optically image a single spin with a resolution down to 38 nm employing only 2-mW total doughnut beam power focused to a diffraction-limited spot of 0.07 μm^2 [83]. The resulting intensity corresponds to ~3 MW cm^{-2}, and can be further reduced by a factor of 20 while keeping a similar resolution by increasing the duration of the doughnut pulse, t_D, from 12.5 to 100 μs [83]. Compared to STED, spin-RESOLFT microscopy technique needs 4 orders of magnitude less power for comparable resolution [67]. It is also interesting to note that the length

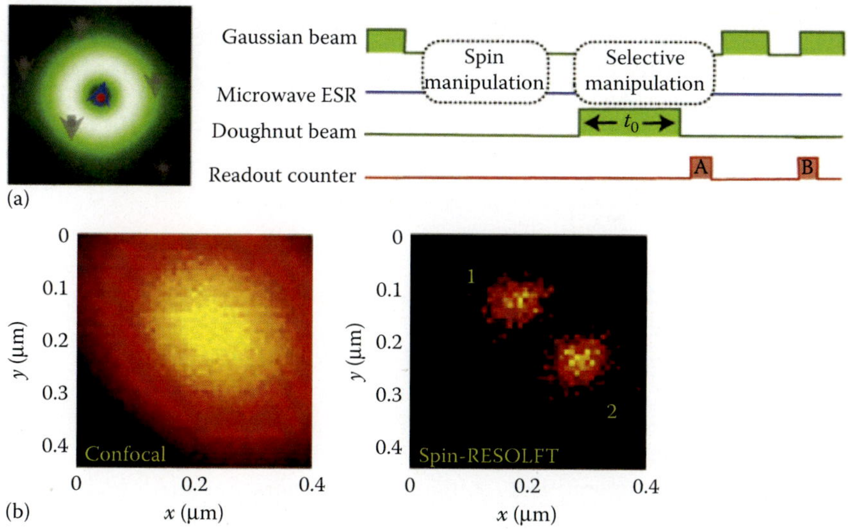

Figure 11.6

(a) Experimental sequence of spin-RESOLFT. NV spins are first optically pumped to $m_s = 0$ with a Gaussian beam at 532 nm followed by microwave manipulation. Application of a 532 nm doughnut beam repolarizes the outer ring to $m_s = 0$, allowing the spin state of the central dark region to be independently read-out. (b) In the confocal image (left) it is not possible to multiple NV^- centers, whereas 2D spin imaging using spin-RESOLFT image right clearly resolves two NV centers separated by ~150 nm. (Reprinted with permission from Maurer, P. C. et al., *Nat. Phys.*, 6, 912, 2010.)

of t_D is limited by the spin-state relaxation time (>100 ms) and the electronic excited-state lifetime (~10 ns) [83].

On one hand, we have deterministic super-resolution imaging techniques, such as STED and spin-RESOLFT, which are capable of resolving NV centers down to 5.8 and 38 nm, respectively [67,83]. However, these approaches require either a high power density for the doughnut-shaped depletion beam (~1 GW cm^{-2}) [67], or due to the serial scanning measurement result in a slow frame rate [83] making problematic imaging of dynamic processes in living cells.

On the other hand, stochastic super-resolution imaging techniques, such as PALM [84,85] and STORM [86], which are based on the localization of sequentially activated photoswitchable fluorophores, are better suited for fast images acquisition. However, they are affected among others by the precise localization of stochastic switching events and the compromise between bleaching and imaging rate. In consequence, the search of fast and noninvasive nonscanning imaging techniques with subdiffraction resolution and accuracy continue. In contrast to PALM and STORM, the widefield super-resolution microscopy using spin-dependent fluorescence in nanodiamonds, referred to as deterministic emitter switch microscopy (DESM), demonstrated the capability to image a 35 × 35 μm^2 surface composed by 100 NV centers in nanodiamonds with a localization precision between 12 and 46 nm, and employing only 90 s [12]. In DESM, the principle is to exploit the dependence of the fluorescence intensity of the NV centers as a function of the spin orientation. In this context, the ODMR technique allows to measure the spin resonance frequencies of all centers within a diffraction-limited spot. If the centers do not overlap, one can distinguish them by microwave excitation in a deterministic way. In consequence, this methodology works properly while nanodiamonds are arbitrarily oriented leading to multiple spin transitions. The basic procedure in DESM to achieve a subdiffraction resolution is to resonantly drive only one ground-state spin transition at a time [12]. Up to now, the world record in far-field-based super-resolution optical microscopy, that is, 8 nm in 2D and 5.8 nm 1D, is held by NV centers embedded in single nanodiamonds [67].

11.5 Nanoscale Sensing of Biological Systems with NV Centers Hosted in Nanodiamonds

As introduced in the Section 11.2 to sensing properties of NV centers hosted in FNDs, various physical properties can be detected with high accuracy. Here we detail two most prominent examples related to biological systems.

Until recently, temperature measurements of biological environments at nanometer-scale resolution presented a grand challenge in metrology. However, using the NV centers hosted in FNDs, Kucsko et al. (2013) demonstrate nanometer-scale temperature probing of the interior of single human embryonic fibroblasts [17]. By inserting nanodiamonds into single living cells, next the FNDs were irradiated with microwaves in order to modulate the electron occupancy of spin states. To determine the changes in the ground-state energy gap authors used using widefield fluorescence measurements that allowed to probe the temperature variations (the microwave frequency that corresponds to the energy difference between spin states) for several FNDs. The temperature inside the living cells has been also modulated by irradiating and heating from the gold nanoparticles also introduced in the cell (see Figure 11.7a and b). Although most of the exciting applications of FNDs are related to fluorescence-based microscopy,

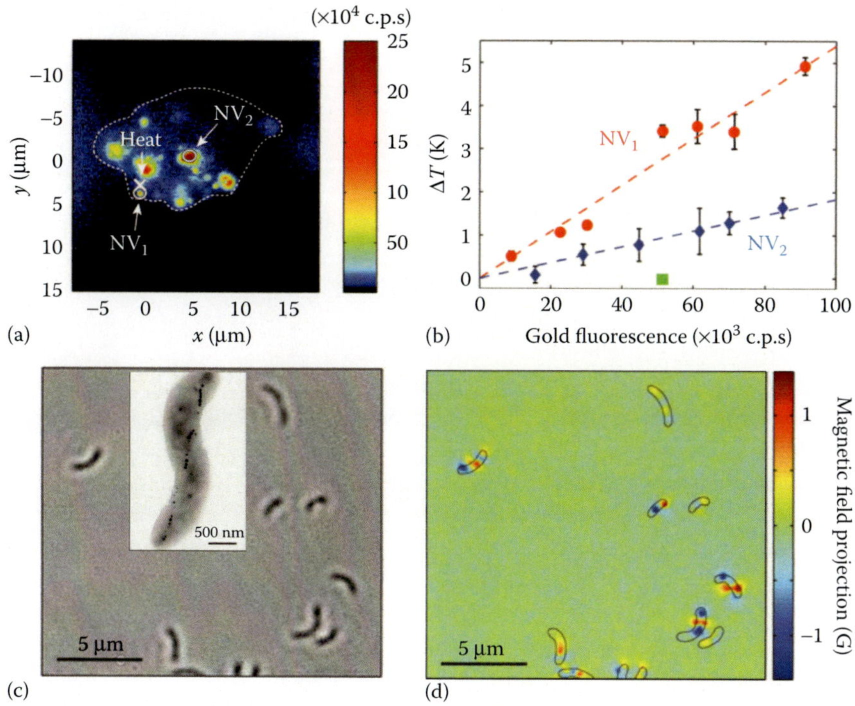

Figure 11.7

(a) Confocal scan of a single living cell obtained using a laser excitation set to 532 nm, with collection at wavelengths greater than 638 nm, which ensures collection of PL signal mostly from FNDs. The cross mark indicates the position of the gold nanoparticle used for heating, and circles represent the location of the nanodiamonds (NV_1 and NV_2) used for thermometry. Color bars indicate the fluorescence in counts per seconds. (b) Measured temperature change at the positions of NV_1 and NV_2 relative to the incident laser power applied to the gold nanoparticle. (a and b: Reprinted with permission from Kucsko, G. et al., *Nature,* 500, 54, 2013.) (c) Widefield optical image of dried MTB on a diamond chip. Inset shows a typical transmission electron microscope (TEM) image of a single MTB. Dark spots inside MTB are magnetite nanoparticles. (d) Stray field distribution recorded with a widefield NV magnetometer—imaged region corresponds to the one shown in (c). (c and d: Reprinted with permission from Le Sage, D. et al., *Nature,* 496, 486, 2013.)

especially in biology, where room temperature functionality and chemistry are crucial parameters, another interesting sensing area is 3D MRI with NV centers hosted in diamonds.

In principle a 1 nm resolution of single electron spins in a solid has been demonstrated [55], while for living cells the achieved resolution has been 400 nm [87]. In contrast, in medical facilities and for biological tissues, the resolution for conventional MRI apparatus is about 1 mm demonstrating huge potential for 3D MRI with NV centers hosted in diamonds. The most prominent example of the magnetic imaging on biological samples has been demonstrated using the magnetic field imaging array consisting of a nanometer-scale layer of nitrogen-vacancy color centers implanted at the surface of a diamond chip. Le Sage et al. have succeed to localize magnetic nanoparticles (magnetosomes) produced in the magnetotactic bacteria (MTB) with subcellular spatial resolution (see Figure 11.7c and d) [87]. Authors suggest that their method can be used to identify potential vertebrate magnetoreceptor cells [88], which should have a magnetic moment that is comparable to or larger than one found in MTBs.

11.6 Manipulation Techniques

To exploit the full FND sensing potential for application in nanoscience and biology, the sensing probe in the form of the single FND has to be positioned and manipulated with high accuracy over the sample. Several exciting techniques such as scanning probe techniques, optical tweezers, and anti-Brownian electrokinetic trap (ABEL) have reached required precision.

11.6.1 Scanning Probe

Thanks to its electronic and optic properties a single NV center embedded in a nanodiamond can be attached to a probe tip, that is, AFM, for atomic-scale resolution in imaging as shown in Figure 11.8. The idea is to bring the scanning probe close to the substrate surface and detect the local fields emanating from the sample through the change in the spin dynamics. The spin-dependent PL of NV center can be collected via far-field optics, whereas optic and microwave excitations is realized as in ensemble level. Alternatively, a fiber-optic waveguide serving as scanning probe can be functionalized such that it delivers microwave signal through an integrated transmission line, performs optic excitation, and collects the fluorescence simultaneously [89,90]. Recently, an engineered NV center hosted in a scanning nanopillar waveguide with long coherence time (~75 μs)

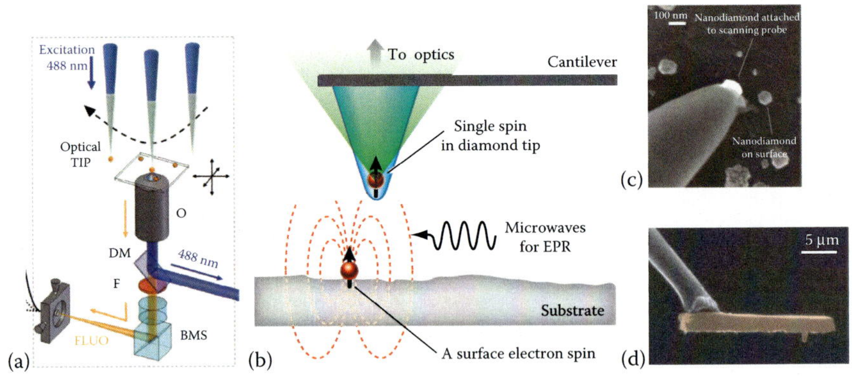

Figure 11.8

Scanning probe-based FND manipulation techniques. (a) In the near-field scanning optical microscopy (NSOM/SNOM) FND has to be first picked by an optical fiber and then is scanned over the magnetic nanostructure. (Reprinted with permission from Cuche, A. et al., *Opt. Express*, 17, 19969, 2009.) (b) Depicts basic principle of all scanning magnetometry techniques. In AFM, a sharp tip with a nitrogen vacancy center at the tip maps out the three-dimensional magnetic field vector above a magnetic nanostructure. Several methods have been developed in order to position/fabricate the FND on the cantilever tip. (c) FND can be picked up from the surface and attached to the cantilever tip. (b,c: Reprinted with permission from Degen, C. L. *Appl. Phys. Lett.*, 92, 2008.) (d) Alternatively, diamond nanopillar probe with a single NV center in its tip can be fabricated from of a single crystalline as shown in SEM image. (Reprinted with permission from Maletinsky, P. et al., *Nat. Nanotechnol.*, 7, 320, 2012.)

and collection efficiency was shown to be a sensitive robust magnetometer imaging magnetic structures as small as 25 nm [91] (see Figure 11.8d).

Similar to conventional MRI technique the scanning probe tip can also be employed for the generation of a magnetic field gradient on NV centers [14]. This modality basically aims to detect the resonance ring of each NV center that appears by scanning the corresponding magnetic field under a fixed microwave excitation. Given that the probe position and hence the gradient profile is known, the NV centers can be localized with the ultimate resolution limited by the resonant ring width. Recently, a spatial resolution of 9 nm has been demonstrated for proximal NV centers in clustered form. In all scanning configurations, the optic data acquired by NV center can be easily correlated to topography mapping achieved with the probe tip for further enhancement in the imaging.

11.6.2 Optical Tweezers

Although the integration of single NV center to a scanning probe tip remarkably improved resolution this approach has some drawbacks regarding experimental realization and operating environment. First, to sense the local fields with high spatial resolution the NV center should be placed close to the probe tip that requires nanoscale positioning and expensive fabrication process. Consequently, a significant effort has been put into the engineering NV defects in diamond. Recently, stable NV centers as close as 2 nm to the diamond surface have been formed without degrading spin coherence time [93]. Yet, this limit is far beyond realization for the nanodiamond integrated to a scanning probe. Also, the probe tip may perturb the environment or the intrinsic characteristics of NV center. Moreover, as this approach is limited by operating environment of the probe tip it is not practical solution for closed microfluidics and bioapplications. Recently, optic-based trapping has been demonstrated as nanoscale positioning tool for nanodiamond hosting NV centers [94] (see Figure 11.9). This technique uses a piezoelectric stage performing 3D scanning of the sample with respect to the focused infrared beam where multiple nanodiamonds are trapped through radiation pressure. The NV centers in FNDs are optically excited by a second green laser while the microwave signal is applied through a nearby antenna. Although the large ensemble of NV centers are randomly oriented in such trap environment, it is shown that the local DC magnetic field sensing is possible. Moreover, a recent work using counterpropagating dual beam technique has achieved the trapping at single particle level [95] (Figure 11.9b). It is also shown that the orientation of the polarized beam on asymmetric nanodiamond can provide the control over the angle between optic axis and the NV quantization axis with high accuracy. This progress facilitates 3D vectorial magnetometery in solution.

11.6.3 ABEL Trap

Despite the fact that laser-based trap is an outstanding manipulation tool for aqueous environment, it is not functional for FNDs of smaller sizes. Given that the force generated in optical tweezers is proportional to the volume of the trapped object particles smaller than 100 nm requires high optic powers (~150 mW) [96]. This is not desirable in bioenvironments as it may affect the properties of the trapped particle as well as the medium. Moreover, as the force arises through a second-order interaction, the applied field must first polarize the object, and then generate the force between the induced dipole and a gradient in the field. ABEL is a novel approach to overcome size limitations in solution phase trapping.

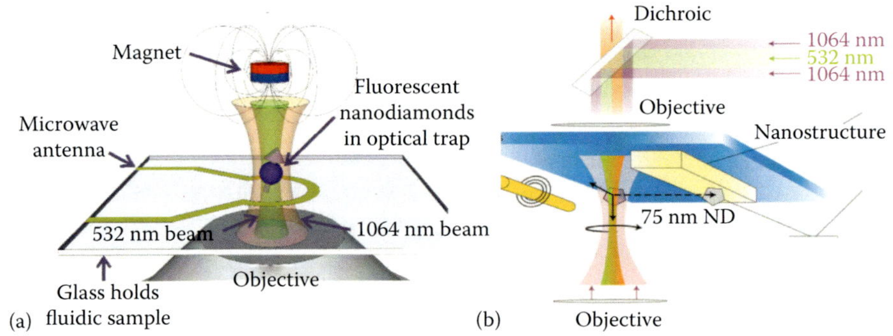

Figure 11.9

Two configurations based on 3D optical trapping optical trapping of FNDs that allow for manipulation of a single-electron spin. (a) Single-beam optical tweezers, photoexcitation, and luminescence detection are integrated on the same setup using a high NA objective. The magnetic field is applied externally, along the axis of the objective. (Reprinted with permission from Horowitz, V. R. et al., *Proc. Natl. Acad. Sci. USA*, 109, 13493, 2012.) (b) The dual beam optical tweezers permits accurate control over the FNDs orientation. (Reprinted with permission from Geiselmann, M. et al., *Nat. Nanotechnol.*, 8, 175, 2013.)

It combines laser-guided SPT with a real-time feedback cancelling the Brownian motion. Unlike optical tweezers, the trap force is not optic-based but electrokinetic, which scales linearly with the particle radius. Recently, single dye molecule of subnanometer size has been successfully trapped by a very low optic excitation power used in the particle tracking [97]. As the trajectories of diffusing particles are not correlated; thus, only one is exposed to the correct feedback, exactly one particle is trapped with this method. Moreover, all surrounding particles around the trap point are subjected to the same feedback therefore applied electrokinetic force on native environment is not perturbative (see Figure 11.10). This also avoids any potential clustering or agglomeration that might occur in optical tweezers for dense dispersions.

Provided these advantages, ABEL trap is an outstanding tool for FND manipulation and positioning. Indeed, more recently this approach has been validated with ESR experiments on 30-nm sized FND containing NV centers [20]. Although the particle has rotationally free behavior, the detection of static magnetic fields through the ODMR spectra has been demonstrated [20].

Moreover, a work by Maclaurin et al. [98] on single diffusing diamond nanocrystal shows the capability of ABEL trap approach for sensing also the fluctuating and oscillating magnetic fields. As well as being a sensitive magnetometer ABEL trap is also compatible to other NV center-based detection schemes such as thermometry and ion concentration measurements. With the advances in fabrication of stable FNDs, this technique will provide a new avenue for 3D high-resolution imaging and sensing in fluidics and physiological environment.

11.7 Conclusion

With this chapter, we attempted to give a short overview of the emerging new probes in the form of FNDs. We focused on their use in nanoscale imaging of biological samples and biological relevant sensing application such as

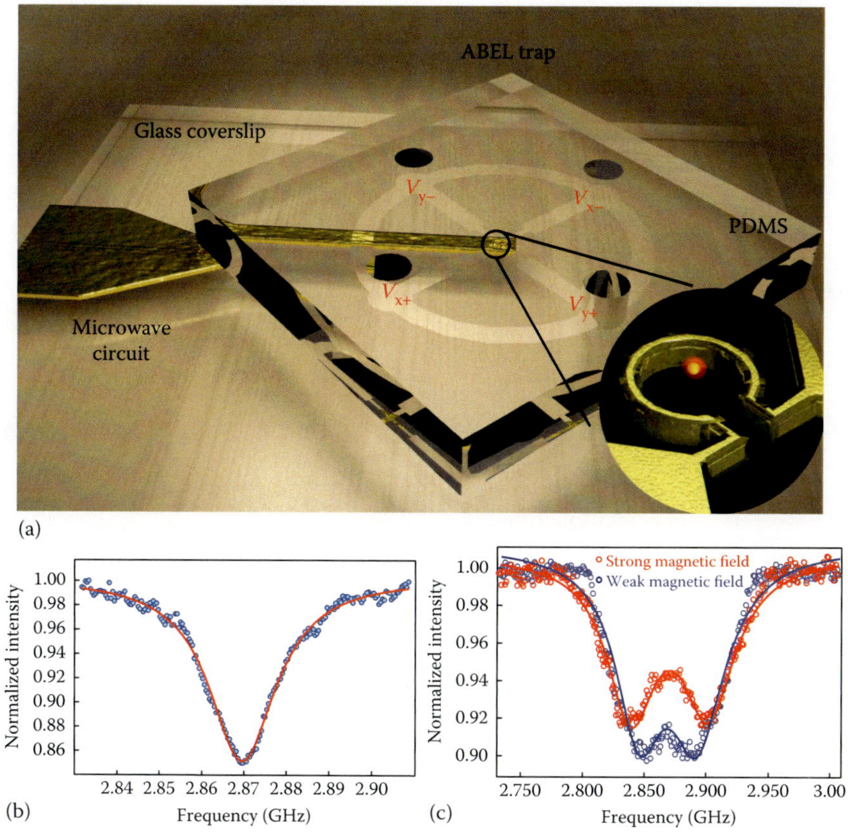

Figure 11.10

(a) Schematic of ABEL trap PDMS microfluidic cell with RF circuit integrated on the glass coverslip. Inset shows a zoom in the trapping area and it indicates the relative location of FND in respect to the area with the enhanced homogeneous magnetic flux density. (b) Typical ODMR spectra of NV defects in ABEL-trapped FND. (c) Optically detected ESR spectra of an ABEL-trapped single FND under the presence of weak and strong static magnetic fields. (From Kayci, M. et al., *Nano Lett.*, 14, 5335, 2014. With permission.)

single-cell thermometry and magnetometery. In relation to the sensing we have also reviewed current nanoscale manipulation techniques that would allow in the future positioning of these ultrasensitive sensors within single living cell. We apologize for any deserving papers that we have wittingly or unwittingly omitted in this chapter.

11.8 List of Acronyms

1D One-dimensional
2D Two-dimensional
3D Three-dimensional
ABEL Anti-Brownian electrokinetic trap
AFM Atomic force microscopy
DESM Deterministic emitter switch microscopy
FND Fluorescent nanodiamond

FP	Fluorescent protein
FWHM	Full width at half maximum
ISC	Intersystem crossing
MRI	Magnetic resonance imaging
MTB	Magnetotactic bacteria
NSOM	Near-field scanning optical microscopy
NV	Nitrogen vacancy
ODMR	Optically detected magnetic resonance
OT	Optical trap
PALM	Photoactivated localization microscopy
PSB	Phonon side bands
QD	Quantum dot
RESOLFT	Reversible saturable optical linear fluorescence transitions
SPT	Single-particle tracking
SQUID	Superconducting quantum interference device
STORM	Stochastic optical reconstruction microscopy
ZFS	Zero field splitting

References

1. Chalfie, M., Y. Tu, G. Euskirchen, W. W. Ward, D. C. Prasher, Green fluorescent protein as a marker for gene-expression. *Science* **263**, 802 (Feb, 1994).
2. Griffin, B. A., S. R. Adams, R. Y. Tsien, Specific covalent labeling of recombinant protein molecules inside live cells. *Science* **281**, 269 (Jul, 1998).
3. Chan, W. C. W., S. M. Nie, Quantum dot bioconjugates for ultrasensitive nonisotopic detection. *Science* **281**, 2016 (Sep, 1998).
4. Miyawaki, A., A. Sawano, T. Kogure, Lighting up cells: Labelling proteins with fluorophores. *Nature Cell Biology*, S1 (Sep, 2003).
5. Giepmans, B. N. G., S. R. Adams, M. H. Ellisman, R. Y. Tsien, Review—The fluorescent toolbox for assessing protein location and function. *Science* **312**, 217 (Apr, 2006).
6. Chang, Y. R., et al., Mass production and dynamic imaging of fluorescent nanodiamonds. *Nature Nanotechnology* **3**, 284 (May, 2008).
7. Yu, S. J., M. W. Kang, H. C. Chang, K. M. Chen, Y. C. Yu, Bright fluorescent nanodiamonds: No photobleaching and low cytotoxicity. *Journal of the American Chemical Society* **127**, 17604 (Dec 21, 2005).
8. Wu, T. J., et al., Tracking the engraftment and regenerative capabilities of transplanted lung stem cells using fluorescent nanodiamonds. *Nature Nanotechnology* **8**, 682 (Sep, 2013).
9. Tisler, J., et al., Highly efficient FRET from a single nitrogen-vacancy center in nanodiamonds to a single organic molecule. *ACS Nano* **5**, 7893 (Oct, 2011).
10. Vaijayanthimala, V., et al., The long-term stability and biocompatibility of fluorescent nanodiamond as an in vivo contrast agent. *Biomaterials* **33**, 7794 (Nov, 2012).
11. Zhang, B. L., et al., Receptor-mediated cellular uptake of folate-conjugated fluorescent nanodiamonds: A combined ensemble and single-particle study. *Small* **5**, 2716 (Dec 4, 2009).

12. Chen, E. H., O. Gaathon, M. E. Trusheim, D. Englund, Wide-field multi-spectral super-resolution imaging using spin-dependent fluorescence in nanodiamonds. *Nano Letters* **13**, 2073 (May, 2013).
13. Han, K. Y., et al., Three-dimensional stimulated emission depletion microscopy of nitrogen-vacancy centers in diamond using continuous-wave light. *Nano Letters* **9**, 3323 (Sep, 2009).
14. Balasubramanian, G., et al., Nanoscale imaging magnetometry with diamond spins under ambient conditions. *Nature* **455**, 648 (Oct 2, 2008).
15. Dolde, F., et al., Electric-field sensing using single diamond spins. *Nature Physics* **7**, 459 (Jun, 2011).
16. Ziem, F. C., N. S. Gotz, A. Zappe, S. Steinert, J. Wrachtrup, Highly sensitive detection of physiological spins in a microfluidic device. *Nano Letters* **13**, 4093 (Sep, 2013).
17. Kucsko, G., et al., Nanometre-scale thermometry in a living cell. *Nature* **500**, 54 (Aug 1, 2013).
18. Shi, F., et al., Protein imaging. Single-protein spin resonance spectroscopy under ambient conditions. *Science* **347**, 1135 (Mar 6, 2015).
19. Aharonovich, I., et al., Diamond-based single-photon emitters. *Reports on Progress in Physics* **74**, (Jul, 2011).
20. Kayci, M., H. C. Chang, A. Radenovic, Electron spin resonance of nitrogen-vacancy defects embedded in single nanodiamonds in an ABEL trap. *Nano Letters* **14**, 5335 (Sep, 2014).
21. Smith, B. R., et al., Five-nanometer diamond with luminescent nitrogen-vacancy defect centers. *Small* **5**, 1649 (Jul 17, 2009).
22. Fu, C. C., et al., Characterization and application of single fluorescent nanodiamonds as cellular biomarkers. *Proceedings of the National Academy of Sciences of the United States of America* **104**, 727 (Jan 16, 2007).
23. Guillois, O., G. Ledoux, C. Reynaud, Diamond infrared emission bands in circumstellar media. *Astrophysical Journal* **521**, L133 (Aug 20, 1999).
24. Goto, M., et al., Spatially resolved 3 μm spectroscopy of Elias 1: Origin of diamonds in protoplanetary disks. *Astrophysical Journal* **693**, 610 (Mar 1, 2009).
25. Vlasov, I. I., et al., Molecular-sized fluorescent nanodiamonds. *Nature Nanotechnology* **9**, 54 (Jan, 2014).
26. Chang, Y. R., et al., Mass production and dynamic imaging of fluorescent nanodiamonds. *Nature Nanotechnology* **3**, 284 (May, 2008).
27. Bradac, C., et al., Observation and control of blinking nitrogen-vacancy centres in discrete nanodiamonds. *Nature Nanotechnology* **5**, 345 (May, 2010).
28. Acosta, V. M., et al., Temperature dependence of the nitrogen-vacancy magnetic resonance in diamond. *Physical Review Letters* **104**, (Feb 19, 2010).
29. Doherty, M. W., et al., Electronic properties and metrology applications of the diamond NV^- center under pressure. *Physical Review Letters* **112**, (Jan 31, 2014).
30. Steinert, S., et al., Magnetic spin imaging under ambient conditions with sub-cellular resolution. *Nature Communications* **4**, (Mar, 2013).
31. Childress, L., R. Walsworth, M. Lukin, Atom-like crystal defects: From quantum computers to biological sensors. *Physics Today* **67**, 38 (Oct, 2014).

32. Chang, I. P., K. C. Hwang, C. S. Chiang, Preparation of fluorescent magnetic nanodiamonds and cellular imaging. *Journal of the American Chemical Society* **130**, 15476 (Nov 19, 2008).
33. Hartl, A., et al., Protein-modified nanocrystalline diamond thin films for biosensor applications. *Nature materials* **3**, 736 (Oct, 2004).
34. Yang, W., et al., DNA-modified nanocrystalline diamond thin-films as stable, biologically active substrates. *Nature materials* **1**, 253 (Dec, 2002).
35. Bumb, A., S. K. Sarkar, N. Billington, M. W. Brechbiel, K. C. Neuman, Silica encapsulation of fluorescent nanodiamonds for colloidal stability and facile surface functionalization. *Journal of the American Chemical Society* **135**, 7815 (May **29**, 2013).
36. Resch-Genger, U., M. Grabolle, S. Cavaliere-Jaricot, R. Nitschke, T. Nann, Quantum dots versus organic dyes as fluorescent labels. *Nature Methods* **5**, 763 (Sep, 2008).
37. Tisler, J., et al., Fluorescence and spin properties of defects in single digit nanodiamonds. *ACS Nano* **3**, 1959 (Jul, 2009).
38. Weber, G., Fluorescence of riboflavin and flavin-adenine dinucleotide. *The Biochemical Journal* **47**, 114 (1950).
39. Gohar, A. V., et al., Subcellular localization-dependent changes in EGFP fluorescence lifetime measured by time-resolved flow cytometry. *Biomedical Optics Express* **4**, 1390 (Aug 1, 2013).
40. Sanchez-Andres, A., Y. Chen, J. D. Muller, Molecular brightness determined from a generalized form of Mandel's Q-parameter. *Biophysical Journal* **89**, 3531 (Nov, 2005).
41. Chen, Y., J. D. Muller, Q. Q. Ruan, E. Gratton, Molecular brightness characterization of EGFP in vivo by fluorescence fluctuation spectroscopy. *Biophysical Journal* **82**, 133 (Jan, 2002).
42. Gaebel, T., et al., Room-temperature coherent coupling of single spins in diamond. *Nature Physics* **2**, 408 (Jun, 2006).
43. Robledo, L., H. Bernien, I. van Weperen, R. Hanson, Control and coherence of the optical transition of single nitrogen vacancy centers in diamond. *Physical Review Letters* **105**, (Oct 21, 2010).
44. Santori, C., D. Fattal, J. Vuckovic, G. S. Solomon, Y. Yamamoto, Indistinguishable photons from a single-photon device. *Nature* **419**, 594 (Oct 10, 2002).
45. He, Y. M., et al., On-demand semiconductor single-photon source with near-unity indistinguishability. *Nature Nanotechnology* **8**, 213 (Mar, 2013).
46. Wilson, A. C., et al., Tunable spin-spin interactions and entanglement of ions in separate potential wells. *Nature* **512**, 57 (Aug 7, 2014).
47. Taylor, J. M., et al., High-sensitivity diamond magnetometer with nanoscale resolution. *Nature Physics* **4**, 810 (Oct, 2008).
48. Maze, J. R., et al., Nanoscale magnetic sensing with an individual electronic spin in diamond. *Nature* **455**, 644 (Oct 2, 2008).
49. Bending, S. J., Local magnetic probes of superconductors. *Advances in Physics* **48**, 449 (Jul–Aug, 1999).
50. Chang, A. M., et al., Scanning Hall-Probe Microscopy of a Vortex and Field Fluctuations in La1.85sr0.15cuo4 Films. *Europhysics Letters* **20**, 645 (Dec 1, 1992).
51. Budker, D., et al., Resonant nonlinear magneto-optical effects in atoms. *Reviews of Modern Physics* **74**, 1153 (Oct, 2002).

52. Mamin, H. J., M. Poggio, C. L. Degen, D. Rugar, Nuclear magnetic resonance imaging with 90-nm resolution. *Nature Nanotechnology* **2**, 301 (May, 2007).
53. Martin, Y., D. W. Abraham, H. K. Wickramasinghe, High-resolution capacitance measurement and potentiometry by force microscopy. *Applied Physics Letters* **52**, 1103 (Mar 28, 1988).
54. Neumann, P., et al., High-precision nanoscale temperature sensing using single defects in diamond. *Nano Letters* **13**, 2738 (Jun, 2013).
55. Grinolds, M. S., et al., Subnanometre resolution in three-dimensional magnetic resonance imaging of individual dark spins. *Nature Nanotechnology* **9**, 279 (Apr, 2014).
56. Vanoort, E., M. Glasbeek, Electric-field-induced modulation of spin echoes of N-V centers in diamond. *Chemical Physics Letters* **168**, 529 (May 18, 1990).
57. Schirhagl, R., K. Chang, M. Loretz, C. L. Degen, Nitrogen-vacancy centers in diamond: Nanoscale sensors for physics and biology. *Annual Review of Physical Chemistry* **65**, 83 (2014).
58. Mohan, N., C. S. Chen, H. H. Hsieh, Y. C. Wu, H. C. Chang, In vivo imaging and toxicity assessments of fluorescent nanodiamonds in Caenorhabditis elegans. *Nano Letters* **10**, 3692 (Sep 8, 2010).
59. Simpson, D. A., et al., In vivo imaging and tracking of individual nanodiamonds in drosophila melanogaster embryos. *Biomedical Optics Express* **5**, 1250 (Apr 1, 2014).
60. Igarashi, R. et al., Real-time background-free selective imaging of fluorescent nanodiamonds in vivo. *Nano Letters* **12**, 5726 (Nov, 2012).
61. Chang, C. C., et al., Exploring cytoplasmic dynamics in zebrafish yolk cells by single particle tracking of fluorescent nanodiamonds. *Advances in Photonics of Quantum Computing, Memory, and Communication V* **8272**, (2012).
62. Huang, Y. A., et al., The effect of fluorescent nanodiamonds on neuronal survival and morphogenesis. *Scientific Reports* **4**, (Nov 5, 2014).
63. Hui, Y. Y., et al., Wide-field imaging and flow cytometric analysis of cancer cells in blood by fluorescent nanodiamond labeling and time gating. *Scientific Reports* **4**, (Jul 4, 2014).
64. McGuinness, L. P., et al., Quantum measurement and orientation tracking of fluorescent nanodiamonds inside living cells. *Nature Nanotechnology* **6**, 358 (Jun, 2011).
65. Manus, L. M., et al., Gd(III)-nanodiamond conjugates for MRI contrast enhancement. *Nano Letters* **10**, 484 (Feb 10, 2010).
66. Deschout, H., et al., Precisely and accurately localizing single emitters in fluorescence microscopy. *Nature Methods* **11**, 253 (Mar, 2014).
67. Rittweger, E., K. Y. Han, S. E. Irvine, C. Eggeling, S. W. Hell, STED microscopy reveals crystal colour centres with nanometric resolution. *Nature Photonics* **3**, 144 (Mar, 2009).
68. Kuhn, S., C. Hettich, C. Schmitt, J. P. H. Poizat, V. Sandoghdar, Diamond colour centres as a nanoscopic light source for scanning near-field optical microscopy. *Journal of Microscopy-Oxford* **202**, 2 (Apr, 2001).
69. Nakayama, Y., et al., Tunable nanowire nonlinear optical probe. *Nature* **447**, 1098 (Jun **28**, 2007).
70. Dutto, F., C. Raillon, K. Schenk, A. Radenovic, Nonlinear optical response in single alkaline niobate nanowires. *Nano Letters* **11**, 2517 (Jun, 2011).
71. Dutto, F., A. Radenovic, Alkaline niobate nanowires as opto-mechanical probes. *Optical Trapping and Optical Micromanipulation Ix* **8458**, (2012).

72. Macias-Romero, C., et al., Probing rotational and translational diffusion of nanodoublers in living cells on microsecond time scales. *Nano Letters* **14**, 2552 (May, 2014).
73. Mor, F. M., A. Sienkiewicz, L. Forro, S. Jeney, Upconversion particle as a local luminescent brownian probe: A photonic force microscopy study. *ACS Photonics* **1**, 1251 (Dec, 2014).
74. Betzig, E., A. Lewis, A. Harootunian, M. Isaacson, E. Kratschmer, Near-field scanning optical microscopy (NSOM)—Development and biophysical applications. *Biophysical Journal* **49**, 269 (Jan, 1986).
75. Durig, U., D. W. Pohl, F. Rohner, Near-field optical-scanning microscopy. *Journal of Applied Physics* **59**, 3318 (May 15, 1986).
76. Betzig, E., J. K. Trautman, T. D. Harris, J. S. Weiner, R. L. Kostelak, Breaking the diffraction barrier—Optical microscopy on a nanometric scale. *Science* **251**, 1468 (Mar 22, 1991).
77. Sonnefraud, Y., et al., Diamond nanocrystals hosting single nitrogen-vacancy color centers sorted by photon-correlation near-field microscopy. *Optics Letters* **33**, 611 (Mar 15, 2008).
78. Einstein, A., Strahlungs-Emission und -Absorption nach der Quantentheorie. *Deutsche Physikalische Gesellschaft* **18**, 318 (1916).
79. Hell, S. W., J. Wichmann, Breaking the diffraction resolution limit by stimulated emission: Stimulated-emission-depletion fluorescence microscopy. *Optics Letters* **19**, 780 (Jun 1, 1994).
80. Klar, T. A., S. W. Hell, Subdiffraction resolution in far-field fluorescence microscopy. *Optics Letters* **24**, 954 (Jul 15, 1999).
81. Hell, S. W., Far-field optical nanoscopy. *Science* **316**, 1153 (May 25, 2007).
82. Arroyo-Camejo S., et al., Stimulated emission depletion microscopy resolves individual nitrogen vacancy centers in diamond nanocrystals. *ACS Nano* **7**, 10912 (Dec 23, 2013).
83. Maurer, P. C., et al., Far-field optical imaging and manipulation of individual spins with nanoscale resolution. *Nature Physics* **6**, 912 (Nov, 2010).
84. Betzig, E., et al., Imaging intracellular fluorescent proteins at nanometer resolution. *Science* **313**, 1642 (Sep 15, 2006).
85. Hess, S. T., T. P. K. Girirajan, M. D. Mason, Ultra-high resolution imaging by fluorescence photoactivation localization microscopy. *Biophysical Journal* **91**, 4258 (Dec, 2006).
86. Rust, M. J., M. Bates, X. W. Zhuang, Sub-diffraction-limit imaging by stochastic optical reconstruction microscopy (STORM). *Nature Methods* **3**, 793 (Oct, 2006).
87. Le Sage, D., et al., Optical magnetic imaging of living cells. *Nature* **496**, 486 (Apr 25, 2013).
88. Eder, S. H. K., et al., Magnetic characterization of isolated candidate vertebrate magnetoreceptor cells. *Proceedings of the National Academy of Sciences of the United States of America* **109**, 12022 (Jul 24, 2012).
89. Cuche, A., et al., Near-field optical microscopy with a nanodiamond-based single-photon tip. *Optics Express* **17**, 19969 (Oct 26, 2009).
90. Fedotov, I. V., et al., Electron spin manipulation and readout through an optical fiber. *Scientific Reports* **4**, (Jul 16, 2014).
91. Maletinsky, P., et al., A robust scanning diamond sensor for nanoscale imaging with single nitrogen-vacancy centres. *Nature Nanotechnology* **7**, 320 (May, 2012).

92. Degen, C. L., Scanning magnetic field microscope with a diamond single-spin sensor. *Applied Physics Letters* **92**, (Jun 16, 2008).
93. Ofori-Okai, B. K., et al., Spin properties of very shallow nitrogen vacancy defects in diamond. *Physical Review B* **86**, (Aug 13, 2012).
94. Horowitz, V. R., B. J. Aleman, D. J. Christle, A. N. Cleland, D. D. Awschalom, Electron spin resonance of nitrogen-vacancy centers in optically trapped nanodiamonds. *Proceedings of the National Academy of Sciences of the United States of America* **109**, 13493 (Aug 21, 2012).
95. Geiselmann, M., et al., Three-dimensional optical manipulation of a single electron spin. *Nature Nanotechnology* **8**, 175 (Mar, 2013).
96. Svoboda, K., S. M. Block, Optical trapping of metallic Rayleigh particles. *Optics Letters* **19**, 930 (Jul 1, 1994).
97. Fields, A. P., A. E. Cohen, Electrokinetic trapping at the one nanometer limit. *Proceedings of the National Academy of Sciences of the United States of America* **108**, 8937 (May 31, 2011).
98. Maclaurin, D., L. T. Hall, A. M. Martin, L. C. L. Hollenberg, Nanoscale magnetometry through quantum control of nitrogen-vacancy centres in rotationally diffusing nanodiamonds. *New Journal of Physics* **15**, (Jan 17, 2013).

SECTION III
Applications in Biology and Medicine

12

Optical Nanoscopy with SOFI

Xuanze Chen, Zhiping Zeng, Xi Zhang, Pingyong Xu, and Peng Xi

12.1 Introduction
12.2 Fluorescent Labels in SOFI
12.3 The Applications of SOFI
12.4 Conclusion and Outlook
References

12.1 Introduction

Usually in optical imaging, we call the specimen under an optical microscope "object" and the image formed at the CCD detector "image." The object can be thought as composing of points, which is dimensionless, or at infinitely high detail. The image, on the contrary, can be seen as composed of so-called "point spread functions" (PSFs). These PSFs are the result of diffraction, which makes a dimensionless point to have a finite size. The resolution of a microscope system is limited by the fact that (1) as the PSFs are with a finite size and (2) when two PSFs get too close with each other simultaneously, the sum of them cannot be discerned.

Consequently, there are two types of super-resolution techniques, which can break the law above:

1. To spatially generate smaller effective PSF
2. To temporally identify and separate the emitters, to avoid them from adding simultaneously

The generation of a smaller PSF includes confocal, STED, RESOLFT, SIM, SSIM, ISM, and so on. Take STED as an example, the peripheral molecules are forced to stay at OFF state by stimulated emission, which is different from spontaneous emission. Thus, a smaller PSF is generated by surrounding a normal excitation PSF with a strong doughnut PSF.

The identification and separation of the emitters temporally includes (f)PALM/STORM, *d*STORM, GSDIM, and so on. In (f)PALM/STORM, single-molecule localization is used, based on the assumption that there is no adjacent emitter molecule within one or several PSF size at each frame. Then, by analyzing the centroid of each PSF, the centers can be precisely localized, with a few molecules a time. As a high-resolution image usually contains many pixels at "ON" state, these procedures are repeated several thousands of times, with mechanisms to effectively switch on randomly other emitting molecules at a time while remaining (forcing) other molecules at OFF (dark) state, to generate one image. Apparently, (f)PALM/STORM requires that the time the molecules stay at ON state be much shorter than at OFF state, so that most of the molecules are at OFF state statistically, or $\tau_{on} \ll \tau_{off}$, to avoid two close-by emitters from being ON simultaneously. This has led to the development of photo-activatable (PA)/switchable probes, to optimize the performance of PALM/STORM.

The current super-resolution techniques, such as STED and PALM/STORM, trade off temporal resolution in exchange for spatial resolution [1]. One simple approach to accelerate these super-resolution techniques is through parallelization detection. For example, 100,000 "doughnuts" have been generated simultaneously to achieve fast super-resolution imaging [2,3]. In PALM/STORM, several algorithms have been developed to fit multiple emitters at one focal volume, such as MLE fitting [4], rapidSTORM [5], DAOSTORM [6], compressed sensing [7], and so on.

SOFI is a very unique super-resolution technique in that it allows much more emitters at ON state within one focal volume. Instead of localizing each of the single molecules, SOFI analyze their fluorescence fluctuation, and generate super-resolution with cumulant analysis.

Similar to STED, SOFI is also a PSF modulation technique. One method to decrease the size of the PSF is to modulate the PSF with itself. This can be done by taking one image of the PSF, which is usually in Gaussian shape, and then multiply it by itself. The full width at half maximum (FWHM) of the PSF^2 will be only $\sqrt{1/2}$ of the original PSF. This seems to be an easy approach to obtain super-resolution: one can multiply one image with itself! This "cheating" magic can be easily done with Photoshop or ImageJ, which has a scientific name in digital image processing: gamma adjustment.

However, if considering two adjacent PSF, one can find that it is not so easy. In fact, when the second PSF is getting to 94% of the FWHM of the first PSF, the dip disappears, making this multiply magic fail to work (Figure 12.1a).

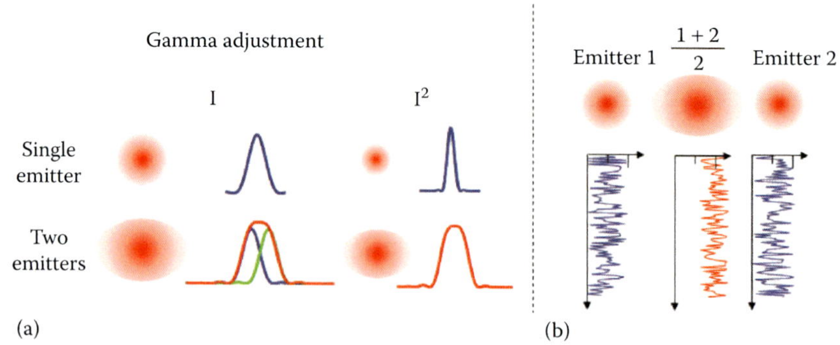

Figure 12.1

(a) Gamma adjustment for single emitter and two close-by emitters. Although the PSF narrowed with this digital image processing technique, the resolution of two close-by emitters are not improved. (b) The blinking of two emitters. When the two emitters get close, the center merged by the two emitters is brighter, but the blinkings are averaged out. Consequently, the close-by emitters can be differentiated. The grid indicates the pixels of the detector.

Although one cannot achieve super-resolution by gamma adjustment, the principle of PSF modulation is a physically valid method. In fact, confocal employs such method to obtain a better resolution, in which a detection pinhole is used to modulate the excitation PSF [8]. Similarly, multiphoton microscopy also employs the principle of self-PSF modulation, by means of the nonlinear process. From Chapter 9 we have learned that, by using the blinking character of fluorescent tags and through the statistical analysis of SOFI, the PSF can be nonlinearly modulated (Chapter 7 Equation 7.3), hence super-resolution can be obtained by increasing the order of the cumulant calculation (Figure 12.1b).

12.2 Fluorescent Labels in SOFI

Instead of the stochastic switching mechanisms in PALM/STORM, SOFI relies on the fluctuation of the fluorescent label. Theoretically, any probe/tag with blinking or fluctuation property can be used for SOFI imaging. However, characteristic optimization of the on/off fluorescent protein (FP)/dye is needed to get a high SOFI signal and the following parameters should be considered: averaged fluorescence intensity, fluctuation range, and photostability. As shown in Equation 9.3 in Chapter 9, the higher the averaged fluorescence intensity and the larger the fluctuation range, the higher the cumulant signal. And, during SOFI imaging, the overall trend of the fluorescence, despite of blinking/fluctuation, should be stable. Otherwise, segmented SOFI should be performed [9].

There are two types of fluorescent labels that have been widely used in SOFI: quantum dots (QDs) and fluctuation FP.

12.2.1 Quantum Dots

QDs are artificial nanoparticles that can emit fluorescence. In recent years, QDs exhibit great potential for fluorescence imaging in life sciences [10–12]. This can be attributed to the remarkable optical properties of QDs, for example, higher fluorescence brightness, superior photostability, blue-shifted absorption spectrum, and narrow fluorescence emission spectra. More importantly, fluorescence intermittency (i.e., blinking) is a significant characteristic of QDs [13–15]. Initially, blinking/fluctuation was treated as a negative effect, so a few surface modification techniques have been developed to suppress it [16,17]. Soon researchers realized that blinking can be utilized as a means for super-resolution [18].

Unlike the PA FPs, QDs have a long ON time and short OFF time. This makes QDs difficult to be applied in PALM/STORM. Yet, since temporally the two blinking molecules or nanoparticles cannot have the same beats, we can use this to extract their location information with statistics. SOFI is a technique developed to utilize the advantage of the blinking mechanism to achieve background-free, contrast-enhanced fast super-resolution imaging [19,20]. Comparing with the pure localization-based techniques, SOFI allows much higher emission density owing to the robust correlation analysis for separating closely spaced emitters with blinking signals [21].

QDs feature a blue-shifted absorption spectrum and narrow, size-dependent emission spectrum [24], thereby facilitating its application in simultaneous excitation and multiplexing fluorescent labeling (Figure 12.2a) [25]. With the application of different colors of QDs, dual-color SOFI has been demonstrated to resolve the spatial relationship between microtubule cytoskeleton and constitutive processing body protein hDcp1a [26].

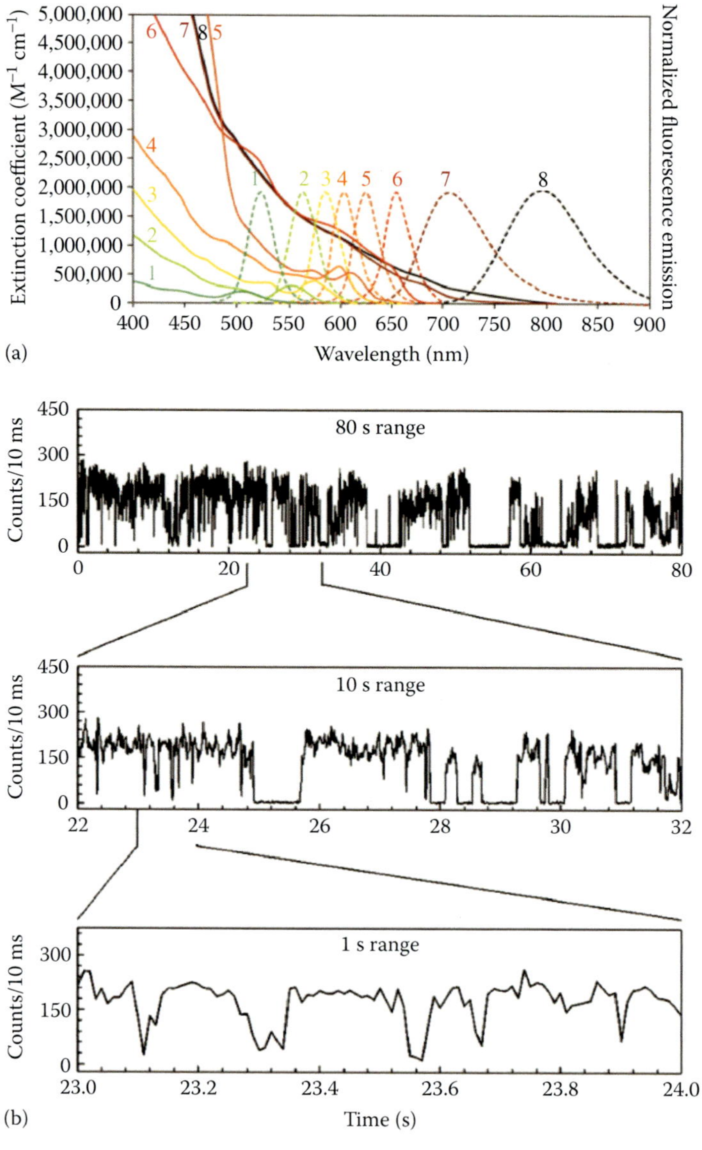

Figure 12.2

The fluorescent emission features of QDs. (a) The fluorescent absorption and emission spectrum of QDs; and (b) the fluctuation of QDs emission can be used in SOFI. (Adapted from Kim, G.B. et al., *Theranostics*, 2, 127–138, 2012; Kuno, M. et al., *J. Chem. Phys.*, 112, 3117–3120, 2000. With permission.)

12.2.2 Fluorescent Proteins

Although QD can be used for live cell imaging after ensembling in polymer to decrease the toxicity, its application is largely limited. The FP plays invaluable roles in the monitoring of cellular systems with optical microscopy, which is awarded the Nobel Prize in Chemistry in 2008 [27]. One major advantage of FP over QD lies on the application in live cell imaging. The FP can be genetically encoded to target the interested protein of the subcellular organelle. Recently, the development of PA/photoswitchable FPs have led to the development of PALM super-resolution microscopy, which is consequently awarded the Nobel Prize in Chemistry in 2014 [28–31]. During the development of PA-FPs, it has been reported that some mutants show strong stochastic blinking or fluctuation feature under one excitation source, such as Dronpa, rsEGFP [32], and pcDronpa2 [33], which makes them ideal for SOFI super-resolution [18]. Recently, through the mutant of mEos3.1 and optimizing toward the SOFI application, we found that the mutant mEos3.1 H62S exhibits very high fluctuation and photostability. We named it Skylan-S (Sky lantern for SOFI). Skylan-S can be used to accurately label multiple subcellular organelles or proteins such as β-actin, caveolin, clathrin and microtubule, and even calcium channel ORAI1. A much larger dynamic range of fluctuation can be obtained from Skylan-S in comparing with Dronpa, leading to the simultaneous super-resolution imaging of both bright and dim structures. It also facilitates real-time imaging of the clathrin-coated pit (CCP) ring structure, benefitting from the high intensity and large fluctuation [34]. Although when SOFI calculates at higher orders than 4, the heterogeneities in blinking statistics and fluorescence intensities may lead to discontinuities in the image structures; Skylan-S with very high averaged fluorescence intensity in the fluctuation state is suitable for computing higher-order SOFI images.

With mutation of FPs, dual-color SOFI is also demonstrated with HeLa cells expressing Lyn-Dronpa for targeting membrane raft, and Kras-rsTagRFP for labeling nonraft microdomains in the plasma membrane [32]. Recently, Dreiklang has also been reported for its application in SOFI live cell super-resolution imaging, through gene transfection to label the vimentin of HeLa cells [35].

In addition to QDs and FPs, organic dyes exhibiting fluctuations can also be employed, such as Alexa647 [35]. Its fluorescence emission features with very low ON/OFF ratio. Alexa647 has been widely used in *d*STORM, through the control of the buffer (Figure 12.3) [36].

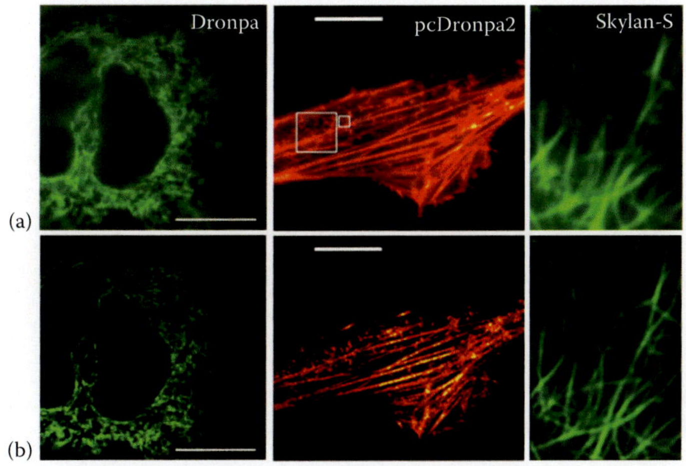

Figure 12.3

The widefield (a) and SOFI (b) imaging with Dronpa, pcDronpa2, and Skylan-S. (Adapted from Dedecker, P. et al., *Proc. Natl. Acad. Sci. U.S.A.*, 109, 10909–10914, 2012; Moeyaert, B. et al., *ACS Nano*, 8, 1664–1673, 2014; Zhang, X. et al., *ACS Nano*, 9, 2659–2667, 2015. With permission.)

12.3 The Applications of SOFI

12.3.1 Three-Dimensional SOFI Techniques

Previously in PALM/STORM, 3D super-resolution imaging are always implemented with sophisticated PSF modulation and following data post-processing, such as astigmatism [37], double helix [38], or with multiplane imaging and detection [35,39]. Taking advantage of its optical sectioning capability, SOFI can provide true 3D PSF improvement [19], enabling 3D imaging [40]. The axial resolution is dependent only on the SOFI post-processing, with large out-of-focus background presented.

Many strategies have been formulated for 3D-SOFI imaging, which can be classified into two approaches:

1. One is the conventional axial scanning mode using widefield or spinning disk confocal illumination. 3D-SOFI images can be straightaway obtained using lamp-based widefiled microscope due to the inherent sectioning capability of SOFI. In addition, the 3D-SOFI imaging can also be obtained with spinning-disk confocal microscopy. Spinning-disk confocal microscopy enables higher SNR and better 3D optical-sectioning ability for deep tissue imaging [41].
2. Another approach is to achieve fast 3D live cell SOFI imaging simultaneously using multiplane technique, which was firstly demonstrated in Reference [35].

Figure 12.4 illustrates the comparison of different imaging modalities. As can be seen, in laser scanning confocal microscopy, during one pixel dwell time there is only one focal spot presented; in spinning disk confocal microscopy, there are multiple focal spots presented, and for widefield microscopy, all the pixels are illuminated. As only the QDs within the focal spots/illumination area can be used for SOFI analysis, the aspect ratio A for laser scanning confocal is only $A = 1/N$, in which N is the pixel number. While for spinning disk confocal, it can be increased

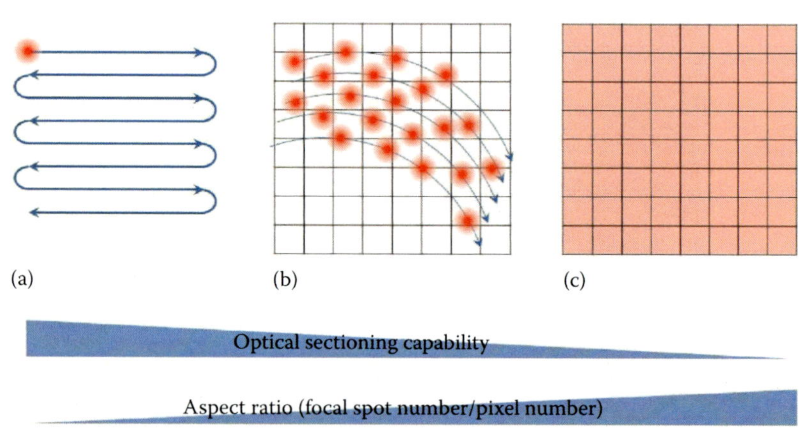

Figure 12.4

The focal spot(s)/area in different imaging modalities during one pixel dwell time: (a) laser scanning confocal; (b) spinning disk confocal; and (c) widefield microscopy. (From Chen, X. et al., *Nano Res.*, 8, 2251–2260, 2015. With permission.)

to $A = M/N$, where M is the number of focal spots that are imaged by the EMCCD simultaneously. Therefore, taking advantage of the confocal optical sectioning capability, and the parallel illumination with 2D detection, spinning disk confocal is better suitable for 3D SOFI super-resolution imaging.

Figure 12.5 illustrates the 3D mitochondria of fixed HeLa cells labeled with DAKAP-Dronpa using a conventional lamp-based widefield microscope [32]. Seven SOFI images collected at 0.5 μm interval are used for the 3D rendering.

The 3D microtubule network in COS7 cells labeled by QD625 were imaged with SD-SOFI in our experiment. As can be seen in Figure 12.6, the 3D resolution are all largely improved through SD-SOFI.

The above mentioned 3D methods require layer by layer 3D stacks, thus it needs a relatively long data acquisition. To improve the temporal resolution for 3D-SOFI, multiplane scheme has been proposed (Figure 12.7). A 3D SOFI super-resolution can be obtained in 0.6 s (325 fps, 200 frames) [42].

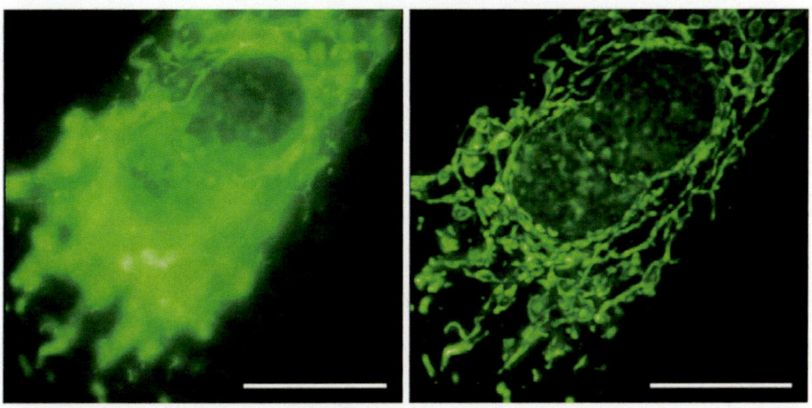

Figure 12.5

3D SOFI with pcDronpa. (From Dedecker, P. et al., *Proc. Natl. Acad. Sci. U.S.A.*, 109, 10909–10914, 2012. With permission.)

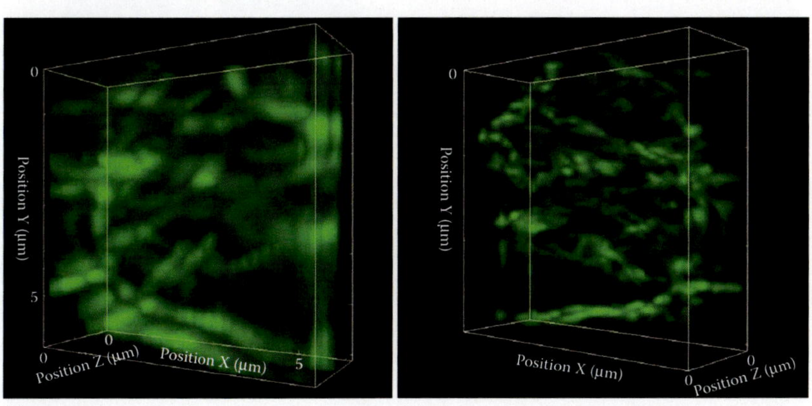

(a) (b)

Figure 12.6

Comparison of the 3D imaging of SD-confocal (a) and SD-SOFI (b). (From Chen, X. et al., *Nano Res.*, 8, 2251–2260, 2015. With permission.)

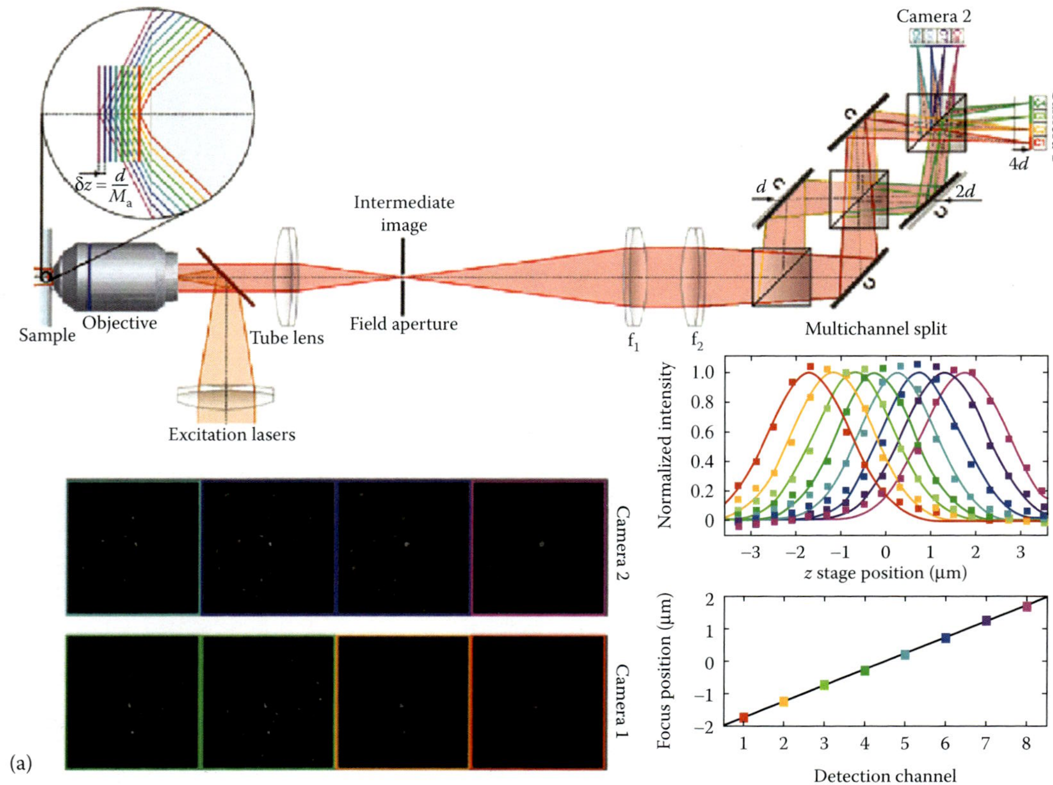

Figure 12.7

Simultaneous detection using multiple focal planes technique (a). (From Geissbuehler, S. et al., *Nat. Commun.*, 5, 2014. With permission.) (*Continued*)

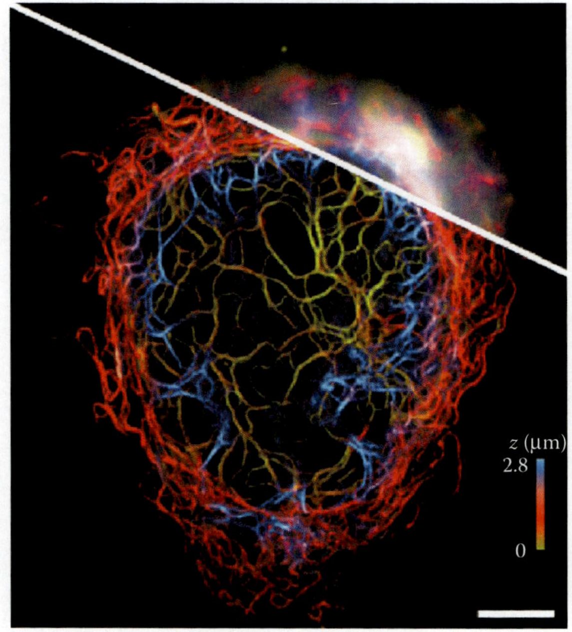

(b)

Figure 12.7 (Continued)

Live cell multiplane 3D-SOFI imaging of HeLa cells expressing vimentin-Dreiklang (b). (From Geissbuehler, S. et al., *Nat. Commun.*, 5, 2014. With permission.)

12.3.2 Fast SOFI Approaches

The increase of the imaging speed is always a war against the noise. Although SOFI can gain super-resolution than conventional microscopy, the stochastic analysis is based on the assumption of infinite frames to be analyzed. When limited number of frames are used in practice, noise correlation/cumulant is nonzero, leading to correlation noise which limits the performance of SOFI and artifacts. There are two approaches to further increase the imaging speed of SOFI: (1) to enhance the blinking of QDs [43] (When more blinking events are detected, the cumulant signal can be amplified.) and (2) to separate the QDs in different domains, so that they are more isolated in each individual domain. When they are closer to single-molecule state, the cross-cumulant noise can be largely decreased, enabling fast SOFI imaging.

Although the blinking of QDs, especially the OFF time can be as short as 0.1 ns [23,44,45], the overall large ON/OFF ratio still pose challenges for effective blinking events to be detected. The blinking ratio can be chemically modified, for example, through thinning the shell of the QDs. The blinking can be significantly enhanced through this approach [43]. With the variant analysis, Watanabe et al. obtained fast super-resolution with temporal and spatial resolution of 80 ms and 90 nm, respectively.

Labeling density is of great importance for ensuring the structural integrity presented by optical fluorescence microscopy [46]. It is the chemical representation of Nyquist–Shannon sampling theorem [47]. For example, to obtain a resolution of 40 nm, the distance between two adjacent emitters should be no more than 20 nm; or we can say the labeling density should be more than 50 (molecules) per micron (50 μm^{-1}) in one dimension, or 2500 μm^{-2} in two dimensions.

The equation below describes the relationship between the labeling density, the ON/OFF ratio, and the emitting molecule density:

$$\text{Emitting molecule density} = \text{Labeling density} \times \frac{\text{ON}}{\text{ON} + \text{OFF}} \text{ratio} \quad (12.1)$$

In PALM/STORM, the emitting molecule density should be at most 1 molecule per PSF area. In contrast, a large labeling density should be applied, in association with sufficient single molecules to be localized, to support the high resolution of PALM/STORM. Therefore, it *employs optical modulation to drive most of the molecules into dark state, leaving only sparse single molecules at ON state (a very small ON/OFF ratio).*

However, because SOFI solely relies on the inherent blinking of the fluorophores to obtain super-resolution, in which the ON/OFF ratio is very high, it poses a great challenge for accurately extracting the fine structure out of the diffraction-limited area. With the increase of the labeling density, the high-order cumulants of SOFI algorithm tend to induce artifacts/discontinuities, leading to a degradation to the image quality.

Inspired by the fact that when the single molecules can be differentiated by their colors (e.g., *m* types of colors), then the labeling density can be decreased *m*-folds for each color channel, we have developed joint tagging SOFI (JT-SOFI). It is aiming at ultrahigh labeling density super-resolution imaging, meanwhile retaining the continuities and integrity of the targets being investigated without compromising on the spatial resolution enhancement, through spectral multiplexing [48–52] and SOFI imaging [53]. Owing to the blue-enhanced absorption spectra and narrow fluorescence emission spectra of QDs, the QDs can be excited simultaneously with the same excitation source, with excessively high spectral encoding capability (Figure 12.8) [25].

Under such circumstances, the labeling density of single color QDs is relatively low which facilitates the accurate separation of single QDs using SOFI algorithm, at relatively low frame numbers. Yet, the overall labeling density is theoretically m-fold increased through the application of m types of QDs, thereby enabling JT-SOFI nanometric imaging with ultrahigh labeling density. By combining the multiple spectral channels, super-resolution images with well-preserved integrity and continuities can be reconstructed, which are capable of revealing subdiffraction-sized structures inside the biological cells in a more genuine perspective at high spatiotemporal resolution.

Experimentally, multiple types of QDs with their fluorescence spectra well separated were jointly immunostained to the same cellular structure. A commercial widefield microscope was utilized for JT-SOFI super-resolution imaging of microtubules jointly labeled with QD525, QD625, and QD705 (Invitrogen). A movie of the blinking QDs-labeled microtubules can be seen in the Supporting Information of Reference [53]. With 4.5 s of data acquisition (100 frames each channel), the spatial resolution can be improved to 85 nm with considerably reduced discontinuities and heterogeneities by employing joint tagging fourth-order SOFI processing (Figure 12.9).

12.3.3 SOFI for FRET and FCS

Before the birth of super-resolution microscopy, the resolution of fluorescent microscopy is limited by the diffraction, which restricts the direct visualization of the dynamics within a focal area. Several indirect methods have therefore been developed, such as Förster resonance energy transfer (FRET) [54] and fluorescence correlation spectroscopy (FCS) [55].

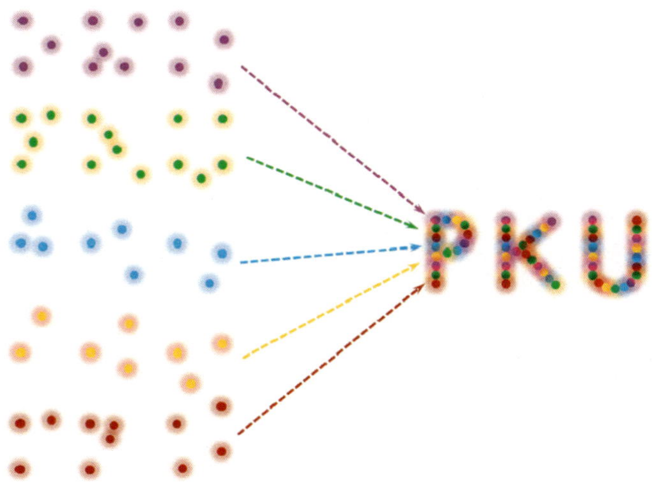

Figure 12.8

Illustration of the JT-SOFI. With different colors of labeling, the individual color channel can be very sparse, yet the combined image contains high labeling density to ensure the faithful super-resolution without under sampling.

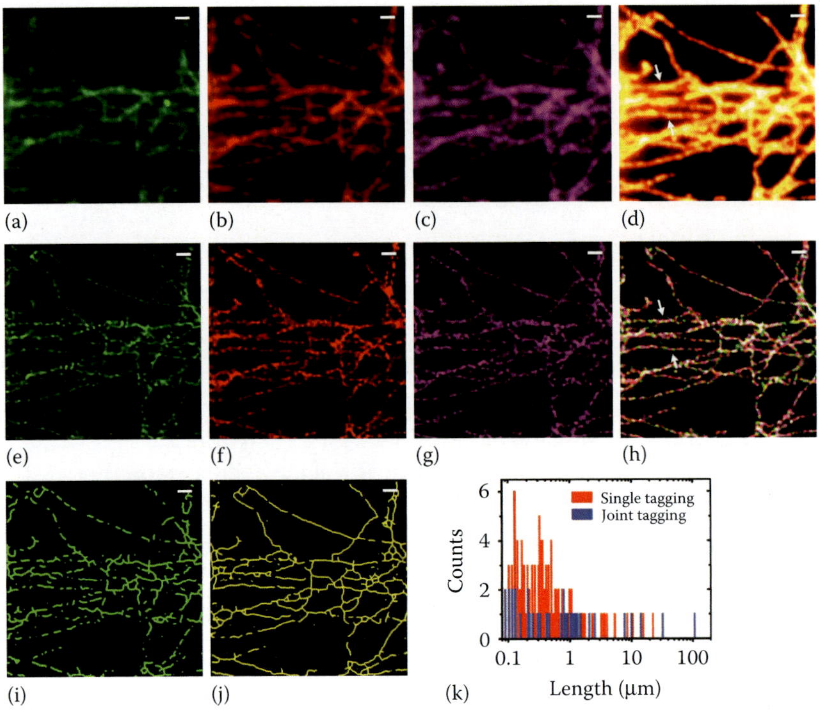

Figure 12.9

(a–k) Comparison of the imaging results of SOFI and JT-SOFI. (From Zeng, Z. et al., *Sci. Rep.*, 5, 8359, 2015. With permission.)

In FRET, when a pair of fluorophores are getting close with each other (typically within 10 nm), the donor molecule can transfer the excited energy directly to the acceptor molecule, so that fluorescence can be emitted from the acceptor. This indicates that an interaction between the molecules linked into the donor and the acceptor has happened. Through the inverse use of this principle, diffuse-assisted FRET SOFI (dSOFI) can turn the donor molecules ON and OFF through FRET effect, creating blinking events in live cell imaging, resulting in SOFI super-resolution imaging (Figure 12.10) [56].

FCS is a technique that calculates the correlation within a focal volume. Because the diffusion of the molecule is randomized, the signal from the focal volume fluctuates. In this regard and in the mathematical form, it is very similar to SOFI. Different from SOFI analysis, which calculates the spatial correlation/cumulants, with a time shift usually set to zero, FCS calculates the temporal correlation. Recently, taking advantage of both techniques, FCS SOFI has been used in the analysis of both structural and dynamics of the nanoporous material [57].

12.4 Conclusion and Outlook

Among the various super-resolution microscopy techniques, SOFI is an "easy" one as it does not require specific instruments. The only requirement for SOFI is that the fluorescence should fluctuate, so that the super-resolution of the

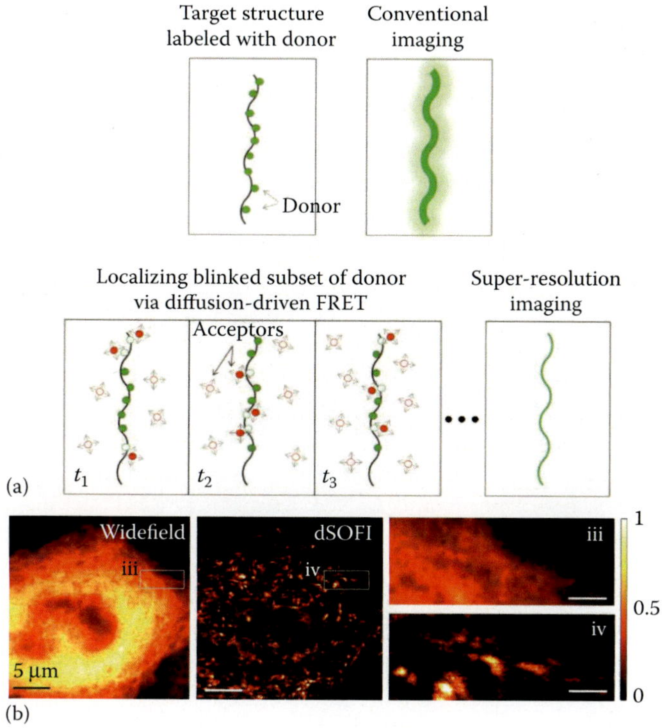

Figure 12.10

Schematic diagram (a) and live cell imaging (b) of dSOFI. (Adapted from Cho, S. et al., *Sci. Rep.*, 3, 1208, 2013. With permission.)

molecules can be addressed with statistical analysis. Consequently, biologists and chemists with little optics background can benefit from this technique, to apply SOFI into probing different cellular world, and to develop better labels for SOFI. In this chapter, we first discussed the excitation and emission features of QDs, which is a widely used inorganic fluorescent dye for SOFI. Then, we elaborated the application of FPs for SOFI, especially for live cell imaging. Third, as the cell has a 3D form, a series of 3D SOFI techniques are compared. Fourth, the methods that improve the temporal resolution of SOFI have been discussed, such as with blinking enhancement, or with joint-tagging. Finally, SOFI with FRET and FCS give us insights of both structural super-resolution and additional dynamic information. We believe that SOFI will find more applications in the dynamic nanoscale interactions of specific macromolecular assemblies in living cells.

However, SOFI should not be constrained on a "optics-free" zone. For example, the application of the multiplane in SOFI has already dramatically boosted the 3D data collection speed of SOFI. In association with optical upgrade, more sophisticated data analysis algorithms can be developed, toward a better resolution, in both spatial and temporal. It is only through efforts from all the aspects that can constantly improve this simple super-resolution technique and widen its application.

References

1. Maglione, M., and S. J. Sigrist, Seeing the forest tree by tree: Super-resolution light microscopy meets the neurosciences, *Nature Neuroscience*, 16, 790–797, 2013.
2. Chen, X., and P. Xi, Hundred-thousand light holes push nanoscopy to go parallel, *Microscopy Research and Technique*, 78, 8–10, 2014.
3. Chmyrov, A., J. Keller, T. Grotjohann, M. Ratz, E. d'Este, S. Jakobs et al., Nanoscopy with more than 100,000 "doughnuts," *Nature Methods*, 10, 737–740, 2013.
4. Huang, F., S. L. Schwartz, J. M. Byars, and K. A. Lidke, Simultaneous multiple-emitter fitting for single molecule super-resolution imaging, *Biomedical Optics Express*, 2, 1377–1393, 2011.
5. Wolter, S., A. Löschberger, T. Holm, S. Aufmkolk, M.-C. Dabauvalle, S. van de Linde et al., rapidSTORM: Accurate, fast open-source software for localization microscopy, *Nature Methods*, 9, 1040–1041, 2012.
6. Holden, S. J., S. Uphoff, and A. N. Kapanidis, DAOSTORM: An algorithm for high-density super-resolution microscopy, *Nature Methods*, 8, 279–280, 2011.
7. Zhu, L., W. Zhang, D. Elnatan, and B. Huang, Faster STORM using compressed sensing, *Nature Methods*, 9, 721–723, 2012.
8. Xi, P., H. Xie, Y. Liu, and Y. Ding, Optical Nanoscopy with Stimulated Emission Depletion, in *Optical Nanoscopy and Novel Microscopy Techniques*, P. Xi, Ed., New York: CRC Press, 2014, pp. 1–22.
9. Dertinger, T., M. Heilemann, R. Vogel, M. Sauer, and S. Weiss, Superresolution optical fluctuation imaging with organic dyes, *Angewandte Chemie*, 122, 9631–9633, 2010.
10. Bentolila, L. A., Y. Ebenstein, and S. Weiss, Quantum dots for in vivo small-animal imaging, *Journal of Nuclear Medicine*, 50, 493–496, 2009.
11. Pathak, S., E. Cao, M. C. Davidson, S. Jin, and G. A. Silva, Quantum dot applications to neuroscience: new tools for probing neurons and glia, *The Journal of Neuroscience*, 26, 1893–1895, 2006.
12. Smith, A. M., X. Gao, and S. Nie, Quantum dot nanocrystals for in vivo molecular and cellular imaging, *Photochemistry and Photobiology*, 80, 377–385, 2004.
13. Xu, Z., and M. Cotlet, Photoluminenscence blinking dynamics of colloidal quantum dots in the presence of controlled external electron traps, *Small*, 8, 253–258, 2012.
14. Gómez, D. E., J. van Embden, J. Jasieniak, T. A. Smith, and P. Mulvaney, Blinking and surface chemistry of single CdSe nanocrystals, *Small*, 2, 204–208, 2006.
15. Frantsuzov, P., M. Kuno, B. Janko, and R. A. Marcus, Universal emission intermittency in quantum dots, nanorods and nanowires, *Nature Physics*, 4, 519–522, 2008.
16. Yuan, C., P. Yu, and J. Tang, Blinking suppression of colloidal CdSe/ZnS quantum dots by coupling to silver nanoprisms, *Applied Physics Letters*, 94, 243108, 2009.
17. Chen, Y., J. Vela, H. Htoon, J. L. Casson, D. J. Werder, D. A. Bussian et al., "Giant" multishell CdSe nanocrystal quantum dots with suppressed blinking, *Journal of the American Chemical Society*, 130, 5026–5027, 2008.

18. Xu, J., J. Chang, Q. Yan, T. Dertinger, M. P. Bruchez, and S. Weiss, Labeling cytosolic targets in live cells with blinking probes, *The Journal of Physical Chemistry Letters*, 4, 2138–2146, 2013.
19. Dertinger, T., R. Colyer, G. Iyer, S. Weiss, and J. Enderlein, Fast, background-free, 3D super-resolution optical fluctuation imaging (SOFI), *Proceedings of the National Academy of Sciences of the United States of America*, 106, 22287–22292, 2009.
20. Dedecker, P., S. Duwé, R. K. Neely, and J. Zhang, Localizer: Fast, accurate, open-source, and modular software package for superresolution microscopy, *Journal of Biomedical Optics*, 17, 126008–126008, 2012.
21. Geissbuehler, S., C. Dellagiacoma, and T. Lasser, Comparison between SOFI and STORM, *Biomedical Optics Express*, 2, 408–420, 2011.
22. Kim, G. B., and Y. P. Kim, Analysis of Protease Activity Using Quantum Dots and Resonance Energy Transfer, *Theranostics*, 2, 127–138, 2012.
23. Kuno, M., D. Fromm, H. Hamann, A. Gallagher, and D. Nesbitt, Nonexponential "blinking" kinetics of single CdSe quantum dots: A universal power law behavior, *The Journal of Chemical Physics*, 112, 3117–3120, 2000.
24. Resch-Genger, U., M. Grabolle, S. Cavaliere-Jaricot, R. Nitschke, and T. Nann, Quantum dots versus organic dyes as fluorescent labels, *Nature Methods*, 5, 763–775, 2008.
25. Han, M., X. Gao, J. Z. Su, and S. Nie, Quantum-dot-tagged microbeads for multiplexed optical coding of biomolecules, *Nature Biotechnology*, 19, 631–635, 2001.
26. Gallina, M. E., J. Xu, T. Dertinger, A. Aizer, Y. Shav-Tal, and S. Weiss, Resolving the spatial relationship between intracellular components by dual color super resolution optical fluctuations imaging (SOFI), *Optical Nanoscopy*, 2, 1–9, 2013.
27. Weiss, P. S., Nobel Prize in Chemistry: Green fluorescent protein, its variants and implications, *ACS Nano*, 2, 1977–1977, 2008.
28. Chudakov, D. M., V. V. Verkhusha, D. B. Staroverov, E. A. Souslova, S. Lukyanov, and K. A. Lukyanov, Photoswitchable cyan fluorescent protein for protein tracking, *Nature Biotechnology*, 22, 1435–1439, 2004.
29. Ando, R., H. Hama, M. Yamamoto-Hino, H. Mizuno, and A. Miyawaki, An optical marker based on the UV-induced green-to-red photoconversion of a fluorescent protein, *Proceedings of the National Academy of Sciences of the United States of America*, 99, 12651–12656, 2002.
30. Sauer, M., Reversible molecular photoswitches: A key technology for nanoscience and fluorescence imaging, *Proceedings of the National Academy of Sciences of the United States of America*, 102, 9433, 2005.
31. Betzig, E., G. H. Patterson, R. Sougrat, O. W. Lindwasser, S. Olenych, J. S. Bonifacino et al., Imaging intracellular fluorescent proteins at nanometer resolution, *Science*, 313, 1642–1645, September 15, 2006.
32. Dedecker, P., G. C. Mo, T. Dertinger, and J. Zhang, Widely accessible method for superresolution fluorescence imaging of living systems, *Proceedings of the National Academy of Sciences of the United States of America*, 109, 10909–10914, 2012.
33. Moeyaert, B., N. Nguyen Bich, E. De Zitter, S. Rocha, K. Clays, H. Mizuno et al., Green-to-red photoconvertible Dronpa mutant for multimodal super-resolution fluorescence microscopy, *ACS Nano*, 8, 1664–1673, 2014.

34. Zhang, X., X. Chen, Z. Zeng, M. Zhang, Y. Sun, P. Xi et al., Development of a reversibly switchable fluorescent protein for super-resolution optical fluctuation imaging (SOFI), *ACS Nano*, 9, 2659–2667, 2015.
35. Geissbuehler, S., A. Sharipov, A. Godinat, N. L. Bocchio, P. A. Sandoz, A. Huss et al., Live-cell multiplane three-dimensional super-resolution optical fluctuation imaging, *Nature Communications*, 5, 5830, 2014.
36. Heilemann, M., S. van de Linde, M. Schüttpelz, R. Kasper, B. Seefeldt, A. Mukherjee et al., Subdiffraction-resolution fluorescence imaging with conventional fluorescent probes, *Angewandte Chemie International Edition*, 47, 6172–6176, 2008.
37. Huang, B., W. Wang, M. Bates, and X. Zhuang, Three-dimensional super-resolution imaging by stochastic optical reconstruction microscopy, *Science*, 319, 810–813, February 8, 2008.
38. Pavani, S. R. P., M. A. Thompson, J. S. Biteen, S. J. Lord, N. Liu, R. J. Twieg et al., Three-dimensional, single-molecule fluorescence imaging beyond the diffraction limit by using a double-helix point spread function, *Proceedings of the National Academy of Sciences of the United States of America*, 106, 2995–2999, 2009.
39. Juette, M. F., T. J. Gould, M. D. Lessard, M. J. Mlodzianoski, B. S. Nagpure, B. T. Bennett et al., Three-dimensional sub-100 nm resolution fluorescence microscopy of thick samples, *Nature Methods*, 5, 527–529, 2008.
40. Dertinger, T., J. Xu, O. F. Naini, R. Vogel, and S. Weiss, SOFI-based 3D superresolution sectioning with a widefield microscope, *Optical Nanoscopy*, 1, 1–5, 2012.
41. Chen, X., Z. Zeng, H. Wang, and P. Xi, Three dimensional multimodal sub-diffraction imaging with spinning-disk confocal microscopy using blinking/fluctuation probes, *Nano Research*, 8, 2251–2260, 2015.
42. Geissbuehler, S., A. Sharipov, A. Godinat, N. L. Bocchio, P. A. Sandoz, A. Huss et al., Live-cell multiplane three-dimensional super-resolution optical fluctuation imaging, *Nature Communications*, 5, 2014.
43. Watanabe, T. M., S. Fukui, T. Jin, F. Fujii, and T. Yanagida, Real-time nanoscopy by using blinking enhanced quantum dots, *Biophysical Journal*, 99, L50–L52, 2010.
44. Cichos, F., C. Von Borczyskowski, and M. Orrit, Power-law intermittency of single emitters, *Current Opinion in Colloid & Interface Science*, 12, 272–284, 2007.
45. Wang, Y., G. Fruhwirth, E. Cai, T. Ng, and P. R. Selvin, 3D super-resolution imaging with blinking quantum dots, *Nano Letters*, 13, 5233–5241, 2013.
46. Ji, N., H. Shroff, H. Zhong, and E. Betzig, Advances in the speed and resolution of light microscopy, *Current Opinion in Neurobiology*, 18, 605–616, 2008.
47. Klein, T., S. Proppert, and M. Sauer, Eight years of single-molecule localization microscopy, *Histochemistry and Cell Biology*, 141, 561–575, 2014.
48. Lemmer, P., M. Gunkel, D. Baddeley, R. Kaufmann, A. Urich, Y. Weiland et al., SPDM: Light microscopy with single-molecule resolution at the nanoscale, *Applied Physics B*, 93, 1–12, 2008.
49. Cremer, C., R. Kaufmann, M. Gunkel, S. Pres, Y. Weiland, P. Müller et al., Superresolution imaging of biological nanostructures by spectral precision distance microscopy, *Biotechnology Journal*, 6, 1037–1051, 2011.

50. Lemmer, P., M. Gunkel, Y. Weiland, P. Müller, D. Baddeley, R. Kaufmann et al., Using conventional fluorescent markers for far-field fluorescence localization nanoscopy allows resolution in the 10-nm range, *Journal of Microscopy*, 235, 163–171, 2009.
51. Cutler, P. J., M. D. Malik, S. Liu, J. M. Byars, D. S. Lidke, and K. A. Lidke, Multi-color quantum dot tracking using a high-speed hyperspectral line-scanning microscope, *PLoS One*, 8, e64320, 2013.
52. van Dijk, T., D. Mayerich, R. Bhargava, and P. S. Carney, Rapid spectral-domain localization, *Optics Express*, 21, 12822–12830, May 20, 2013.
53. Zeng, Z., X. Chen, H. Wang, N. Huang, C. Shan, H. Zhang et al., Fast super-resolution imaging with ultra-high labeling density achieved by joint tagging super-resolution optical fluctuation imaging, *Scientific Reports*, 5, 8359, 2015.
54. Kenworthy, A. K., Imaging protein-protein interactions using fluorescence resonance energy transfer microscopy, *Methods*, 24, 289–296, 2001.
55. Magde, D., E. Elson, and W. W. Webb, Thermodynamic fluctuations in a reacting system—Measurement by fluorescence correlation spectroscopy, *Physical Review Letters*, 29, 705, 1972.
56. Cho, S., J. Jang, C. Song, H. Lee, P. Ganesan, T.-Y. Yoon et al., Simple super-resolution live-cell imaging based on diffusion-assisted Forster resonance energy transfer, *Scientific Reports*, 3, 1208, 2013.
57. Kisley, L., R. Brunetti, L. J. Tauzin, B. Shuang, X. Yi, A. W. Kirkeminde et al., Characterization of porous materials by fluorescence correlation spectroscopy super-resolution optical fluctuation imaging (fcsSOFI), *ACS Nano*, 9, 9158–9166, 2015.

13

Super-Resolution Fluorescence Microscopy of the Nanoscale Organization of RNAP and DNA in *E. coli*

Christoph Spahn, Ulrike Endesfelder, and Mike Heilemann

13.1 Introduction
13.2 Spatial Organization of Transcription in *E. coli*
13.3 Spatial Organization of DNA Replication and Segregation in *E. coli*
13.4 Correlative Super-Resolution Imaging of RNAP, Membrane, and DNA in *E. coli*
13.5 Experimental Design of SMLM Studies in *E. coli*
13.6 Protocols
Acknowledgments
References

13.1 Introduction

Cellular processes in bacteria are orchestrated by the spatiotemporal organization of proteins into functional units. The size and composition of such functional units is sensitively regulated in the bacterial cell, and among other factors depend on the cell cycle stage as well as external stimuli or growth conditions. Revealing parameters such as size, protein content, stoichiometry, and dynamics is crucial to understand the functionality of molecular machines, as well as their regulation at the molecular level.

Small cellular structures with a size of tens of nanometers are typically not accessible for visualization with light microscopy, because diffraction of light limits the spatial resolution to about $\lambda/2$ or 200–300 nm. The development of microscopic techniques that bypass this barrier, often referred to as "super-resolution microscopy" (Heilemann, 2010), has paved the way to investigate tiny structures down to the molecular level. In particular, single-molecule localization microscopy (SMLM) techniques achieve a near-molecular resolution, and uniquely allow extracting quantitative information on protein copy numbers and dynamics. Briefly, these techniques employ photoswitchable fluorescent probes (Fürstenberg and Heilemann, 2013), stochastic activation of a subset of fluorophores to realize single-molecule detection, and position determination with a precision of a few nanometers. A pointillistic reconstruction of single-molecule coordinates into an artificial image provides maps of proteins at the near molecular level with a spatial resolution approaching 20 nm. In addition, the post-analysis of SMLM data provides access to quantitative information such as protein subpopulations and heterogeneities that are otherwise hidden by ensemble averaging. Applying SMLM to bacterial cells, protein copy numbers and distributions (Biteen et al., 2008, Endesfelder et al., 2013, Greenfield et al., 2009, Spahn et al., 2014) as well as dynamics of cellular processes in living cells were studied (Badrinarayanan et al., 2012, Bakshi et al., 2013, Persson et al., 2013, Sanamrad et al., 2014).

In this chapter, we review quantitative single-molecule super-resolution imaging of RNA polymerase (RNAP) in *E. coli*. We first describe how protein copy numbers of RNAP were determined for single bacterial cells and under different growth conditions. We then demonstrate how the organization of RNAP in nanoscale units can be determined from single-molecule data, and how this data can be interpreted in the context of the *E. coli* genome. In a second part, we present an experimental strategy for high-density labeling of the bacterial DNA and super-resolution imaging of the chromosome. We introduce sequential super-resolution imaging of the bacterial chromosome and RNAP in fixed cells. Finally, we present correlative single-molecule tracking of RNAP in live cells, and super-resolution imaging of the chromosome. Detailed information on the experimental design and optimized protocols are listed in the final section.

13.2 Spatial Organization of Transcription in *E. coli*

13.2.1 Single-Molecule Super-Resolution Microscopy of the RNAP Organization

RNAP is the key enzyme orchestrating transcription within the bacterial cell. *E. coli* possesses only a single type of RNAP, whose "core" is composed of the four subunits $\alpha_2\beta\beta'$. Transcription initiation is globally controlled by sigma (σ) factors (Mooney et al., 2009, Mukhopadhyay et al., 2001), which mediate homeostasis and facilitate a fast adaption to new conditions or stress situations. In addition, *E. coli* expresses more than 300 other transcription factors, which illustrates the complexity of transcriptional regulation in a bacterial cell (Browning and Busby, 2004). Recruited to a promoter region RNAP forms an extremely stable elongation complex processing the DNA with an average speed of 25–50 bp/s (Golding et al., 2005, Neidhardt, 1987). It has been shown that transcriptional organization is very sensitive to environmental cues like chemical composition of the growth medium. In rich medium *E. coli* cells grow rapidly since they are supplied by all nutrients. Here, the ribosomal operons exhibit 60–80 active RNAPs (Miller et al., 1970), but most of the other active genes are associated with only one RNAP (French and Miller, 1989). This leads to the conclusion that nearly all active genes are transcribed by less than one RNAP at a time and mostly less than once per cell cycle. In minimal media, essential substrates for cell growth are not available and have to be produced by the cell itself. The genes for the necessary proteins are transcribed by the transcription machinery, and it has been shown that most of these genes are transcribed by less than one RNAP at a time (Bon et al., 2006).

This variation in transcriptional activity also influences DNA replication and generation time of *E. coli*. The *E. coli* strain MG1655 possesses a single 4.6 Mbp chromosome including seven ribosomal operons and requires between 18 and 100 min for division, depending on the growth condition (Blattner et al., 1997, Neidhardt, 1987, Nomura, 1999). As a complete chromosome replication takes ~40 min (Michelsen et al., 2003), *E. coli* cells maintain multiple overlapping replication cycles during fast growth. Depending on the doubling time and therefore on the number of overlapping cell cycles and the progress of each replication cycle, cells possess different numbers of ribosomal operons. How this interplay of replication and transcription, especially for fast growth conditions, is synchronized and maintained, is largely unknown.

Confocal fluorescence microscopy studies of, by the green fluorescent protein (GFP)-tagged RNAP reported a spatial pattern of RNAP (Cabrera and Jin, 2003) that changes with cell growth conditions. By introducing a photoswitchable or photoactivatable fluorescent protein (FP) such as yGFP (Bakshi et al., 2012), PAmCherry1 (Endesfelder et al., 2013), or mEos2 (Bakshi et al., 2013, Endesfelder et al., 2013), these structures were studied with SMLM.

Copy numbers can be extracted by imaging a bacterial strain in which the endogenous RNAP is tagged with a quantitative photoactivatable FP such as PAmCherry1 (Endesfelder et al., 2013). Applying appropriate imaging conditions and subsequent image analysis strategies can account for specific fluorophore properties and avoid overcounting. The calibration with biochemical data allowed estimating the detection efficiency of PAmCherry1-RNAP in *E. coli* to around 50%–60% (Endesfelder et al., 2013) (for details in quantification strategies see also the practical Section 13.5). RNAP copy numbers were determined to 3321 (s.d. 1014) at fast growth (LB medium) and 1292 (s.d. 599) at slow growth (M9 medium) (Endesfelder et al., 2013) (Figure 13.1).

The analysis of the cellular distribution of RNAP in rich medium (LB) reveals that the underlying banding structure increases linearly with cell size and is regularly spaced (Figure 13.2). At the beginning of the cell cycle, two bands are visible (group 1) which then expand and divide into four bands (group 2 and 3) as the bacteria grow larger. *E. coli* in group 3 also visibly exhibit a septum in the white light transmission images. The number of bands of each bacterium corresponds to the number of replicating chromosomes (three on average in LB medium) which together are reported to divide themselves into distinct regions (Ishihama, 2000). The RNAP banding pattern thus mirrors the organization of DNA, which was also confirmed by DAPI co-staining of DNA. FRAP measurements determined the transcriptionally active RNAP fraction to 50%–80% (Endesfelder et al., 2013). Within these banded regions RNAP is believed to further form clusters of different size. A large number of RNAP molecules are engaged on a small number of ribosomal operons (~70 RNAPs per operon and 7 operons per chromosome). The high-density areas of RNAP (Figure 13.2) suggest that the ribosomal operons do adopt a compact conformation (each 4.4 kb operon ~1.5 μm when linear). For minimal medium (M9), DAPI staining suggests that the cellular DNA, consisting of 1–2 chromosome copies, is distributed fairly uniformly throughout the ellipse-like nucleoid (Cabrera and Jin, 2003). The analysis of the banding patterns typically shows the presence of one (group 1) or sometimes two copies (group 2) of the bacterial chromosome.

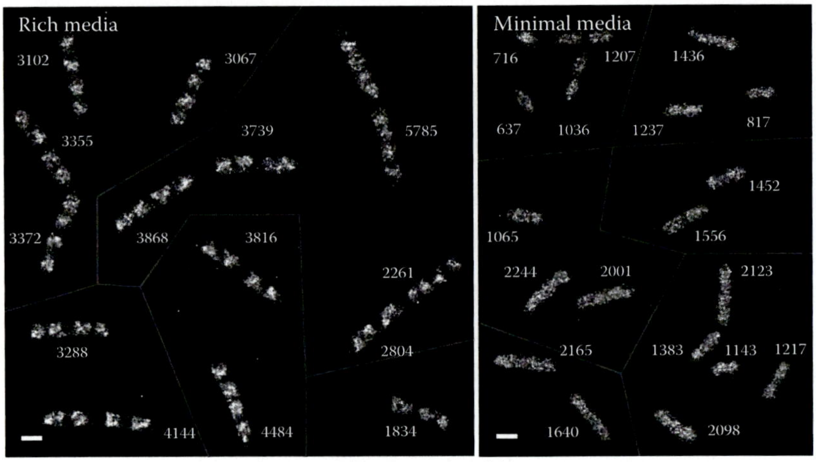

(a) (b)

Figure 13.1

Super-resolution protein map of the distribution of RNAP in *E. coli*. *E. coli* strain MG1655 *rpoC-PAmCherry1 Amp^R*, in which the chromosomal β'-subunit is tagged with PAmCherry1, was grown at 32°C in (a) LB (rich medium) or (b) in M9 (a minimal medium) to OD600 of 0.4, fixed, and imaged using SMLM. The numbers next to the cells indicate molecular localizations. Images were rendered by blurring the localization position with a 2D Gaussian to reflect the achieved spatial resolution. Each panel was assembled from different images (gray lines; scale bars 1 μm). (Reprinted from *Biophys. J.*, 105, Endesfelder, U. et al., Multiscale spatial organization of RNA polymerase in *Escherichia coli*, 172–181, Copyright 2013, with permission from Elsevier.)

13.2.2 Clustering of RNAP at the Nanoscale

Next to absolute protein copy numbers, SMLM provides spatial coordinates of proteins. Cluster analysis of single-molecule positions yields information on the size, frequency, and protein content of clusters. In case of high copy numbers and small volumes like in *E. coli*, the clustering algorithm must be able to cope with high densities. Here, density-based cluster analysis like density-based spatial clustering analysis with noise (DBSCAN) (Ester et al., 1997) or OPTICS (Ankerst et al., 1999) can be applied. These algorithms are based on the density distribution of points in a dataset. Points with greater than *MinPts* neighbors within a radius ε are connected to their neighbors; regions of adjacent connected points are considered to be clustered. Points without connection are registered to be unclustered protein. The sensitivity of the algorithm critically depending on the parameter is ε, which must be sufficiently large to not miss clusters. Here, a sophisticated choice of ε can be made by determining the effective localization precision of the dataset (Endesfelder et al., 2014) from a control sample, for example, cytosolic monomeric PAmCherry1 fluorophores. Cluster analysis of the distribution of RNAP in cells growing in LB medium revealed an average number of 6.1 large clusters per chromosome with 80–100 copies of RNAP. The cluster diameter was determined to around 160 nm. The RNAP density within the cluster was calculated to be 687 (s.d. 76) localizations/μm^2 (M9) and 706 (s.d. 119) localizations/μm^2 (LB) (Endesfelder et al., 2013) (Figure 13.3).

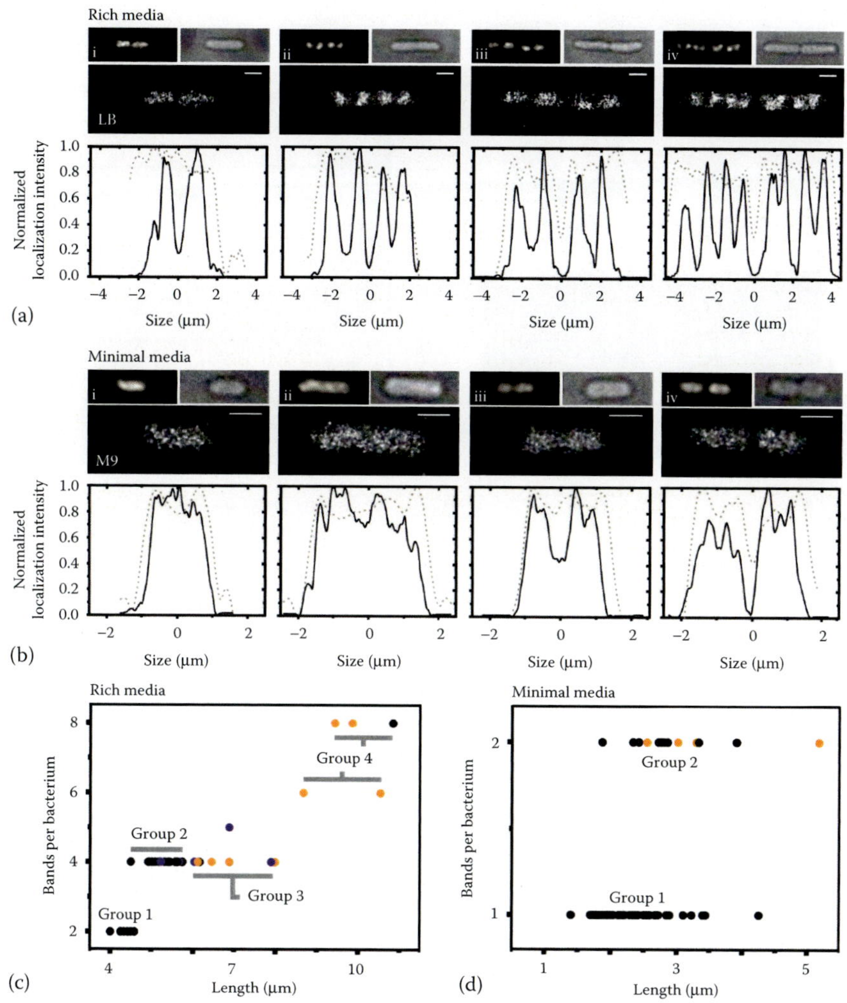

Figure 13.2

RNAP molecules are organized in bands. *E. coli* cells expressing β′-PAmCherry1 were grown in LB (a and b) or M9 medium (c and d), fixed, and imaged using SMLM and bright field microscopy. (a) Three views of four fields are shown (large images—SMLM; top-left—bright field image; top-right—reconstruction of a conventional fluorescence image using single-molecule data) above the corresponding RNAP localization densities (SMLM data, solid black line) and smoothed intensity profiles (from bright field image, dashed gray line). Fields i–iv include cells of different length (scale bar 500 nm). (b) Number of bands per cell dependent on cell length. Cells lacking a septum are shown in black; cells that begin to build up or contain a fully formed septum are shown in blue and orange, respectively. (c and d) Analysis as in (a, b) for cells grown in minimal medium. (Reprinted from *Biophys. J.*, 105, Endesfelder, U. et al., Multiscale spatial organization of RNA polymerase in *Escherichia coli*, 172–181, Copyright 2013, with permission from Elsevier.)

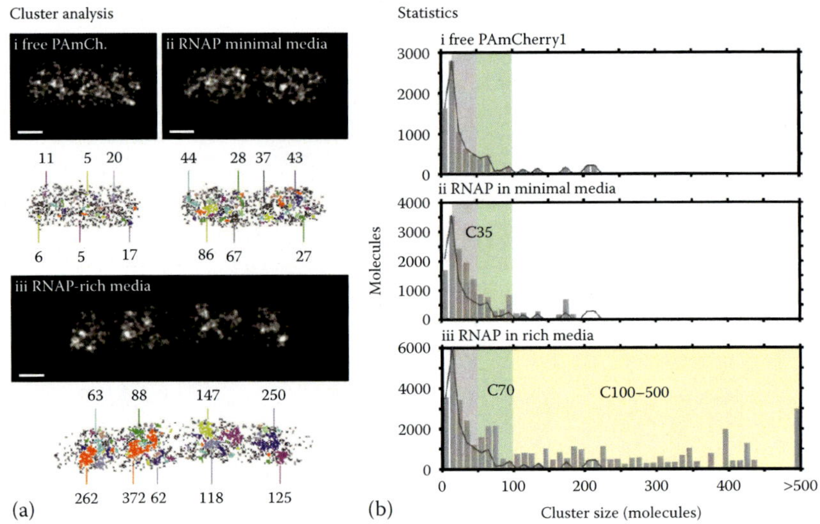

Figure 13.3

Clustering analysis of RNAP in *E. coli* cells expressing free PAmCherry1 (i) or β′-PAmCherry1 (ii, iii) grown in rich medium (LB) (i, iii) or minimal medium (M9) (ii) using DBSCAN. (a) SMLM images are shown above localizations plotted as crosses (crosses in one cluster share the same randomly chosen color) (scale bars 500 nm). (b) Cluster sizes observed for cells shown in (a). The solid gray line shows the frequency distribution for free PAmCherry1 (i), normalized to the maximum frequency in each panel. Blue, green, and yellow backgrounds highlight regions containing clustered molecules. (Reprinted from *Biophys. J.*, 105, Endesfelder, U. et al., Multiscale spatial organization of RNA polymerase in *Escherichia coli*, 172–181, Copyright 2013, with permission from Elsevier.)

This ultrastructural analysis in *E. coli* grown in rich medium (L9) revealed a dense packing of RNAP in "hot spots" of transcription. There is reason to speculate that these clusters correspond to sites transcribing rrn operons: about 70% of active RNAP are closely packed to the rrn operons (Neidhardt, 1987). The identified clusters contain at least 70 RNAPs, which is consistent to earlier results from electron microscopy that determined 80–90 RNAP per single rrn operon (Miller et al., 1970). Moreover, there are ~6 clusters per chromosome (overlapping clusters are included), which is close to seven rrn operons in the *E. coli* genome. When reducing rrn transcription by stringent response, RNAP is redistributed to other genes and no large clusters are observed (Endesfelder et al., 2013). A similar observation was made for cells treated with rifampicin, which inhibits transcription (Endesfelder et al., 2013).

13.3 Spatial Organization of DNA Replication and Segregation in *E. coli*

13.3.1 Visualization of Chromosomal DNA Using SMLM

E. coli possesses a single circular chromosome, which is replicated bidirectionally by two independent replisomes starting from the origin of replication (*oriC*) toward the terminator region (*ter*) (Reyes-Lamothe et al., 2012). A complete chromosomal replication process (C-period) requires about 40 min (Michelsen et al., 2003). However, *E. coli* is able to divide much faster under optimal growth conditions (e.g., within 20 min doubling time in rich medium). This is facilitated by overlapping rounds of replication, the so-called multifork replication. Astonishingly, other processes involving DNA such as transcription or chromosome segregation co-occur during multifork replication, raising the question of how the chromosome is organized in order to orchestrate such a fast and precise cell cycle.

While tagging and tracking of genetic loci using fluorescent repressor–operator systems (FROS) like Lac or Tet systems (Lau et al., 2003) or the *parS*-ParB system (Li et al., 2003) gave valuable insight into general chromosome organization and dynamics, these techniques are most useful under conditions where the generation time exceeds the time required for DNA replication. It was found that chromosomes segregate in the same order as genes are placed along the replichores (Youngren et al., 2014), and that the chromosome consist of domains with different mobility (Espeli et al., 2008). Nucleoid dynamics and structure were indirectly investigated during slow and fast growth using fusion proteins of FPs and nucleoid-associated proteins such as HU or Fis, which are involved in chromosome organization and thus almost completely decorate the chromosome during the cell cycle (Dorman, 2013, Fisher et al., 2013, Hadizadeh Yazdi et al., 2012). Structured nucleoid morphology could be observed in living cells during fast growth (Hadizadeh Yazdi et al., 2012).

In 2009, Ferullo and colleagues (Ferullo et al., 2009) applied a protocol for direct DNA labeling in order to investigate cell cycle synchronization in *E. coli*. They used a highly specific reaction of an alkyne and azide group, a type of chemical reaction which is typically referred to as click chemistry (Salic and Mitchison, 2008) (Figure 13.4a). With this approach, cells can be pulse-labeled after they were exposed to medium containing the thymidine analogue 5-ethynyl-2′-deoxiuridine (EdU). Due to the high-density labeling of DNA during replication and the possibility of attaching any desired fluorophore carrying an azide group, this approach is perfectly suited to prepare samples for SMLM (Zessin et al., 2012). Using this labeling strategy, the nucleoid of fixed *E. coli* cells at different stages of the cell cycle was visualized using direct stochastic optical reconstruction microscopy (*d*STORM) (Heilemann et al., 2008, Spahn et al., 2014).

The super-resolved images shown in Figure 13.4 represent snapshots of the nucleoid at the time point of fixation. Since the nucleoid is a very dynamic entity (Fisher et al., 2013, Hadizadeh Yazdi et al., 2012), fixation facilitates to resolve the chromosomal ultrastructure with a high resolution (localization precision 13 ± 1 nm; [Spahn et al., 2014]), which would otherwise not be possible due to chromosome movement and remodeling.

13.3.2 Cell Cycle Dependent DNA Structure Analysis

Many processes in *E. coli* such as replication initiation or septum formation are related to the volume or the length of the cell cylinder (Chien et al., 2012). By sorting SMLM by cell length, the segregation process throughout the cell cycle can be reconstructed (Figure 13.4b). During this process the nucleoid undergoes strong remodeling and exhibits different substructures such as small fibers (Figure 13.4b, 3.5 μm) or helical structures (Figure 13.4b, 6.6 μm). Other structures (crescents, squares, "horse shoes," Figure 13.5) tend to appear at specific length intervals, indicating a strong organization of the bacterial chromosome. Interestingly,

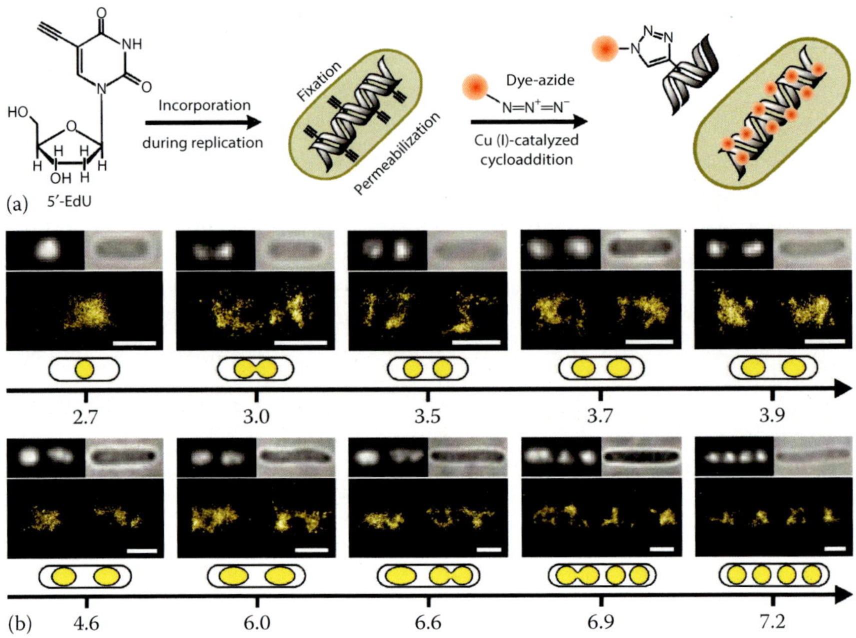

Figure 13.4

Labeling of chromosomal DNA in *E. coli* using click chemistry. (a) 5-ethynyl-2′-deoxyuridine (EdU) is added to the growth medium and is incorporated into nascent DNA during replication. After fixation and permeabilization of the cells, dye azides are added and covalently attached to the alkyne group in a copper-catalyzed Huisgens 1,3-dipolar cycloaddition. (b) Using this labeling technique, chromosomal DNA of fast grown *E. coli* was labeled with Alexa Fluor 647, facilitating *d*STORM imaging of the bacterial nucleoid. The super-resolved images reveal spatial information hidden in the widefield images (upper left), for example, DNA fibers with diameters of 50–100 nm (3.5 μm). Sister chromosomes exhibit asymmetrical replication/segregation behavior, as depicted by the cartoons below the *d*STORM images (scale bars 1 μm). (Reprinted from *J. Struct. Biol.*, 185, Spahn, C. et al., Super-resolution imaging of *Escherichia coli* nucleoids reveals highly structured and asymmetric segregation during fast growth, 243–249, Copyright 2014, with permission from Elsevier.)

bacteria with two or more nucleoids show asymmetry between the respective sister chromosomes, indicating that one sister chromosome has proceeded further in the second replication round. This manifests in the existence of bacteria with an odd amount of clearly distinguishable nucleoids (e.g., 6.9 μm; see cartoons in Figure 13.4b, which depict this asymmetric behavior schematically [Spahn et al., 2014]).

More information from super-resolved DNA images can be extracted by measuring the dimensions of the nucleoid (length and width) as well as internucleoid centroid distances. We found that centroid distances between sister nucleoids and their expansion along the bacterial length axis increase linearly with respect to the cell length (Spahn et al., 2014). At the same time, the chromosome remains in close proximity to the membrane throughout the whole cell cycle. Whether this effect is caused by physical/entropic phenomena such as radial confinement (Fisher et al., 2013) or molecular crowding (Jun and Wright, 2010), or biological models such as the transertion model (Woldringh and Nanninga, 2006) or an interplay between these and other segregation models, is still unclear.

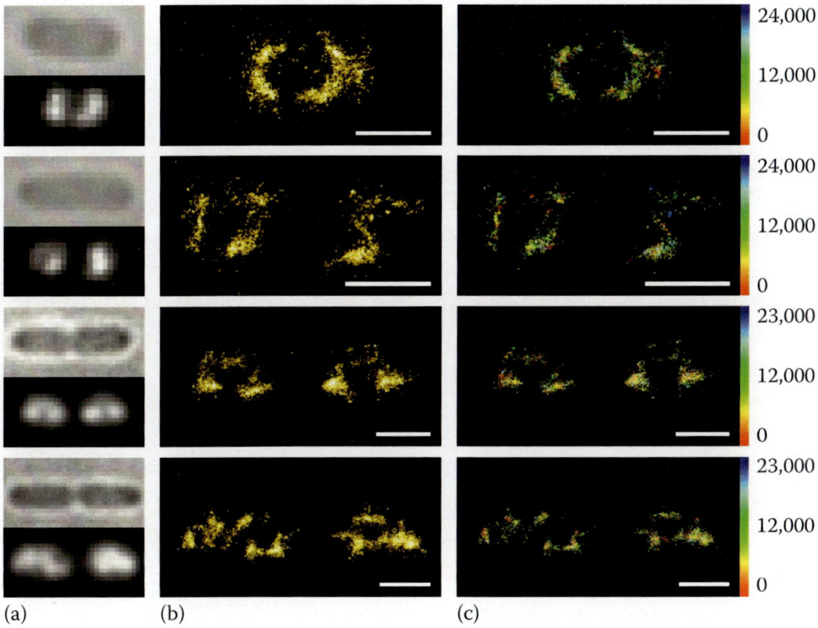

Figure 13.5

Highly regular nucleoid structures at different stages of the *E. coli* cell cycle. (a) Bright-light (top) and diffraction-limited (bottom) images of the highly resolved nucleoids in (b) demonstrate the strong nucleoid remodeling during the replication/segregation process. (c) Single-molecule localizations were color-coded dependent on the frame number as a control for blinking artifacts or residual DNA movement. The uniform color distribution highlights that these artifacts can be excluded, and that fixation "captures" the DNA in the cytosolic protein meshwork (scale bars 1 μm). (Reprinted from *J. Struct. Biol.*, 185, Spahn, C. et al., Super-resolution imaging of *Escherichia coli* nucleoids reveals highly structured and asymmetric segregation during fast growth, 243–249, Copyright 2014, with permission from Elsevier.)

13.4 Correlative Super-Resolution Imaging of RNAP, Membrane, and DNA in *E. coli*

The development of quantitative SMLM to determine RNAP copy numbers and nanoscale patterns as well as high-density labeling and super-resolution imaging of the bacterial nucleoid paves the way toward correlated imaging of both DNA and RNAP in the same *E. coli* cell. However, combining PALM imaging of photoactivatable or photoconvertible FPs with labeling protocols and imaging conditions for organic dyes is not trivial. First, many well established protocols for click chemistry labeling require copper ions as catalysts, which chemically degrade FPs. Second, imaging buffers required for organic fluorophores affect the photophysical properties of FPs by for example, introducing nonfluorescent dark states (Endesfelder et al., 2011). Furthermore, sequential imaging can induce photobleaching of the second fluorophore while imaging the first fluorophore.

A solution to these constraints is a sequential labeling and imaging approach, where fixed or live cells are immobilized on the glass surface and covered with a hydrophilic gel matrix to prevent movement or detachment during post-labeling and imaging. In this section, we present two workflows for sequential imaging that allow structural and dynamical investigation of proteins with respect to DNA.

The spatial organization of these targets with respect to the geometry of cell cylinder is visualized by PAINT-imaging of the bacterial membrane.

13.4.1 Structural Studies of RNAP, Membrane and Chromosomal DNA

Due to the low temporal resolution of SMLM, visualization of different molecule classes in fast growing bacterial cells at near-molecular resolution is only possible in fixed cells. Here, a sequential workflow allows the combination of quantitative PALM imaging with other SMLM imaging methods.

Different approaches have been published for relocating cells on the same or different microscopes, by using relative positions of sample intrinsic or extrinsic fiducial markers (Gunkel et al., 2014, Tam et al., 2014). While adherent mammalian cells can be used for these workflows right after fixation, bacteria have to be immobilized and it has to be assured that their position and orientation does not change during sample transfer and post-labeling procedures. One solution is to add a hydrophilic gel matrix (Zessin et al., 2013) to immobilized bacteria before imaging (Figure 13.6).

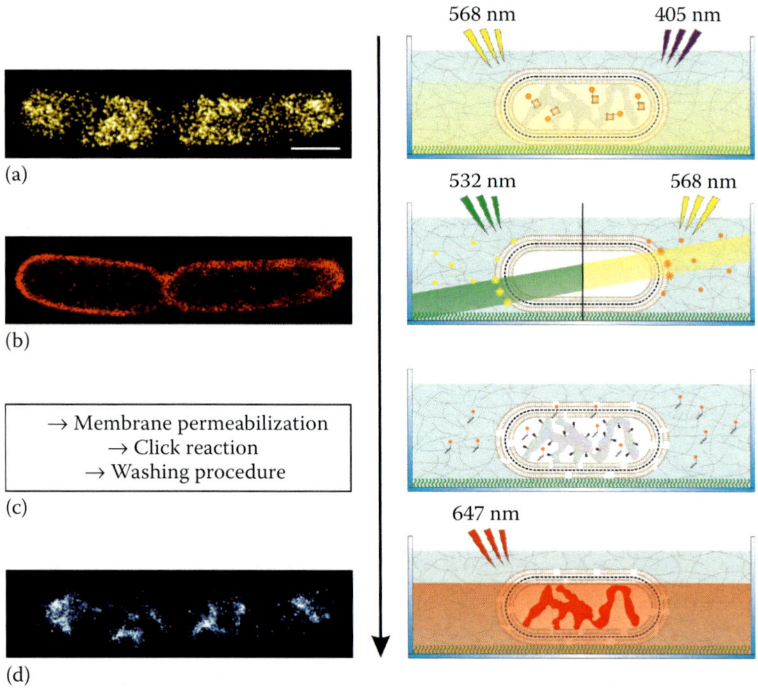

Figure 13.6

Sequential (a) PALM, (b) PAINT, and (c) *d*STORM imaging of fixed *E. coli* cells grown in LB medium. Fixed cells are immobilized on poly-L-lysine coated chamber slides and embedded into a hydrophilic gel matrix to maintain position and shape. RNAP-PAmCherry1 is imaged first using activation at 405 nm and read-out at 568 nm. After all RNAP-PAmCherry1 were imaged, 100–500 pM R6G or Nile red were added to the chamber to stain and visualize the bacterial membrane via PAINT imaging. Membrane permeabilization and click chemistry labeling of DNA can be performed in the porous gel matrix, and DNA was visualized using *d*STORM (scale bar 1 μm). (Reproduced with permission from Spahn, C. et al., Correlative super-resolution imaging of RNA polymerase distribution and dynamics, bacterial membrane and chromosomal structure in *Escherichia coli*, *Methods Appl. Fluoresc.*, 3, 014005. Copyright 2015, Institute of Physics.)

Single-molecule imaging of RNAP tagged with a photoactivatable FP is conducted in a nonredox buffer and allows extracting information on copy numbers. DNA is then post-labeled via click chemistry, and a suitable imaging buffer is added for photoswitching of organic fluorophores. This sequential procedure allows obtaining super-resolved images of both RNAP and the bacterial nucleoid.

This experimental workflow can be further extended by imaging the cell boundaries using synthetic fluorophores that reversibly bind to the membrane while dramatically increasing their quantum yield, for example, Nile red or Rhodamine 6G (Lew et al., 2011) (Figure 13.6). This requires a membrane-conserving fixation procedure, for example, using methanol-free formaldehyde.

The overlay of the super-resolution images allows investigating the spatial relationship between proteins and DNA (Figure 13.7). PAINT images can be used as a calibration for the cell borders in bright field images (Figure 13.7e). Furthermore, cell centroids, orientations, and lengths of single bacteria can be used for rotational alignment and size sorting, which allows reconstruction of a growing cell out of many differently sized fixed cells.

The sequential imaging of RNAP and DNA reveals distinct bands, which populate similar areas within the cell and with respect to the long axis. This is

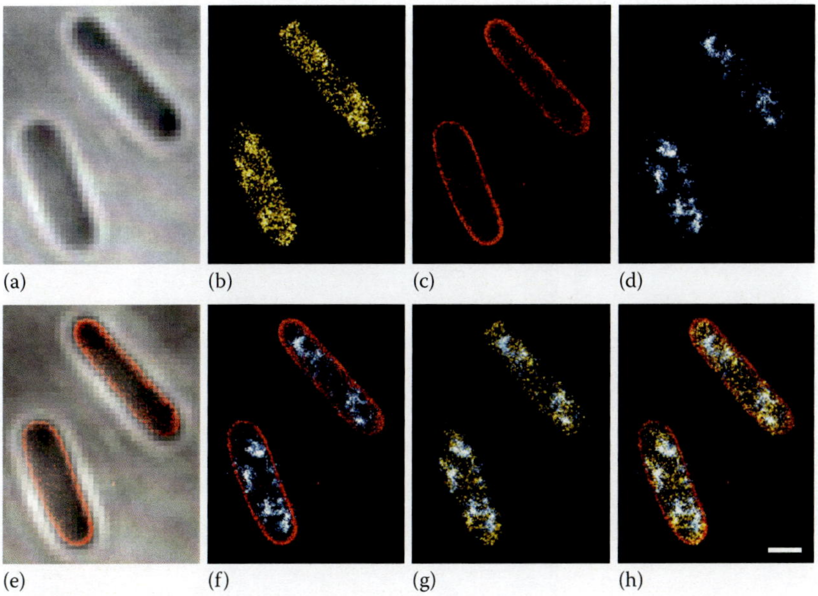

Figure 13.7

Overlay of bright field (a) and super-resolution images of RNAP (b), cell membrane (c), and DNA (d). An overlay of (a) and (c) can be used to extract the cell boundaries (e). The structure of the nucleoid(s) and its positioning within the bacterial cell (f) can be investigated by combining DNA and membrane images. The spatial relationship between RNAP and dense chromatin (g) reveals that RNAP preferentially surrounds the dense nucleoid regions where active genes are exposed. Finally, all highly resolved channels can be overlaid (h) (scale bar 1 μm). (Reproduced with permission from Spahn, C. et al., Correlative super-resolution imaging of RNA polymerase distribution and dynamics, bacterial membrane and chromosomal structure in *Escherichia coli*, *Methods Appl. Fluoresc.*, 3, 014005. Copyright 2015, Institute of Physics.)

explained by RNAP largely bound to DNA in order to mediate transcription. However, RNAP appears to be mostly located at the periphery of the dense chromatin and in small gaps between the condensed chromosome regions. One can assume that active genes are exposed at outer parts of the condensed chromatin, where they are accessible to RNAP, or is even guided towards the nucleoid surface via RNAP (Stracy et al., 2015). This state is maintained during the whole cell cycle, assuring that an equal amount of the RNAP pool is passed to the daughter cells upon cell division.

13.4.2 Correlative Single-Molecule Tracking of RNAP and Super-Resolution Imaging of the Bacterial Chromosome

Super-resolution images provide structural information on the spatial organization of proteins at the nanoscale. However, the temporal resolution inversely scales with the labeling density of fluorophores, which limits following protein dynamics in a live cell at the same time. A hybrid approach of sequential single-molecule tracking in live cells, fast fixation, and subsequent super-resolution imaging in a fixed cell allows correlating the two modes of single-molecule imaging and was first demonstrated in mammalian cells (Balint et al., 2013).

We developed an experimental protocol that allows extracting dynamic information on RNAP from single-molecule trajectories, and ultrastructural information on the chromosome (Figure 13.8). This approach allows to associate different diffusional states of RNAP (e.g., actively transcribing vs. stagnant) to specific positions on the bacterial nucleoid.

Correlating trajectories of RNAP to the underlying chromosome structure showed that RNAP mainly populates the edges of the bacterial nucleoid, transcribing active genes at the nucleoid periphery. This might be caused by

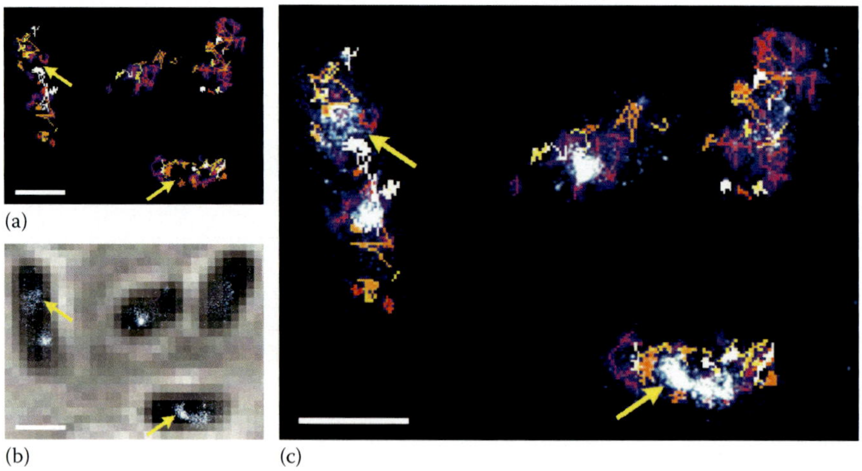

Figure 13.8

Correlative single-molecule tracking and *d*STORM imaging. (a) Tracking of RNAP in live cells reveals areas which are less explored by RNAP (yellow arrows). (b) Subsequent *d*STORM imaging of the chromosomal DNA. (c) Overlay of single-molecule trajectories and the chromosome reveals that dense chromatin is located in these areas, indicating that it is less accessible for RNAP (scale bar 1 μm). (Reproduced with permission from Spahn, C. et al., Correlative super-resolution imaging of RNA polymerase distribution and dynamics, bacterial membrane and chromosomal structure in *Escherichia coli*, *Methods Appl. Fluoresc.*, 3, 014005. Copyright 2015, Institute of Physics.)

the entropic effect of phase separation between DNA and cytosolic proteins (Woldringh, 2002), but still requires further investigation.

13.5 Experimental Design of SMLM Studies in *E. coli*

SMLM delivers fluorescence images with near-molecular spatial resolution. However, given the experimental procedure underlying an SMLM experiment, it is justified to review sources and consequences of experimental artifacts. What is the localization precision or the effective overall resolution of the data? As a fluorescence technique, one does observe the biomolecule of interest via a fluorescent marker. How efficient is the labeling, and how does the introduction of a fluorescent tag impair protein function and localization? Can fixation artifacts be excluded, for example, does the fixation preserve the natural organization of the intracellular target molecules, and are the fluorescent properties of the label retained? Are these structures also present in a living cell? Finally, how reliably can molecules be counted?

In order to answer these questions, suitable control experiments need to be designed and are discussed within this section.

13.5.1 Tagging the Biomolecule of Interest

Recording of growth curves of recombinant *E. coli* strains and determination of the doubling time is important to assure cell health. This can be performed by measuring the optical density (absorption, A) under experimental conditions and calculate the mass doubling time (t_d) from the growth rate k using an exponential fit (Equations 13.1 and 13.2).

$$A(t) = A_0 \times e^{k \times t} \tag{13.1}$$

$$t_d = \frac{\ln 2}{k} \tag{13.2}$$

MG1655 *rpoC–PamCherry1* exhibits normal growth behavior (Figure 13.9). A visual inspection for normal cell morphologies as length, width, localization of division points, and so on can further confirm the unperturbed native growth behavior.

Western blots of whole cell lysates report on the tagging stoichiometry and whether the FP is cleaved off. Targeting the lysate of the MG1655 *rpoC–PamCherry1 Amp*^{R} with an antibody targeting RNAP confirms that >90% of RNAP is tagged. A second experiment targeting with an anti-mCherry antibody shows that >90% of the FP is attached to RNAP.

It is advisable to confirm the spatial structures and organizations of biomolecules using different fluorescent labeling strategies as well as correlative imaging methods. Photophysical properties of specific fluorophores that depend on the redox potential of the environment, pH, laser intensity, or labeling chemistry need to be explored (Annibale et al., 2010, Dempsey et al., 2011, Durisic et al., 2014, Endesfelder et al., 2011). The FP mEos2 expressed in the cytosol of *E. coli* cells exhibits blinking (Endesfelder et al., 2013). Similar blinking characteristics are observed for mEos2 in reducing buffers, as needed for two-color imaging with organic fluorophores (Endesfelder et al., 2011). Furthermore, possible dimerization of FPs, maturation kinetics, and activation efficiencies play an important role in quantification studies.

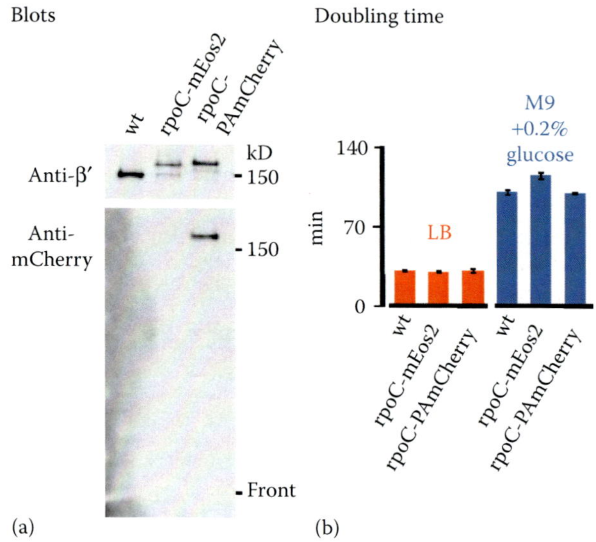

Figure 13.9

Labeling and growth controls of genetic *rpoC* fusions. (a) Western blots are reported on labeling efficiency and potential fusion protein cleavage. Probing with anti-β′ antibody (upper panel) reveals that wild-type β′ migrates with an apparent molecular weight of ~155 kDa, while >90% of β′-PAmCherry1 migrates slower, consistent with an additional mass contributed by the tagged fluorescent protein (~27 kDa). Probing with anti-mCherry antibody shows that >90% of the fluorescent protein is attached to β′. (b) Tagging β′ with PAmCherry1 has no effect on growth rate. Growth rate was determined by measuring OD 600 increase during logarithmic phase (OD 600 < 0.4). The PAmCherry1 tag on β′ causes no change in cell doubling time in either media (i.e., <1%). Tagging β′ with mEos2 causes some growth defect in minimal media. (Reprinted from *Biophys. J.*, 105, Endesfelder, U. et al., Multiscale spatial organization of RNA polymerase in Escherichia coli, 172–181, Copyright 2013, with permission from Elsevier).

13.5.2 Sample Preparation

Chemical fixation can introduce artifacts on protein localization, and appropriate controls are important. Here, different experiments can be set up to check for either structural damage caused by fixative, residual movement after fixation/labeling, or comparison to the native structure in living cells. Commonly two different types of fixatives are used for microscopic studies: crosslinking reagents such as formaldehyde and glutaraldehyde, or denaturing reagents like methanol. Glutaraldehyde and formaldehyde fixation delivered similar structures. However, glutaraldehyde has to be quenched with sodium borohydrate in order to prevent strong autofluorescence in the red channel. Denaturing fixation led to different, less-defined nucleoid structures, possibly caused by dehydration/rehydration during fixation process and a less rigid fixation. It further has to be investigated whether fixation impairs the function of the FP. High concentrations of fixative (e.g., 2.4% formaldehyde + 0.04% glutaraldehyde) might destroy the fluorophore, leading to an underestimation of copy numbers (Figure 13.10a). Reliable counting also presumes that the experiment is performed until all molecules are photoconverted and read out (Figure 13.10b).

Residual movement, sample drift during the imaging procedure, and blinking artifacts can be investigated easily by color coding the localizations according to their temporal appearance (Figure 13.5). Due to the high labeling density and

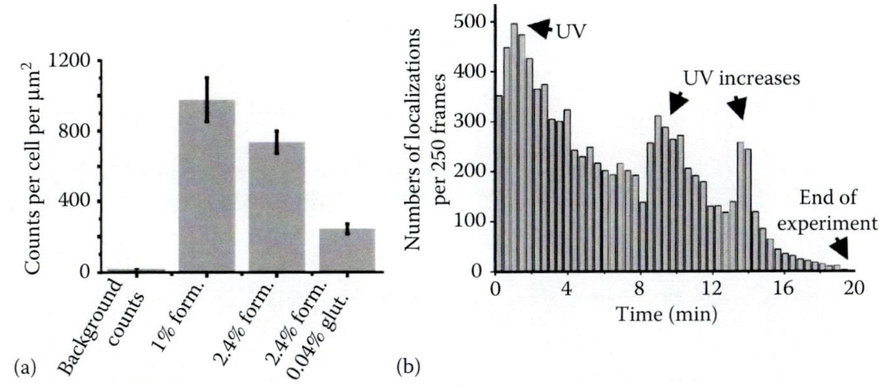

Figure 13.10

(a) Histogram of single PAmCherry1 localizations (binning of 250 frames; 1 frame equals 100 ms). The initial localization density of ~2 localizations per frame decreases as the number of remaining unactivated molecules decreases. The localization density increases again when the intensity of UV activation light is increased (black arrows), but eventually decreases to negligible values, after which imaging is stopped. Few fluorophores (<2) are simultaneously active in the same cell, and as a result, the probability of double spots within the diffraction-limited area is very small. Given sufficient imaging time, nearly all fluorescent PAmCherry1 molecules will be identified and counted. (b) *E. coli* cells expressing PAmCherry1-RNAP were grown in LB medium at 32°C, fixed using different fixation mixtures based on formaldehyde (columns 2–4), and glutaraldehyde (column 4), and imaged. Wild-type MG1655 cells (expressing no fluorescent proteins) produce a very low number of false positive localizations (column 1) due to the presence of autofluorescence. (Reprinted from *Biophys. J.*, 105, Endesfelder, U. et al., Multiscale spatial organization of RNA polymerase in Escherichia coli, 172–181, Copyright 2013, with permission from Elsevier).

multiple blinking events per dye molecule under *d*STORM conditions, we expect an equal color distribution along the whole nucleoid. Rainbow-like structures indicate either sample drift or residual movement.

However, this visualization is no proof for the nativity of the observed nucleoid structures. Time-lapse imaging of nucleoid-associated proteins under similar conditions (e.g., in agarose pads) can be performed in order to observe the chromosome structure over the cell cycle in living cells (Figure 13.11). RNAP tagged with GFP (Cabrera and Jin, 2003) in cells exhibiting similar doubling times as the click-labeled cells are a suitable control experiment. Live-cell images of rpoC-GFP and the highly resolved nucleoid images show similar structures, which also appear in the same chronological order.

13.5.3 Imaging Procedure

A crucial parameter for successful imaging is that the spot density per image is low enough to ensure that single-molecule PSFs are identified by the post-processing software. To assure this spatial separation, the majority of all fluorophores except a small subset have to reside in the nonfluorescent dark state at a given time.

Typically, a series of 4,000–20,000 images with a frame rate between 10 Hz and 2 kHz is recorded with an electron multiplied charge coupled device (EMCCD) camera. The appropriate choice is dependent on the target (e.g., density, dimensionality), the labeling density, and the switching properties of the fluorophore. Briefly, for structural studies, the frame rate should be adjusted for the average time τ_{on} a fluorophore resides in its fluorescent state, which by

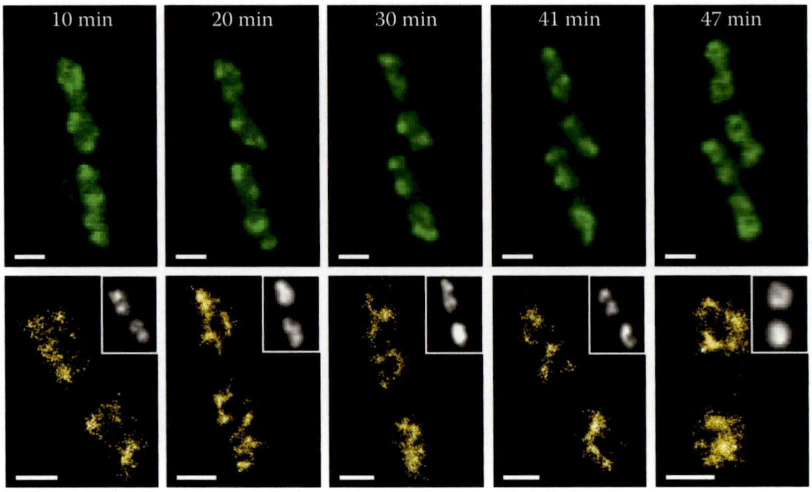

Figure 13.11

RNAP live-cell imaging and *d*STORM imaging of the nucleoid structure using rpoC-GFP (DJ2599 [Cabrera and Jin, 2003]). DJ2599 cultures were synchronized, released from starvation, and imaged on pre-warmed agarose pads. After recovery from starvation, cells begin to replicate (top panel) while showing RNAP structures similar to the diffraction-limited fluorescence images of fixed *E. coli* labeled with Alexa Fluor 647 (insets in lower panel). The respective *d*STORM images are shown in the lower panel (scale bars 1 μm). (Reprinted from *J. Struct. Biol.*, 185, Spahn, C. et al., Super-resolution imaging of *Escherichia coli* nucleoids reveals highly structured and asymmetric segregation during fast growth, 243–249, Copyright 2014, with permission from Elsevier.)

itself can be altered changing the irradiation intensity (van de Linde et al., 2011). For tracking schemes the fluorescence emission should be spread for at least 10 imaging frames. The EMCCD camera should be operated in the most sensitive mode, applying optimal cooling for reduced thermal background noise, the highest dynamical range settings, and the fastest read-out option (e.g., in frame transfer mode, which shortens read-out periods in which the camera is insensitive as much as possible) and an optimal EM gain. Pulsing the UV activation in PALM or sptPALM reduces autofluorescence and photodamage. The number of detected photons per fluorophore depends on pH, concentration of reducing agents, temperature, and chemical additives (e.g., COT) (Dempsey et al., 2011, Olivier et al., 2013). The quantum yield of fluorophores can be increased by choosing for example D_2O instead of conventional water (Klehs et al., 2014).

Quantitative imaging requires that all fluorophores were activated and detected (Figure 13.10). Experimental controls are needed to determine the overall detection efficiency, which can be affected by bleaching prior to the experiment, incomplete maturation, or sample fixation. In addition, the fluorescence emission of a single fluorophore can be spread over several consecutive imaging frames. These multiple events should be corrected into a single localization using a tracking filter that groups multiple spots from the same molecule in consecutive frames. Whether a fluorophore is reliably activated and bleached without entering a reversible dark state depends on the chemical nanoenvironment (Annibale et al., 2010, Endesfelder et al., 2011, 2013, Winterflood and Ewers, 2014). By the use of cytosolic fluorophores (expressed on a low level from a low copy plasmid) undesired blinking characteristics in the bacterial environment can be explored.

13.5.4 Post-Processing

The characteristic of SMLM data is that the raw data is a list of single-molecule localizations. Thus, imaged-based analysis as used for conventional fluorescence microscopy require the generation of an intensity-like image. However, one should be aware that the list of single-molecule localizations can contain artifacts. These can occur if background signal is recognized as a single emitter, or if multiple emitters are interpreted as a single fluorophore. For single-emitter fitting software, the spot density is the most crucial factor and should not exceed 0.2–0.6 spots per μm^2 (Wolter et al., 2011). The spatial localization precision is not much affected by an increased spot density as it decreases only by a few nanometers and either identifies a spot or directly discards it. The stochastic precision and recall decay exponentially with increasing spot density (without multispot analysis) (Wolter et al., 2011).

With SMLM, colocalization analysis is getting closer to molecular resolution. However, this requires reliable drift correction and channel alignment. Highly regular structures are best treated with feature-based drift correction using a sliding-window algorithm (Mlodzianoski et al., 2011). Such algorithms are not applicable to low-density samples or samples with diffuse localizations, for example, for the distribution of proteins in membranes. In this case, fiducial markers can be used for drift correction and channel alignment (Zessin et al., 2013). When imaging immobilized bacteria in different channels, images can also be registered using fast Fourier transformation analysis. This analysis performs well if a sufficient number (4+) of differently orientated bacteria are in the field of view (Spahn et al., 2015).

13.6 Protocols

In this last section, optimized protocols for the different sample preparations and imaging modes are listed.

13.6.1 Preparation of Fixed RNAP Samples

Growth: Grow *E. coli* shaking at 200 rpm at 32°C to OD 0.4.

Fixation: Mix and incubate 100 mM phosphate buffer, pH 7.5, 1% PFA, and bacteria in liquid culture at OD 0.4 for 30 min at room temperature. Wash in phosphate buffer, pH 7.5 at room temperature, spin down using a centrifuge (commonly 5.000× g for 3 min), remove supernatant, and resuspend in phosphate buffer pH 7.5. Excess formaldehyde can be quenched by adding 50 mM ammonium chloride to the washing buffer.

Immobilization on cover slip surface: Clean glass surface (0.5% HF or KOH etching or plasma cleaning). Incubate with 0.01% (w/v) poly-L-lysine solution for 10–20 min, take off, and air dry in dust free environment. Incubate suspension of fixed bacteria and wash with phosphate buffer, pH 7.5. Incubate with 1% PFA in phosphate buffer, pH 7.5 for 5 min, and wash afterward. Sample can be stored until imaging in PBS (or phosphate buffer, pH 7.5) at 4°C.

13.6.2 Fluorescent Labeling of Chromosomal DNA via Click Chemistry

Growth: *E. coli* working culture is inoculated from overnight culture, usually by diluting 1:200 into fresh medium (OD_{600} ~ 0.025). Grow until early to mid-exponential phase and add EdU to an end concentration of ~10 μM. Depending on the growth condition (e.g., minimal medium without further added nucleotides), EdU might impair cell division at this concentration. Here, a lower EdU concentration and/or addition of other nucleotides (ACGU) reduce the negative effects. Depending on the incubation time in EdU-containing medium, pulse labeling or whole chromosome labeling can be performed. While cultures should be synchronized for pulse labeling experiments (e.g., stringent response [Ferullo and Lovett, 2008]), whole chromosome labeling can be performed by adjusting the EdU incubation time to the mass doubling time determined for the applied growth conditions. We usually incubate cultures for 1.5 mass doubling times.

Fixation, permeabilization, and click reaction: Fixation is performed as described in 13.6.1. In case of DNA visualization only, samples can be permeabilized and click-labeled in solution. Permeabilization is performed using 0.5% Triton-X 100 in phosphate buffer for 30 min at room temperature. After washing the sample once, the click reaction buffer (Qu et al., 2011) is added for 30 min at room temperature.

Immobilization on cover slip surface is performed as described in Section 13.6.1.

13.6.3 Sample Preparation for Correlative Sequential Approaches

13.6.3.1 Correlative Multicolor Imaging for Structural Analysis

Sequential PALM, PAINT, and *d*STORM imaging requires some changes in the previously described protocols. In case of additional membrane visualization using PAINT imaging, fixation has to be performed using methanol-free formaldehyde. After fixation, cells are immobilized on cleaned glass surfaces as described in Section 13.6.1, but without post-fixation step. At the same time, extracellular matrix (ECM) is thawed on ice and diluted with pre-chilled phosphate buffer to a final protein concentration of ~3–8 mg/ml. After immobilization, buffer is removed and 30 μl ECM is added to the sample (use chilled tips to prevent immediate polymerization). Incubate on ice for ~5 min and move the sample to 37°C for at least 10 min in order to facilitate matrix polymerization. Prevent a complete drying-out of the sample. The matrix is then fixed using prewarmed formaldehyde solution (1%–2% in phosphate buffer) for 30 min, washed with phosphate buffer containing 50 mM ammonium chloride. The sample is ready for PALM imaging after washing once with the final storage buffer.

After PALM imaging, Nile red or R6G is added to the ECM-coated sample at 100–300 pM concentration. After PAINT imaging, the sample is washed once and permeabilization is performed on the slide as already described for bacterial suspensions. The click reaction is also performed as described in Section 13.6.2; however, the reaction time might be increased to obtain a higher labeling density. The presence of the ECM layer requires more washing steps than the DNA labeling approach in solution (3–5 washing steps of at least 10 min each).

The sample is then transferred onto the microscope again and *d*STORM imaging is performed in PBS + 100 mM MEA pH 8.5.

13.6.3.2 Correlative sptPALM and dSTORM Imaging

Grow *E. coli* cells overnight in EZ-rich defined medium (EZ-RDM) and inoculate 1:200 as working culture. Add EdU (10 μM) to the culture medium at the desired growth phase. Depending on the experiment, we usually add EdU during early to mid-exponential phase (OD_{600} ~ 0.2). After immobilization, cells were washed once with EZ-RDM and covered with 300 µl EdU-containing RZ-RDM. It has to be assured that cells are incubated for more than 1.5 mass doubling times until they are fixed after sptPALM imaging (preincubation in the shaker + immobilization procedure + sptPALM imaging time). After sptPALM imaging, cells are immediately fixed on the slide by using high concentration of formaldehyde (4%) for 30 min. After quenching of excess formaldehyde, 30 µl ECM is added and all subsequent steps are similar as described above.

Acknowledgments

The authors acknowledge funding by the German Science Foundation (EXC 115 and SFB 902).

References

Ankerst, M., Breunig, M. M., Kriegel, H. P. and Sander, J. 1999. OPTICS: Ordering points to identify the clustering structure. *Sigmod Record,* 28(2), June 1999, 49–60.

Annibale, P., Scarselli, M., Kodiyan, A. and Radenovic, A. 2010. Photoactivatable fluorescent protein mEos2 displays repeated photoactivation after a long-lived dark state in the red photoconverted form. *J Phys Chem Lett,* 1, 1506–1510.

Badrinarayanan, A., Reyes-Lamothe, R., Uphoff, S., Leake, M. C. and Sherratt, D. J. 2012. In vivo architecture and action of bacterial structural maintenance of chromosome proteins. *Science,* 338, 528–531.

Bakshi, S., Dalrymple, R. M., Li, W., Choi, H. and Weisshaar, J. C. 2013. Partitioning of RNA polymerase activity in live *Escherichia coli* from analysis of single-molecule diffusive trajectories. *Biophys J,* 105, 2676–2686.

Bakshi, S., Siryaporn, A., Goulian, M. and Weisshaar, J. C. 2012. Superresolution imaging of ribosomes and RNA polymerase in live *Escherichia coli* cells. *Mol Microbiol,* 85, 21–38.

Balint, S., Verdeny Vilanova, I., Sandoval Alvarez, A. and Lakadamyali, M. 2013. Correlative live-cell and superresolution microscopy reveals cargo transport dynamics at microtubule intersections. *Proc Natl Acad Sci USA,* 110, 3375–3380.

Biteen, J. S., Thompson, M. A., Tselentis, N. K., Bowman, G. R., Shapiro, L. and Moerner, W. E. 2008. Super-resolution imaging in live Caulobacter crescentus cells using photoswitchable EYFP. *Nat Methods,* 5, 947–949.

Blattner, F. R., Plunkett, G., 3rd, Bloch, C. A., Perna, N. T., Burland, V., Riley, M., Collado-Vides, J., et al. 1997. The complete genome sequence of *Escherichia coli* K-12. *Science,* 277, 1453–1462.

Bon, M., McGowan, S. J. and Cook, P. R. 2006. Many expressed genes in bacteria and yeast are transcribed only once per cell cycle. *FASEB J,* 20, 1721–1723.

Browning, D. F. and Busby, S. J. 2004. The regulation of bacterial transcription initiation. *Nat Rev Microbiol,* 2, 57–65.

Cabrera, J. E. and Jin, D. J. 2003. The distribution of RNA polymerase in *Escherichia coli* is dynamic and sensitive to environmental cues. *Mol Microbiol,* 50, 1493–1505.

Chien, A. C., Hill, N. S. and Levin, P. A. 2012. Cell size control in bacteria. *Curr Biol,* 22, R340–R349.

Dempsey, G. T., Vaughan, J. C., Chen, K. H., Bates, M. and Zhuang, X. 2011. Evaluation of fluorophores for optimal performance in localization-based super-resolution imaging. *Nat Methods,* 8, 1027–1036.

Dorman, C. J. 2013. Genome architecture and global gene regulation in bacteria: Making progress towards a unified model? *Nat Rev Microbiol,* 11, 349–355.

Durisic, N., Laparra-Cuervo, L., Sandoval-Alvarez, A., Borbely, J. S. and Lakadamyali, M. 2014. Single-molecule evaluation of fluorescent protein photoactivation efficiency using an in vivo nanotemplate. *Nat Methods,* 11, 156–162.

Endesfelder, U., Finan, K., Holden, S. J., Cook, P. R., Kapanidis, A. N. and Heilemann, M. 2013. Multiscale spatial organization of RNA polymerase in *Escherichia coli. Biophys J,* 105, 172–181.

Endesfelder, U., Malkusch, S., Flottmann, B., Mondry, J., Liguzinski, P., Verveer, P. J. and Heilemann, M. 2011. Chemically induced photoswitching of fluorescent probes—A general concept for super-resolution microscopy. *Molecules,* 16, 3106–3118.

Endesfelder, U., Malkusch, S., Fricke, F. and Heilemann, M. 2014. A simple method to estimate the average localization precision of a single-molecule localization microscopy experiment. *Histochem Cell Biol,* 141, 629–638.

Espeli, O., Mercier, R. and Boccard, F. 2008. DNA dynamics vary according to macrodomain topography in the *E. coli* chromosome. *Mol Microbiol,* 68, 1418–1427.

Ester, M., Kriegel, H. P. and Sander, J. 1997. Spatial data mining: A database approach. *Advances in Spatial Databases,* 1262, 47–66.

Ferullo, D. J., Cooper, D. L., Moore, H. R. and Lovett, S. T. 2009. Cell cycle synchronization of *Escherichia coli* using the stringent response, with fluorescence labeling assays for DNA content and replication. *Methods,* 48, 8–13.

Ferullo, D. J. and Lovett, S. T. 2008. The stringent response and cell cycle arrest in *Escherichia coli. PLoS Genet,* 4, e1000300.

Fisher, J. K., Bourniquel, A., Witz, G., Weiner, B., Prentiss, M. and Kleckner, N. 2013. Four-dimensional imaging of *E. coli* nucleoid organization and dynamics in living cells. *Cell,* 153, 882–895.

French, S. L. and Miller, O. L., Jr. 1989. Transcription mapping of the *Escherichia coli* chromosome by electron microscopy. *J Bacteriol,* 171, 4207–4216.

Fürstenberg, A. and Heilemann, M. 2013. Single-molecule localization microscopy-near-molecular spatial resolution in light microscopy with photoswitchable fluorophores. *Phys Chem Chem Phys,* 15, 14919–14930.

Golding, I., Paulsson, J., Zawilski, S. M. and Cox, E. C. 2005. Real-time kinetics of gene activity in individual bacteria. *Cell,* 123, 1025–1036.

Greenfield, D., Mcevoy, A. L., Shroff, H., Crooks, G. E., Wingreen, N. S., Betzig, E. and Liphardt, J. 2009. Self-organization of the *Escherichia coli* chemotaxis network imaged with super-resolution light microscopy. *PLoS Biol,* 7, e1000137.

Gunkel, M., Flottmann, B., Heilemann, M., Reymann, J. and Erfle, H. 2014. Integrated and correlative high-throughput and super-resolution microscopy. *Histochem Cell Biol,* 141, 597–603.

Hadizadeh, Y. N., Guet, C. C., Johnson, R. C. and Marko, J. F. 2012. Variation of the folding and dynamics of the *Escherichia coli* chromosome with growth conditions. *Mol Microbiol,* 86, 1318–1333.

Heilemann, M. 2010. Fluorescence microscopy beyond the diffraction limit. *J Biotechnol,* 149, 243–251.

Heilemann, M., Van De Linde, S., Schuttpelz, M., Kasper, R., Seefeldt, B., Mukherjee, A., Tinnefeld, P. and Sauer, M. 2008. Subdiffraction-resolution fluorescence imaging with conventional fluorescent probes. *Angew Chem Int Ed Engl,* 47, 6172–6176.

Ishihama, A. 2000. Functional modulation of *Escherichia coli* RNA polymerase. *Annu Rev Microbiol,* 54, 499–518.

Jun, S. and Wright, A. 2010. Entropy as the driver of chromosome segregation. *Nat Rev Microbiol,* 8, 600–607.

Klehs, K., Spahn, C., Endesfelder, U., Lee, S. F., Furstenberg, A. and Heilemann, M. 2014. Increasing the brightness of cyanine fluorophores for single-molecule and superresolution imaging. *ChemPhysChem,* 15, 637–641.

Lau, I. F., Filipe, S. R., Soballe, B., Okstad, O. A., Barre, F. X. and Sherratt, D. J. 2003. Spatial and temporal organization of replicating *Escherichia coli* chromosomes. *Mol Microbiol,* 49, 731–743.

Lew, M. D., Lee, S. F., Ptacin, J. L., Lee, M. K., Twieg, R. J., Shapiro, L. and Moerner, W. E. 2011. Three-dimensional superresolution colocalization of intracellular protein superstructures and the cell surface in live Caulobacter crescentus. *Proc Natl Acad Sci USA,* 108, E1102–E1110.

Li, Y., Youngren, B., Sergueev, K. and Austin, S. 2003. Segregation of the *Escherichia coli* chromosome terminus. *Mol Microbiol,* 50, 825–834.

Michelsen, O., Teixeira De Mattos, M. J., Jensen, P. R. and Hansen, F. G. 2003. Precise determinations of C and D periods by flow cytometry in *Escherichia coli* K-12 and B/r. *Microbiology,* 149, 1001–1010.

Miller, O. L., Jr., Hamkalo, B. A. and Thomas, C. A., Jr. 1970. Visualization of bacterial genes in action. *Science,* 169, 392–395.

Mlodzianoski, M. J., Schreiner, J. M., Callahan, S. P., Smolkova, K., Dlaskova, A., Santorova, J., Jezek, P. and Bewersdorf, J. 2011. Sample drift correction in 3D fluorescence photoactivation localization microscopy. *Opt Express,* 19, 15009–15019.

Mooney, R. A., Davis, S. E., Peters, J. M., Rowland, J. L., Ansari, A. Z. and Landick, R. 2009. Regulator trafficking on bacterial transcription units in vivo. *Mol Cell,* 33, 97–108.

Mukhopadhyay, J., Kapanidis, A. N., Mekler, V., Kortkhonjia, E., Ebright, Y. W. and Ebright, R. H. 2001. Translocation of sigma(70) with RNA polymerase during transcription: Fluorescence resonance energy transfer assay for movement relative to DNA. *Cell,* 106, 453–463.

Neidhardt, F. C. 1987. Escherichia coli *and* Salmonella typhimurium: *Cellular and Molecular Biology*, Washington, DC: American Society for Microbiology.

Nomura, M. 1999. Engineering of bacterial ribosomes: Replacement of all seven *Escherichia coli* rRNA operons by a single plasmid-encoded operon. *Proc Natl Acad Sci USA,* 96, 1820–1822.

Olivier, N., Keller, D., Gonczy, P. and Manley, S. 2013. Resolution doubling in 3D-STORM imaging through improved buffers. *PLoS One,* 8, e69004.

Persson, F., Linden, M., Unoson, C. and Elf, J. 2013. Extracting intracellular diffusive states and transition rates from single-molecule tracking data. *Nat Methods,* 10, 265–269.

Qu, D., Wang, G., Wang, Z., Zhou, L., Chi, W., Cong, S., Ren, X., Liang, P. and Zhang, B. 2011. 5-Ethynyl-2′-deoxycytidine as a new agent for DNA labeling: Detection of proliferating cells. *Anal Biochem,* 417, 112–121.

Reyes-Lamothe, R., Nicolas, E. and Sherratt, D. J. 2012. Chromosome replication and segregation in bacteria. *Annu Rev Genet,* 46, 121–143.

Salic, A. and Mitchison, T. J. 2008. A chemical method for fast and sensitive detection of DNA synthesis in vivo. *Proc Natl Acad Sci USA,* 105, 2415–2420.

Sanamrad, A., Persson, F., Lundius, E. G., Fange, D., Gynna, A. H. and Elf, J. 2014. Single-particle tracking reveals that free ribosomal subunits are not excluded from the *Escherichia coli* nucleoid. *Proc Natl Acad Sci USA,* 111, 11413–11418.

Spahn, C., Cella-Zannacchi, F., Endesfelder, U. and Heilemann, M. 2015. Correlative super-resolution imaging of RNA polymerase distribution and dynamics, bacterial membrane and chromosomal structure in *Escherichia coli. Methods Appl Fluoresc,* 3, 014005.

Spahn, C., Endesfelder, U. and Heilemann, M. 2014. Super-resolution imaging of *Escherichia coli* nucleoids reveals highly structured and asymmetric segregation during fast growth. *J Struct Biol,* 185, 243–249.

Stracy, M., Lesterlin, C., Garza de Leon, F., Uphoff, S., Zawadzki, P., Kapanidis, A. N. 2015. Live-cell superresolution microscopy reveals the organization of RNA polymerase in the bacterial nucleoid. *Proc Natl Acad Sci USA,* 112, E4390–E4399.

Tam, J., Cordier, G. A., Borbely, J. S., Sandoval Alvarez, A. and Lakadamyali, M. 2014. Cross-talk-free multi-color STORM imaging using a single fluorophore. *PLoS One,* 9, e101772.

van de Linde, S., Loschberger, A., Klein, T., Heidbreder, M., Wolter, S., Heilemann, M. and Sauer, M. 2011. Direct stochastic optical reconstruction microscopy with standard fluorescent probes. *Nat Protoc,* 6, 991–1009.

Winterflood, C. M. and Ewers, H. 2014. Single-molecule localization microscopy using mCherry. *ChemPhysChem,* 15, 3447–3451.

Woldringh, C. L. 2002. The role of co-transcriptional translation and protein translocation (transertion) in bacterial chromosome segregation. *Mol Microbiol,* 45, 17–29.

Woldringh, C. L. and Nanninga, N. 2006. Structural and physical aspects of bacterial chromosome segregation. *J Struct Biol,* 156, 273–283.

Wolter, S., Endesfelder, U., van de Linde, S., Heilemann, M. and Sauer, M. 2011. Measuring localization performance of super-resolution algorithms on very active samples. *Opt Express,* 19, 7020–7033.

Youngren, B., Nielsen, H. J., Jun, S. and Austin, S. 2014. The multifork *Escherichia coli* chromosome is a self-duplicating and self-segregating thermodynamic ring polymer. *Genes Dev,* 28, 71–84.

Zessin, P. J., Finan, K. and Heilemann, M. 2012. Super-resolution fluorescence imaging of chromosomal DNA. *J Struct Biol,* 177, 344–348.

Zessin, P. J., Krüger, C. L., Malkusch, S., Endesfelder, U. and Heilemann, M. 2013. A hydrophilic gel matrix for single-molecule super-resolution microscopy. *Opt Nanoscopy,* 2, 1–8.

14

Correlative Live-Cell and Super-Resolution Microscopy and Its Biological Applications

Melike Lakadamyali

14.1 Introduction
14.2 Challenges with Live-Cell Super-Resolution Microscopy
14.3 General Concept of Correlative Live-Cell and Super-Resolution Microscopy
14.4 Biological Application of Correlative Imaging to Cargo Transport
14.5 Technical Developments in Correlative Imaging
14.6 Conclusions and Future Perspectives
References

14.1 Introduction

Light microscopy has been a revolutionary method for studying biological processes for several reasons. Light is noninvasive, allowing the study of cellular and subcellular dynamics in real time. In addition, the large toolbox of fluorescent probes and labeling strategies that are available, allow the detection of specific subcellular components with exquisite molecular specificity and in multiple colors. These capabilities render light microscopy superior to electron microscopy, which is only suited for imaging fixed cells and suffers from poor molecular specificity. However, the diffraction limit has, until recently, presented an impenetrable barrier to achieving spatial resolution beyond ~200–300 nm in x–y and ~500 nm in z using light. The low spatial resolution has been a major obstacle for observing the make-up of many biological structures (viruses, chromatin, cytoskeleton) and the subcellular organization and distribution of multiprotein complexes, which are smaller than the diffraction limit.

In recent years, remarkable advances have been made in breaking the diffraction limit with the development of stimulated emission depletion microscopy, stimulated emission depletion (STED) in 2000,[1] followed by nonlinear structured illumination in 2005[2] and finally single-molecule localization-based methods (stochastic optical reconstruction microscopy, STORM; fluorescence photoactivated localization microscopy, PALM/fPALM) in 2006.[3–5] These methods have improved the spatial resolution of light microscopy by one order of magnitude (~20 nm in x–y and 50 nm in z). Since their first introduction, all of these methods have undergone enormous technological developments. For example, they have been extended to multicolor[6–10] and 3D imaging,[11–14] including imaging in thick samples.[15] The concepts and methodological developments behind these microscopy techniques is the subject of earlier chapters. One of the main advantages of light microscopy, its compatibility with live-cell imaging, however, still remains a major challenge. While all of these methods have been applied to image biological dynamics at nanoscale spatial resolution,[15–29] live-cell super-resolution microscopy is still in its early days. This chapter will review some of the main challenges associated with imaging fast dynamic processes at high resolution and describe an alternative approach termed "correlative live-cell and super-resolution microscopy"[30] that generates information-rich datasets and allows interpreting dynamic information in the context of super-resolution data. In particular, the biological application of this method to study cargo transport dynamics in the crowded cellular environment will be described. Finally, recent technical developments that automate and streamline the method[31] and extend it to multicolor super-resolution imaging[32] will be discussed.

14.2 Challenges with Live-Cell Super-Resolution Microscopy

Live-cell imaging has several extra requirements over fixed cell imaging. First, it requires that the acquisition speed is faster than the dynamics of the biological process to be studied. Second, the ability to fluorescently label intracellular proteins with ease in living cells is highly important. Finally, phototoxicity during imaging is a major consideration for live-cell imaging. Therefore, use of low laser power densities is essential for cell viability.

In principle, there is no fundamental restriction that prevents the concept of super-resolution microscopy to be applied to living cells. However, there are certain technical limitations, taking into consideration parameters such as temporal resolution, phototoxicity, and ease of intracellular labeling. These technical challenges are summarized below. For a more detailed review of live-cell super-resolution imaging and its biological applications, the reader is directed to Lakadamyali.[33]

In the case of STED, the temporal resolution is determined by the speed at which the focal spot can be scanned across the sample and the size of the imaging area. Therefore, high temporal resolution can be achieved at the expense of field-of-view and/or spatial resolution. The dependence of speed on the field-of-view can be improved by using the parallel detection scheme in which several doughnuts are simultaneously scanned.[19] In this case, an impressive temporal resolution of <1 s can be achieved in a large field of view and with relatively high spatial resolution. Live-cell STED imaging has been carried out in several biological systems including imaging neurons deep inside the brain tissue of living mice.[17] However, the relatively high laser powers needed to induce stimulated emission and switch off the fluorescence of molecules using the STED depletion beam means that photobleaching and phototoxicity can become potential problems. This can limit the imaging duration as well as the overall choice of fluorophores to those that are bright and photostable. This problem can be alleviated with the use of reversible saturable optical linear fluorescence transition (RESOLFT),[34,35] in which fluorescent probes are switched off from long-lived states compared to the short excited lifetime exploited in STED, therefore lowering the intensity requirement for the depletion beam. However, the temporal resolution of RESOLFT is limited by the switching kinetics of the photoswitchable fluorescent proteins and currently this method cannot achieve as high a temporal resolution as STED.

In the case of single-molecule localization-based methods (STORM/PALM), the temporal resolution is limited by the time needed to acquire enough fluorophore localizations to satisfy the Nyquist criterion for a given spatial resolution. Nyquist criterion states that the density of imaged labels should be such that the distance between the individual localizations is half the desired spatial resolution.[36] The temporal resolution is thus ultimately limited by how fast the fluorophores can be switched between on and off states, the camera frame rate, and the field-of-view. Fluorescent proteins provide easy intracellular labeling in living cells; however, the slow switching kinetics of fluorescent proteins and their low photon output limit both the temporal (~20–40 s) and the spatial (~30–50 nm) resolution.[27] Organic fluorophores are typically brighter than fluorescent proteins and they can be switched to dark states very fast by using high laser powers without compromising photon output. Therefore, high temporal resolution (~0.5–1 s) can be achieved while maintaining a high spatial resolution (~20 nm).[24] However, intracellular labeling using organic fluorophores is highly challenging and high laser powers needed to increase their switching rate can introduce phototoxicity. The recent development of new membrane permeable, bright fluorophores,[37] which can be used in conjunction with genetically encoded tags such as SNAP-, CLIP- and HALO-tags that specifically target these fluorophores to the protein of interest[24,38–40] holds promise to alleviate the intracellular labeling problem.

To summarize, live-cell super-resolution imaging requires finding a delicate balance between several parameters such as spatial resolution, temporal resolution, field-of-view, and the duration (length) of imaging. Future development of new fatigue-resistant photoswitchable fluorescent probes with faster photoswitching kinetics, higher photon output, and higher photostability will derive progress in this field in the future and open new doors for live-cell super-resolution microscopy. However, with the currently available toolbox, achieving nanoscale spatial resolution with millisecond temporal resolution is still highly challenging. Many biological processes, such as microtubule-dependent cargo transport, take place at faster timescales than the typical temporal resolution that can be achieved with live-cell super-resolution microscopy, obscuring their observation in living cells. One approach to circumvent this problem is to use correlative live-cell and super-resolution microscopy.[30]

14.3 General Concept of Correlative Live-Cell and Super-Resolution Microscopy

Correlative live-cell and super-resolution microscopy involves sequentially imaging the same cell under two different imaging modalities: (1) live-cell imaging/single particle tracking and (2) super-resolution microscopy (Figure 14.1).[30,41,42] In general terms, the target of interest (organelle, protein complex, or molecule) is first labeled with a fluorescent marker and a conventional fluorescence time-lapse movie of its dynamics is recorded with high temporal resolution (millisecond scale). The brightness and photostability of the label used for tagging the target of interest determines the achievable temporal resolution, which is limited by photobleaching. Endo/lysosomal vesicles and mitochondria can be readily tagged with markers such as lysotracker and mitotracker, providing a temporal resolution of ~250 ms with several minutes of imaging time. Quantum dots, fluorescent nanodiamonds, and fluorescent beads get internalized and delivered into endo/lysosomal vesicles when added to cells and can be used to follow the transport of these vesicles with higher temporal resolution (~50–100 ms) due to their high photostability. A single-particle tracking routine,[43] which in general terms finds the centroid position of each fluorescent spot and links them together across the frames, can be used to generate dynamic trajectories of the target objects with high spatial resolution (10–20 nm depending on the brightness of the object that is being tracked).

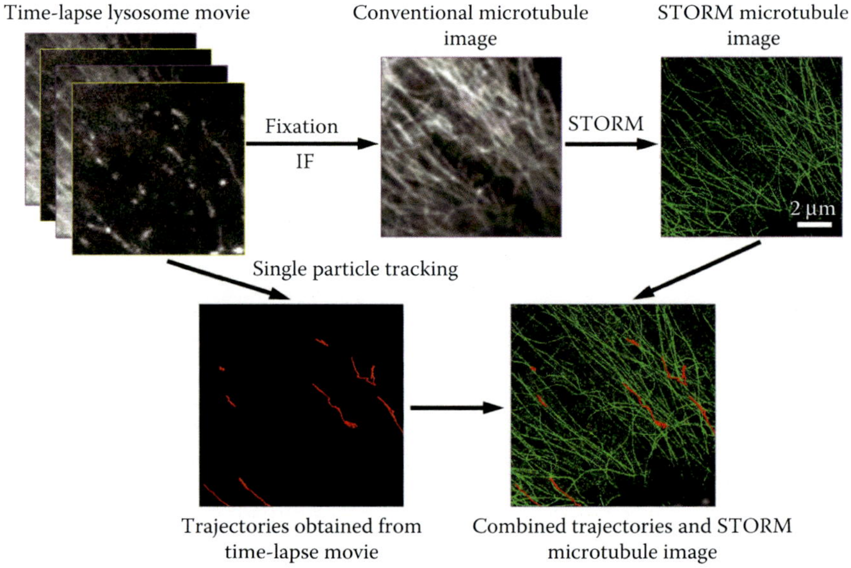

Figure 14.1

Workflow of correlative live-cell and super-resolution microscopy. A conventional time-lapse movie is first recorded at high frame rate (~10–20 Hz). The sample is then fixed on the microscope stage, the structure of interest is labeled with super-resolution compatible fluorophores using an appropriate method such as immunofluorescence (IF) and a super-resolution image (e.g., STORM microtubule image) is recorded. Single-particle tracking provides high spatiotemporal resolution trajectories of subcellular dynamics. These trajectories can be aligned with the super-resolution image using fiduciary markers and a registration algorithm such as affine transformation. (Reprinted with permission from Balint, S. et al., *Proc. Natl. Acad. Sci. USA*, 110, 3375–3380, 2013.)

Once the live-cell time-lapse movie is completed, the sample is subsequently fixed *in situ* (on the microscope stage) at a time point of interest. Once optimized, fixation is typically fast and structures can be preserved as they appear in the final frame of the time-lapse movie. After fixation, further sample preparation steps are carried out to label the structure to be imaged at high resolution using appropriate labeling methods (e.g., immunostaining with primary antibodies and secondary antibodies labeled with super-resolution compatible fluorophores). Once the sample preparation is complete, a super-resolution image of the target structure can be recorded. Finally, the trajectories obtained from the time-lapse movie can be mapped onto this super-resolution image. It is important to achieve precise registration of the two images. Image alignment can be achieved with the help of fiduciary markers, such as fluorescent beads, that appear and that can be localized in both images. The centroid positions of these fiduciary markers provide coordinates that can be used in a registration algorithm such as affine transformation, which can typically align the two images with a final registration uncertainty of ~9–10 nm.[30,32] The registration uncertainty can be determined as the average distance between the transformed centroid positions of the fiduciary markers from one movie and the centroid positions of these from the other movie (Equation 14.1):

$$\text{Registration Error} = \frac{1}{N}\sum_{i=1}^{N}\sqrt{\Delta x_i^2 + \Delta y_i^2} \tag{14.1}$$

where:

Δx_i is the shift between the x-centroid positions

Δy_i is the shift between the y-centroid positions of the ith fiduciary marker in the two images

14.4 Biological Application of Correlative Imaging to Cargo Transport

14.4.1 Cargo Transport and Cellular Roadblocks

An exemplary system in which correlative live-cell and super-resolution microscopy can give novel insights is the transport of vesicles and organelles by motor proteins along microtubule tracks. The cytoplasm of mammalian cells is highly crowded, with several types of vesicles, organelles, proteins, and cytoskeletal filaments densely packed together. This cytoplasmic crowding is precisely the reason why active transport with motor proteins (dynein and kinesin) is essential, since diffusion becomes a highly inefficient mechanism for transporting large protein complexes, vesicles, and organelles over long distances. Motor proteins such as dynein and kinesin are nanoscale molecular machines that convert the energy from ATP hydrolysis to mechanical motion.[44–46] They bind to cargos such as vesicles and organelles with their "head" domain and to microtubules with their "leg" domain. They take defined hand-over-hand steps ranging from 8 to 74 nm along the microtubule and exert forces ranging from 0.1 to 8 pN depending on the motor,[45–51] therefore executing directed, processive motion that translocates the vesicle from one place to another. Dynein transports cargos toward the minus end of the microtubule located at the microtubule organizing center (MOC), whereas most kinesins transport cargos in the opposite direction, toward the microtubule plus end.[52,53]

Cytoplasmic crowding can pose challenges to motor protein-mediated transport. For example, vesicles transported by motors can run into roadblocks and traffic jams, which can hinder their directed motion.[54,55] Nonetheless, motors are able to deliver their cargo efficiently and rapidly through the cytoplasm. To achieve this task, several motors must coordinate their action to navigate a complex cytoskeleton, overcoming traffic jams and roadblocks.[54,55] The mechanisms by which motors overcome or avoid roadblocks can give important insights into neurodegenerative diseases such as Alzheimer's, in which transport may become disrupted due to accumulation of roadblocks along the microtubule tracks.[56,57]

Studying cargo transport dynamics in the context of cellular roadblocks requires both high temporal and high spatial resolution. Given that motor protein-mediated transport can reach speeds of 1–2 μm/s, for accurate determination of transport dynamics (such as pauses) an exposure time of maximum 100–250 ms is desirable. In addition, a large field of view (at least $40 \times 40\ \mu m^2$) allows tracking several cargos over large enough distances such that sufficiently large statistics can be accumulated for quantitative analysis and interpretation. Finally, the imaging time should be long enough (in the scale of minutes) to capture sufficiently long trajectories. These parameters (acquisition time, field of view, imaging length) are mostly incompatible with what can be readily achieved using live-cell super-resolution microscopy. On the other hand, super-resolution microscopy is essential for resolving individual microtubules and roadblocks along the microtubules (e.g., microtubule intersections or microtubule-associated proteins, MAPs), since the cytoskeleton is organized into a highly complex and dense three-dimensional (3D) network. In this case, imaging the cargo transport dynamics using conventional time-lapse microscopy and correlating the trajectories to the individual microtubules visualized at high resolution after fixation is ideally suited to put these dynamics into the context of the microtubule network.

14.4.2 Correlative Imaging of Cargo Dynamics at Microtubule–Microtubule Intersections

Filament intersections, such as those between two or more microtubules can represent both roadblocks to cargo transport and switching points that can divert transport in different directions. Therefore, microtubule intersections can potentially affect the efficiency of cargo transport. The impact of microtubule intersections on the movement of individual motors and motor-decorated beads was previously studied using *in vitro* reconstituted microtubules deposited on top of each other to generate intersecting microtubules.[58,59] These studies constitute a first starting point to understand how filament intersections can impact cargo transport, however, their implication for intracellular transport is not clear. The microtubule network has a more complex architecture inside the cell than what has been achieved in these *in vitro* experiments. For example, it is not clear if the microtubule switching observed in these studies can happen inside the cell, in which the intersecting microtubules may not necessarily be touching each other. In addition, likely both polarity motors (dynein and kinesin) are present on the cargo simultaneously[60,61] and their activity and their numbers can be regulated through mechanisms that are not present *in vitro*.[62]

Balint et al. set out to study the impact of microtubule intersections on cargo transport inside living cells.[30] Lysosomes were chosen as the cargo of interest, since they can be easily labeled using lysotracker and they exhibit complex, directed motion along microtubules. This motion could be tracked in live-cell time-lapse movies using single particle tracking, which provided high spatio-temporal resolution trajectories of lysosome dynamics inside the cell. These trajectories exhibited periods of directed motion with mean square displacement >1.5 and an average speed of ~0.5 μm/s, which is typical for microtubule and motor protein-dependent motion. In order to align the trajectories with the microtubule network, the cells were fixed and the microtubules labeled and imaged at high resolution. It is essential that the dynamic changes of the microtubule network are slower than the imaging duration of the time-lapse movie. Otherwise, the microtubule network will rearrange during the time-lapse movie and the end point image of the microtubules will not constitute a faithful representation of where the microtubules were during the time-lapse movie. The growing, shrinking, and buckling dynamics of the microtubules could be reduced substantially (to minutes timescales) by treating the cells with low concentration of taxol and nocodazole without significantly affecting lysosome transport.

After overcoming these technical challenges, the trajectories could be reliably aligned with the 2D super-resolution images of the microtubule network (Figure 14.2) and their dynamics at microtubule intersections could be studied. Microtubule intersections were identified as points in which two or more microtubules crossed each other in the super-resolution images. Lysosomes exhibited four distinct behaviors when they approached a microtubule intersection. The majority of lysosomes (48.6%) slowed down and paused (no net displacement for 1 s or longer) when they arrived at an intersection point between multiple microtubules. A large percentage of lysosomes (31.5%) could also pass through the intersection unhindered and continued to move on the same microtubule. A small percentage of lysosomes (14.5%) switched to the intersecting microtubule while reversing transport direction and moving backward on the same

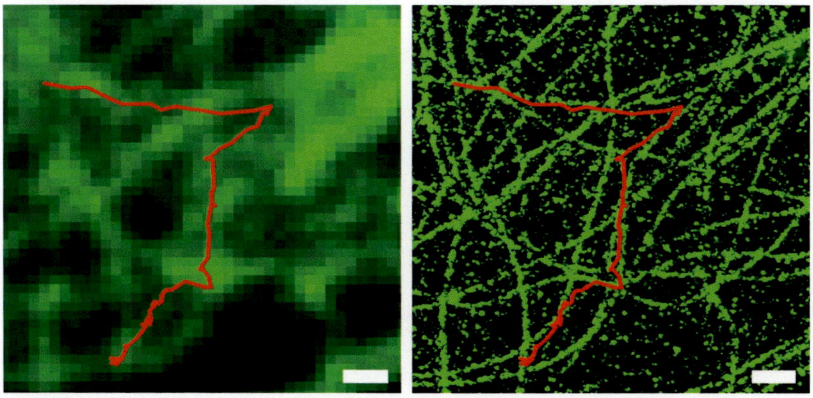

Figure 14.2

Full trajectory (red line) of a lysosome (not displayed) mapped on top of the end-point image of microtubules, either with conventional fluorescence microscopy (b) or with the correlative live-cell and super-resolution imaging approach (a). Scale bar: 1 μm. (Reprinted with permission from Oddone, A. et al., Super-resolution imaging with single-molecule localization, in *Cell Membrane Nanodomains: From Biochemistry to Nanoscopy*, 2014.)

microtubule was rare (5.4%). Pausing was temporary and did not constitute an "end-point" to lysosome transport. In majority of the cases (83%), lysosomes once again moved after the pausing period by either switching to the intersecting microtubule or eventually passing through the intersection. Interestingly, when an anterograde (or retrograde) moving organelle switched to the intersecting microtubule either as a primary behavior or after pausing, the transport direction was maintained as anterograde or as retrograde in the majority of cases. Previously, it was shown that both polarity motors (dynein and kinesin) are simultaneously present on the membrane of purified lysosomes.[60] Taken together, the maintenance of transport directionality indicates that likely only one motor-type is maintained active at a time. This regulation to maintain one motor-type active can potentially improve transport efficiency by ensuring that the cargo continues to move overall in a given direction, even after track switching.

14.4.3 Cargo Dynamics and the 3D Microtubule Architecture

Since a large number of lysosomes paused at microtubule intersections, this observation brought up the possibility that the axial separation of the microtubules along the z-axis may determine whether the intersection constitutes a roadblock or not. To test this possibility, the trajectories were correlated to 3D super-resolution images of the microtubule network. Several approaches have been developed to improve the spatial resolution in the z-dimension. In the case of single-molecule localization-based super-resolution methods, the astigmatism approach is relatively easy to implement and provides high resolution (~50 nm) along the z-axis.[11] This approach takes advantage of the fact that when a cylindrical lens is present in the imaging path, single molecules that are further from the focal plane will have an elliptical point spread function (PSF). The ellipticity will be along one direction (e.g., x-axis) when a molecule is above the focal plane and along the other direction (e.g., y-axis) when it is below the focal plane. Therefore, the z-position of the molecule along the focal plane can be precisely determined from the width of the image PSF along the x- and y-axes while the x–y position of the molecule can be determined from the centroid position of the image PSF as usual.

Using this astigmatism approach, it was possible to measure the axial separation of microtubules as the peak-to-peak distance between individual microtubule images and correlate the lysosome trajectories to these 3D super-resolution images (Figure 14.3). Analysis of lysosome trajectories at microtubule intersections showed that the pausing events correlated with intersections having a small axial separation. These results indicate that the microtubule network geometry is the main determinant of the amount of hindrance the intersection constitutes to the directed transport of cargo. However, the motors could eventually overcome these obstructions with high fidelity by either switching to the intersecting microtubule or passing through the intersection after the initial pause. Given the size of the lysosomes (~350 nm), the mechanism by which they can pass through a tight intersection remains to be determined. One possibility could be that there is a high degree of flexibility for the lysosome to change its shape to "squeeze" through the intersection. Another possibility could be that the motors move the lysosome away from the intersecting microtubule by changing their position on the microtubule surface. Future experiments will be able to distinguish between these and other possibilities.

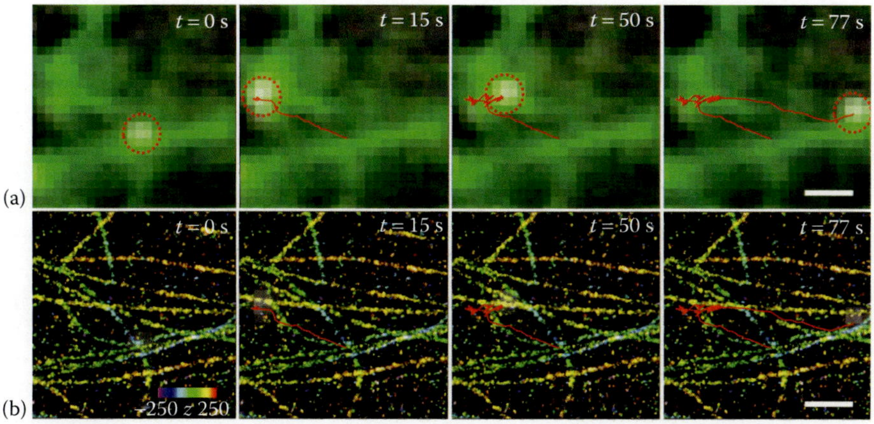

Figure 14.3

Correlative live-cell and STORM imaging of cargo transport. (a) Conventional two-color time-lapse images of lysosome (white) and microtubules (green). The red line shows the transport trajectory of the lysosome obtained with single particle tracking. (b) The same field-of-view but with the conventional image of microtubules replaced by the 3D STROM image (color coding shows *z*-scale according to the *z*-color bar). The transport trajectory of lysosome could be mapped precisely to the STORM image of individual microtubules. Scale bar: 500 nm. (From Lakadamyali, M., *ChemPhysChem,* 15, 630–636, 2014. With permission.)

14.5 Technical Developments in Correlative Imaging

14.5.1 Automated Correlative Imaging with Microfluidics

In principle, the correlative approach can be extended to study other subcellular processes in which it is necessary to interpret dynamic information in the context of nanoscale structural information or nanoscale quantitative information (such as counting protein copy numbers).[63–65] However, fixation and the subsequent labeling prior to super-resolution imaging require precise delivery and removal of fluid from the sample, which remains on the microscope stage for the duration of the experiment. This procedure, when performed manually, is imprecise, labor-intensive, and time consuming. Automating the sample preparation steps in between the two imaging modalities can therefore improve the throughput and ease-of-use of the method, rendering it more versatile. With this goal in mind, Tam et al. designed a PDMS-based microfluidic device for automated fluid delivery (Figure 14.4) and validated it for correlative live-cell super-resolution imaging.[31]

A simple design, with a single-layered microfluidic chip containing external valves was preferred to reduce cost and allow the microfluidic device to be easily transported on and off the microscope stage. In order to parallelize the imaging, eight independent imaging chambers were incorporated into a single microfluidic device, in which mammalian cells can be seeded and grown. A multiplexed design was developed for reagent delivery that can be easily modified to give flexibility to experimental strategy. Instead of assigning each imaging chamber to its own dedicated set of reagent reservoirs, all imaging chambers were designed to have access to a common set of reagent reservoirs (Figure 14.4), allowing the numbers of reservoirs and/or imaging chambers to be increased or decreased independently of upstream or downstream components.

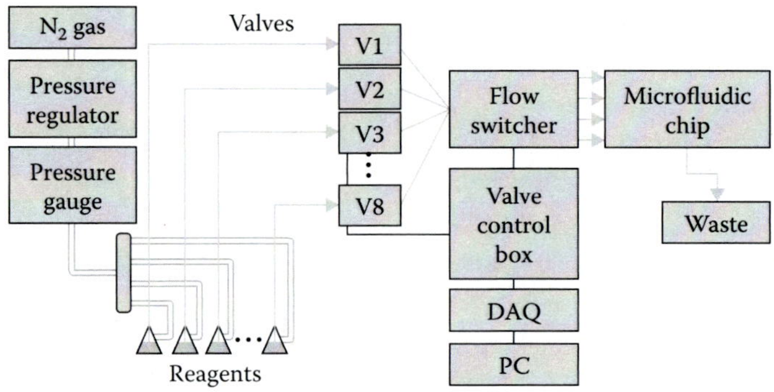

Figure 14.4

Schematic drawing of the computer-controlled fluid injection system. The number of reagent reservoirs can be adjusted by adding or removing valves. Pressurized air is coupled to reagents which are injected through the system using solenoid pinch valves. (From Tam et al., *PLoS One*, 9, e115512, 2014. With permission.)

One of the biggest challenges of adapting the correlative live-cell and super-resolution microscopy to microfluidic imaging chambers was the optimization of sample labeling and imaging protocols. Sample preparation is especially important in the case of super-resolution imaging, since the improved resolution requires an even more stringent criterion for the preservation and labeling of sample features. In particular, high labeling densities must be achieved using higher-than-typical antibody concentrations since the labeling density is directly linked to the spatial resolution through the Nyquist criterion.[36] In addition, the imaging buffer conditions are crucial for inducing the desired photoswitching properties in the case of small organic fluorophores.[41,42] It is difficult to achieve these conditions inside a microfluidic device for multiple reasons. First, cells must be injected through small diameter tubes, which lead to a dramatic decrease in the density of cells in suspension. Second, PDMS has significantly different material properties than polystyrene or glass, which are typically used in imaging chambers. Therefore, direct implementation of established protocols for cell seeding result in very low cell concentrations and to the failure of cells to establish a confluent monolayer. In addition, established protocols for immunostaining lead to poor labeling densities and low spatial resolution. Finally, the commonly used composition of imaging buffers lead to the complete absence of photoswitching in fluorophores. These difficulties could be overcome by systematically testing and optimizing the concentration of different reagents used. For example, the concentration of cells had to be increased by 30-fold during cell seeding and the concentration of antibodies by about 5-fold to achieve desired results. Once optimized, these labeling and imaging conditions resulted in comparable image resolution to manually labeled samples.

The versatility of the microfluidic devices was demonstrated by correlating the dynamics of mitochondria to their morphology at high resolution as well as to the distribution of proteins on the mitochondrial membrane. This proof-of-principle application served as a validation of the microfluidic-assisted correlative live-cell and super-resolution microscopy. Importantly, it was possible to carry out a total of 28 correlative live-cell super-resolution experiments in a time period of about 65 h with high quality data in 21 out of 28 experiments (75% success rate), a dramatic improvement in throughput and success rate, while the immunostaining was carried out completely unsupervised.

The microfluidic-assisted, high-throughput correlative imaging enabled investigation of the relationships between dynamics, size, and protein distribution at the level of individual organelles for a large population of mitochondria. A total of 577 mitochondria could be identified, which were clearly discernable both in the live-cell type-lapse movie showing mitochondrial dynamics up to the point of fixation, as well as in the super-resolution image and therefore could be correlated. Mitochondria showed diverse dynamics, which could be broadly assigned into three categories: (i) stationary; (ii) undergoing a slow, but dynamic motion (dynamic-slow); or (iii) undergoing a fast, microtubule-dependent translation across the cell (dynamic-fast). Mitochondrial size determined from super-resolution images was correlated to these mitochondrial dynamics determined from the live-cell videos. The correlation clearly showed that mitochondria that were dynamic slow or interacting were larger in size when compared to all other categories. Furthermore, dynamic fast mitochondria were on average smaller than stationary mitochondria. Together, these results suggest that mitochondrial size, dynamics, and interactions are not independent parameters, but rather, related to each other. Large mitochondria may constitute a substantial load to motor proteins due to their size, slowing down their motion or large mitochondria may be hindered more in their transport along microtubules due to roadblocks.

Finally, to demonstrate that this kind of streamlined and automated correlative imaging and analysis can be applied to screen changes in mitochondrial dynamics, morphology, and protein distribution, cells stably expressing a GFP-tubulin marker were analyzed. Since dynamic mitochondria are likely associated with microtubules, overexpression of GFP-tagged tubulin could in principle alter the correlations between mitochondrial morphology and dynamics that were observed in the wild-type cells. However, overall, the distribution of mitochondrial dynamics was similar between wild-type and transfected cells and there were no statistically significant differences when comparing mitochondrial morphology between dynamic categories. Therefore, protein overexpression did not alter this particular organelle in these stably transfected cells.

In summary, the microfluidic imaging devices provide a versatile platform for automated correlative microscopy that has the capability to image with both high temporal resolution (millisecond range) and high spatial resolution (10–20 nm range). This platform adds to a growing list of recent "systems microscopy" approaches[66–69] that address the need for automation in advanced microscopy and can be applied to investigate biological phenomena at very high spatial and temporal scales with a wide range of potential applications which include cargo transport[30] and mitochondrial motility.

14.5.2 Multicolor Imaging Using the Correlative Approach

Recent years have seen a large amount of development in multicolor super-resolution microscopy. There are two general approaches for extending super-resolution to multiple colors. One approach uses fluorophore pairs in which the same reporter is coupled to different activators.[6] In this case, the fluorescence of the reporter can be activated into the "on" state by exciting the activator with the corresponding wavelength laser light. The color is thus determined based on the wavelength of the activating laser light, that is, by using alternating pulses of activation laser light with different wavelengths, it is possible to color-code the resulting localizations based on when they turn on during the imaging cycle.[6,8] This approach is free from chromatic aberrations and the need for image registration since all colors are acquired in the same image channel. However, it is prone to color crosstalk,[7] since fluorophores can also undergo spontaneous activation, independent of the activation laser, or alternately, fluorophores can be activated by the "wrong" activation laser. A second approach uses spectrally distinct reporter dyes coupled to the same (or different) activator dyes.[7] A variation of this second approach uses spectrally distinct photoswitchable reporter fluorophores alone without an activator dye (*d*STORM).[9,10,70,71] The advantage of this second approach is that color crosstalk can be reduced or completely eliminated. However, due to chromatic aberrations, sophisticated algorithms may be required to properly align the multiple images at the nanoscale level.[72] More importantly, photoswitchable fluorophores must fulfill important criteria in terms of their photon output and photoswitching properties (e.g., duty cycle)[9] in order to generate images at high spatial resolution. Most fluorophores do not fulfill these criteria and there is very limited availability of spectrally distinct photoswitchable fluorophores with favorable photophysical properties.[9] A detailed analysis of a large number of photoswitchable fluorophores showed that AlexaFluor647 outperforms most fluorophores leading to images with the highest resolution.[9] Since the best fluorophore can only be used once, differences in the duty cycle and brightness of different fluorophores can impact the relative resolution of the images in the different color channels. Finally, it is often difficult to find one optimal imaging buffer compatible with all fluorophores. The performance of the imaging buffer can also decrease over time as more colors are acquired.

These caveats in multicolor super-resolution microscopy can be overcome by extending the correlative imaging concept to multicolor microscopy. In this case, the correlation is done over multiple, sequentially acquired super-resolution images instead of multiple imaging modalities. Fiduciary markers such as fluorescent beads can be used as described above to precisely align the different images into one multicolor image with a final image registration precision of 10 nm using simple, commonly employed registration algorithms such as affine transformation (Figure 14.5). Importantly, since the structures of interest are labeled and imaged sequentially, one single fluorophore can be used to label the different structures and acquire all the images. After each imaging session is completed, the residual fluorescence from that particular target can be fully quenched by treating the cells with sodium borohydride, therefore successfully avoiding any color crosstalk. The best performing fluorophore can therefore be used across all images, leading to uniform spatial resolution over all the different colors and the need for only one single imaging buffer. Importantly, a large number of targets can be labeled sequentially without significantly altering the structural properties of the different targets to be imaged. Finally, using this approach, the number of colors that can be acquired is no longer limited by the number of available spectrally distinct activator and/or reporter fluorophores. One limitation can be the number of available primary antibodies of different species that can be orthogonally recognized by fluorophore-labeled secondary antibodies. However, this is not a specific limitation of sequential multicolor imaging and can potentially be overcome with the use of labeled primary antibodies or labeled oligos.[73]

The sequential labeling can be carried out *in situ* on the microscope stage and can potentially also be automated with the use of microfluidic devices described above.

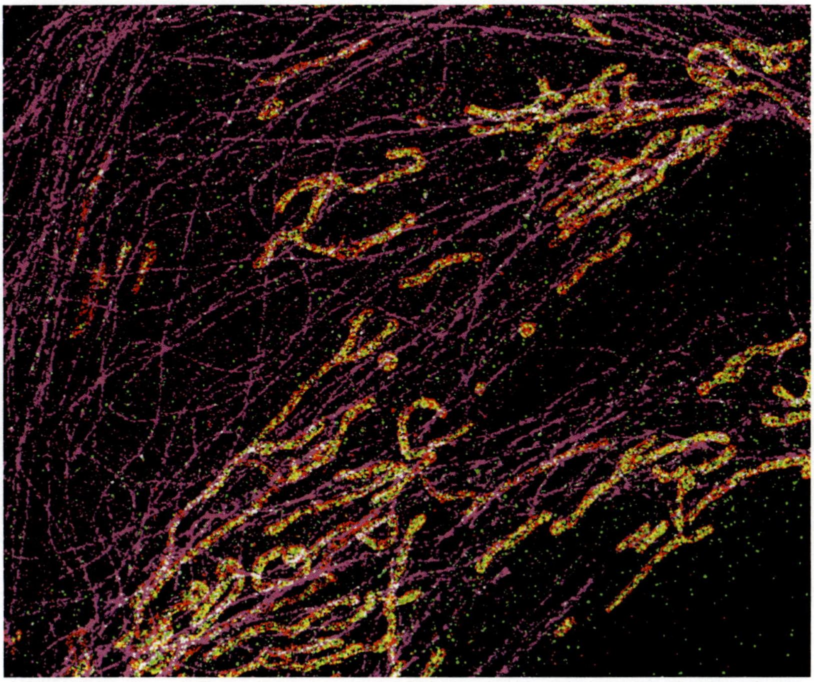

Figure 14.5

Three-color image of microtubules (magenta), mitochondrial outer membrane protein Tom20 (red) and mitochondrial inner membrane protein (ATP-synthase, green)-imaged sequentially using the same fluorophore activator-reporter pair (AlexaFluor405-AlexaFluor647). (From Tam et al., *PLoS One*, 9, e115512, 2014. With permission.)

Alternatively, to free microscope time, the sample can be removed from the microscope stage in between the imaging sessions for sequential labeling. In this case, it is essential to locate the same region of interest multiple times. The use of high magnification objectives in STORM makes this task rather challenging since the field of view being imaged is relatively small (typically $40 \times 40\ \mu m^2$ or smaller). In correlative microscopy, this challenge is overcome with the help of a "finder grid".[74] However, glass-bottom chambers that are readily available and used by most laboratories for fluorescence microscopy applications typically do not include a finder grid. To simplify the method for locating the same region of interest, a "virtual grid" can instead be used with the help of a precision motorized stage with a readout for the stage coordinates in the X and Y directions. In this case, a "virtual grid" is generated by recording the coordinates of two reference points (e.g., P_1 and P_2, which can be the corner points of the sample chamber) as well as the coordinates of the region of interest (C) during the first imaging session. At the start of the next imaging session, the two reference points are relocated and the new coordinates of these two points are recorded as P_1' and P_2'. These reference points can then be used to re-locate the region of interest, C. The rotation angle between day 1 and day 2 is calculated as

$$\Delta\theta = \cos^{-1}\left(\frac{\overrightarrow{P_1P_2}\cdot\overrightarrow{P_1'P_2'}}{\left\|P_1P_2\right\|\left\|P_1'P_2'\right\|}\right)$$

If the sample is mounted with care, $\Delta\theta$ can be maintained below 0.05°. The new location of the region, C', relative to the corner point P_1' can then be calculated as

$$C' = \left\|\overrightarrow{P_1C}\right\| \angle(\theta + \Delta\theta), \text{ where } \theta = \tan^{-1}\left(\frac{\left.\overrightarrow{P_1C}\right|_y}{\left.\overrightarrow{P_1C}\right|_x}\right).$$

This procedure allows locating the same region of interest within 5 μm. To locate the region more precisely, images of fiduciary markers (fluorescent beads) can be acquired and compared to images of these same markers from the first imaging session.

In summary, the simple strategy of sequential labeling and imaging eliminates a large number of technical problems and enables crosstalk-free, multicolor STORM with the best performing fluorophore. The "virtual grid" approach makes it easy to re-locate the same region of interest multiple times (or across multiple microscopes), freeing up microscope time and simplifying the sample preparation.

14.6 Conclusions and Future Perspectives

The all-optical correlative live-cell and super-resolution imaging method circumvents one of the major limitations of super-resolution microscopy and allows relating fast subcellular dynamics to nanoscale information. When applied to cargo transport, it has revealed important insights about the dynamics of cargos at roadblocks such as microtubule intersections. Thanks to microfluidic-assisted automation, which streamlines the method and improves its throughput, in the future, it would be possible to screen the impact of other roadblocks such as MAPs or other organelles on cargo dynamics. An interesting question remains, which is how motor proteins eventually overcome a roadblock such as tight microtubule intersections, to continue transporting large cargos along the same microtubule. Future extension of the method to 3D single particle tracking can help determine if motors can move around the microtubule track to avoid roadblocks. In addition, it would be possible to relate quantitative information at the nanoscale level, such as motor protein copy numbers that may be determined by methods such as single-molecule counting,[63,64] to cargo transport dynamics such as speed, processivity, and run length. Understanding the precise molecular mechanisms of active transport can help give important clues into how these processes fail in neurodegenerative diseases such as Alzheimer's and amyotrophic lateral sclerosis (ALS).[75]

In its current form, correlative live-cell and super-resolution microscopy has been used to relate dynamics to super-resolution images acquired using single-molecule localization-based methods, in particular STORM. However, there is no fundamental limitation that precludes extending this method to correlating dynamics with STED images. It is also important to emphasize that this all-optical correlative method is not limited to studying cargo transport and it should have a wide range of applications in biology, where putting dynamics into the context of nanoscale ultrastructural or molecular information is important. Other examples of biological systems in which the correlative analysis may generate novel information include gene dynamics and chromatin structure, as well as endocytosis, sorting, and endo/lysosomal subdomain protein organization.

The automation should in particular help with more widespread application of the method to diverse biological questions.

Overall, in the short term, this method should help bridge the gap between dynamic live-cell imaging and super-resolution microscopy and improve our understanding of how multiprotein complexes or subcellular structures remodel dynamically in time.

References

1. Klar, T. A., S. Jakobs, M. Dyba, A. Egner, and S. W. Hell. Fluorescence Microscopy with Diffraction Resolution Barrier Broken by Stimulated Emission. *Proc Natl Acad Sci USA* 97(15):8206–8210, 2000.
2. Gustafsson, M. G. Nonlinear Structured-Illumination Microscopy: Wide-Field Fluorescence Imaging with Theoretically Unlimited Resolution. *Proc Natl Acad Sci USA* 102(37):13081–13086, 2005.
3. Betzig, E., G. H. Patterson, R. Sougrat et al. Imaging Intracellular Fluorescent Proteins at Nanometer Resolution. *Science* 313(5793):1642–1645, 2006.
4. Hess, S. T., T. P. Girirajan, and M. D. Mason. Ultra-High Resolution Imaging by Fluorescence Photoactivation Localization Microscopy. *Biophys J* 91(11):4258–4272, 2006.
5. Rust, M. J., M. Bates, and X. Zhuang. Sub-Diffraction-Limit Imaging by Stochastic Optical Reconstruction Microscopy (Storm). *Nat Methods* 3(10):793–795, 2006.
6. Bates, M., B. Huang, G. T. Dempsey, and X. Zhuang. Multicolor Super-Resolution Imaging with Photo-Switchable Fluorescent Probes. *Science* 317(5845):1749–1753, 2007.
7. Bates, M., G. T. Dempsey, K. H. Chen, and X. Zhuang. Multicolor Super-Resolution Fluorescence Imaging Via Multi-Parameter Fluorophore Detection. *ChemPhysChem* 13(1):99–107, 2012.
8. Lakadamyali, M. High Resolution Imaging of Neuronal Connectivity. *J Microsc* 248(2):111–116, 2012.
9. Dempsey, G. T., J. C. Vaughan, K. H. Chen, M. Bates, and X. Zhuang. Evaluation of Fluorophores for Optimal Performance in Localization-Based Super-Resolution Imaging. *Nat Methods* 8(12):1027–1036, 2011.
10. Lampe, A., V. Haucke, S. J. Sigrist, M. Heilemann, and J. Schmoranzer. Multi-Colour Direct Storm with Red Emitting Carbocyanines. *Biol Cell* 104(4):229–237, 2012.
11. Huang, B., W. Wang, M. Bates, and X. Zhuang. Three-Dimensional Super-Resolution Imaging by Stochastic Optical Reconstruction Microscopy. *Science* 319(5864):810–813, 2008.
12. Pavani, S. R., M. A. Thompson, J. S. Biteen et al. Three-Dimensional, Single-Molecule Fluorescence Imaging Beyond the Diffraction Limit by Using a Double-Helix Point Spread Function. *Proc Natl Acad Sci USA* 106(9):2995–2999, 2009.
13. Shtengel, G., J. A. Galbraith, C. G. Galbraith et al. Interferometric Fluorescent Super-Resolution Microscopy Resolves 3D Cellular Ultrastructure. *Proc Natl Acad Sci USA* 106(9):3125–3130, 2009.
14. Jia, S., J. C. Vaughan, and X. Zhuang. Isotropic 3D Super-Resolution Imaging with a Self-Bending Point Spread Function. *Nat Photonics* 8:302–306, 2014.

15. Cella Zanacchi, F., Z. Lavagnino, M. Perrone Donnorso et al. Live-Cell 3D Super-Resolution Imaging in Thick Biological Samples. *Nat Methods* 8(12):1047–1049, 2011.
16. Benke, A. and S. Manley. Live-Cell Dstorm of Cellular DNA Based on Direct DNA Labeling. *ChemBioChem* 13(2):298–301, 2012.
17. Berning, S., K. I. Willig, H. Steffens, P. Dibaj, and S. W. Hell. Nanoscopy in a Living Mouse Brain. *Science* 335(6068):551, 2012.
18. Biteen, J. S., M. A. Thompson, N. K. Tselentis et al. Super-Resolution Imaging in Live Caulobacter Crescentus Cells Using Photoswitchable Eyfp. *Nat Methods* 5(11):947–949, 2008.
19. Chmyrov, A., J. Keller, T. Grotjohann et al. Nanoscopy with More Than 100,000 'Doughnuts.' *Nat Methods* 10:737–740, 2013.
20. Cox, S., E. Rosten, J. Monypenny et al. Bayesian Localization Microscopy Reveals Nanoscale Podosome Dynamics. *Nat Methods* 9(2):195–200, 2012.
21. Fiolka, R., L. Shao, E. H. Rego, M. W. Davidson, and M. G. Gustafsson. Time-Lapse Two-Color 3d Imaging of Live Cells with Doubled Resolution Using Structured Illumination. *Proc Natl Acad Sci USA* 109(14):5311–5315, 2012.
22. Hess, S. T., T. J. Gould, M. V. Gudheti et al. Dynamic Clustered Distribution of Hemagglutinin Resolved at 40 Nm in Living Cell Membranes Discriminates between Raft Theories. *Proc Natl Acad Sci USA* 104(44):17370–17375, 2007.
23. Huang, F., T. M. Hartwich, F. E. Rivera-Molina et al. Video-Rate Nanoscopy Using Scmos Camera-Specific Single-Molecule Localization Algorithms. *Nat Methods* 10(7):653–658, 2013.
24. Jones, S. A., S. H. Shim, J. He, and X. Zhuang. Fast, Three-Dimensional Super-Resolution Imaging of Live Cells. *Nat Methods* 8(6):499–508, 2011.
25. Nagerl, U. V., K. I. Willig, B. Hein, S. W. Hell, and T. Bonhoeffer. Live-Cell Imaging of Dendritic Spines by Sted Microscopy. *Proc Natl Acad Sci USA* 105(48):18982–18987, 2008.
26. Rossy, J., D. M. Owen, D. J. Williamson, Z. Yang, and K. Gaus. Conformational States of the Kinase Lck Regulate Clustering in Early T Cell Signaling. *Nat Immunol* 14(1):82–89, 2013.
27. Shroff, H., C. G. Galbraith, J. A. Galbraith, and E. Betzig. Live-Cell Photoactivated Localization Microscopy of Nanoscale Adhesion Dynamics. *Nat Methods* 5(5):417–423, 2008.
28. Westphal, V., S. O. Rizzoli, M. A. Lauterbach et al. Video-Rate Far-Field Optical Nanoscopy Dissects Synaptic Vesicle Movement. *Science* 320(5873):246–249, 2008.
29. Wombacher, R., M. Heidbreder, S. van de Linde et al. Live-Cell Super-Resolution Imaging with Trimethoprim Conjugates. *Nat Methods* 7(9):717–719, 2010.
30. Balint, S., I. Verdeny Vilanova, A. Sandoval Alvarez, and M. Lakadamyali. Correlative Live-Cell and Superresolution Microscopy Reveals Cargo Transport Dynamics at Microtubule Intersections. *Proc Natl Acad Sci USA* 110(9):3375–3380, 2013.
31. Tam, J., G. A. Cordier, S. Balint et al. A Microfluidic Platform for Correlative Live-Cell and Super-Resolution Microscopy. *PLoS One* 9(12):e115512, 2014.
32. Tam, J., G. A. Cordier, J. S. Borbely, A. Sandoval Alvarez, and M. Lakadamyali. Cross-Talk-Free Multi-Color Storm Imaging Using a Single Fluorophore. *PLoS One* 9(7):e101772, 2014.

33. Lakadamyali, M. Super-Resolution Microscopy: Going Live and Going Fast. *ChemPhysChem* 15(4):630–636, 2014.
34. Hell, S. W. Toward Fluorescence Nanoscopy. *Nat Biotechnol* 21(11):1347–1355, 2003.
35. Hofmann, M., C. Eggeling, S. Jakobs, and S. W. Hell. Breaking the Diffraction Barrier in Fluorescence Microscopy at Low Light Intensities by Using Reversibly Photoswitchable Proteins. *Proc Natl Acad Sci USA* 102(49):17565–17569, 2005.
36. Lakadamyali, M., H. Babcock, M. Bates, X. Zhuang, and J. Lichtman. 3D Multicolor Super-Resolution Imaging Offers Improved Accuracy in Neuron Tracing. *PLoS One* 7(1):e30826, 2012.
37. Grimm, J. B., B. P. English, J. Chen et al. A General Method to Improve Fluorophores for Live-Cell and Single-Molecule Microscopy. *Nat Methods* 12:244–250, 2015.
38. Klein, T., A. Loschberger, S. Proppert et al. Live-Cell Dstorm with Snap-Tag Fusion Proteins. *Nat Methods* 8(1):7–9, 2011.
39. Lee, H. L., S. J. Lord, S. Iwanaga et al. Superresolution Imaging of Targeted Proteins in Fixed and Living Cells Using Photoactivatable Organic Fluorophores. *J Am Chem Soc* 132(43):15099–15101, 2010.
40. van de Linde, S., M. Heilemann, and M. Sauer. Live-Cell Super-Resolution Imaging with Synthetic Fluorophores. *Annu Rev Phys Chem* 63:519–540, 2012.
41. Oddone, A., I. V. Vilanova, J. Tam, S. Balint, and M. Lakadamyali. Super-Resolution Imaging with Single-Molecule Localization. In Cambi, A., and Lidke, D. S., eds. *Cell Membrane Nanodomains: From Biochemistry to Nanoscopy*, Boca Raton, FL: CRC Press, 2014.
42. Oddone, A., I. V. Vilanova, J. Tam, and M. Lakadamyali. Super-Resolution Imaging with Stochastic Single-Molecule Localization: Concepts, Technical Developments, and Biological Applications. *Microsc Res Tech* 77(7):502–509, 2014.
43. Chenouard, N., I. Smal, F. de Chaumont et al. Objective Comparison of Particle Tracking Methods. *Nat Methods* 11(3):281–289, 2014.
44. Kreis, T. and R. Vale. *Guidebook to the Cytoskeletal and Motor Proteins*. 2nd edn. New York: Oxford University Press, 1999.
45. Schliwa, M. *Molecular Motors*. Weinheim, Germany: Wiley-VCH, 2003.
46. Vale, R. D. The Molecular Motor Toolbox for Intracellular Transport. *Cell* 112(4):467–480, 2003.
47. Yildiz, A., J. N. Forkey, S. A. McKinney et al. Myosin V Walks Hand-Over-Hand: Single Fluorophore Imaging with 1.5-Nm Localization. *Science* 300(5628):2061–2065, 2003.
48. Yildiz, A., M. Tomishige, R. D. Vale, and P. R. Selvin. Kinesin Walks Hand-Over-Hand. *Science* 303(5658):676–678, 2004.
49. Visscher, K., M. J. Schnitzer, and S. M. Block. Single Kinesin Molecules Studied with a Molecular Force Clamp. *Nature* 400(6740):184–189, 1999.
50. Clancy, B. E., W. M. Behnke-Parks, J. O. Andreasson, S. S. Rosenfeld, and S. M. Block. A Universal Pathway for Kinesin Stepping. *Nat Struct Mol Biol* 18(9):1020–1027, 2011.
51. Leidel, C., R. A. Longoria, F. M. Gutierrez, and G. T. Shubeita. Measuring Molecular Motor Forces in Vivo: Implications for Tug-of-War Models of Bidirectional Transport. *Biophys J* 103(3):492–500, 2012.

52. Roberts, A. J., T. Kon, P. J. Knight, K. Sutoh, and S. A. Burgess. Functions and Mechanics of Dynein Motor Proteins. *Nat Rev Mol Cell Biol* 14(11):713–726, 2013.
53. Hirokawa, N., Y. Noda, Y. Tanaka, and S. Niwa. Kinesin Superfamily Motor Proteins and Intracellular Transport. *Nat Rev Mol Cell Biol* 10(10):682–696, 2009.
54. Lakadamyali, M. Navigating the Cell: How Motors Overcome Roadblocks and Traffic Jams to Efficiently Transport Cargo. *Phys Chem Chem Phys* 16(13):5907–5916, 2014.
55. Ross, J. L., M. Y. Ali, and D. M. Warshaw. Cargo Transport: Molecular Motors Navigate a Complex Cytoskeleton. *Curr Opin Cell Biol* 20(1):41–47, 2008.
56. Kanaan, N. M., G. F. Pigino, S. T. Brady et al. Axonal Degeneration in Alzheimer's Disease: When Signaling Abnormalities Meet the Axonal Transport System. *Exp Neurol* 246:44–53, 2013.
57. Stokin, G. B., and L. S. Goldstein. Axonal Transport and Alzheimer's Disease. *Annu Rev Biochem* 75:607–627, 2006.
58. Ross, J. L., H. Shuman, E. L. Holzbaur, and Y. E. Goldman. Kinesin and Dynein-Dynactin at Intersecting Microtubules: Motor Density Affects Dynein Function. *Biophys J* 94(8):3115–3125, 2008.
59. Vershinin, M., B. C. Carter, D. S. Razafsky, S. J. King, and S. P. Gross. Multiple-Motor Based Transport and Its Regulation by Tau. *Proc Natl Acad Sci USA* 104(1):87–92, 2007.
60. Hendricks, A. G., E. Perlson, J. L. Ross et al. Motor Coordination via a Tug-of-War Mechanism Drives Bidirectional Vesicle Transport. *Curr Biol* 20(8):697–702, 2010.
61. Gross, S. P., M. Vershinin, and G. T. Shubeita. Cargo Transport: Two Motors Are Sometimes Better Than One. *Curr Biol* 17(12):R478–R486, 2007.
62. Mallik, R. and S. P. Gross. Molecular Motors: Strategies to Get Along. *Curr Biol* 14(22):R971–R982, 2004.
63. Durisic, N., L. Laparra-Cuervo, A. Sandoval-Alvarez, J. S. Borbely, and M. Lakadamyali. Single-Molecule Evaluation of Fluorescent Protein Photoactivation Efficiency Using an In Vivo Nanotemplate. *Nat Methods* 11(2):156–162, 2014.
64. Durisic, N., L. L. Cuervo, and M. Lakadamyali. Quantitative Super-Resolution Microscopy: Pitfalls and Strategies for Image Analysis. *Curr Opin Chem Biol* 20:22–28, 2014.
65. Ricci, M. A., C. Manzo, M. G. Parajo, M. Lakadamyali, and M. P. Cosma. Chromatin Fibers Are Formed by Heterogeneous Groups of Nucleosomes In Vivo. *Cell* 160:1145–1158, 2015.
66. Conrad, C. and D. W. Gerlich. Automated Microscopy for High-Content Rnai Screening. *J Cell Biol* 188(4):453–461, 2010.
67. Conrad, C., A. Wunsche, T. H. Tan et al. Micropilot: Automation of Fluorescence Microscopy-Based Imaging for Systems Biology. *Nat Methods* 8(3):246–249, 2011.
68. Lock, J. G. and S. Stromblad. Systems Microscopy: An Emerging Strategy for the Life Sciences. *Exp Cell Res* 316(8):1438–1444, 2010.
69. Neumann, B., T. Walter, J. K. Heriche et al. Phenotypic Profiling of the Human Genome by Time-Lapse Microscopy Reveals Cell Division Genes. *Nature* 464(7289):721–727, 2010.

70. Endesfelder, U., S. Malkusch, B. Flottmann et al. Chemically Induced Photoswitching of Fluorescent Probes—A General Concept for Super-Resolution Microscopy. *Molecules* 16(4):3106–3118, 2011.
71. van de Linde, S., A. Loschberger, T. Klein et al. Direct Stochastic Optical Reconstruction Microscopy with Standard Fluorescent Probes. *Nat Protoc* 6(7):991–1009, 2011.
72. Annibale, P., M. Scarselli, M. Greco, and A. Radenovic. Identification of the Factors Affecting Co-Localization Precision for Quantitative Multicolor Localization Microscopy. *Opt Nanoscopy* 1:9, 2012.
73. Jungmann, R., M. S. Avendano, J. B. Woehrstein et al. Multiplexed 3D Cellular Super-Resolution Imaging with DNA-Paint and Exchange-Paint. *Nat Methods* 11(3):313–318, 2014.
74. Schwartz, C. L., V. I. Sarbash, F. I. Ataullakhanov, J. R. McIntosh, and D. Nicastro. Cryo-Fluorescence Microscopy Facilitates Correlations between Light and Cryo-Electron Microscopy and Reduces the Rate of Photobleaching. *J Microsc* 227(Pt 2):98–109, 2007.
75. Maday, S., A. E. Twelvetrees, A. J. Moughamian, and E. L. Holzbaur. Axonal Transport: Cargo-Specific Mechanisms of Motility and Regulation. *Neuron* 84(2):292–309, 2014.

15

SAX Microscopy and Its Application to Imaging of 3D-Cultured Cells

Katsumasa Fujita, Kentaro Mochizuki, and Nicholas I. Smith

15.1 Introduction
15.2 The Principle of SAX Microscopy
15.3 The Spatial Resolution in SAX Microscopy
15.4 The Frequency Response of Fluorescence Emission
15.5 SAX Imaging of 3D-Cultured Cells
15.6 Conclusions
References

15.1 Introduction

By the recent development of super-resolution techniques, the spatial resolution of optical microscopy is no longer limited by the wave nature of light. The common concept that has enabled us to break the diffraction limit is utilizing the switching, blinking, or nonlinear property of fluorescence emission of probes that label a sample. Localization microscopy, such as photoactivated localization microscopy (PALM) or stochastic optical reconstruction microscopy (STORM), relies on the photoactivation, photoconversion, and transition to dark state of fluorescent probes to separately image the emitters in a sample (Hofmann et al. 2005, Betzig et al. 2006, Hess et al. 2006, Rust et al. 2006, Fölling et al. 2008). Stimulated emission depletion (STED) microscopy utilizes the saturation effect in stimulated emission to make the size of the donut hole small far beyond the diffraction limit (Hell and Wichmann 1994, Klar and Hell 1999).

It has been known that the nonlinear optical property of a sample can be utilized to improve the spatial resolution. Two-photon excitation of fluorescence and harmonic generation are well known optical phenomena that have been applied to biological imaging (Gannaway and Sheppard 1978, Denk et al. 1990). However, they do not provide a spatial resolution beyond the limit of conventional optical microscopy, since a long-wavelength excitation light, which is typically at the near-infrared region, is used for sample illumination. Therefore, the effective spatial resolution is equivalent or even worse than that in optical microscopy using visible light.

Although conventional two-photon excitation and harmonic generation were not applied to super-resolution imaging in practice, the nonlinear relation between the illumination intensity and the induced optical effect can essentially provide the spatial resolution beyond the limit given by the microscope optics and the wavelength used for sample illumination. Therefore, there are several attempts for inducing nonlinear optical responses by using visible light. As one of such approaches, saturation effects seen in fluorescence excitation have been utilized for improvement of the spatial resolution (Enderlein 2005, Fujita et al. 2007, Humpolíčková et al. 2009). Fluorophores have a nonzero lifetime of the excitation state (~ns) and, therefore, the excitation efficiency becomes nonlinear when a large number of photons are incident to the fluorophores. Transitions to triplet or a dark state of fluorophores also result in a nonlinear fluorescence response, since the number of excitable fluorophore decreases due to the transition induced during excitation and emission cycles in continuous light irradiation. Those phenomena have been utilized in laser-scanning fluorescence microscopy, and fluorescence imaging with a sub-diffraction-limited resolution has been demonstrated (Enderlein 2005, Fujita et al. 2007, Humpolíčková et al. 2009).

In this chapter, the principle of saturated excitation (SAX) microscopy and its application to 3D cell culture imaging are introduced. In addition to the resolution improvement, nonlinearity-based optical microscopy has an advantage in observation of thick samples. The localization of fluorescence signal in an illumination focal volume, which is well investigated in two-photon excitation microscopy, suppresses fluorescence emission from out-of-focus planes and maintains a high image contrast even in the observation of a deep part of a sample. This advantage can also be expected in the nonlinearity-based super-resolution techniques. All the biological systems in nature consist of varieties of 3D structures from cellular to organ levels, and it is important to image these structures in 3D in order to study biology or medicine in more natural conditions. For those purposes, the nonlinearity-based super-resolution technique can be an effective tool owing to its advantage in the depth imaging.

15.2 The Principle of SAX Microscopy

SAX microscopy is based on laser-scanning confocal microscopy, in which fluorescent probes in a sample are excited by a focused laser light and fluorescence emission from the probes is detected while the laser focus scans the sample (Fujita et al. 2007). SAX microscopy utilizes the saturation effect seen in fluorescence excitation. Figure 15.1 shows a relationship between the excitation intensity and fluorescence signal of Rhodamine 6G solution. As seen in Figure 15.1, the excitation-emission relation becomes nonlinear at the high intensity of excitation light. This nonlinear fluorescence response can be seen prominently at high-intensity region, especially at the center of the laser focus. Therefore, extracting the nonlinear fluorescence signal can resolve structures smaller than the focal size.

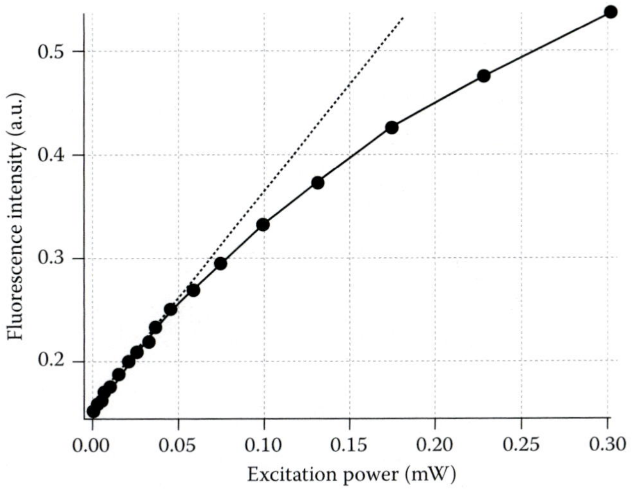

Figure 15.1

Fluorescence excitation and emission relationship measured by exciting Rhodamine B solution with light at a wavelength of 488 nm.

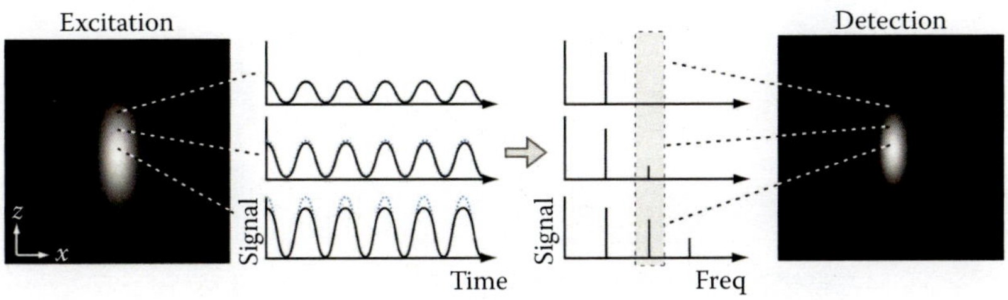

Figure 15.2

Schematic to show the resolution improvement in SAX microscopy.

Figure 15.2 shows the imaging scheme of SAX microscopy. In SAX microscopy, the excitation intensity is temporally modulated at a frequency (f) and the fluorescence signal is demodulated at the harmonic frequencies ($2f$, $3f$, ...). Since the saturation in fluorescence excitation causes a distortion in the fluorescence modulation, fluorescence signals induced under the saturation condition have high-harmonic frequencies, which can be simply extracted by lock-in detection.

Figure 15.3 shows an optical setup of SAX microscopy using the temporal modulation of excitation for extracting the nonlinear fluorescence response. The optical setup is based on that of a typical confocal microscope, and the introduction of an excitation modulator and a lock-in amplifier realizes a SAX microscope. In Figure 15.3, two acousto-optic modulators (AOMs) are used to produce excitation modulation at a single frequency. The two AOMs are driven at different frequencies and cause Doppler shift to the incident beam at the diffraction. The interference of beams diffracted by each AOM presents the excitation intensity modulated at a single frequency. This modulated laser beam can also be used to confirm the linearity of the measurement system. The fluorescence signal is detected by a photomultiplier tube and sent to a lock-in amplifier for

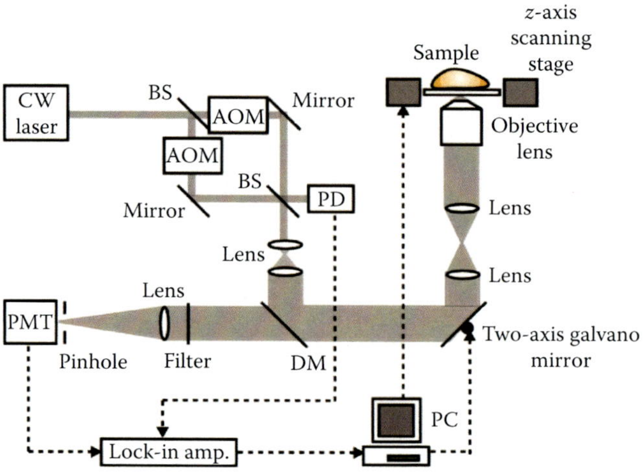

Figure 15.3

Optical setup of SAX microscopy.

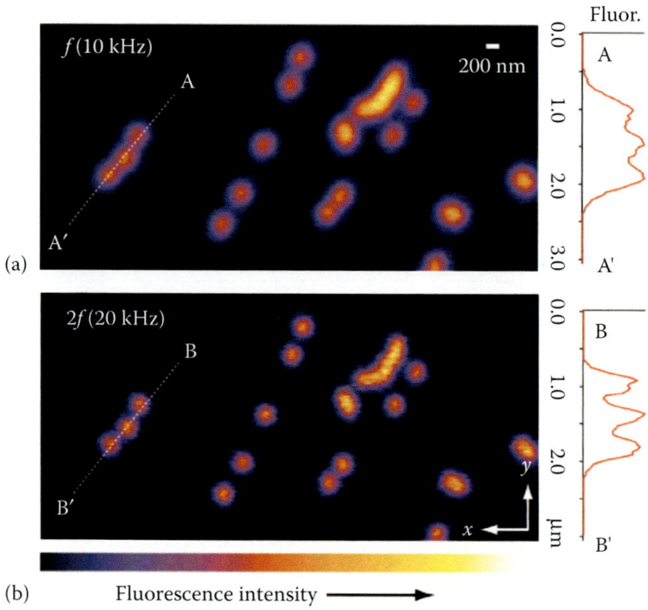

Figure 15.4

Fluorescence images of fluorescence beads measured by SAX microscopy with the demodulation at (a) the fundamental (*f*) and (b) the second harmonic frequency (2*f*). The diameter of a bead is approximately 200 nm. (Reprinted with permission from Fujita, K. et al., *Phys. Rev. Lett.*, 99, 228105. Copyright 2007 by the American Physical Society.)

demodulation. Changing the demodulation frequency switches the modes of fluorescence imaging from confocal to SAX microscopy.

Figures 15.4 through 15.6 shows fluorescence images obtained by SAX microscopy and conventional confocal microscopy for comparisons (Fujita et al. 2007, Yamanaka et al. 2013a). These experimental results show the clear improvement of the spatial resolution by introducing SAX in confocal microscopy.

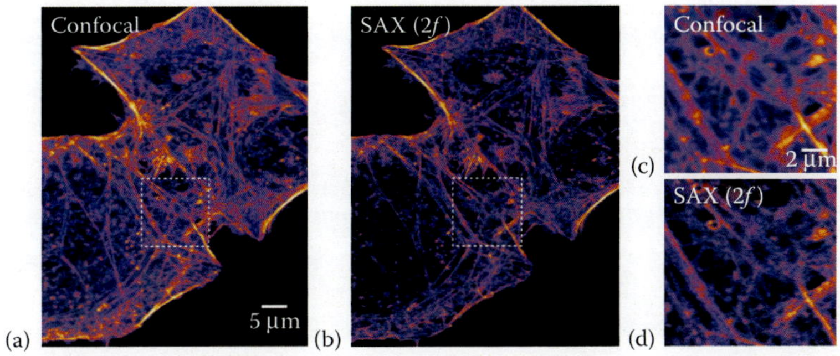

Figure 15.5

Fluorescence images of actin in HeLa cells stained with ATTO Rho6G and obtained by (a) confocal and (b) SAX microscopy, with the demodulation at the second harmonic frequency (2*f*). (c) and (d) show the magnified views of the boxed areas in (a) and (b), respectively.

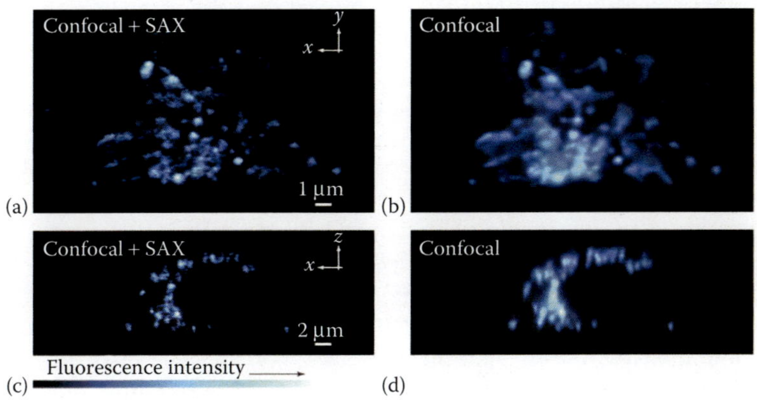

Figure 15.6

Fluorescence images of Golgi apparatus expressing EGFP in the focal plane (*x*-*y*) and along the optical axis (*x*-*z*), obtained by (a, c) SAX and (b, d) confocal microscopy.

Any fluorescent probe shows the saturation effect in excitation and can be basically used in SAX microscopy (Yamanaka et al. 2008). However, the high excitation intensity that introduces the saturation may cause strong photobleaching. Therefore, the photostability of the fluorescent probe is an important factor for better imaging quality. It is confirmed that ATTO Rho6G and Rhodamine 6G are effective in observing fixed cells (Yamanaka et al. 2013b) and EGFP for live cells (Yamanaka et al. 2013a).

Since the signal localization is in the 3D space, SAX microscopy is also effective in 3D imaging of samples. Figure 15.7 shows a comparison of fluorescence images obtained by conventional confocal and SAX microscopy (Yonemaru et al. 2014). The 3D structures in the cell can be better observed with higher contrast in the SAX microscopy image.

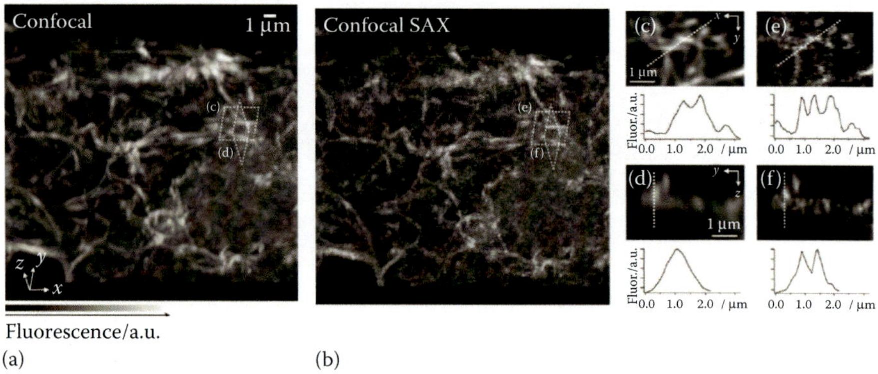

Figure 15.7

Projection images of actin in HeLa cells stained with phalloidin-conjugated ATTO Rho6G. The images were constructed from 3D image data sets measured by (a) confocal and (b) SAX microscopy. (c–f) show the images and the line profiles of the *x*-*y* and *y*-*z* planes in (a) and (b) for a comparison of the spatial resolutions. (From Yonemaru, Y. et al.: Saturated excitation microscopy with optimized excitation modulation. ChemPhysChem. 2014. 15. 743. Copyright Wiley-VCH Verlag GmbH & Co. KGaA. Reproduced with permission.)

15.3 The Spatial Resolution in SAX Microscopy

The spatial resolution in SAX microscopy can be described using a point spread function (PSF). Similar to confocal microscopy, the effective PSF of SAX microscopy is given as the product of excitation and detection PSFs (Equation 15.1) (Gu 1996). The detection PSF is determined by the detection optics (the same PSF for conventional fluorescence imaging). On the other hand, the excitation PSF presents a complicated form, since the excitation efficiency is nonlinear to the excitation intensity, and the nonlinearity is valiant in the 3D space. Therefore, it is important to start from considering the fluorescence response under various excitation intensities.

$$h_{\mathrm{SAX}} = h_{\mathrm{ex}} h_{\mathrm{det}} \tag{15.1}$$

To describe the nonlinear fluorescence response that contributes to the resolution improvement, we model fluorescence excitation and emission by using a five-level model that describes the transition of fluorophores between different energy states (Figure 15.8a) (Kawano et al. 2011). In this model, the fluorescence signal is proportional to the population of the single-excitation state (S_1) and the excitation intensity proportionally increases the transition from the ground state (S_0) to the single-excitation state (S_1). In addition to these transition essentials to fluorescence emission, there are several pathways that fluorophores can take, that are, the transition to triple-excitation state (T_1 and T_n), excitation to the upper states ($S_{>1}$ and S_n), and photobleaching. The rates for these transitions were well investigated for Rhodamine 6G (Eggeling et al. 2005), and we discuss the imaging property of SAX microscopy for Rhodamine 6G by using the five-level model.

Equation 15.2 describes the populations of molecules at the ground and the excitation states in the five-level model in Figure 15.8a. The relationship between excitation and emission can be calculated by solving the equations under different excitation intensities (Eggeling et al. 2005). Figure 15.8b shows a result of

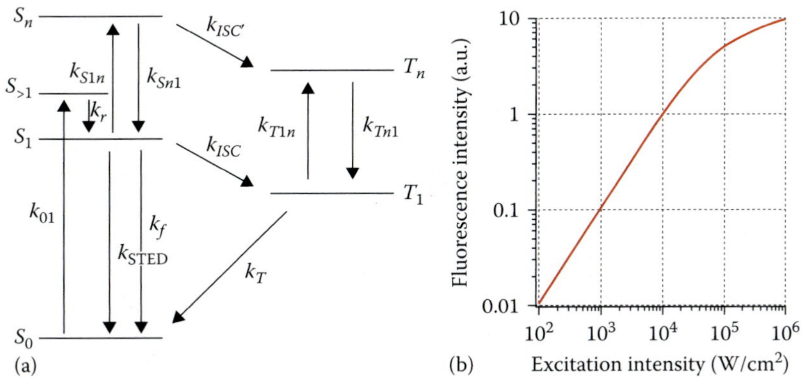

Figure 15.8

(a) Five-level molecular electronic state model. S_0: the ground state, S_1: the singlet first excitation state, $S_{>1}$: the vibrational state of first singlet state, S_n: the nth singlet state, T_1: the first triplet state, T_n: the nth triplet state. (b) The relationship between fluorescence and excitation intensity for Rhodamine 6G molecules calculated using the five-level molecular electronic state model.

fluorescence response calculated for excitation of Rhodamine 6G. A linear fluorescence response is seen at the low excitation intensity, and the slope of the curve becomes smaller at the high excitation owing to the saturation effect.

$$\begin{aligned}
\frac{d}{dt}S_0 &= -k_{01}(t)S_0 + (k_{SE}+k_f)S_1 + k_T T_1 \\
\frac{d}{dt}S_1 &= -(k_{SE}+k_f+k_{ISC}+k_{1Sn}+k_b)S_1 + k_r S_{>1} + k_{Sn1}S_n \\
\frac{d}{dt}S_{>1} &= k_{01}(t)S_0 - k_r S_{>1} \\
\frac{d}{dt}T_1 &= k_{ISC}S_1 - (k_T + k_{1Tn} + k_b)T_1 + k_{Tn1}T_n \\
\frac{d}{dt}T_n &= k_{ISC}S_n + k_{1Tn}T_1 - (k_{Tn1}+k_{bn})T_n
\end{aligned} \tag{15.2}$$

Fluorescence signals under modulated excitation can be calculated by introducing temporally modulated excitation intensity into the rate equations. Applying Fourier transform to the modulated fluorescence signal, the relationships between the excitation and demodulated fluorescence signals are obtained, as show in Figure 15.9a. The calculation results show that the harmonic demodulation can extract the nonlinear responses given by the excitation saturation. The higher the modulation frequency becomes, the higher the order of the nonlinear signal that is extracted, which indicates that the spatial resolution can be improved further by demodulating the fluorescence signal at a higher frequency. The relationships of demodulated fluorescence signal and excitation measured for Rhodamine 6G solution (Figure 15.9b) also confirm that the high frequency demodulation can extract high-order nonlinear components from fluorescence signal induced under the excitation saturation.

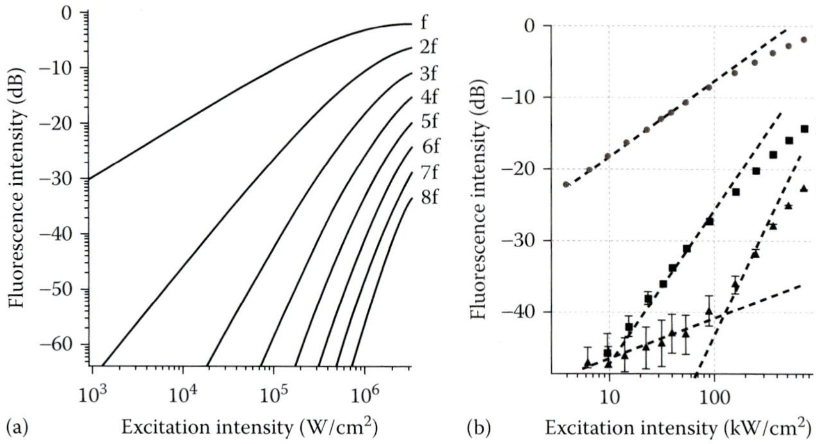

Figure 15.9

(a) Calculated and (b) experimentally measured relationship between the excitation intensity and the demodulated fluorescence signals for Rhodamine 6G molecules excited by a CW laser with a wavelength of 532 nm. The lateral axis shows the average of the modulated excitation intensity. The modulation frequency (*f*) is 10 kHz for both plots. (Reprinted with permission from Fujita, K. et al., *Phys. Rev. Lett.*, 99, 228105. Copyright 2007 by the American Physical Society; Yamanaka, M. et al, *J. Biomed. Opt.*, 13, 050507. Copyright 2008 by Society of Photo Optical Instrumentation Engineers.)

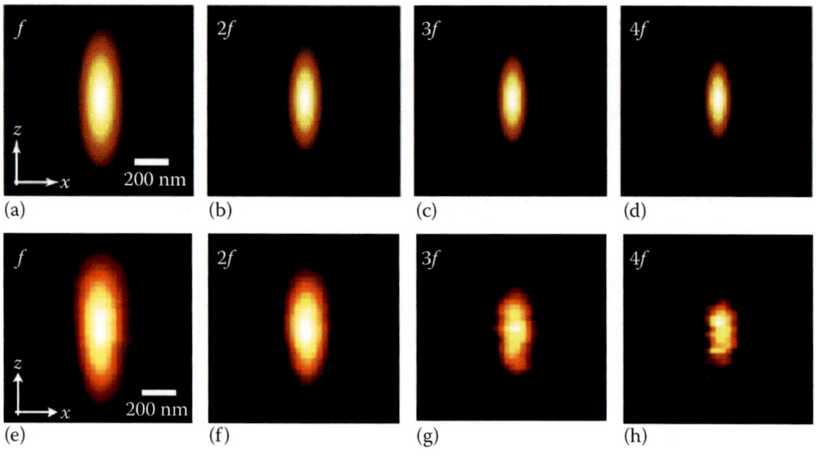

Figure 15.10

(a) Calculated effective PSFs in SAX microscopy, demodulated at the fundamental (*f*) and the harmonic frequencies (2*f*, 3*f*, and 4*f*). The excitation intensities for *f*, 2*f*, 3*f*, and 4*f* are 3.5, 9, 17, and 25 kW/cm^2, respectively. (b) Fluorescence images of a single fluorescent nanodiamond (diameter: ~100 nm). The images were measured by demodulating at the fundamental frequency (e: 10 kHz) and the harmonic frequencies (f: 20 kHz, g: 30 kHz, and h: 40 kHz).

Figure 15.10a shows the effective PSFs estimated by calculating the demodulated signal intensity at each position around the laser focus (Fujita et al. 2007, Yamanaka et al. 2011). As indicated above, the excitation efficiency of the demodulated fluorescence around the laser focus shows the strong dependence on the demodulation frequency, and the resolution improvement becomes more

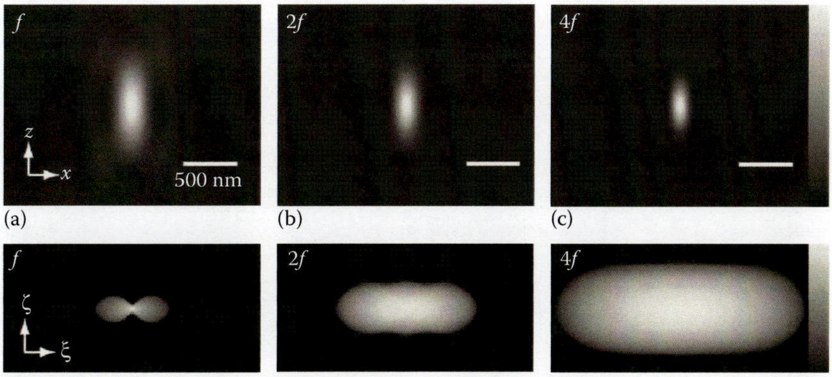

Figure 15.11

Excitation PSFs with demodulation at frequencies of (a) f, (b) $2f$, and (c) $4f$, along with their corresponding OTFs (d–f). The OTFs are shown with logarithmic gray scale. (Copyright 2011 The Japan Society of Applied Physics.)

prominent with demodulation at higher frequency. This result simply reflects the fact that the nonlinear fluorescence response is induced at positions with high excitation intensity, which, in this case, is the center of the laser focus.

Above calculations indicate that the spatial resolution can be improved by simply increasing the demodulation frequency. However, in practical conditions, the signal-to-noise ratio (SNR) in fluorescence detection determines the upper limit of the spatial resolution. As shown in Figure 15.10a, the higher harmonic signal can be a few magnitudes smaller than the signal at the fundamental frequency. So far, by using fluorescent probes with strong photostability, such as fluorescent nanodiamonds, and a long-time exposure, the 4th-order nonlinear response has been extracted and used for imaging (Figure 15.10b) (Yamanaka et al. 2011). For organic fluorescent probes, $3f$ would be the limit for demodulation frequency in practical conditions for imaging biological samples with organic probes.

To clarify the theoretical limit of the spatial resolution, considering the optical transfer function (OTF) of the imaging system is essential. The OTF of SAX microscopy can be given by applying Fourier transform to the PSF. Figure 15.9 shows the excitation OTFs calculated with different demodulation frequencies. The OTFs show that the cut-off frequency is expanded by increasing the demodulation frequency. The excitation OTF with demodulation at $2f$ and $4f$ has the cut-off frequency about twice and four times larger than that a conventional excitation OTF. In practical conditions, it is difficult to resolve sample structures to this limit owing to the limitation in the SNR. However, these results indicate that the deconvolution technique can improve the image contrast of the structures smaller than the diffraction limit. Note that Figure 15.11 shows only the excitation OTF, and the effective OTF is obtained by convolving the detection OTF, which is equivalent to Figure 15.10d, by assuming a same wavelength for excitation and emission.

15.4 The Frequency Response of Fluorescence Emission

Since fluorescent molecules exhibit the transitions between energy states in nano to a few hundred microseconds, the fluorescence signal under temporally modulated excitation shows different behaviors at different modulation frequencies.

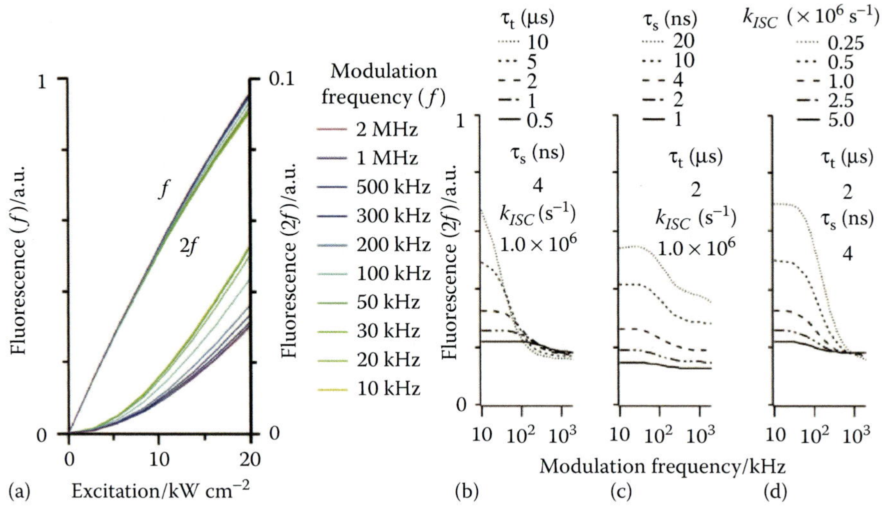

Figure 15.12

(a) Relationship between the averaged excitation intensity and fluorescence signal demodulated at fundamental (f) and 2nd-order harmonic ($2f$) frequency for a single Rhodamine 6G molecule (modulation frequencies: 10 kHz–2 MHz). (b–d) Relationships between modulation frequency and fluorescence intensity demodulated at $2f$ with different lifetimes of (b) the excited triplet (τ_t) state and (c) the excited singlet state (τ_s), and (d) different intersystem crossing rates (k_{ISC}). The excitation intensity used for all simulation was 10 kW/cm^2. (From Yonemaru, Y. et al.: Saturated excitation microscopy with optimized excitation modulation. *ChemPhysChem*. 2014. 15. 743. Copyright Wiley-VCH Verlag GmbH & Co. KGaA. Reproduced with permission.)

Therefore, estimating the dependence of harmonic signal intensity on the modulation frequency is necessary to increase the quality of SAX images. As reported by Gatzogiannis et al. (2011), the amplitude and phase of harmonic signals can be affected by the modulation frequency of excitation.

Figure 15.12a shows the fluorescence responses with fundamental (f) and second harmonic ($2f$) frequencies, calculated at different modulation frequencies (f) (Yonemaru et al. 2014). This calculation result illustrates that the strength of the second harmonic signal shows a clear dependence on the modulation frequency, and the modulation frequencies can be categorized into two groups according to the strength of the harmonic signals. Figures 15.12b, c, and d show the plots of the second harmonic signals, calculated to know the effects of the singlet and triplet excitation-state lifetimes and intersystem crossing. The comparison of the calculation results indicates that the frequency response is strongly affected by the triplet lifetime, and the modulation frequency can be determined using it. The signal strength is also affected by the singlet excitation-state lifetime and the efficiency of intersystem crossing, which are also important parameters in choosing a fluorescent probe for SAX imaging. Figure 15.13 shows the fluorescence responses measured by using Rhodamine 6G, ATTO 488, ATTO Rho6G, and phalloidin-conjugated ATTO Rho6G solution. The experimental results also show frequency responses similar to the calculation results.

Photobleaching is one of the important factors that determine the quality of fluorescence images. Figure 15.14 shows the comparison of photobleaching efficiencies given at different modulation frequencies. This experimental result did

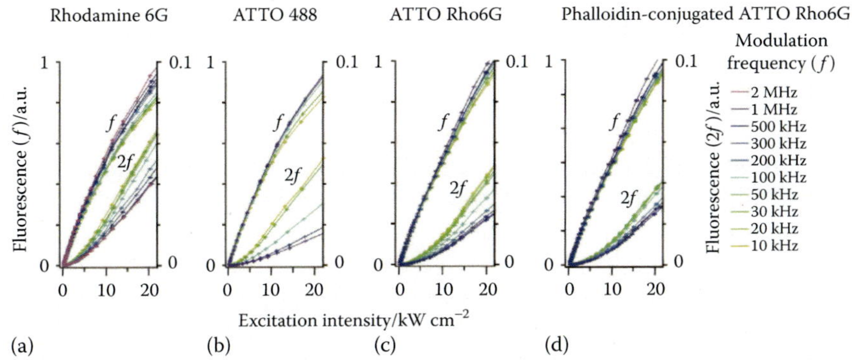

Figure 15.13

Experimental result of relationships between the averaged excitation intensity and the fluorescence signal demodulated at f and $2f$ with (a) Rhodamine 6G, (b) ATTO 488, (c) ATTO Rho6G, and (d) phalloidin-conjugated ATTO Rho6G solution (10 μM). (From Yonemaru, Y. et al.: Saturated excitation microscopy with optimized excitation modulation. *ChemPhysChem*. 2014. 15. 743. Copyright Wiley-VCH Verlag GmbH & Co. KGaA. Reproduced with permission.)

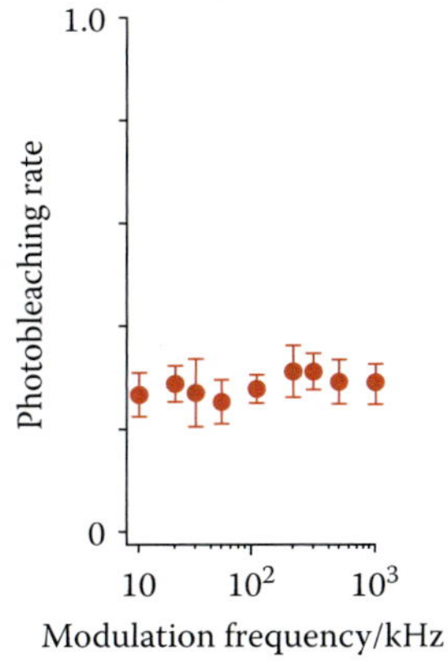

Figure 15.14

Relationship between photobleaching rate and modulation frequency. The photobleaching rate was calculated from the signal intensities of the first confocal image (I_1) and the second confocal image (I_2) as $(I_1-I_2)/I_1$. (From Yonemaru, Y. et al.: Saturated excitation microscopy with optimized excitation modulation. *ChemPhysChem*. 2014. 15. 743. Copyright Wiley-VCH Verlag GmbH & Co. KGaA. Reproduced with permission.)

not show a strong dependence of photobleaching on the modulation frequency in the examined range, indicating that the photobleaching efficiency is primarily determined by the light exposure. This actually gives a trade-off between the strength of harmonic signals and photobleaching, because the lower modulation frequency, which typically gives stronger harmonic signals, requires a longer exposure time, resulting in significant photobleaching. Therefore, it is important to choose a modulation frequency by considering this trade-off.

15.5 SAX Imaging of 3D-Cultured Cells

Cell culturing is inevitable in current biological researches. Typically, cells are cultured on a hard substrate, such as glass or plastic, and the cellular structure or responses to stimuli, such as protein expression and localization and its dynamics, are observed by optical microscopy. The observation results provide tremendous information about biological activities of cells and are used to clarify hypotheses of biology and medicine.

The recent study reported that the morphology of cells in culture is quite important to derive a reasonable answer from observation results. One of the important examples is that maturation of differentiated ES or iPS cells is well promoted when they form cell spheroids (Takayama et al. 2013). This fact clearly mentions that the morphology or surrounding environments affect cell activities and functions. It is also reported that there are differences in cell proliferation, viability, gene expression, and drug metabolism between cells cultured on a hard substrate (Pampaloni et al. 2007) (2D culture, e.g., Figure 15.15a) and those cultured in a gel (3D culture, e.g., Figure 15.15b). Therefore, in the near future, there will be a demand of high-resolution imaging of 3D-cultured cells in order to understand cellular activities in conditions close to the nature.

SAX microscopy utilizes nonlinear fluorescence responses induced by saturated excitation and can provide high image contrast, even from samples with large thickness and high fluorescence background. As described above, the noise by fluorescence signal from out-of-focus planes can be strongly suppressed by the harmonic demodulation. Figure 15.16 shows the comparison of the edge responses for the axial direction theoretically and experimentally obtained by conventional confocal and SAX microscopy (Yamanaka et al. 2013b); the depth discrimination capability of SAX microscopy does not show strong dependence on the pinhole size, indicating that the inherent background rejection effectively works to improve the image contrast of thick samples. Figure 15.17 shows the *x*-*z* images of fixed HeLa cells observed by conventional confocal and SAX microscopy with different pinhole sizes. The comparison clearly demonstrates the improvement of axial resolution and contrast by SAX microscopy. Even

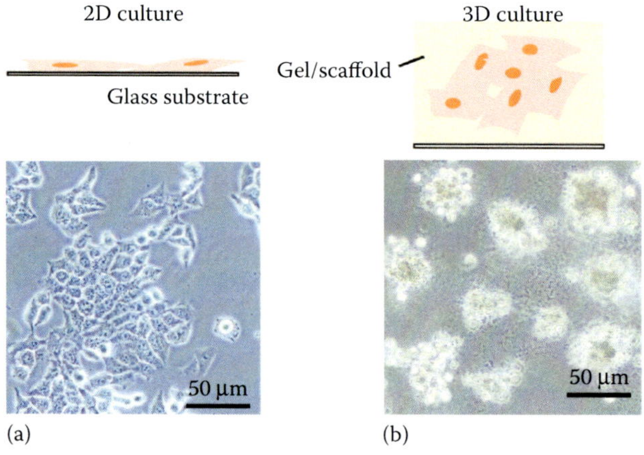

Figure 15.15

Schematic showing the differences between (a) 2D- and (b) 3D-cultured cells.

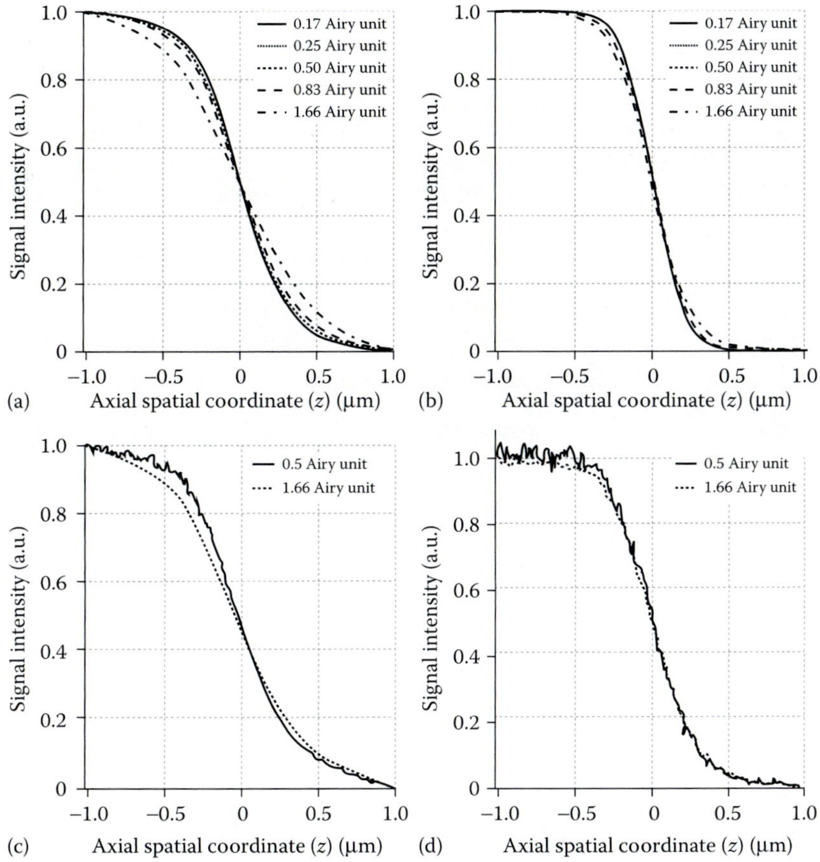

Figure 15.16

(a,b) Calculated edge responses for a border between a Rhodamine 6G solution and a glass substrate in (a) confocal and (b) SAX microscopy, with demodulations at $2f$. The calculation was performed for different pinhole diameters (0.17, 0.25, 0.5, 0.83, and 1.66 Airy units). (c,d) Experimental results for the edge response in (c) confocal and (d) SAX microscopy, with harmonic demodulation at $2f$ (20 kHz). The edge responses were measured with pinhole sizes of 0.5 and 1.66 Airy units. (Reprinted with permission from Yamanaka, M. et al, *J. Biomed. Opt.*, 18, 126002, 2013. Copyright 2008 by Society of Photo Optical Instrumentation Engineers.)

without a pinhole, SAX microscopy still exhibits the resolution in the axial direction, which is provided by the nonlinear nature, similar to two-photon excitation.

The images in Figure 15.18 show the results of observing 3D-cultured cells. The cells were cultured in a 3D matrix (BD matrigel) and the cells formed cell clusters; these cells were observed by the layer-by-layer scanning from the bottom. The 3D projections of the volume images are shown in the figures. The comparison of SAX and confocal images shows that the details of intracellular structures in the cell clusters are visualized with higher resolution and contrast in the observations by SAX microscopy. These observations also show the clear difference of the protein distribution in the cell clusters and the cells cultured on glass substrate. For more detailed comparison, the x-z cross section images obtained by conventional confocal and SAX microscopy are shown in Figure 15.19.

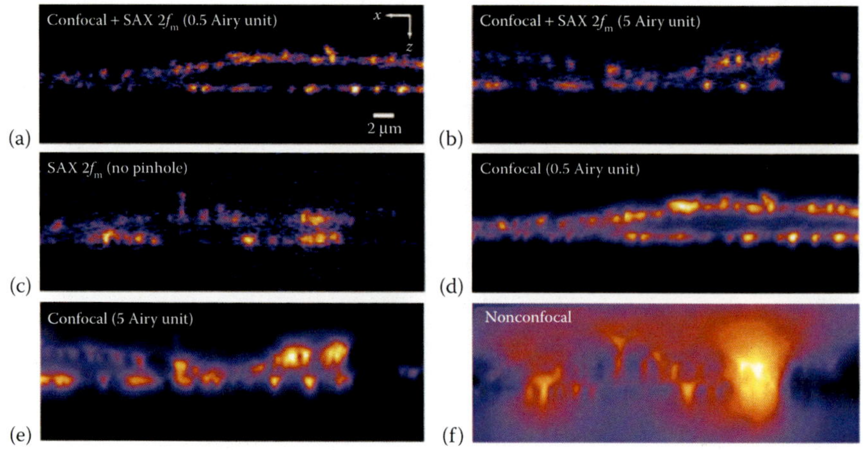

Figure 15.17

Fluorescence images of HeLa cells stained with ATTO Rho6G phalloidin by (a–c) SAX microscopy demodulated at 2*f*, (d, e) confocal microscopy, and (f) nonconfocal microscopy. (Reprinted with permission from Yamanaka, M. et al, *J. Biomed. Opt.*, 18, 126002, 2013. Copyright 2008 by Society of Photo Optical Instrumentation Engineers.)

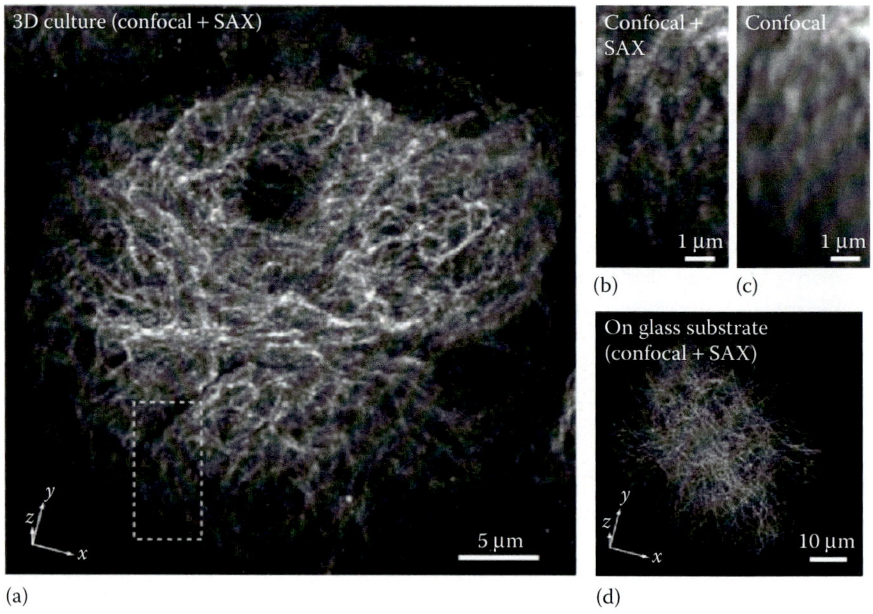

Figure 15.18

3D projection of HeLa cells in 3D culture. (a) SAX image of alpha-tubulin. (b) The magnification of the boxed area in (a). (c) Confocal image of the same area in (b). (d) SAX image of alpha-tubulin in HeLa cells on the glass substrate. *(Continued)*

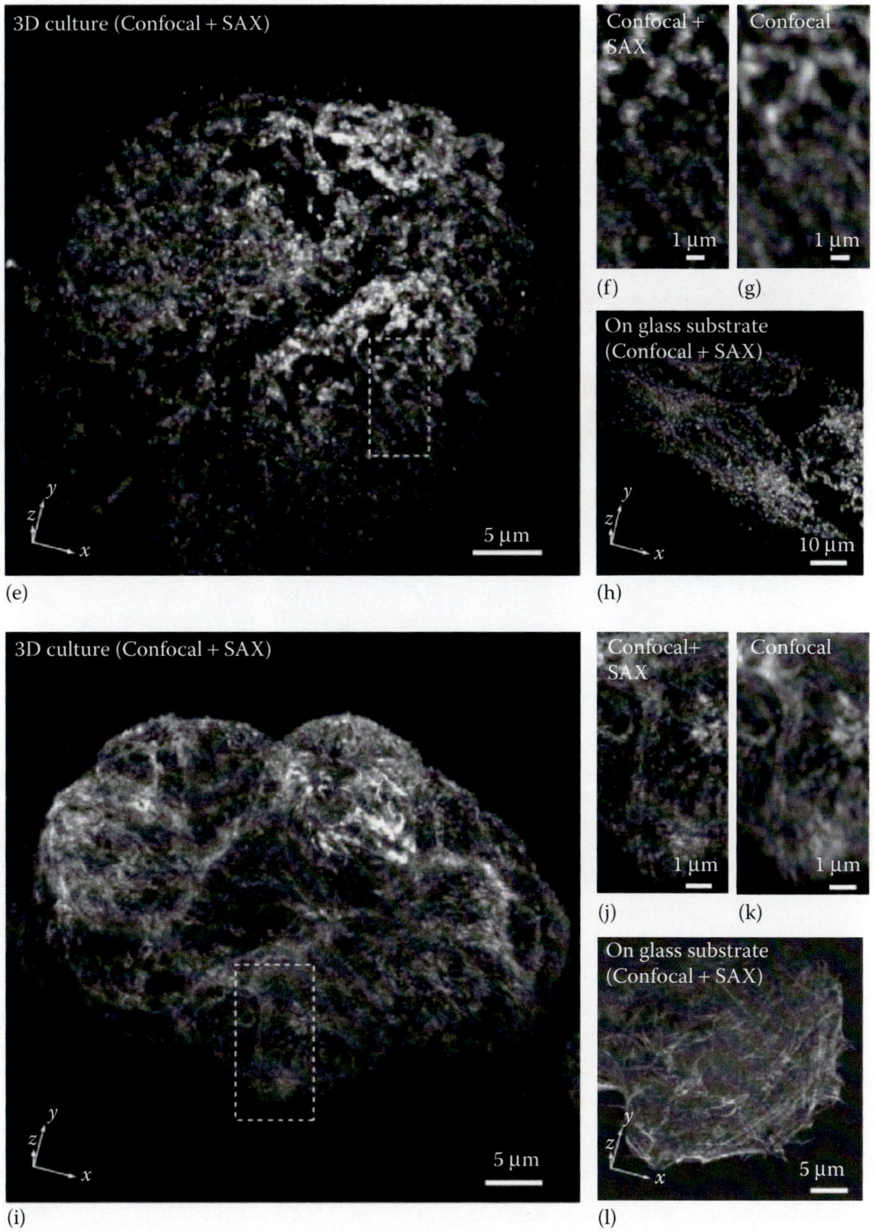

Figure 15.18 (Continued)

3D projection of HeLa cells in 3D culture. (e–l) Same data series for mitochondria and actin in HeLa cells. The SAX images were obtained with demodulation at the 2nd-harmonic frequency (100 kHz). (Reprinted with permission from Yamanaka, M. et al, *J. Biomed. Opt.*, 18, 126002, 2013. Copyright 2008 by Society of Photo Optical Instrumentation Engineers.)

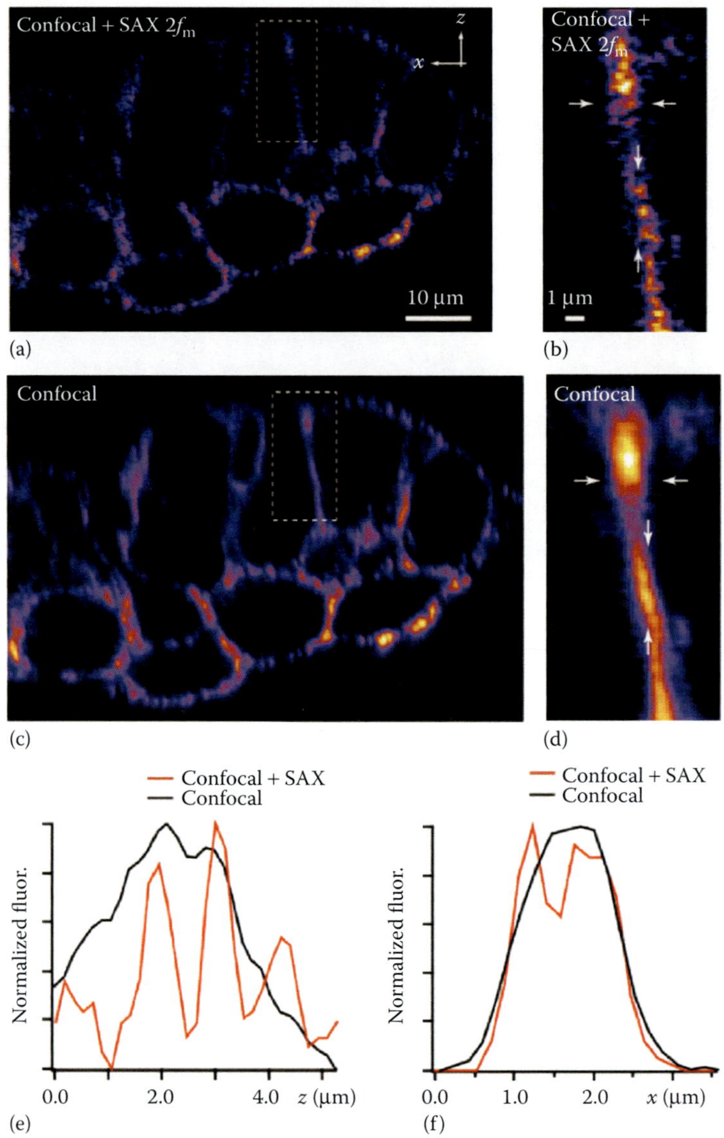

Figure 15.19

Fluorescence images of HeLa cells cultured in 3D matrix and the line profiles of the structures. (a) SAX and (c) confocal images of actin in HeLa cells along the optical axis (*x-z*). (b) and (d) Magnifications of the boxed area in (a) and (c). The line profiles for the structure indicated by the arrows in the adjacent images are shown in (e) and (f). The sample was stained with phalloidin-conjugated ATTO 488, and the excitation intensities for SAX and confocal microscopy were 12 and 1.2 kW/cm^2, respectively. A silicone-oil-immersion objective lens (1.3 NA) was used, with a pinhole size corresponding to 0.5 Airy units. Low-pass filtering was applied to the images. (Reprinted with permission from Yamanaka, M. et al, *J. Biomed. Opt.*, 18, 126002, 2013. Copyright 2008 by Society of Photo Optical Instrumentation Engineers.)

The sample was a cluster of HeLa cells, of which actin was stained by ATTO 488 phalloidin. The observation was performed from the bottom of the images. The comparison of the images in Figure 15.19 clearly shows the improvement of the spatial resolution in the SAX image throughout the imaging depth. The spatial resolution is degraded due to the spherical aberration for both imaging modes. Even in such a condition, the effect of saturated excitation in the resolution improvement is clearly observed.

15.6 Conclusions

In this chapter, the principle of SAX microscopy and the application to observation of 3D-cultured cells are described. Although the resolution improvement in SAX microscopy is limited to about twice of the conventional techniques, the strong suppression of the background fluorescence signal contributes effectively in the observation of the intracellular structures in detail. Another advantage of SAX microscopy is the simple configuration of the microscope system. The installation of the light modulator and a lock-in amplifier can easily realize a SAX microscope. When using a signal acquisition board (A/D board) fast enough to record the fluorescence modulation, the lock-in amplifier is not required, since the demodulation can also be done by a computer that controls a set of galvanometer mirrors.

The concept can also be applied to two-photon excitation microscopy, which realizes super-resolution in imaging deep pats of tissue samples. Basically, any saturable optical effect provides a chance to achieve the spatial resolution beyond the diffraction limit. So far, the saturations of light scatterings from metallic nanoparticles (Chu et al. 2014a, 2014b) and photo-induced acoustic waves (Danielli et al. 2014) have been utilized for super-resolution imaging.

Other than the saturation-based techniques mentioned above, techniques using Förster resonance energy transfer and photo-induced charge separation are proposed to induce the nonlinear responses of fluorescent probes, in which the repetitive single-photon excitation is employed (Hänninen et al. 1996, Chen and Cheng 2009, Mochizuki et al. 2015). Although there is no demonstration of 3D imaging so far, those techniques offer lower excitation intensity in the observation and are expected for 3D imaging with less photodamage. Since the resolution improvement is inherently provided by the probes, no signal or image processing with multiple images is required, which leads to the advantage in the imaging speed.

References

Betzig, E., Patterson, G. H., Sougrat, R. et al., Imaging intracellular fluorescent proteins at nanometer resolution, *Science* 313(2006):1642–1645.

Chen, J., Cheng, Y., Far-field superresolution imaging with dual-dye-doped nanoparticles, *Opt. Lett.* 34(2009):1831–1833.

Chu, S.-W., Su, T.-Y., Oketani, R. et al., Measurement of a saturated emission of optical radiation from gold nanoparticles: Application to an ultrahigh resolution microscope, *Phys. Rev. Lett.* 112(2014a):017402.

Chu, S.-W., Wu, H.-Y., Huang, Y.-T. et al., Saturation and reverse saturation of scattering in a single plasmonic nanoparticle, *ACS Photonics* 1(2014b):32–37.

Danielli, A., Maslov, K., Garcia-Uribe, A., Winkler, A. M., Li, C., Wang, L., Chen, Yu., Dorn II, G. W., Wang, L. V., Label-free photoacoustic nanoscopy, *J. Biomed. Opt.* 19(2014):086006.

Denk, W., Strickler J. H., Webb, W. W., Two-photon laser scanning fluorescence microscopy, *Science* 248(1990):73–76.

Eggeling, C., Volkmer, A., Seidal, C. A. M., Molecular photobleaching kinetics of rhodamine 6G by one- and two-photon induced confocal fluorescence microscopy, *ChemPhysChem* 6(2005):791–804.

Enderlein, J., Breaking the diffraction limit with dynamic saturation optical microscopy, *Appl. Phys. Lett.* 87(2005):094105.

Fujita, K., Kobayashi, M., Kawano, S., Yamanaka, M., Kawata, S., High-resolution confocal microscopy by saturated excitation of fluorescence, *Phys. Rev. Lett.* 99(2007):228105.

Fölling, J., Bossi, M., Bock, H. et al., Fluorescence nanoscopy by ground-state depletion and single-molecule return, *Nat. Methods* 5(2008):943–945.

Gannaway, J. N., Sheppard, C. J. R., Second-harmonic imaging in the scanning optical microscope, *Opt. Quant. Electron.* 10(1978):435–439.

Gatzogiannis, E., Zhu, X., Kao, Y.-T., Min, W., Observation of frequency-domain fluorescence anomalous phase advance due to dark-state hysteresis, *J. Phys. Chem. Lett.* 2(2011):461–466.

Gu, M., *Principles of Three Dimensional Imaging in Confocal Microscopes* (Singapore: World Scientific Publishing, 1996).

Hänninen, P. E., Lehtelä, L., Hell, S. W., Two- and multiphoton excitation of conjugate-dyes using a continuous wave laser, *Opt. Comm.* 130(1996):29–33.

Hell, S. W., Wichmann, J., Breaking the diffraction resolution limit by stimulated emission: Stimulated-emission-depletion fluorescence microscopy, *Opt. Lett.* 19(1994):780–782.

Hess, S. T., Girirajan, T. P. K., Mason, M. D., Ultra-high resolution imaging by fluorescence photoactivation localization microscopy, *Biophys. J.* 91(2006):4258–4272.

Hofmann, M., Eggeling, C., Jakobs, S., Hell, S. W., Breaking the diffraction barrier in fluorescence microscopy at low light intensities by using reversibly photoswitchable proteins, *Proc. Natl. Acad. Sci. USA* 102(2005):17565–17569.

Humpolíčková, J., Benda, A., Enderlein, J., Optical saturation as a versatile tool to enhance resolution in confocal microscopy, *Biophys. J.* 97(2009):2623–2629.

Kawano, S., Smith, N. I., Yamanaka, M., Kawata, S., Fujita, K., Determination of the expanded optical transfer function in saturated excitation imaging and high harmonic demodulation, *Appl. Phys. Express* 4(2011):042401.

Klar, T. A., Hell, S. W., Subdiffraction resolution in far-field fluorescence microscopy, *Opt. Lett.* 24(1999):954–956.

Mochizuki, K., Shi, L., Mizukami, S. et al., Nonlinear fluorescence imaging by photoinduced charge separation, *Jpn. J. Appl. Phys.* 54(2015):042403.

Pampaloni, F., Reynaud, E. G., Stelzer, E. H. K., The third dimension bridges the gap between cell culture and live tissues, *Nat. Rev. Mol. Cell Biol.* 8(2007):839–845.

Rust, M. J., Bates, M., Zhuang, X., Sub-diffraction-limit imaging by stochastic optical reconstruction microscopy (STORM), *Nat. Methods* 3(2006):793–795.

Takayama, K., Kawabata, K., Nagamoto, Y., Kishimoto, K., Tashiro, K., Sakurai, F., Tachibana, M., Kanda, K., Hayakawa, T., Kusuda Furue, M., Mizuguchi, H., 3D spheroid culture of hESC/hiPSC-derived hepatocyte-like cells for drug toxicity testing, *Biomaterials* 34(2013):1781–1789.

Yamanaka, M., Kawano, S., Fujita, K., Smith, N. I., Kawata, S., Beyond the diffraction-limit biological imaging by saturated excitation microscopy, *J. Biomed. Opt* 13(2008):050507.

Yamanaka, M., Saito, K., Smith, N. I. et al., Saturated excitation of fluorescent proteins for subdiffraction-limited imaging of living cells in three dimensions, *Interface Focus* 3(2013a):20130007.

Yamanaka, M., Tzeng, Y.-K., Kawano, S. et al., SAX microscopy with fluorescent nanodiamond probes for high-resolution fluorescence imaging, *Biomed. Opt. Express* 2(2011):1946–1954.

Yamanaka, M., Yonemaru, Y., Kawano, S. et al., Saturated excitation microscopy for sub-diffraction-limited imaging of cell clusters, *J. Biomed. Opt.* 18(2013b):126002.

Yonemaru, Y., Yamanaka, M., Smith, N. I., Kawata, S., Fujita, K., Saturated excitation microscopy with optimized excitation modulation, *ChemPhysChem* 15(2014):743–749.

16

Quantitative Super-Resolution Microscopy for Cancer Biology and Medicine

Alec Peters, Andrew Nickerson, Li-Jung Lin, and Xiaolin Nan

16.1 Introduction
16.2 Quantitative Super-Resolution Fluorescence Microscopy
16.3 Visualizing Raf Multimer Formation in Cells with Quantitative PALM
16.4 Imaging Ras-Raf Interactions at the Nanoscale with BiFC-PALM
16.5 Concluding Remarks
Acknowledgment
References

16.1 Introduction

16.1.1 The Ras-Raf-MAPK Signaling Pathway in Cancer

The Ras-Raf-mitogen-activated protein kinase (MAPK) signaling pathway is a key regulator of cell growth and proliferation (Cox and Der 2010; Wellbrock et al. 2004), and its constitutive activation is among the most frequent abnormalities in cancer (Cox and Der 2010; Roberts and Der 2007; Schubbert et al. 2007) (Figure 16.1). Activation of the MAPK pathway can result from different upstream signals such as ligand binding to receptor tyrosine kinases (RTKs), activated Raf kinases, and GTP-bound Ras, all of which are highly pursued drug targets in cancer (Downward 2003; Roberts and Der 2007; Takashima and Faller 2013). However, efforts to inhibit these molecular targets for cancer therapy have yielded only limited success, urging more thorough investigations of the mechanisms underlying the signaling events in the MAPK pathway (McCormick 2007; Takashima and Faller 2013).

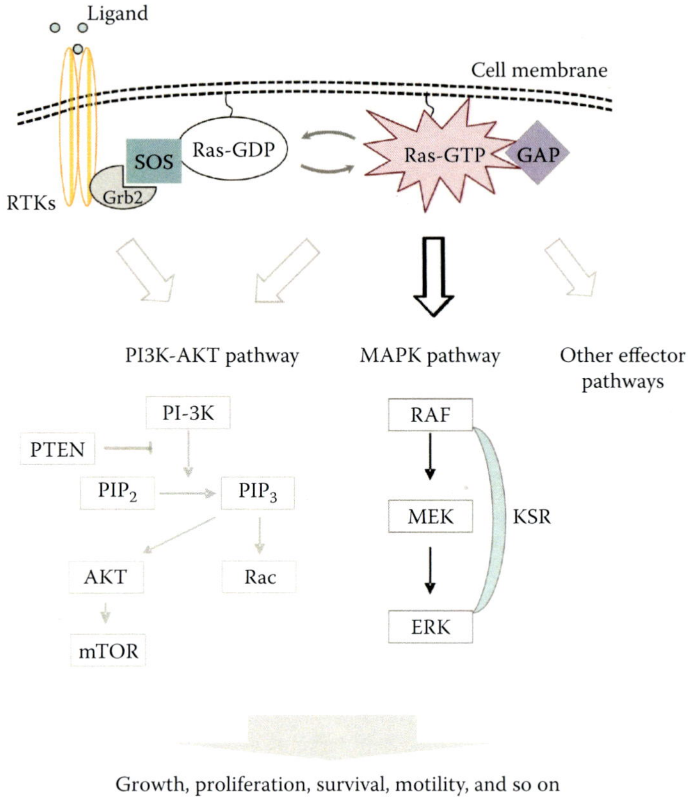

Figure 16.1

The Ras-Raf-MAPK signaling pathway.

RAS is by far the most frequently mutated oncogene in human cancers, with ~1/3 of all human tumors harboring activating *RAS* mutations (Schubbert et al. 2007; Roberts and Der 2007). The gene product, Ras, is a small GTPase residing on the inner leaflet of the plasma membrane, acting as a molecular switch by cycling between an inactive GDP-bound form and an active GTP-bound form. Binding of GTP causes conformational changes in the Ras protein that enable binding and activation of downstream effectors such as Raf and phosphoinositide 3-kinase (PI3K), through which Ras regulates many key aspects of cell physiology, including growth, proliferation, and survival, among others (Cox and Der 2010).

Mammals ubiquitously express four main Ras isoforms, H-, N-, KRas 4A, and KRas 4B, where KRas 4A and 4B are splicing variants; the 4B isoform is commonly referred to simply as KRas. These isoforms have nearly identical G-domains and primarily differ in the last ~20 amino acids at the C-terminus, also known as the hypervariable region (HVR) (Figure 16.2a). The last four amino acids in the HVR form a CAAX motif (C = cysteine; A = aliphatic residue; X = any amino acid), which undergoes posttranslational prenylation. Additional lipid modifications to other cysteine residues within the HVR can take place, depending on the sequence of each Ras isoform (Karnoub and Weinberg 2008). For example, HRas is dually palmitoylated at residues C181 and C184 and NRas is palmitoylated only at C181, whereas KRas is not palmitoylated (Figure 16.2b). These lipid modifications facilitate membrane targeting and may correlate with

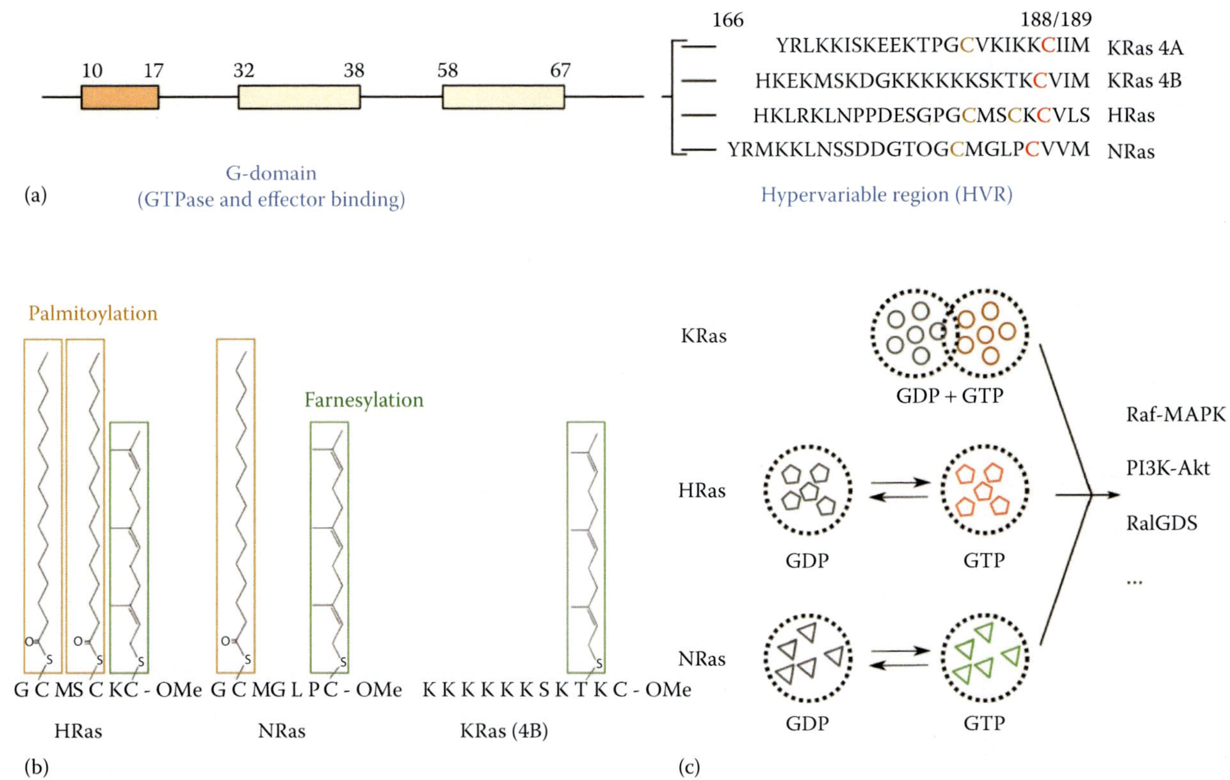

Figure 16.2

Domain structure (a), posttranslational modifications (b), and nanocluster-dependent signaling (c) of major Ras isoforms (H-, N-, and KRas).

the isoform-specific membrane partitioning, spatial clustering, and signaling specificity of Ras (Henis et al. 2009; Rocks et al. 2005) (Figure 16.2c).

Oncogenic mutations of Ras (including single amino acid substitutions at residues 12, 13, and 61) increase the population of Ras-GTP in cells and can lead to the deregulation of cell growth and proliferation, a key event in tumorigenesis (Gibbs et al. 1984; McCormick 1992). It is not surprising then that these mutations are commonly associated with human cancer. For example, *RAS* mutations are found in more than 95% of pancreatic ductal adenocarcinomas, which has a 5-year survival rate less than 5% after the initial diagnosis. Molecular agents to inhibit Ras activity may prove effective for a variety of cancers; currently, however, such agents with clinical efficacy are nonexistent, leaving mutant Ras an intractable therapy target (Cox et al. 2014).

Like *RAS*, *RAF* (specifically *BRAF*) is another oncogene in the MAPK pathway frequently associated with human cancers (Roberts and Der 2007; Wellbrock et al. 2004). For example, activating mutations of *BRAF* (such as the V600E mutation) are found in approximately 67% of melanomas (Davies et al. 2002). The three isoforms of Raf kinases, namely A-, B-, and CRaf, possess three conserved regions termed CR1, CR2, and CR3 (Figure 16.3a). The N-terminal CR1 region contains the Ras-binding domain (RBD) and a cysteine-rich domain. The CR2 region consists of regulatory residues, and the C-terminal CR3 region is the kinase (catalytic) domain. In resting conditions, wild-type Raf proteins are mostly held in an autoinhibitory conformation through intramolecular interactions. Raf is activated by binding (via the RBD) to active, GTP-loaded Ras at the cell membrane, followed by phosphorylation of residues in the CR2 region. Intriguingly, membrane localization, forced dimerization, or N-terminal truncation of Raf are each sufficient to activate Raf (at least in part) (Figure 16.3b). Somatic mutation of the kinase domain (CR3) also activates BRaf in particular, and is common in melanomas as well as in colorectal and thyroid cancers (Davies et al. 2002; Roberts and Der 2007). Of note, *BRAF* and *RAS* mutations are almost mutually exclusive in human tumors, possibly due to the functional redundancy in these mutations (Fernández-Medarde and Santos 2011).

Constitutively, active BRaf can lead to deregulation of the MAPK pathway and tumorigenesis through the phosphorylation of downstream effectors such as MEK and ERK (Collisson et al. 2012; Davies et al. 2002). Therapeutic strategies for tumors with mutant *BRAF* include small-molecule kinase inhibitors of Raf (RAFi) (Rahman et al. 2014; Roberts and Der 2007). To date, however, this strategy is only moderately effective in treating tumors with the $BRAF^{V600E/K}$ mutations, due to both initial and acquired resistance (Callahan et al. 2012; Lito et al. 2013; Solit and Rosen 2014). Tumors with $BRAF^{V600E}$ are initially sensitive to RAFi but can become insensitive to the treatment in a matter of weeks; the growth of $BRAF^{WT}$ tumors, on the other hand, is accelerated by the treatment, an adverse effect that has been shown to be mediated by paradoxical activation of the Raf-MAPK pathway (Figure 16.3b) (Hatzivassiliou et al. 2010; Poulikakos et al. 2010; Solit and Rosen 2014). In some cases, RAFi treatment even leads to the appearance of new tumors. All these factors limit the efficacy of Raf-targeting therapeutics and urge the development of more effective therapies.

16.1.2 Regulation of Ras and Raf Signaling through Multimer Formation

The limited success in therapeutic targeting of the Ras-Raf-MAPK signaling pathway in human cancers drives the search for additional mechanisms that regulate the signaling activities of Ras and Raf. Attempts to better understand

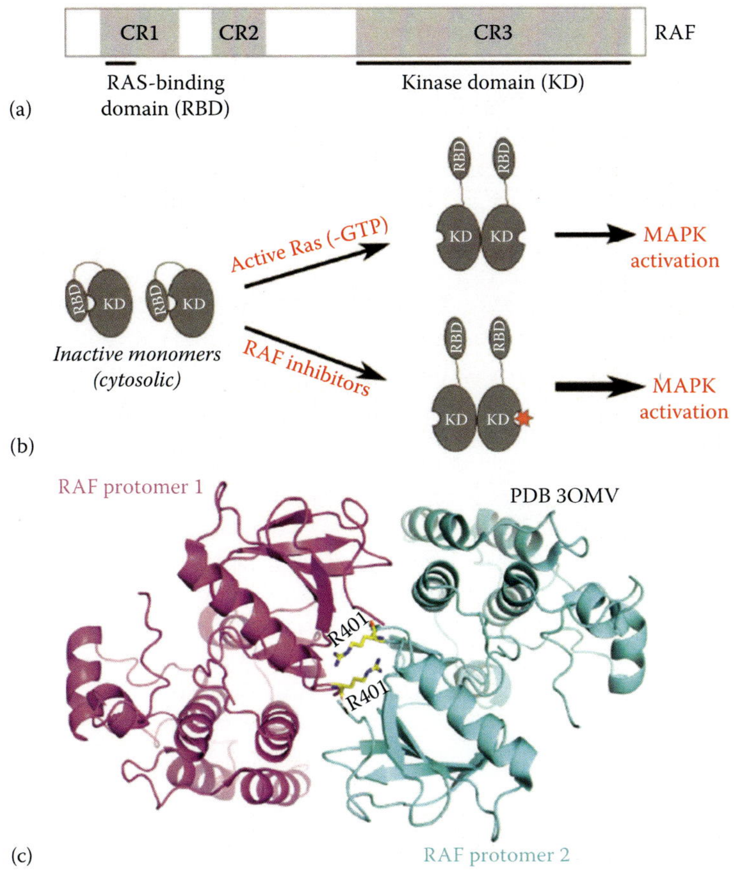

Figure 16.3

Domain structure (a), dimerization-dependent signaling (b), and kinase domain dimer structure (c) of the Raf kinase.

the cellular regulation of Ras and Raf have led to the discovery of spatial mechanisms, for example, the formation of higher-order structures (i.e., multimers), in determining the signaling activity and outcome of Ras-Raf-MAPK (Prior et al. 2003; Rajakulendran et al. 2009).

Raf kinases have been suggested to function as dimers in the early 1990s (Farrar et al. 1996; Luo et al. 1996), but the association between Raf dimer formation and cell signaling has not been appreciated until recently (Hatzivassiliou et al. 2010; Poulikakos et al. 2010, 2011; Solit and Rosen 2014). Both biochemical and structural studies have started to converge on a dimerization-dependent activation mechanism of Raf. Artificial dimerization of Raf was shown to activate Raf kinase and subsequently MAPK (namely MEK and ERK) (Farrar et al. 1996; Luo et al. 1996). Immunoprecipitation (IP) studies suggested the existence of Raf dimers in the presence of active Ras (Weber et al. 2001). Furthermore, X-ray crystallography studies of the BRaf catalytic domain (CatB) revealed a well-defined dimer interface between the CatB molecules (Rajakulendran et al. 2009). Amino acid residues at the dimer interface were identified to be critical to Raf dimerization (Figure 16.3b and c), and single amino acid substitutions of these residues (such as R401H in CRaf and R509H in BRaf) were shown to profoundly affect the kinase activity of Raf (Rajakulendran et al. 2009). The dimerization of Raf is also implicated in RAFi-mediated activation of Raf signaling in $BRAF^{WT}$ tumors (Hatzivassiliou et al. 2010; Poulikakos et al. 2010) and in the acquired resistance to RAFi in $BRAF^{V600E}$ tumors (Poulikakos et al. 2011).

Along with Raf, Ras has been shown to form higher-order structures termed nanoclusters on the cell membrane, and the formation of these nanoclusters appears to be an important event in the activation of Ras (Figure 16.2c) (Henis et al. 2009; Janosi et al. 2012). Ras nanoclusters were first revealed by immunoelectron microscopy (immuno-EM), where the plasma membranes of cells transiently overexpressing GFP-Ras were peeled off, followed by labeling with gold-conjugated antibodies specific to GFP and imaging with EM. Immuno-EM studies not only indicated that Ras forms higher-order structures on the plasma membrane but also suggested that the different Ras isoforms localize and form these higher-order structures in different areas of the membrane (Plowman et al. 2005; Prior et al. 2003; Tian et al. 2007).

Despite the strong evidence, the existence and involvement of Ras and Raf multimers in cellular processes have not been fully validated. X-ray crystallography studies utilize purified and truncated proteins, which may behave differently than the wild-type protein in whole cells. Immunoprecipitation (IP) experiments detect protein-protein interactions (PPIs) but do not measure the stoichiometry of these interactions. Immuno-EM has been extensively used to characterize Ras nanocluster formation on membrane peel-offs but not in whole or living cells. The lack of methodologies with which to resolve individual PPIs in whole cells on the nanometer scale is one of the main reasons that the relevance of Ras and Raf multimers in cell signaling has been obscure.

16.1.3 Super-Resolution Microscopy as a New Tool for Biological Imaging

The recent advent of single-molecule super-resolution imaging (SRM) techniques, such as photoactivated localization microscopy (PALM) (Betzig et al. 2006; Hess et al. 2006), stochastic optical reconstruction microscopy (STORM) (Rust et al. 2006; van de Linde et al. 2011), and their derivatives, overcome resolution barriers of conventional light microscopy to allow for the localization of individual fluorescent molecules on the nanometer scale. Both PALM and STORM operate using subdiffraction localization of individual fluorescent probes that stochastically switch on and off. PALM or STORM probes typically emit sufficient photons per switching cycle to allow localization to ~10 nm precision. Essentially, PALM and STORM imaging are simultaneous molecular enumeration and localization processes and are potentially capable of directly resolving protein complexes such as receptor clusters and Raf dimers in cells at detection sensitivities far greater than those of more conventional methods such as IP, while keeping the cell relatively intact when compared with immuno-EM.

The use of fluorescent tagging strategies in PALM and STORM also allows for the combination of existing fluorescence techniques with SRM. For example, bimolecular fluorescence microscopy has been widely used to visualize PPIs (Liu et al. 2014; Nickerson et al. 2014), an essential aspect of cell signaling; it can potentially be combined with PALM to image PPIs such as the interaction between Ras and Raf to determine their precise cellular location and protein stoichiometry.

In this chapter, PALM, in combination with spatial analysis, is shown to be capable of directly resolving protein homodimers and oligomers in whole cells at the nanometer spatial scale. This quantitative PALM approach was used to study Raf and Ras multimerization (Nan et al. 2013), suggesting that Raf forms dimers (and occasionally higher-order structures) when activated by mutant Ras, artificial membrane localization, or RAFis. PALM was combined with BiFC to visualize multimers formed by Ras-Raf complexes on the cell membrane (Nickerson et al. 2014). These results suggest that multimer formation is a key event in Ras and Raf signaling and demonstrate the utility of quantitative SRM techniques in studying basic and translational cancer medicine.

16.2 Quantitative Super-Resolution Fluorescence Microscopy

16.2.1 Considerations for Quantitative SRM

The goal of quantitative SRM is to both accurately localize and determine the stoichiometry (i.e., count) of the biological molecules of interest. This poses strict requirements on the biochemical and photophysical properties of the fluorescent probes. PALM and STORM use special probes such as photoactivated fluorescent proteins (PA-FPs) and photoswitchable organic fluorophores to achieve single-molecule localization on densely labeled biological samples (Dempsey et al. 2011; Fernández-Suárez and Ting 2008). The photophysical properties of the fluorescent probe will govern how each fluorescently tagged molecule appears in the imaging process and hence how authentically the final reconstructed image represents the underlying cellular structure of interest (Ha and Tinnefeld 2012).

To understand the challenges for quantitative super-resolution imaging, it is helpful to go over the PALM and STORM imaging process. PALM and STORM image acquisition is an iterative process in which a small, random subset of fluorescent molecules is activated to emit fluorescence at a time. Each activated fluorescent molecule produces a diffraction-limited and isolated image on the detector, the centroid of which can be localized to within 10–20 nm. Fluorescence from each activated fluorescent molecule is collected until the fluorescence is extinguished by permanent photobleaching or the molecule entering transient dark states (Annibale et al. 2011). Successive rounds of activation, imaging, and (apparent) photobleaching continue until all molecules in the sample are depleted. For quantitative, stoichiometry measurements with PALM and STORM, emission (also referred to as "localization") events from individual, fluorescently tagged target molecules should be spatially and/or temporally isolated from each other. When multiple target molecules reside in close proximity, such as those in a single nanocluster, it becomes difficult to distinguish them spatially even at ~10 nm localization precision; however, if the localization events from these molecules are temporally separated (i.e., emitting at distinct times during the image acquisition process), then it is still possible to assign individual (or a group of) localization events to individual molecules (Figure 16.4a).

The scenario described above poses requirements on the photophysical properties of the fluorescent probe, in addition to the generic requirements for SRM probes such as having a high photon yield. The probe should also be monomeric, so that false multimer signals do not arise from self-aggregation of probes. Last but not least, the fluorescent probe should not have long-lived dark states; that

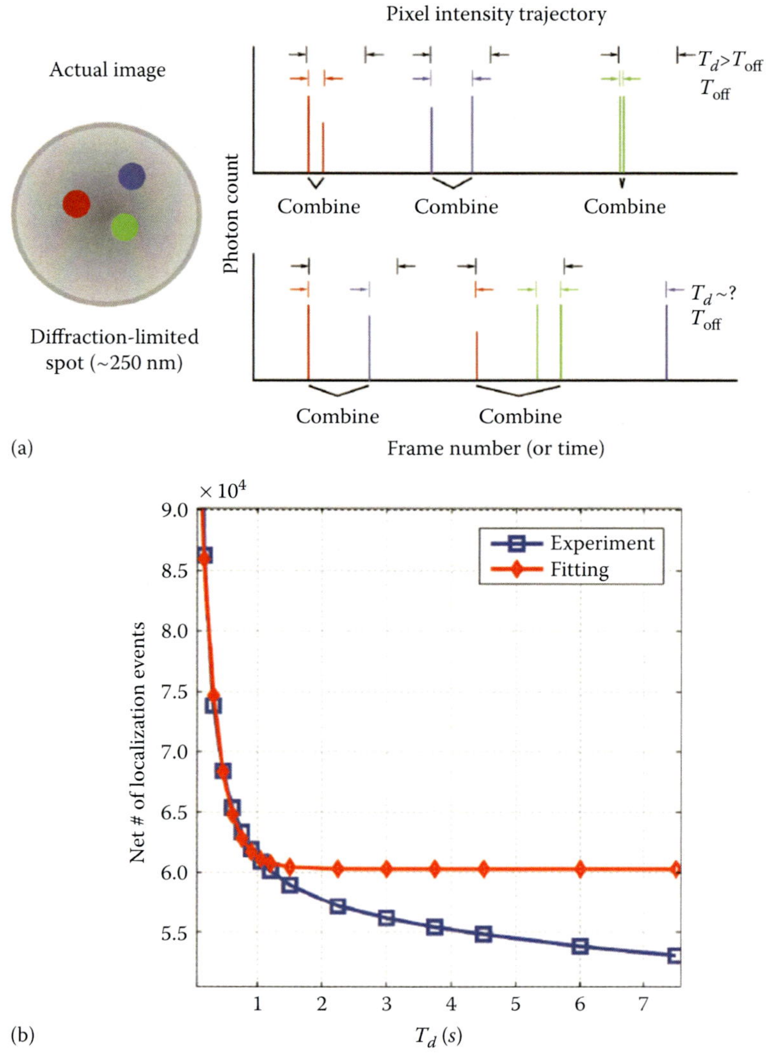

Figure 16.4

Quantitative SRM imaging on a densely labeled sample. (a) Localization events from fluorescent molecules residing in the same diffraction-limited spot (left) may be distinguishable if their dark state lifetime (T_{off}) is well defined and is much shorter than the average interval between emission events from individual molecules (a, top right). By combining localization events within T_d ($>T_{off}$) and a certain distance (e.g., 100 nm) into a single localization event, each combined localization event can be considered as one molecule. In the presence of long-lived dark states (T_{off} is large), however, this treatment becomes problematic, as indicated by the wrong grouping scheme of localization events from different molecules (bottom right); (b) Estimating T_{off} of PAmCherry1 by varying T_d and counting the resulting ("net") number of localization events.

is, the molecule should not switch on again after an extended period in a dark state. Ideally, the fluorescent probe should switch on only once—that is, emitting fluorescence continuously once activated until permanently photobleached.

Unfortunately, this is rarely achieved with current-generation PALM/STORM probes. To date, no organic fluorophores and few fluorescent proteins have been reported to be suitable for quantitative super-resolution microscopy. Emission from most PALM/STORM probes after initial photoactivation or photoswitching is interrupted by transient dark states (the fluorophore blinks); sometimes this can cycle tens or even hundreds of times. This complicates analysis, since during the dark periods of one fluorescent molecule, another molecule in close proximity may be switched on and may start to emit fluorescence. If two probes happen to be at a distance less than the localization accuracy and emit fluorescence in successive frames, it becomes difficult to distinguish the two as individual and separate fluorophores. This can be partially rectified by adjusting the photoactivation illumination intensity, so that the average time between switching of two closely spaced fluorescent probes is long compared with the lifetime of the dark states (T_{off}). In these circumstances, multiple emission events from the same molecule can be identified and combined by allowing the events to be separated by certain predictable transient durations of dark periods (T_d) (Figure 16.4a, right top). Evidently, this solution would fail if a significant population of the probe has long-lived dark states (Figure 16.4a, right bottom) (Annibale et al. 2011).

Few PALM/STORM probes, all of which are fluorescent proteins, have been reported to perform reasonably well for quantitative super-resolution microscopy (Durisic et al. 2014; Lee et al. 2012; Nan et al. 2013; Puchner et al. 2013; Sengupta et al. 2011). One such protein is PAmCherry1; this PA-FP has been a popular PALM probe in many studies, owing to its relatively high photon yield, monomeric nature, and irreversible photoactivation in bulk measurements (Subach et al. 2009). The utility of PAmCherry1 as a probe for quantitative PALM imaging was tested by expressing it in BHK21 cells and imaging the cells after fixation. PAmCherry1 was observed to enter transient dark states (blink), and the average lifetime of the dark states (T_{off}) was measured to be ~0.26 ± 0.03 s (Figure 16.4b). The interval of time between emissions of two PAmCherry1 molecules that are within 80 nm was estimated to be about 20 s. This suggests that there is a very low chance of observing two separate (but spatially close) PAmCherry1 molecules as one. To compensate for the blinking, multiple emission events from one PAmCherry1 molecule were combined if they occurred within ~1.3 s (i.e., 5 times the T_{off}) and ~80 nm (i.e., 4 times the localization error) before the final image was reconstructed.

The resulting composite PALM image showed that PAmCherry1 molecules existed as monomeric fluorophores randomly scattered throughout the cytoplasm, as expected. An example image is shown in Figure 16.5a, where each dot represents the center of mass of one PAmCherry1 molecule. Visual inspection of the image indicates that the PAmCherry1 molecules predominantly exist as monomers with very few "clusters" (most of which can be attributed to the random scattering of the molecules). A Ripley's K test (Dixon 2006; Ripley 1977) was used to quantitate clustering by counting the average number of PAmCherry1 particles found within a given distance to a specified particle and comparing this with predictions from a random spatial pattern. The test suggested that ~6% of the PAmCherry1 molecules were in dimers or higher-order structures (Figure 16.5b). These apparent PAmCherry1 clusters may result either from the presence of long-lived dark states that cause a single PAmCherry1 protein to give rise to multiple localization events or from a slight multimer formation propensity of PAmCherry1. In either case, the level of clustering is sufficiently small to permit the use of PAmCherry1 in quantitative PALM imaging such as assessing

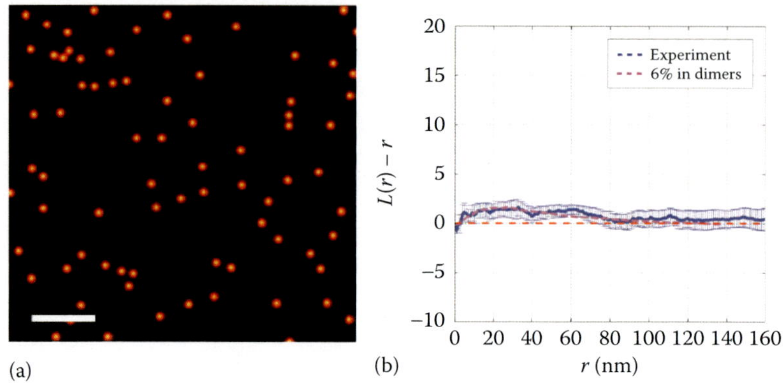

Figure 16.5

Imaging PAmCherry1 monomers expressed in BHK21 cells with PALM (a). Ripley's K test shows a slight clustering (multimerization) potential, and (b) simulations suggest that ~6% of PAmCherry1 molecules appear in clusters (multimers) in the PALM images. Scale bar: 200 nm.

biologically induced multimer formation. These results suggest that PAmCherry1 is a useful, albeit not ideal, fluorescent reporter molecule for quantitative PALM and may be used to measure protein multimerization in the cell.

16.2.2 Resolving Protein Multimers using Quantitative PALM

The efficacy of PAmCherry1 as a probe for quantitative PALM imaging of protein multimer formation was examined using artificial PAmCherry1-PAmCherry1 dimers, which are tandem PAmCherry1 peptides genetically fused by a flexible linker sequence. The PAmCherry1 dimer was overexpressed in BHK21 cells and was imaged using the same methodology as used to examine PAmCherry1 monomers. The resulting images suggest that a much larger fraction (when compared with PAmCherry1 monomer images) of the fluorescent molecules existed as dimers in cells (Figure 16.6a). Compared with about 6% of dimers in PAmCherry1 monomer images, PALM images of the PAmCherry1 artificial dimers are estimated (via visual inspection and a Ripley's K test) to have 77 ± 5% of fluorescent molecules existing as dimers or multimers (Figure 16.6b). Here, it is important to note that all multimers were assumed to be dimers in estimating the percentage of molecules in multimers (clusters), because the Ripley's K test is not as effective in quantitating cluster size distributions, despite its sensitivity to spatial clustering.

To measure the cluster size distribution of fluorescent probes in PALM images, a new spatial analysis algorithm based on DBSCAN (Ester et al. 1996) was developed. DBSCAN has been extensively used in spatial analysis, for example, in identifying individual "clusters" in PALM images (Figure 16.6c). DBSCAN alone, however, was inaccurate in reporting the cluster size distribution when there is significant positional noise in the image, as shown on simulated data (Nan et al. 2013). A new method combining computer simulations with DBSCAN, coined simulation-aided DBSCAN (SAD), was used to assess particle density distribution in PALM images. SAD addresses the problem of positional noise in an image by using computer-generated data sets that contain the same positional noise as the PALM image being analyzed. Both data sets are then analyzed with

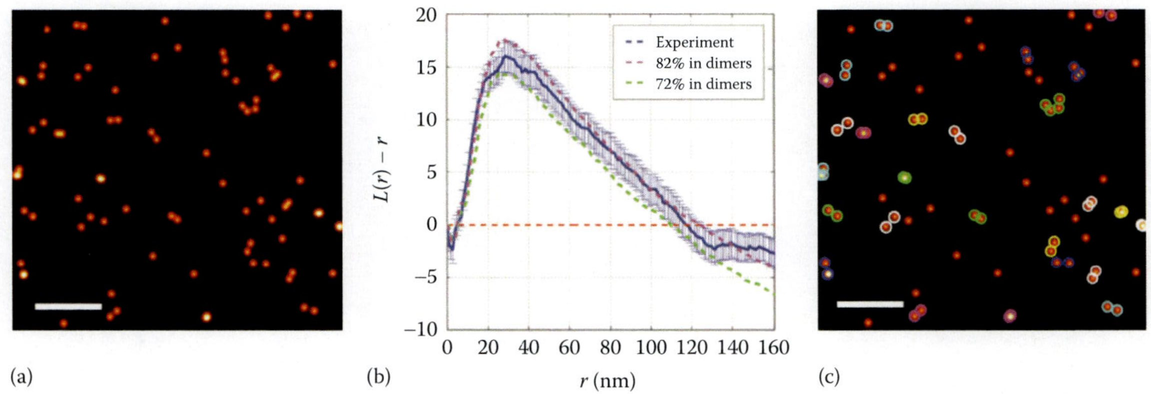

Figure 16.6

Direct observation of PAmCherry1-PAmCherry1 artificial dimers with PALM. (a) PALM image of PAmCherry1–PAmCherry1 artificial dimers expressed in a BHK21 cell. (b) Ripley K test on PALM images of PAmCherry1–PAmCherry1. (c) Density-based spatial clustering analysis with noise (DBSCAN) output of the image shown in (a) with a cluster diameter threshold of 37 nm. Scale bars: 200 nm.

DBSCAN, and the computer simulation parameters are changed over successive iterations, until the simulated and PALM data sets match (Figure 16.7a). At this point, the cluster size distribution of PA-FPs in the PALM sample is presumed to represent the actual data, as demonstrated with simulated data sets (Figure 16.7b and c).

SAD was used to analyze PALM images of PAmCherry1 monomer and artificial dimer, and the technique robustly returns similar cluster size distributions to the measured distributions in the PALM images. Consistent with previous data, PALM images of PAmCherry1 monomer, when analyzed with SAD, show that about 6 ± 3% of the molecules exist as "dimers" and a negligible amount exist as higher-order structures, consistent with previous measurements (Figure 16.7d). PALM images of PAmCherry1-PAmCherry1 artificial dimers were then analyzed with SAD, and the calculated number of dimers was found to contain ~74% of the total molecules and trimers contained ~4% of the total molecules (Figure 16.7e). The SAD results largely agree with previous analyses done using the Ripley's K test, even with increasing fluorophore densities.

These results were used to estimate the fluorescent probe recovery rate (detection efficiency) of PAmCherry1 in PALM images. If probe recovery rate is x, the probability of detecting a dimer in the PALM images is x^2, assuming that the two fluorophores in the dimer are activated independently. The probability of a dimer being detected as a monomer (i.e., the probability of detecting one of the fluorophores and not the other) is $x(1-x)$. The ratio between these two is $x\hat{}2/(x*(1-x))$ or $x/(1-x)$. Using SAD, this ratio was measured to be about 0.74 in PAmCherry1 images, and therefore, the detection efficiency of PAmCherry1 in PALM images is about 74%. As this is far below 100%, subsequent PALM imaging results were corrected for the detection efficiency.

This study demonstrates that PALM imaging, coupled with quantitative spatial analysis, provides high enough resolution and detection efficiency to directly visualize protein monomers and multimers in whole cells expressing a suitable fluorescent probe such as PAmCherry1.

16.3 Visualizing Raf Multimer Formation in Cells with Quantitative PALM

16.3.1 Raf Multimer Formation in the Presence of Mutant Ras

PAmCherry1 was fused to CRaf and expressed in BHK21 cells to analyze the distribution of PAmCherry1-CRaf in whole cells. The cells were transiently overexpressing either PAmCherry1-CRaf alone or with oncogenic $KRas^{G12D}$. Since Raf activation typically takes place at the cell membrane (Rizzo et al. 2000), a total internal reflection fluorescence (TIRF) illumination scheme was used during PALM imaging to visualize only fluorescent molecules on or within ~100 nm of the plasma membrane.

Cells expressing PAmCherry1-CRaf exhibited a fibroblast-like morphology (typical of BHK21 cells) and had very low signal intensity in TIRF PALM images (Figure 16.8a). PALM images suggested that the membrane density of PAmCherry1 was only about 20 molecules per μm^2 in these cells, with minimal clustering (Figure 16.8b). These results are consistent with CRaf existing primarily as a cytosolic monomer in resting cells with wild-type Ras, and therefore, the artificial PAmCherry1-CRaf constructs have a low propensity to form multimers.

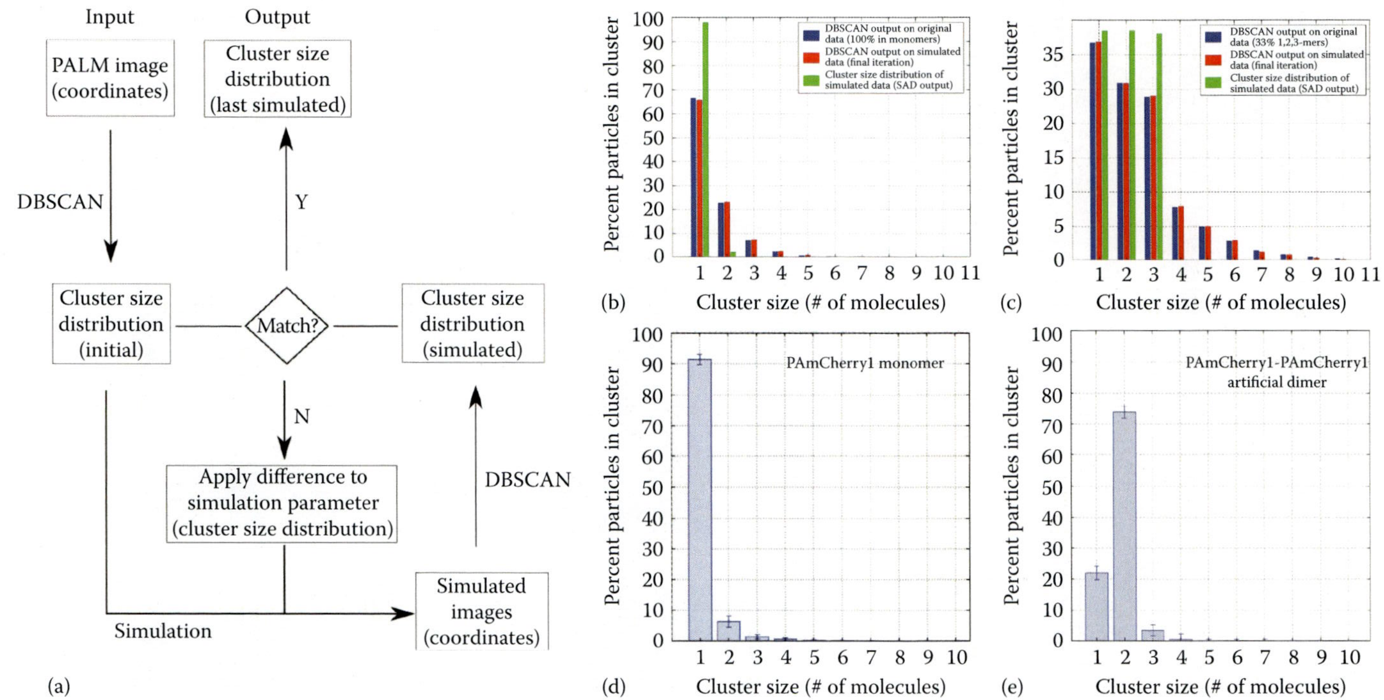

Figure 16.7

Simulation-aided DBSCAN (SAD). (a) Algorithm design of SAD; (b) Testing SAD with simulated PALM data set that contains 100% monomers; (c) Testing SAD with simulated PALM data set that contains 33% of 1-, 2-, and 3-mers. In both cases, particle densities were set at 100 per μm^2 and positional noise from a real PALM data set was added; (d) SAD output of PALM images of PAmCherry1 monomers; (e) SAD output of PALM images of PAmCherry1-PAmCherry1 artificial dimers.

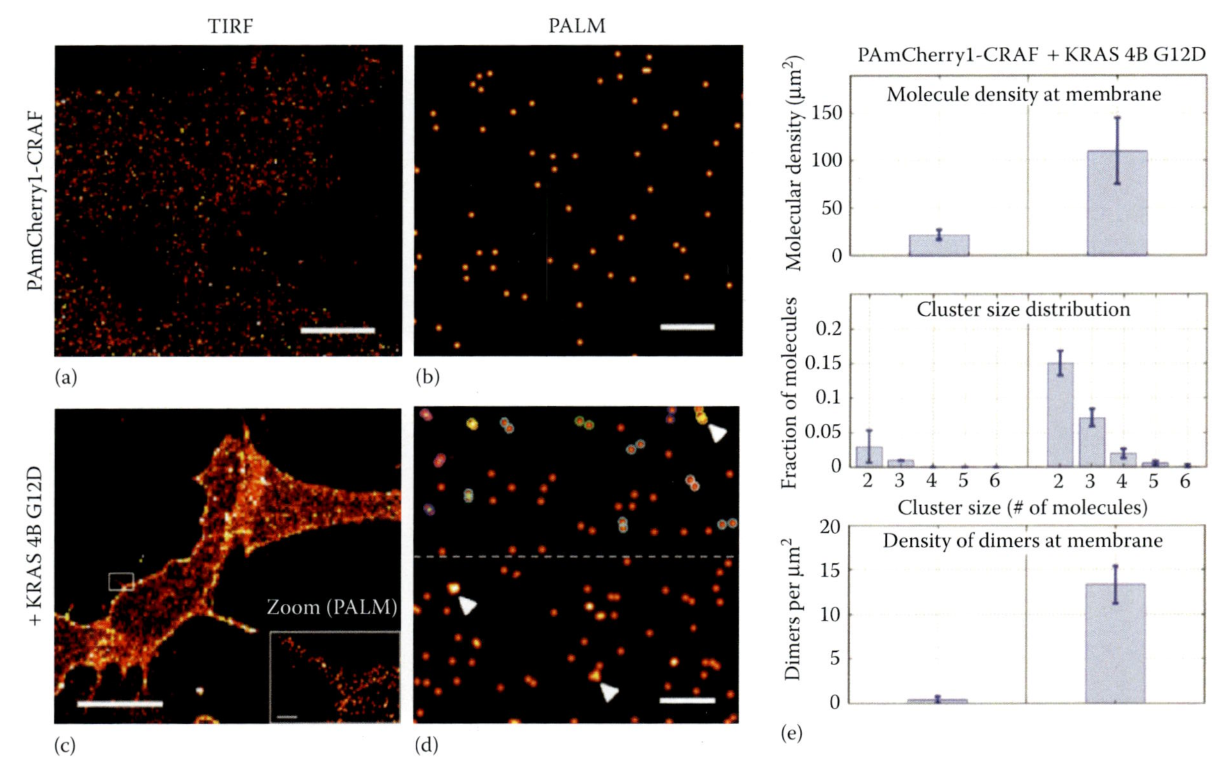

Figure 16.8

Visualizing CRaf multimers in cells with quantitative PALM. (a) TIRF and (b) PALM images of PAmCherry1-CRaf expressed alone in BHK21 cells; (c) TIRF and (d) PALM images of PAmCherry1-CRaf co-expressed with $KRAS^{G12D}$. Individual clusters were marked based on DBSCAN results in the upper half of the image (above the dotted line). Arrows indicate higher-order multimers. Inset: zoom-in PALM image of the area in the white box showing a membrane protrusion; (e) Comparisons between PAmCherry1-cRaf expressed alone (left column) and co-expressed with $KRas^{G12D}$ (right column) on membrane localization (top panel), cluster size distribution (middle panel), and density of dimers (bottom panel). Scale bars (a, c): 5 μm; (c): Inset, and (b, d): 200 nm.

BHK21 cells expressing PAmCherry-1-CRaf along with oncogenic $KRas^{G12D}$ exhibited different cell morphologies and PALM data than those expressing PAmCherry1-CRaf alone (Figure 16.8c). Notably, the cells had sharp membrane protrusions that resemble actin-rich filopodia or invadopodia (Figure 16.8c, inset) (Murphy and Courtneidge 2011); these protrusions are about 100–200 nm wide, which are below the diffraction limit of conventional light microscopy and difficult to study with conventional microscopy. Furthermore, PALM imaging of these cells revealed that CRaf exists as clusters (multimers) on the plasma membrane, and a large percentage of these clusters are shown to be dimers when analyzing the images with DBSCAN (Figure 16.8d, above the dotted line). SAD analyses indicated that a large fraction of the PAmCherry1-CRaf molecules exist as dimers. A small fraction of PAmCherry1-CRaf molecules existed as higher-order multimers (Figure 16.8d, arrows), which have not been previously reported in IP and X-ray crystallography studies (Rajakulendran et al. 2009); the functional significance of the higher-order multimers is still unclear.

Overall, the expression of oncogenic KRas increased both the clustering density and the membrane localization of CRaf molecules when compared with expressing CRaf with wild-type KRas. The membrane density of CRaf increased from 0.5 to 13 multimers per μm^2 (Figure 16.8e). This sharp increase in CRaf multimer density on Ras-mediated activation is consistent with a switch-like function of the Ras GTPase, suggesting that the formation of CRaf multimers is important in Ras-mediated activation of the Raf-MAPK pathway.

Given that CRaf dimers have functional significance in Ras-mediated MAPK activation, the protein domain(s) of CRaf that mediate dimerization were explored. Previous studies indicated that CatC, a truncated CRaf without the CR1 and CR2 regulatory regions but with its catalytic domain (Figure 16.3), is a constitutively active kinase. This hyperactivity is likely due to the loss of N-terminal autoinhibition, followed by spontaneous dimerization (Cutler et al. 1998; Stanton and Cooper 1987).

To provide evidence of CatC dimerization in cells, PALM images of BHK21 cells expressing either PAmCherry1-CatC or PAmCherry1-$CatC^{R401H}$ were acquired (Figure 16.9a). The R401H mutation was selected because previous studies suggested that this mutation disrupts CatC and CRaf dimerization (see also Figure 16.3c). PALM imaging of the PAmCherry1-CatC and -$CatC^{R401H}$ constructs was completed using relaxed TIRF excitation to allow for visualization of both membrane-bound and cytosolic fluorophores.

Consistent with previous studies and observations, PALM imaging at all expression levels showed that a large fraction of cytosolic PAmCherry1-CatC molecules exist as multimers, whereas PAmCherry1-$CatC^{R401H}$ exist almost entirely as monomers (Figure 16.9a–c). Biochemical data suggest that $CatC^{R401H}$ does not activate the MAPK pathway (via ERK phosphorylation), whereas the wild-type CatC does (Poulikakos et al. 2010; Hatzivassiliou et al. 2010). These data suggest that residue 401 on CatC does indeed play a role in both CatC dimerization and signaling activity.

Further analyses of the PALM images with SAD also revealed that unlike full-length CRaf, CatC primarily forms dimers, and at higher expression levels, it forms a small fraction of tetramers (i.e., dimer of dimers) but essentially no trimers (Figure 16.9c). This suggests that cytosolic CatC primarily forms dimers, and this dimerization is independent of Ras binding or membrane localization. Given that CatC is the CRaf kinase domain, this evidence suggests that the fundamental functional unit of CRaf is also likely a dimer.

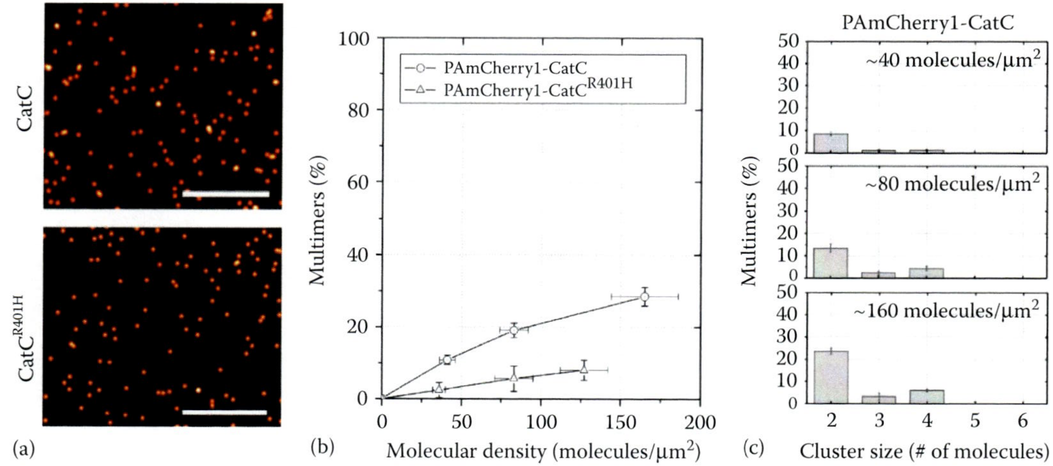

Figure 16.9

CatC dimerization and its dependence on an intact dimer interface. (a) PALM image of PAmCherry1-CatC WT (upper) and R401H mutant (lower). (b) Comparison on clustered fractions of PAmCherry1-CatC WT (circles) and R401H mutant (triangles) measured at different molecular densities. (c) Cluster size distributions of CatC WT measured with SAD in PALM images at different molecular densities. Scale bars: 500 nm.

16.3.2 Raf Multimer Formation upon Forced Membrane Localization

Differences in clustering behavior between CRaf (in the presence of $KRas^{G12D}$) and its truncated version CatC may indicate differences in multimerization dynamics between the two systems. The CR1 and CR2 regions of Raf have been shown to both inhibit Raf catalytic activity (of the CatC or CR3 domain) and contain an RBD that binds Ras-GTP, and in doing so, they release the kinase from an inhibited state (Cutler et al. 1998). Therefore, it is likely that the dimers, trimers, and multimers of CRaf in the presence of mutant KRas observed in PALM experiments are driven by Ras-GTP and that the dimers and tetramers of CatC primarily reflect the binding stoichiometry between the kinase domains of Raf in the absence of CR1 and CR2.

Electron microscopy studies support this view, showing that Ras forms membrane clusters (Abankwa et al. 2007). Under highly overexpressing conditions, a typical Ras cluster is ~20 nm in size and can contain up to eight Ras molecules; formation of these Ras nanoclusters appears to be critical for Ras signaling and dependent on posttranslational modifications to the C-terminal HVR (including the CAAX motif) (Abankwa et al. 2007; Plowman et al. 2005). Fusion of the KRas HVR to the C-terminus of CRaf yields a chimeric CRaf-CAAX protein that localizes to the cell membrane and has constitutive kinase activity (Leevers et al. 1994). Based on the established connection between Raf dimerization and kinase activation, this implies that the CAAX motif could drive CRaf dimerization and clustering on the membrane, similar to driving Ras nanocluster formation. Moreover, when Raf binds Ras-GTP, the resulting complex may exist as clusters, as Ras forms clusters on the membrane via its CAAX motif.

Findings from PALM studies corroborate the role of the CAAX motif (HVR) in Ras and Raf clustering and activation. Tagging of CRaf-CAAX and mutant $CRaf^{401H}$-CAAX with PAmCherry1 yields chimeric proteins with similar biological activities as their nontagged analogs (Figure 16.10a). Both chimeric proteins showed similar membrane clustering and cluster size distributions as revealed by PALM, and these clusters mimicked those of CRaf (specifically, PAmCherry1-CRaf in the presence of $KRas^{G12D}$) and not of CatC in membrane localization, cluster size, and distribution of clusters on the membrane (Figure 16.10b and c). These observations strongly support the hypothesis that the Ras HVR mediates CRaf-CAAX and likely CRaf:$KRas^{G12D}$ complex clustering on the cell membrane.

Lastly, biochemical data also indicate that multimer formation alone may not be sufficient to activate CRaf and the MAPK pathway, since PAmCherry1-$CRaf^{R401H}$-CAAX formed nanoclusters but displayed much less activation of the MAPK pathway when compared with its wild-type counterpart, as indicated by the level of ERK phosphorylation (Figure 16.10a). This may indicate that distinct mechanisms mediate Raf oligomerization and activation.

16.3.3 Raf Multimer Formation Induced by Raf Kinase Inhibitors

Multimer formation has also been implicated in resistance of human cancers to small kinase inhibitors of Raf (RAFi). Despite the centrality of Raf-MAPK signaling in many types of human cancers, including melanoma, where ~2/3 of all tumors harbor activating *BRAF* mutations, RAFi treatment results in resistance after a brief, initial response in patients with $BRAF^{V600E}$ (Lito et al. 2013; Rahman et al. 2014; Solit and Rosen 2014). One of the mechanisms through which the

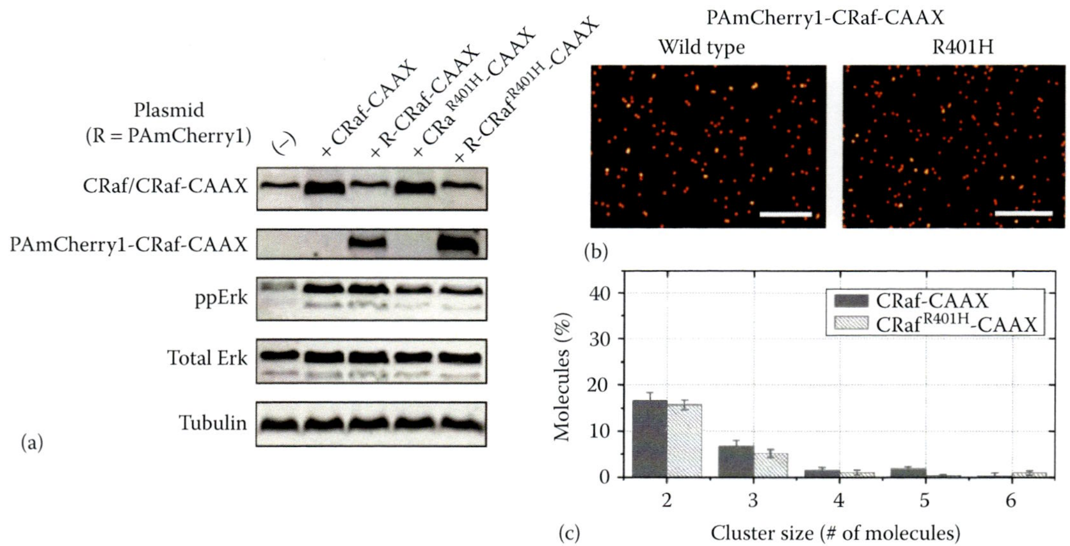

Figure 16.10

Biological analysis and clustering of PAmCherry1-CRaf-CAAX WT and R401H mutant. (a) BHK21 cells transfected with CRAF-CAAX (lane 2), PAmCherry1-CRaf-CAAX (lane 3), and R401H mutants (lanes 4 and 5) were harvested and assayed for protein expression and MAPK pathway activation by Western blotting; (b) PALM images of BHK21 cells transfected with PAmCherry1-CRaf-CAAX WT (left) and R401H mutant (right). (c) Comparison of cluster size distributions between PAmCherry1-CRaf-CAAX WT and R401H mutant measured with PALM and SAD. Scale bars: 500 nm.

BRAF^{V600E} tumor cells develop resistance to RAFi is to express an alternatively spliced form of *BRAF*^{V600E}, the p61 *BRAF*^{V600E}, which constitutively multimerizes and activates MAPK signaling (Poulikakos et al. 2011). Moreover, in patients with wild-type *BRAF*, RAFis paradoxically promote tumor growth and can even induce new tumors (Oberholzer et al. 2012; Su et al. 2012). This has been shown to be through activation of Raf-MAPK signaling by RAFi-induced Raf multimerization, a process apparently dependent on both CRaf and Ras-GTP (Hatzivassiliou et al. 2010; Poulikakos et al. 2010). In these studies, Raf dimerization was again detected with IP.

To better understand the mechanisms of RAFi-induced Raf multimer formation, PALM has also been used to visualize the spatial organization of PAmCherry1-CRaf on the membrane after treating cells with RAFis. Two different RAFis, GDC-0879 (Hansen et al. 2008) and PLX-4032 (Yang et al. 2010), were used to treat cells expressing PAmCherry1-CRaf; both compounds were potent Raf inhibitors *in vitro*. Quantitative PALM imaging clearly indicates that both GDC-0879 and PLX-4032 caused significant membrane localization (Figure 16.11a and b) and multimer formation (Figure 16.11b and c) of CRaf. In particular, GDC-0879 caused a massive accumulation of CRaf on the membrane, from ~20 molecules/μm^2 in control cells to ~80 molecules/μm^2 in treated cells, even when used at a low concentration (1 μM). Around 15% of the membrane-bound CRaf molecules were in multimers, most of which exist as dimers, resulting in greater than 12 Raf dimers/μm^2. By contrast, the PLX-4032 compound induced only a modest increase in membrane-bound CRaf, to ~30 molecules/μm^2 at 1 μM. At 10 μM concentration, PLX-4032 was able to induce ~50 CRaf molecules/μm^2 to the membrane, of which less than 10% were in dimers. Hence, GDC-0879 is significantly more potent in inducing both membrane localization and multimerization of CRaf.

The above observations are consistent with previous biochemical measurements, where GDC-0879 caused a more profound upregulation of Raf-MAPK pathway activation (Hatzivassiliou et al. 2010). The other compound, PLX-4032 (Vemurafenib), had a much milder adverse effect and was approved by the FDA in 2011 for treating advanced melanoma with *BRAF*^{V600E} mutations (Chapman et al. 2011). This study further demonstrates the utility of quantitative super-resolution microscopy in revealing key molecular details in translational cancer research.

16.4 Imaging Ras-Raf Interactions at the Nanoscale with BiFC-PALM

16.4.1 BiFC-PALM Based on Split PAmCherry1

Data presented in the previous sections have shown that both Ras and Raf form dimers or multimers on the cell membrane when effective MAPK signaling occurs. Since the functional unit of Raf is a dimer and Raf binds to Ras-GTP on the membrane, the role of Ras is likely to multimerize Raf to activate the kinase and initiate downstream signaling. An interesting prediction from this hypothesis is that an active Ras-Raf complex should at least contain two copies of Ras and two copies of Raf, suggesting a new model of Ras-Raf signaling that has not been proposed or validated before. This new Ras-Raf interaction model would ideally be validated through quantitative imaging of both Ras and Raf. To date, however, PA-FPs suited for quantitative PALM imaging mostly emit in the orange or red channel and one

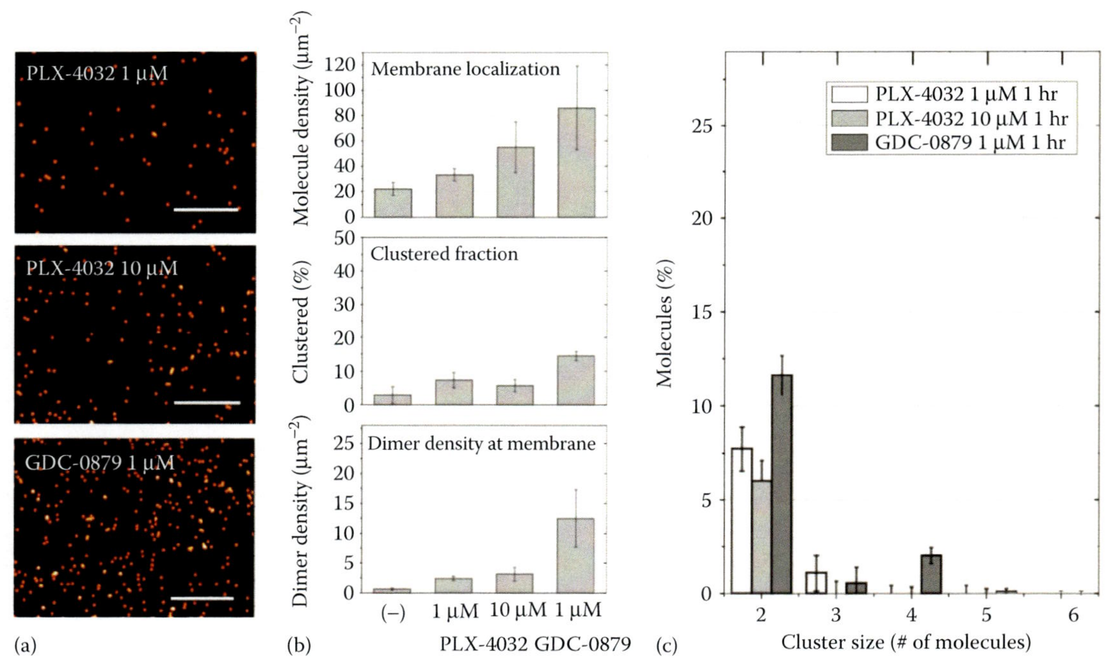

Figure 16.11

RAFi-induced membrane localization and dimerization of CRaf. (a) PALM images of PAmCherry1-CRaf protein expressed in BHK21 cells that were serum-starved and treated with 1 μM and 10 μM PLX-4032 (Vemurafenib) and 1 μM GDC-0879 for 1 h at 37°C; (b) Comparison of membrane localization (top), clustered fraction (middle), and dimer concentration (bottom) under different conditions; (c) Comparison of the cluster size distribution under different conditions. Scale bar: 500 nm.

that emits in a different channel is yet to be identified. This makes it impractical to perform dual-color, quantitative PALM imaging of Ras and Raf at once.

As an alternative to dual-color co-localization, bimolecular fluorescence complementation (BiFC) has been used in place of co-localization for detecting PPIs in cells (Kerppola 2006). BiFC involves the splitting of an FP into two nonfluorescent fragments, and when attached to interacting proteins and brought into proximity through the interaction, the two fragments can spontaneously refold into a complete FP with restored fluorescence (Figure 16.12a). In BiFC, fluorescence signal results from only a pair of interacting proteins (which can be specific or nonspecific) and emits in only one channel. As such, BiFC imaging can be performed on a standard fluorescence microscope, making it a more direct and simplified approach than other techniques.

As demonstrated in 2014, BiFC can be combined with PALM for imaging PPIs at the nanometer and single-molecule scales (Nickerson et al. 2014; Liu et al. 2014). This combination, named BiFC-PALM, was possible because both BiFC and PALM use FPs. Specifically, by splitting an FP that is suited for PALM imaging (e.g., a PA-FP) into two fragments that are compatible with BiFC, the reconstituted protein resulting from PPIs may retain the photophysical properties similar to the original FP and thus may allow for PALM imaging of the PPIs. Until recently, Dronpa was the only PALM-compatible FP split for BiFC (Lee et al. 2010), but its photophysical properties are suboptimal for PALM. Our laboratory first demonstrated the use of split PAmCherry1, the PA-FP that has been used throughout this chapter for quantitative PALM imaging, for BiFC-PALM (Nickerson et al. 2014).

Reconstitution of the fluorescent protein in BiFC is dependent on the split site. Different split sites have been tested for mCherry, the parent protein of PAmCherry1, by Fan et al. (2008), and the best site was determined to be between amino acids 159 and 160. This site is within a loop region between beta sheets 7 and 8 of the FP chromophore (Figure 16.12b), a common site for splitting many FP proteins for BiFC (Kerppola 2008). Sequence alignment also reveals that mCherry and PAmCherry1 are identical around residues 159/160. Based on these commonalities, splitting PAmCherry1 around residues 159/160 may generate a functional BiFC pair of fragments.

To test the utility of this split site for PAmCherry1 BiFC, the two fragments, RN (residues 1–159) and RC (methionine plus residues 160–236), were genetically fused to DmrA and DmrC, two peptides used in the Clontech iDimerize system. The DmrA and DmrC peptides interact and form a 1:1 heterodimer only on addition of a small-molecule heterodimerizer. An N-terminal myristoylation signal was also attached to DmrA to target the peptide to the cell membrane. This helps localize the expected BiFC signal to a distinct cellular compartment (the cell membrane) for easy validation. It also helps reduce the number of BiFC configurations to be tested from eight to four: DmrA-RN/DmrC-RC, DmrA-RN/RC-DmrC, DmrA-RC/DmrC-RN, and DmrA-RC/RN-DmrC (Figure 16.12c). A flexible (GGGGS) × 2 linker sequence (Chen et al. 2013) was used in all fusion constructs to facilitate RN-RC complementation.

On addition of the heterodimerizer, two of the four tested configurations showed strong signal in TIRF images (Figure 16.12d, right panels) and the others had none at all. Importantly, the signal appeared to be predominantly localized to the cell membrane, which was expected given the membrane targeting sequence attached to DmrA. As a control, when the heterodimerizer was withheld from the

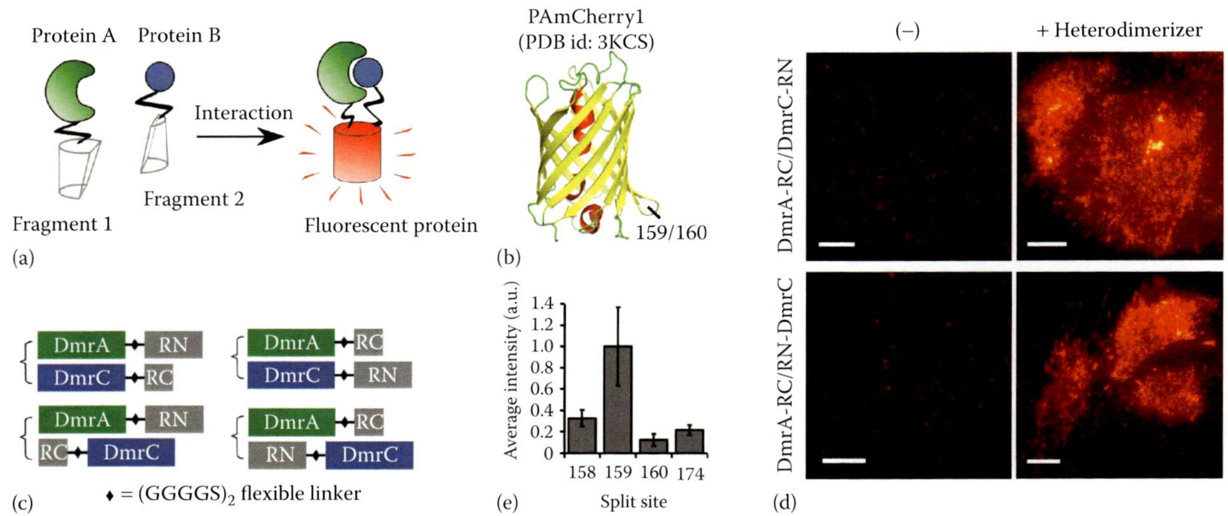

Figure 16.12

Split PAmCherry1 for BiFC. (a) Crystal structure of PAmCherry1 with the 159/160 split site for BiFC indicated; (b) Artificial dimerization system for testing PAmCherry1 BiFC; (c) Four tested configurations for PAmCherry1 BiFC with the inducible heterodimerization system. RN = PAmCherry1 N-terminal fragment (residues 1–159), RC = PAmCherry1 C-terminal fragment (residues 160–236); (d) Total internal reflection fluorescence (TIRF) images of heterodimerizer-induced BiFC signals in DmrA-RC/DmrC-RN (top right panel), and DmrA-RC/RN-DmrC (bottom right panel); (e) Average BiFC signal intensities at four different split sites: 158, 159, 160, and 174 (n = 5–10). Intensities are normalized to 159 split site, and error bars are SEM. Scale bar: 10 μm in (d).

working configurations, the BiFC signal was absent (Figure 16.12d, left panels), suggesting that the signal was a result of the interaction and not of the spontaneous complementation between the RN and RC fragments.

Using the same scheme as above, additional split sites were tested, including those between residues 158 and 159, 160 and 161, and between 174 and 175 (located in another loop between beta sheets 8 and 9). Although these splitting schemes did yield some BiFC signals on addition of the heterodimerizer, the signals were much lower than that observed with the 159/160 split site (Figure 16.12e). Hence, the site 159/160 is likely optimal for PAmCherry1 BiFC.

Similar to many other BiFC systems, it typically takes 1–2 h for a significant PAmCherry1 BiFC signal to develop, most of which would likely be the time needed for chromophore formation and maturation. In addition, similar to most other BiFC systems, chromophore complementation between RN and RC appears to be irreversible, as removal of the heterodimerizer did not reduce BiFC signal at least in a 2-h period of incubation in regular culture medium (Kerppola 2006). Compared with other BiFC systems, however, PAmCherry1 BiFC did not require that the cells be incubated at low temperatures (25°C or even 4°C); incubation at low temperatures for extended durations could have adverse effects on cell physiology. Interestingly, BiFC based on the parent protein, mCherry, was reported to require incubation at low temperatures (Fan et al. 2008), suggesting significant differences in BiFC kinetics between the two FPs, despite their high degrees of homology in sequence and structure.

16.4.2 Nanoscopic Imaging of Ras-Raf Complexes with BiFC-PALM

To combine BiFC with PALM, the reconstituted PA-FP needs to retain some important photophysical properties of the original PA-FP, namely photoactivation (or photoswitching) with high contrast ratio and good single-molecule photon yield (brightness), among others. This was the case for split PAmCherry1. Specifically, reconstituted PAmCherry1 exhibited fast photoactivation and a high contrast ratio in both fixed and living cells. Clear, single-molecule images could be easily acquired on PAmCherry1 BiFC samples owing to the low spontaneous activation rate of the fluorophore, as with the original PAmCherry1 (Figure 16.13a). Both the reconstituted PAmCherry1 and the original PAmCherry1 emitted ~600 photons on average per localization event, with an actual localization precision of ~18 nm (Figure 16.13b). Lastly, reconstituted PAmCherry1 exhibited a dark state lifetime of 0.26 ± 0.05 s, identical to that of the original PAmCherry1 (Nan et al. 2013; Nickerson et al. 2014).

Similarities in these properties make reconstituted PAmCherry1 a suitable probe for quantitative PALM imaging of PPIs. For example, the artificially induced DmrA/DmrC complexes were selectively imaged with PALM to reveal their nanoscale distribution on the membrane. As shown in Figure 16.13c, although DmrA/DmrC complexes are distributed across the entire membrane area, the distribution is clearly not random. Instead, the complexes often form higher-order structures (clusters), some of which are as large as 4–5 membered. By contrast, these features are unresolved with conventional fluorescence microscopy (Figure 16.13c, bottom right).

Next, the spatial organization of Ras-Raf complexes was investigated using a similar approach. RN or RC was genetically fused to the N-terminus of KRas^{G12D}, an active mutant of KRas, since tagging at the C-terminus of Ras would disrupt its posttranslational modification and correct membrane targeting. This reduces the number of possible Ras-Raf BiFC complexes from

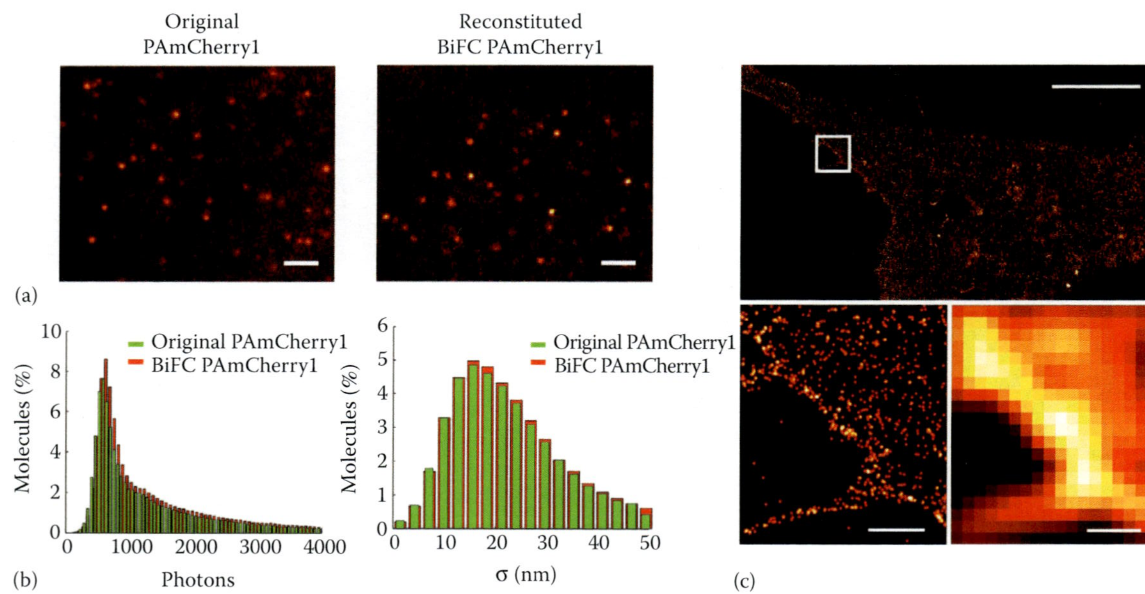

Figure 16.13

Nanoscopic imaging of protein-protein interactions (PPIs) in cells with BiFC-PALM based on split PAmCherry1. (a) Single molecules images of the original PAmCherry1 (left) and BiFC-reconstituted PAmCherry1 (right); (b) Comparing the photon number per localization event (left) and the localization precision (right) between the original PAmCherry1 (green) and BiFC-reconstituted PAmCherry1 (red); (c) Imaging the DmrA/DmrC complexes in cells using BiFC-PALM. Bottom panels, PALM (left) and the corresponding low-resolution (right) images of the boxed area in the PALM image of the whole cell (top). Scale bars: 2 μm (a), 5 μm (c, top), and 500 nm (c, bottom).

eight to four, RN-KRas/RC-CRaf, RC-KRas/RN-CRaf, RN-KRas/CRaf-RC, and RC-KRas/CRaf-RN, as in the case of DmrA-DmrC BiFC. To simplify the experiment, only the RBD of CRaf (amino acids 51–131) (Vojtek et al. 1993) was used instead of the full-length CRaf (Figure 16.14a), in part because the irreversibility of BiFC would otherwise result in constant signaling of Raf.

Of the four BiFC configurations tested, two (RN-KRas/RC-RBD and RN-KRas/RBD-RC) yielded strong BiFC signals, whereas the other two (RC-KRas/RN-RBD and RC-KRas/RBD-RN) gave no or a weak signal. Results from one of these configurations, the RN-KRas/RBD-RC pair, are shown in Figure 16.14b. Differences in BiFC efficiency among the different configurations for both Ras-Raf and DmrA-DmrC interaction pairs may indicate differences in structural orientations of the fusion proteins as well as in BiFC kinetics. Furthermore, the lack of BiFC signal in some of the configurations also suggests that spontaneous interactions between RN and RC, should they exist, have only a negligible contribution to BiFC.

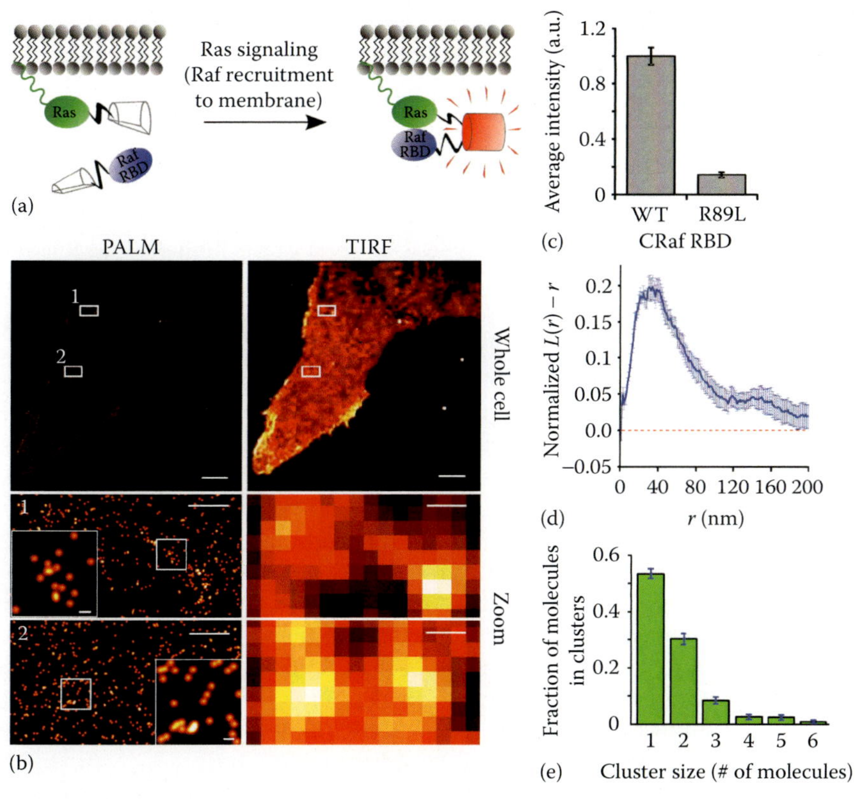

Figure 16.14

Imaging individual KRas G12D/CRaf RBD complexes with BiFC-PALM. (a) KRas G12D and CRaf RBD (CRaf residues 51–131) BiFC scheme; (b) PALM images (left) and low-resolution TIRF representations (right) of a U2OS cell expressing RN-KRas G12D and CRaf RBD-RC. Bottom panels are zoomed-in views of boxed areas 1 and 2 in the whole cell views, respectively; (c) Average BiFC signal intensity with wild-type CRaf RBD and the R89L mutant ($n = 5$–8); (d) Ripley's K-test analysis of 10 sampled areas from cell in B. Dashed line represents random monomeric distribution; (e) Average cluster size distribution within the same 10 sampled areas as in d. All error bars are SEM in (c) and standard deviations in (d) and (e). Scale bars in (b): 5 μm top, 500 nm middle and bottom, and 50 nm insets.

To further ensure that the observed BiFC signal was specific to Ras-Raf interactions, an R89L mutation in the CRaf RBD was introduced to the RC-RBD construct and tested for BiFC. The R89L mutation was reported to disrupt Raf binding to Ras (Fabian et al. 1994). Consistently, this mutation diminished RN-KRas/RC-RBD BiFC by ~10 fold (Figure 16.14c), confirming that the observed BiFC signal reflected specific interactions between Ras and Raf.

PALM imaging of the RN-KRas/RBD-RC pair reveals a similar, heterogeneous distribution pattern of Ras-Raf (RBD) complexes on the cell membrane (Figure 16.14b, left panels). Although the low-resolution TIRF image of the same cell (Figure 16.14b, right panels) shows a largely homogeneous distribution of the complexes, individual clusters intermixed with monomers of RN-KRas/RBD-RC BiFC complex can be clearly observed in the high-resolution PALM image. The apparent diameter of these clusters is ~30 nm, based on the Ripley's K test, which likely represents the resolution of PALM imaging rather than the actual dimensions of the clusters (Figure 16.14d).

SAD analysis shows that the clusters are mostly dimers and trimers (Figure 16.14e), resembling the observed CRaf multimers in the presence of $KRas^{G12D}$ (Figure 16.8). This was expected because CRaf is cytosolic when active and most of the CRaf molecules found on the membrane should be bound to $KRas^{G12D}$; hence, PALM images of membrane-bound PAmCherry1-CRaf would be equivalent to those of PAmCherry1-(Ras/Raf) BiFC complexes.

Data presented above confirm that Ras-Raf complexes can exist as multimers, where each multimer would comprise multiple copies of Ras and multiple copies of Raf. The dominance of dimers implies that most of the higher-order Ras-Raf complexes would contain two Ras and two Raf molecules.

16.4.3 Tracking Individual Ras-Raf Complexes using BiFC-PALM

Single-molecule-tracking (smt)-PALM is a powerful technique for analyzing the diffusion dynamics of biological molecules in cells (Manley et al. 2008). Each smt-PALM experiment could yield a large number of diffusion trajectories, enabling sophisticated statistical analyses such as variational Bayesian single-particle tracking (vbSPT) (Persson et al. 2013) to help reveal otherwise hidden diffusion and/or interaction states. Successful combination of BiFC with PALM thus extends smt-PALM to studying the diffusion of individual protein complexes in living cells. In particular, PAmCherry1 has been commonly used in smt-PALM experiments for its high contrast ratio (hence low spontaneous fluorescence background) and brightness (Subach et al. 2009). Moreover, the fact that PAmCherry1 BiFC performs well at 37°C makes it fully compatible with live-cell imaging under physiological conditions. Thus, the PAmCherry1 BiFC system described above could be readily used for smt-PALM.

Smt-PALM tracking of individual Ras-Raf complexes was performed using the same RN-KRas/RBD-RC BiFC configuration as described in the previous section. Throughout the imaging experiments, cells expressing the BiFC constructs were maintained at 37°C on a temperature-controlled stage and in CO_2-independent growth medium (which replaced regular growth medium right after the cells were taken out of the incubator). A low level of 405 nm light (2.5–5 W/cm^2) was used to illuminate the sample, activating a subset of PAmCherry1 molecules at any given time, which were imaged continuously (Figure 16.15a). In addition, TIRF illumination was used for both the activation (405 nm) and imaging (561 nm) lasers to confine the observation volume to the cell membrane. With

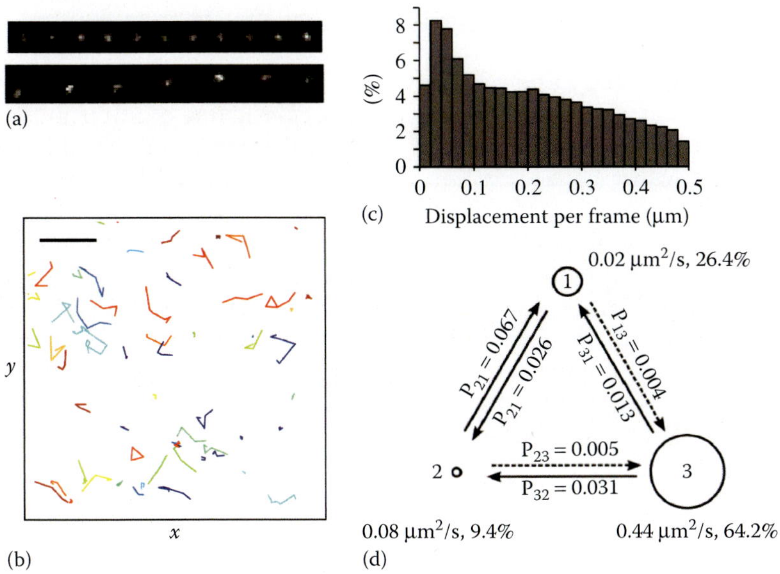

Figure 16.15

Tracking individual KRas G12D/CRaf RBD complexes with BiFC and smt-PALM. (a) Displacement trajectories at 50 ms intervals of PAmCherry1-tagged KRas G12D/CRaf RBD BiFC complexes showcasing an immobile molecule (top) and a mobile molecule (bottom); (b) *x-y* plot of a subset of tracked molecules; (c) Histogram showing the distribution of displacement per frame (50 ms); (d) Variational Bayesian single-particle tracking (vbSPT) analysis of the diffusion trajectories showing three possible diffusion states and the transition probabilities between states within the frame acquisition time (50 ms). Scale bar: 1 μm in (b).

this configuration, typically more than 10,000 trajectories could be generated per cell (Figure 16.15b). Of note, only molecules that appeared in more than two successive frames were considered when constructing diffusion trajectories.

Analyses of the Ras-Raf BiFC diffusion trajectories further support a heterogeneous distribution of the complexes on the membrane. First, in the *x-y* positional plot of the molecules (Figure 16.15b), it immediately became obvious that the complexes diffuse at multiple rates; some are highly motile, whereas others are stationary. A more clear view of the existence of multiple diffusion rates could be obtained from the histogram of displacement per frame, which is equivalent to an "instantaneous diffusion rate" (Figure 16.15c). This is a similar observation to those made previously on individual Ras proteins; however, the throughput of previous studies was much lower because PALM or smt-PALM had not been introduced at the time.

A more quantitative analysis of the diffusion trajectories using vbSPT indicated that at least three diffusion states of the Ras-Raf complexes exist on the membrane (Figure 16.15d). A mobile state with an average diffusion constant (D) of 0.44 $\mu m^2/s$ is the dominant (abundance ~60%) form. Another major diffusion state is an immobile state, with a low diffusion constant (average $D \sim 0.02\ \mu m^2/s$); the diffusion rate of this state is so low that the measured displacements per frame on this population were of the order of 10–20 nm within the frame acquisition time, that is, 25–50 ms. This measured "displacement" is most likely dominated by the localization error rather than the actual, physical movement of the molecules (complexes). Hence, this state appears to be entirely stationary at the

25–50 ms time scales, indicating a strong binding to structural elements such as the cortical actin or the integrin complexes that reside on or are proximal to the cell membrane. Lastly, about 10% of the complexes are in an intermediate diffusion state (average $D \sim 0.08\ \mu m^2/s$), but the nature of this population is currently unclear. Importantly, vbSPT analysis also reveals that the three states interconvert with each other at certain probabilistic rates. Together, information contained within the vbSPT output plot not only confirms the presence of multiple diffusion rates of Ras-Raf complexes but also quantitatively measures the abundance, lifetime, and conversion rates of the states; such information would otherwise be inaccessible if a sufficient number of trajectories have not been acquired with smt-PALM. In combination with biochemical manipulations, BiFC with smt-PALM may yield further information as to how the Ras-Raf interactions are spatially and temporally regulated in a living cell.

16.5 Concluding Remarks

Multimer formation has been suggested as a key step in the activation of the Raf kinase and subsequent downstream effectors in the MAPK pathway (Farrar et al. 1996; Luo et al. 1996; Rajakulendran et al. 2009). Before the advent of SRM, the existence of Raf multimers and their involvement in cell signaling could not be fully demonstrated. Quantitative PALM imaging results directly show the formation of CRaf multimers (predominantly dimers) in whole mammalian cells, which can arise through interactions between the CatC domain of CRaf (when the N-terminal regulatory domain is removed) and/or through Ras-Ras interactions (when Raf is bound to Ras-GTP). In the latter case, multimer formation appears to be dominated by the Ras HVR. Similarly, when Raf is artificially localized to the membrane by fusion to the Ras HVR, the chimeric protein also forms multimers on the membrane, which could explain the constitutive kinase activity of Raf-CAAX. Based on the data presented in this chapter and in other literature, the following working models are proposed to describe Raf multimer formation under various activating conditions (Figure 16.16).

That being the case, studies involving $CRaf^{R401H}$-CAAX indicate that multimer formation may not be sufficient to fully activate Raf and the MAPK pathway, since this chimeric protein localizes to the membrane and forms multimers to a similar extent to the CRaf-CAAX fusion protein. Furthermore, although interactions between Ras-GTP and Raf may be needed to activate the MAPK pathway in normal cells, certain oncogenic variants of Raf, such as $BRaf^{V600E}$, appear to activate the MAPK pathway without multimer formation (Poulikakos et al. 2011). Indeed, the $BRaf^{V600E;R509H}$ double mutant has been shown to activate MAPK. Clearly, more investigations are needed to fully elucidate the details of intra- and intermolecular interactions involved in Raf kinase activation.

Existing techniques have helped generate comprehensive lists of the molecular components that make up biological systems such as a cell and have provided critical insight into cellular processes. How the different components interact in a cellular space and carry out the processes and how these interactions and processes may go awry in cancer remain poorly understood. An important aspect lacking in most existing measurement technologies is the capability to visualize molecular orientations and interactions in intact biological samples with sufficient resolutions. In this perspective, recent and still rapidly evolving SRM techniques have combined advantages of electron microscopy (high resolution)

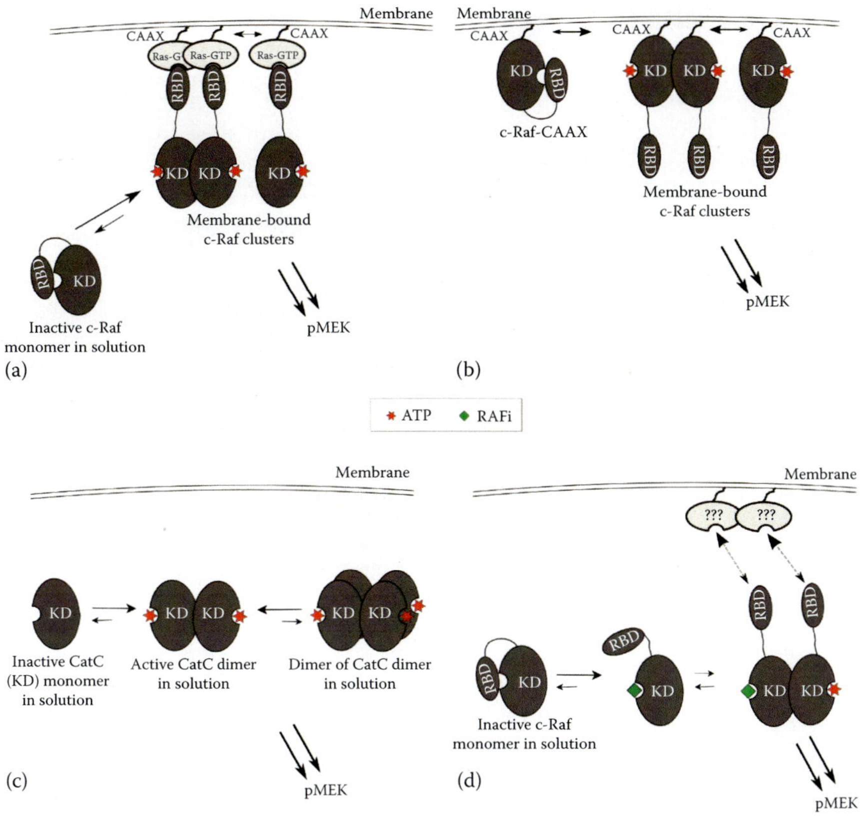

Figure 16.16

Working models of CRaf dimerization and multimerization (clustering) under various activating conditions. (a) Ras-GTP clusters on the membrane and recruits Raf, driving the latter into multimers and activating the kinase; (b) Raf-CAAX localizes to the membrane and clusters (forms multimers) similar to Ras, resulting in Raf multimerization and activation; (c) Catalytic domain of Raf (e.g., CatC) spontaneously forms dimers and dimer of dimers independent of Ras; (d) RAFi induces heterodimerization between a RAFi-bound Raf molecule and a RAFi-free (ATP-bound) Raf molecule, a process that may also depend on Raf binding to Ras-GTP.

and fluorescence microscopy (high specificity, sensitivity, and live-cell compatibility); they offer a unique opportunity to examine biology with unprecedented details and precision.

In this chapter, the utility of PALM to detect individual proteins with high spatial and stoichiometric resolutions led to the direct visualization of Raf and Ras-Raf multimers in cells under various signaling conditions and mutational states. The findings both corroborate previous reports and yield novel insight into the regulation and biological function of these protein multimers. More significantly, using a similar approach and carefully controlled protein expression system, the Ras GTPase was shown to signal as a dimer in activating the Raf-MAPK pathway, suggesting that both Ras and Raf may be inhibited through disruption of the dimers (Nan et al. 2015). Evidently, the new biological insight from SRM imaging studies will provide useful guidance for efforts on improving targeted cancer therapy.

Acknowledgment

Work described in this chapter was supported by start-up funds to X.N. from OHSU and by NIH 5U54CA143836-05 (PI: J. Liphardt). Research in the Nan lab was also supported by the Damon Runyon Cancer Research Foundation, the M. J. Murdock Charitable Trust, the FEI company, and the Prospect Creek Foundation. The authors also thank Drs. Joe W. Gray, Steven Chu, Frank McCormick, and Eric A. Collisson for helpful discussions and Dr. Tao Huang, Dr. Tanja M. Tamgüney, and Sophia Lewis for their contributions to the work.

References

Abankwa, Daniel, Alemayehu A Gorfe, and John F Hancock. 2007. Ras Nanoclusters: Molecular Structure and Assembly. *Seminars in Cell & Developmental Biology* 18(5):599–607. doi:10.1016/j.semcdb.2007.08.003.

Annibale, Paolo, Stefano Vanni, Marco Scarselli, Ursula Rothlisberger, and Aleksandra Radenovic. 2011. Quantitative Photo Activated Localization Microscopy: Unraveling the Effects of Photoblinking. *PLoS One* 6(7). doi:10.1371/journal.pone.0022678.

Betzig, Eric, George H Patterson, Rachid Sougrat, O Wolf Lindwasser, Scott Olenych, Juan S Bonifacino, Michael W Davidson, Jennifer Lippincott-Schwartz, and Harald F Hess. 2006. Imaging Intracellular Fluorescent Proteins at Nanometer Resolution. *Science (New York, N.Y.)* 313(5793):1642–1645. doi:10.1126/science.1127344.

Callahan, Margaret K, Raajit Rampal, James J Harding, Virginia M Klimek, Young Rock Chung, Taha Merghoub, Jedd D Wolchok et al., 2012. Progression of RAS-Mutant Leukemia during RAF Inhibitor Treatment. *The New England Journal of Medicine* 367(24):2316–2321. doi:10.1056/NEJMoa1208958.

Chapman, Paul B, Axel Hauschild, Caroline Robert, John B Haanen, Paolo Ascierto, James Larkin, Reinhard Dummer et al., 2011. Improved Survival with Vemurafenib in Melanoma with BRAF V600E Mutation. *The New England Journal of Medicine* 364(26):2507–2516. doi:10.1056/NEJMoa1103782.

Chen, Xiaoying, Jennica L Zaro, and Wei Chiang Shen. 2013. Fusion Protein Linkers: Property, Design and Functionality. *Advanced Drug Delivery Reviews* 65(10):1357–1369 (Elsevier B.V.). doi:10.1016/j.addr.2012.09.039.

Collisson, Eric A, Christy L Trejo, Jillian M Silva, Shenda Gu, James E Korkola, Laura M Heiser, Roch-Philippe Charles et al., 2012. A Central Role for RAF→MEK→ERK Signaling in the Genesis of Pancreatic Ductal Adenocarcinoma. *Cancer Discovery* 2(8):685–693. doi:10.1158/2159-8290.CD-11-0347.

Cox, Adrienne D and Channing J Der. 2010. Ras History: The Saga Continues. *Small GTPases* 1(1):2–27. doi:10.4161/sgtp.1.1.12178.

Cox, Adrienne D, Stephen W Fesik, Alec C Kimmelman, Ji Luo, and Channing J Der. 2014. Drugging the Undruggable RAS: Mission Possible? *Nature Reviews Drug Discovery* 13(11):828–851 (Nature Publishing Group). doi:10.1038/nrd4389.

Cutler, Richard E, Robert M Stephens, Misty R Saracino, and Deborah K Morrison. 1998. Autoregulation of the Raf-1 Serine/threonine Kinase. *Proceedings of the National Academy of Sciences of the United States of America* 95(16):9214–9219. doi:10.1073/pnas.95.16.9214.

Davies, Helen, Graham R Bignell, Charles Cox, Philip Stephens, Sarah Edkins, Sheila Clegg, Jon Teague et al., 2002. Mutations of the BRAF Gene in Human Cancer. *Nature* 417(6892):949–954. doi:10.1038/nature00766.

Dempsey, Graham T, Joshua C Vaughan, Kok Hao Chen, Mark Bates, and Xiaowei Zhuang. 2011. Evaluation of Fluorophores for Optimal Performance in Localization-Based Super-Resolution Imaging. *Nature Methods* 8(12):1027–1036 (Nature Publishing Group, a division of Macmillan Publishers Limited. All Rights Reserved). doi:10.1038/nmeth.1768.

Dixon, Philip M. 2006. Ripley's K Function. *Encyclopedia of Environmetrics.* doi:10.1002/9780470057339.var046.

Downward, Julian. 2003. Targeting RAS Signalling Pathways in Cancer Therapy. *Nature Reviews Cancer* 3(1):11–22. doi:10.1038/nrc969.

Durisic, Nela, Lara Laparra-Cuervo, Angel Sandoval-Álvarez, Joseph Steven Borbely, and Melike Lakadamyali. 2014. Single-Molecule Evaluation of Fluorescent Protein Photoactivation Efficiency Using an in Vivo Nanotemplate. *Nature Methods*, January. doi:10.1038/nmeth.2784.

Ester, Martin, Hans P Kriegel, Jorg Sander, and Xiaowei Xu. 1996. A Density-Based Algorithm for Discovering Clusters in Large Spatial Databases with Noise. *Second International Conference on Knowledge Discovery and Data Mining*, 226–231. doi:10.1.1.71.1980.

Fabian, John R, Anne B Vojtek, Jonathan A Cooper, and Deborah K Morrison. 1994. A Single Amino Acid Change in Raf-1 Inhibits Ras Binding and Alters Raf-1 Function. *Proceedings of the National Academy of Sciences of the United States of America* 91(13):5982–5986.

Fan, Jin-Yu, Zong-Qiang Cui, Hong-Ping Wei, Zhi-Ping Zhang, Ya-Feng Zhou, Yun-Peng Wang, and Xian-En Zhang. 2008. Split mCherry as a New Red Bimolecular Fluorescence Complementation System for Visualizing Protein-Protein Interactions in Living Cells. *Biochemical and Biophysical Research Communications* 367:47–53. doi:10.1016/j.bbrc.2007.12.101.

Farrar, Michael A, J Alberol-Ila, Roger M Perlmutter, and José Alberola-Lla. 1996. Activation of the Raf-1 Kinase Cascade by Coumermycin-Induced Dimerization. *Nature* 383(6596):178–181. doi:10.1038/383178a0.

Fernández-Medarde, Alberto and Eugenio Santos. 2011. Ras in Cancer and Developmental Diseases. *Genes & Cancer* 2(3):344–358. doi:10.1177/1947601911411084.

Fernández-Suárez, Marta and Alice Y Ting. 2008. Fluorescent Probes for Super-Resolution Imaging in Living Cells. *Nature Reviews Molecular Cell Biology* 9(12):929–943. doi:10.1038/nrm2531.

Gibbs, Jackson B, Irving S Sigal, Martin Poe, and Edward M Scolnick. 1984. Intrinsic GTPase Activity Distinguishes Normal and Oncogenic Ras p21 Molecules. *Proceedings of the National Academy of Sciences of the United States of America* 81(18):5704–5708.

Ha, Taekjip and Philip Tinnefeld. 2012. Photophysics of Fluorescent Probes for Single-Molecule Biophysics and Super-Resolution Imaging. *Annual Review of Physical Chemistry* 63(1):595–617. doi:10.1146/annurev-physchem-032210-103340.

Hansen, Joshua D, Jonas Grina, Brad Newhouse, Mike Welch, George Topalov, Nicole Littman, Michele Callejo et al., 2008. Potent and Selective Pyrazole-Based Inhibitors of B-Raf Kinase. *Bioorganic & Medicinal Chemistry Letters* 18(16):4692–4695. doi:10.1016/j.bmcl.2008.07.002.

Hatzivassiliou, Georgia, Kyung Song, Ivana Yen, Barbara J Brandhuber, Daniel J Anderson, Ryan Alvarado, Mary J C Ludlam et al., 2010. RAF Inhibitors Prime Wild-Type RAF to Activate the MAPK Pathway and Enhance Growth. *Nature* 464(7287):431–435. doi:10.1038/nature08833.

Henis, Yoav I, John F Hancock, and Ian A Prior. 2009. Ras Acylation, Compartmentalization and Signaling Nanoclusters (Review). *Molecular Membrane Biology* 26(1):80–92. doi:10.1080/09687680802649582.

Hess, Samuel T, Thanu P K Girirajan, and Michael D Mason. 2006. Ultra-High Resolution Imaging by Fluorescence Photoactivation Localization Microscopy. *Biophysical Journal* 91(11):4258–4272 (Elsevier). doi:10.1529/biophysj.106.091116.

Janosi, Lorant, Zhenlong Li, John F Hancock, and Alemayehu A Gorfe. 2012. Organization, Dynamics, and Segregation of Ras Nanoclusters in Membrane Domains. *Proceedings of the National Academy of Sciences of the United States of America* 109(21):8097–8102. doi:10.1073/pnas.1200773109.

Karnoub, Antoine E, and Robert A Weinberg. 2008. Ras Oncogenes: Split Personalities. *Nature Reviews Molecular Cell Biology* 9(7):517–531. doi:10.1038/nrm2438.

Kerppola, Tom K. 2006. Design and Implementation of Bimolecular Fluorescence Complementation (BiFC) Assays for the Visualization of Protein Interactions in Living Cells. *Nature Protocols* 1:1278–1286. doi:10.1038/nprot.2006.201.

Kerppola, Tom K. 2008. Bimolecular Fluorescence Complementation (BiFC) Analysis as a Probe of Protein Interactions in Living Cells. *Annual Review of Biophysics* 37(January):465–487. doi:10.1146/annurev.biophys.37.032807.125842.

Lee, Sang-Hyuk, Jae Yen Shin, Antony Lee, and Carlos Bustamante. 2012. Counting Single Photoactivatable Fluorescent Molecules by Photoactivated Localization Microscopy (PALM). *Proceedings of the National Academy of Sciences of the United States of America* 109(43):17436–17441. doi:10.1073/pnas.1215175109.

Lee, You Ri, Jong-Hwa Park, Soo-Hyun Hahm, Lin-Woo Kang, Ji Hyung Chung, Ki-Hyun Nam, Kwang Yeon Hwang, Ick Chan Kwon, and Ye Sun Han. 2010. Development of Bimolecular Fluorescence Complementation Using Dronpa for Visualization of Protein-Protein Interactions in Cells. *Molecular Imaging and Biology* 12:468–478. doi:10.1007/s11307-010-0312-2.

Leevers, Sally J, Hugh F Paterson, and Christopher J Marshall. 1994. Requirement for Ras in Raf Activation Is Overcome by Targeting Raf to the Plasma Membrane. *Nature* 369(6479):411–414. doi:10.1038/369411a0.

Lito, Piro, Neal Rosen, and David B Solit. 2013. Tumor Adaptation and Resistance to RAF Inhibitors. *Nature Medicine* 19(11):1401–1409. doi:10.1038/nm.3392.

Liu, Zhen, Dong Xing, Qian Peter Su, Yun Zhu, Jiamei Zhang, Xinyu Kong, Boxin Xue et al., 2014. Super-Resolution Imaging and Tracking of Protein–protein Interactions in Sub-Diffraction Cellular Space. *Nature Communications* 5(July):1–8 (Nature Publishing Group). doi:10.1038/ncomms5443.

Luo, Zhijun, Guri Tzivion, Peter J Belshaw, Demetrios Vavvas, Mark Marshall, and Joseph Avruch. 1996. Oligomerization Activates c-Raf-1 through a Ras-Dependent Mechanism. *Nature* 383(6596):181–185. doi:10.1038/383181a0.

Manley, Suliana, Jennifer M Gillette, George H Patterson, Hari Shroff, Harald F Hess, Eric Betzig, and Jennifer Lippincott-Schwartz. 2008. High-Density Mapping of Single-Molecule Trajectories with Photoactivated Localization Microscopy. *Nature Methods* 5(2):155–157.

McCormick, Frank. 1992. Coupling of Ras p21 Signalling and GTP Hydrolysis by GTPase Activating Proteins. *Philosophical Transactions of the Royal Society of London. Series B, Biological Sciences* 336(1276):43–47; discussion 47–48. doi:10.1098/rstb.1992.0042.

McCormick, Frank. 2007. Success and Failure on the Ras Pathway. *Cancer Biology & Therapy* 6(10):1654–1659.

Murphy, Danielle A, and Sara A Courtneidge. 2011. The 'Ins' and 'Outs' of Podosomes and Invadopodia: Characteristics, Formation and Function. *Nature Reviews Molecular Cell Biology* 12(7):413–426. doi:10.1038/nrm3141.

Nan, Xiaolin, Eric A Collisson, Sophia Lewis, Jing Huang, Tanja M Tamgüney, Jan T Liphardt, Frank McCormick, Joe W Gray, and Steven Chu. 2013. Single-Molecule Superresolution Imaging Allows Quantitative Analysis of RAF Multimer Formation and Signaling. *Proceedings of the National Academy of Sciences of the United States of America* 110(46):18519–18524. doi:10.1073/pnas.1318188110.

Nan, Xiaolin, Tanja M Tamgüney, Eric A Collisson, Li-Jung Lin, Cameron Pitt, Jacqueline Galeas, Sophia Lewis, Joe W Gray, Frank McCormick, and Steven Chu. Ras-GTP Dimers Activate the Mitogen-Activated Protein Kinase (MAPK) Pathway. *Proceedings of the National Academy of Sciences of the United States of America*, 2015. doi: 10.1073/pnas.1509123112.

Nickerson, Andrew, Tao Huang, Li-Jung Lin, and Xiaolin Nan. 2014. Photoactivated Localization Microscopy with Bimolecular Fluorescence Complementation (BiFC-PALM) for Nanoscale Imaging of Protein-Protein Interactions in Cells. *PloS One* 9(6):e100589. doi:10.1371/journal.pone.0100589.

Oberholzer, Patrick A, Damien Kee, Piotr Dziunycz, Antje Sucker, Nyam Kamsukom, Robert Jones, Christine Roden et al., 2012. RAS Mutations Are Associated with the Development of Cutaneous Squamous Cell Tumors in Patients Treated with RAF Inhibitors. *Journal of Clinical Oncology* 30(3):316–321. doi:10.1200/JCO.2011.36.7680.

Persson, Fredrik, Martin Lindén, Cecilia Unoson, and Johan Elf. 2013. Extracting Intracellular Diffusive States and Transition Rates from Single-Molecule Tracking Data. *Nature Methods* 10(3):265–269. doi:10.1038/nmeth.2367.

Plowman, Sarah J, Cornelia Muncke, Robert G Parton, and John F Hancock. 2005. H-Ras, K-Ras, and Inner Plasma Membrane Raft Proteins Operate in Nanoclusters with Differential Dependence on the Actin Cytoskeleton. *Proceedings of the National Academy of Sciences of the United States of America* 102(43):15500–15505. doi:10.1073/pnas.0504114102.

Poulikakos, Poulikos I, Chao Zhang, Gideon Bollag, Kevan M Shokat, and Neal Rosen. 2010. RAF Inhibitors Transactivate RAF Dimers and ERK Signalling in Cells with Wild-Type BRAF. *Nature* 464(7287):427–430. doi:10.1038/nature08902.

Poulikakos, Poulikos I, Yogindra Persaud, Manickam Janakiraman, Xiangju Kong, Charles Ng, Gatien Moriceau, Hubing Shi et al., 2011. RAF Inhibitor Resistance Is Mediated by Dimerization of Aberrantly Spliced BRAF(V600E). *Nature* 480(7377):387–390. doi:10.1038/nature10662.

Prior, Ian A, Cornelia Muncke, Robert G Parton, and John F Hancock. 2003. Direct Visualization of Ras Proteins in Spatially Distinct Cell Surface Microdomains. *Journal of Cell Biology* 160(2):165–170. doi:10.1083/jcb.200209091.

Puchner, Elias M, Jessica M Walter, Robert Kasper, Bo Huang, and Wendell A Lim. 2013. Counting Molecules in Single Organelles with Superresolution Microscopy Allows Tracking of the Endosome Maturation Trajectory. *Proceedings of the National Academy of Sciences* 110(40):16015–16020. doi:10.1073/pnas.1309676110.

Rahman, Md Atiqur, Ali Salajegheh, Robert Anthony Smith, and AK-Y Lam. 2014. BRAF Inhibitors: From the Laboratory to Clinical Trials. *Critical Reviews in Oncology/Hematology* 90(3):220–232 (Elsevier Ireland Ltd). doi:10.1016/j.critrevonc.2013.12.008.

Rajakulendran, Thanashan, Malha Sahmi, Martin Lefrançois, Frank Sicheri, and Marc Therrien. 2009. A Dimerization-Dependent Mechanism Drives RAF Catalytic Activation. *Nature* 461(7263):542–545. doi:10.1038/nature08314.

Ripley, Brian D. 1977. Modelling Spatial Patterns. *Journal of the Royal Statistical Society. Series B (Methodological)*, 172–212.

Rizzo, Mark A, Kuntala Shome, Simon C Watkins, and Guillermo Romero. 2000. The Recruitment of Raf-1 to Membranes Is Mediated by Direct Interaction with Phosphatidic Acid and Is Independent of Association with Ras. *The Journal of Biological Chemistry* 275(31):23911–23918. doi:10.1074/jbc.M001553200.

Roberts, Patrick J, and Der, Channing J. 2007. Targeting the Raf-MEK-ERK Mitogen-Activated Protein Kinase Cascade for the Treatment of Cancer. *Oncogene* 26(22):3291–3310. doi:10.1038/sj.onc.1210422.

Rocks, Oliver, Anna Peyker, Martin Kahms, Peter J Verveer, Carolin Koerner, Maria Lumbierres, Jürgen Kuhlmann, Herbert Waldmann, Alfred Wittinghofer, and Philippe IH H Bastiaens. 2005. An Acylation Cycle Regulates Localization and Activity of Palmitoylated Ras Isoforms. *Science (New York, N.Y.)* 307(5716):1746–1752. doi:10.1126/science.1105654.

Rust, Michael J, Mark Bates, and Xiaowei Zhuang. 2006. Sub-Diffraction-Limit Imaging by Stochastic Optical Reconstruction Microscopy (STORM). *Nature Methods* 3(10):793–795. doi:10.1038/nmeth929.

Schubbert, Suzanne, Kevin Shannon, and Gideon Bollag. 2007. Hyperactive Ras in Developmental Disorders and Cancer. *Nature Reviews Cancer* 7(4):295–308. doi:10.1038/nrc2109.

Sengupta, Prabuddha, Tijana Jovanovic-Talisman, Dunja Skoko, Malte Renz, Sarah L Veatch, and Jennifer Lippincott-Schwartz. 2011. Probing Protein Heterogeneity in the Plasma Membrane Using PALM and Pair Correlation Analysis. *Nature Methods* 8(11):969–975 (Nature Publishing Group, a division of Macmillan Publishers Limited. All Rights Reserved). doi:10.1038/nmeth.1704.

Solit, David B, and Neal Rosen. 2014. Towards a Unified Model of RAF Inhibitor Resistance. *Cancer Discovery* 4(1):27–30. doi:10.1158/2159-8290.CD-13-0961.

Stanton, Vincent P, and Geoffrey M Cooper. 1987. Activation of Human Raf Transforming Genes by Deletion of Normal Amino-Terminal Coding Sequences. *Molecular and Cellular Biology* 7(3):1171–1179. doi:10.1128/MCB.7.3.1171.Updated.

Su, Fei, Amaya Viros, Carla Milagre, Kerstin Trunzer, Gideon Bollag, Olivia Spleiss, Jorge S Reis-Filho et al., 2012. Mutations in Cutaneous Squamous-Cell Carcinomas in Patients Treated with BRAF Inhibitors. *New England Journal of Medicine*. doi:10.1056/NEJMoa1105358.

Subach, Fedor V, George H Patterson, Suliana Manley, Jennifer M Gillette, Jennifer Lippincott-Schwartz, and Vladislav V Verkhusha. 2009. Photoactivatable mCherry for High-Resolution Two-Color Fluorescence Microscopy. *Nature Methods* 6(2):153–159. doi:10.1038/nmeth.1298.

Takashima, Asami and Douglas V Faller. 2013. Targeting the RAS Oncogene. *Expert Opinion on Therapeutic Targets* 17(5):507–531. doi:10.1517/14728222.2013.764990.Targeting.

Tian, Tianhai, Angus Harding, Kerry Inder, Sarah Plowman, Robert G Parton, and John F Hancock. 2007. Plasma Membrane Nanoswitches Generate High-Fidelity Ras Signal Transduction. *Nature Cell Biology* 9(8):905–914. doi:10.1038/ncb1615.

Van de Linde, Sebastian, Anna Löschberger, Teresa Klein, Meike Heidbreder, Steve Wolter, Mike Heilemann, and Markus Sauer. 2011. Direct Stochastic Optical Reconstruction Microscopy with Standard Fluorescent Probes. *Nature Protocols* 6(7):991–1009 (Nature Publishing Group, a division of Macmillan Publishers Limited. All Rights Reserved). doi:10.1038/nprot.2011.336.

Vojtek, Anne B, Stanley M Hollenberg, and Jonathan A Cooper. 1993. Mammalian Ras Interacts Directly with the Serine/threonine Kinase Raf. *Cell* 74(1):205–214. doi:10.1016/0092-8674(93)90307-C.

Weber, Christoph K, Joseph R Slupsky, H Andreas Kalmes, and Ulf R Rapp. 2001. Active Ras Induces Heterodimerization of cRaf and BRaf. *Cancer Research* 61(9):3595–3598.

Wellbrock, Claudia, Maria Karasarides, and Richard Marais. 2004. The RAF Proteins Take Centre Stage. *Nature Reviews Molecular Cell Biology* 5(11):875–885. doi:10.1038/nrm1498.

Yang, Hong, Brian Higgins, Kenneth Kolinsky, Kathryn Packman, Zenaida Go, Raman Iyer, Stanley Kolis et al., 2010. RG7204 (PLX4032), a Selective BRAFV600E Inhibitor, Displays Potent Antitumor Activity in Preclinical Melanoma Models. *Cancer Research* 70(13):5518–5527. doi:10.1158/0008-5472.CAN-10-0646.

17

Distribution, Organization, and Dynamics of EGF Receptor in the Plasma Membrane Studied by Super-Resolution Imaging

Susana Rocha, Herlinde De Keersmaecker, Eduard Fron, Mitsuhiro Abe, Asami Makino, Hiroshi Uji-i, Johan Hofkens, Toshihide Kobayashi, Atsushi Miyawaki, and Hideaki Mizuno

17.1 Introduction
17.2 Super-Resolution Imaging of Membrane Components
17.3 Single-Particle Tracking of Receptor Molecules
17.4 Closing Remarks
Acknowledgments
References

17.1 Introduction

The lipid molecules are heterogeneously distributed in the plasma membrane, forming clusters or aggregates often referred to as membrane rafts (Simons and Ikonen 1997). Ever since this heterogeneity was described, there has been an increased interest in the organization of the biological membranes. By microscopic approaches, such as single-particle tracking and fluorescence correlation spectroscopy (Hancock 2006), lipid rafts are now seen as small (10–200 nm), heterogeneous, highly dynamic, sterol- and sphingolipid-enriched domains that compartmentalize cellular processes (Pike 2006). There is increasing evidence that clustering of distinct cholesterol- and sphingolipid-enriched membrane microdomains plays an important role in transmembrane signaling in a variety of mammalian cells. Many receptors, including tumour necrosis factor-α receptors, death receptors, insulin receptors, and integrins, as well as postreceptor signaling molecules, may be localized within these microdomains to form signaling platforms, that is, plasma membrane regions where receptors and downstream molecules co-exist, promoting molecular interactions (Simons and Toomre 2000). Fluorescence microscopy has been used to visualize these membrane microdomains (Gaus et al. 2003; Sugii et al. 2003; Ishitsuka et al. 2004; Reid et al. 2004; Kiyokawa et al. 2005). However, owing to their small size below the diffraction limit of light, conventional fluorescence microscopy is unable to elucidate detailed structure of such domains. Recently, we developed new fluorescent probes against cholesterol- and sphingomyelin-enriched membrane rafts on the plasma membrane (Mizuno et al. 2011), applicable for a super-resolution microscopic modality, photoactivated localization microscopy (PALM) (Betzig et al. 2006). This approach made it possible to observe membrane rafts with a resolution of 10–20 nm, circumventing the diffraction limit.

In addition to the lipid components of plasma membrane, super-resolution microscopy can be used to visualize the distribution and organization of membrane proteins. Using direct stochastic optical reconstitution microscopy (*d*STORM), another super-resolution microscopic modality (Rust et al. 2006; Heilemann et al. 2008), we can visualize the distribution of the epidermal growth factor (EGF) receptor (EGFR) at the nanoscale. The EGF signaling is involved in diverse physiological processes, and EGFR activation is linked to multiple downstream signaling pathways. Proper regulation of EGF signaling is critical for several biological phenomena, such as survival, proliferation, differentiation, and locomotion. Aberrant spatiotemporal regulation of its pathway is involved in different cancer types (Yarden 2001). Overexpression of EGFR is observed in various types of cancer cells, and hence, drugs that antagonize the EGF signaling pathway have been used as anticancer drugs (Citri and Yarden 2006). The EGFR is localized in the plasma membrane and forms a dimer on EGF binding, which leads to transautophosphorylation of the receptor C-terminal tail. Phosphorylated EGFR forms a signaling complex with downstream adaptor and effector molecules, including Grb2, PLCγ, Shc, and c-Cbl (Bogdan and Klämbt 2001). It has been proposed that EGFR and some downstream molecules co-exist in microdomains before EGF binding, which promotes molecular interaction. Clustering of receptor molecules could also be important for EGFR signaling activated by ligands other than EGF (Linggi and Carpenter 2006) and could control signaling differences between different EGFR/ErbB family members (Lemmon 2009, Sorkin and Goh 2009). Furthermore, the distribution of the receptor in the plasma membrane can be regulated by other factors such as protein–protein interactions. It is also known that microdomains play an important role in EGFR endocytosis on stimulation, which is a major downregulation pathway (Pike et al. 2005). These molecular interactions are also likely to affect the dynamics of the receptor molecules.

In addition to acquiring super-resolution images, the visualization of single emitters by PALM/*d*STORM grants the possibility to analyze molecular dynamics by single-particle tracking (SPT). This technique is based on image acquisition of single particles as a function of time by fluorescence microscopy. In the last decades, SPT has become an essential tool for studying dynamics in biological processes (Kusumi et al. 2014). Tracking of individual particles provides a better spatial and temporal resolution than other methods to analyze molecular dynamics, such as fluorescence recovery after photobleaching (FRAP), and more importantly, since there is no averaging, heterogeneity of the processes is retained. In this chapter, we introduce PALM/*d*STORM and SPT to investigate distribution, organization, and dynamics of EGFR in the plasma membrane.

17.2 Super-Resolution Imaging of Membrane Components

17.2.1 Indirect Immunostaining for *d*STORM Imaging of EGFR

In order to perform imaging by super-resolution microscopy, EGFR molecules need to be labeled fluorescently. Here, we introduce two methods to label EGFR with Alexa 647 for *d*STORM imaging: indirect immunostaining and agonist labeling.

In the first method, we use an antibody against the EGFR ectodomain, Ab-5 (mouse monoclonal antibody, GR15; Calbiochem) as a primary antibody, and Alexa Fluor 647 goat antimouse IgG (H+L) (Life Technologies) as a secondary antibody. Since our target molecules are localized in the plasma membrane, pretreatments that affect membrane structure such as permeabilization should be avoided. Therefore, we have chosen a primary antibody that recognizes the extracellular region of the receptor. Figure 17.1 shows EGFR molecules in the plasma membrane of HeLa cells stained by this method. Even by confocal microscopy, clusters of EGFR distributing on both the apical and basal membranes are visible (Figure 17.1a).

As reported by (Heilemann et al. 2008), Alexa 647 fluorophores can be switched reversibly between a fluorescent and a dark state with high efficiency by using a "switching buffer": phosphate-buffered saline (PBS, pH 7.4) containing oxygen scavenger (0.5 mg/mL glucose oxidase, 40 mg/mL catalase, and 10 w/v% glucose) and 50 mM β-mercaptoethylamine (MEA). Exploiting the switching behavior of Alexa 647, detailed information concerning the nanoscale distribution of the EGFR molecules was obtained by *d*STORM. By using total internal reflection (TIRF) mode on a wide-field microscope, the illumination was restricted to a few hundred nanometers above the glass coverslip. This means that fluorophores present at the bottom membrane of the cells are exclusively excited, eliminating the background fluorescence caused by out-of-focus molecules (Figure 17.1b). The specimen was simultaneously illuminated at 635 and 489 nm, and the laser power was adjusted to ensure that only a subset of fluorophores is activated at any time in the field of view. Precise determination of particle localization was achieved by fitting of a 2D Gaussian function to the point spread function of the detected molecules (Equation 17.1).

$$I = I_0 \cdot \exp\left[-\frac{1}{2}\left[\left(\frac{x-x_0}{s_x}\right)^2+\left(\frac{y-y_0}{s_y}\right)^2\right]\right]+b \tag{17.1}$$

where s is the diffraction limit resolution of the optics and b is the background on the region used for the fitting. The localization precision ($\sigma_{x,y}$) depends on s, pixel size (q), photon counts from the molecule (N), and b (Equation 17.2) (Thompson et al. 2002).

$$\sigma_{x,y}^2 = \frac{s_{x,y}^2+(q^2/12)}{N}+\frac{8\pi s^4 b^2}{q^2 N^2} \tag{17.2}$$

The determined positions were then used to reconstruct the *d*STORM images with a precision of 18 nm (Figure 17.1c). On the super-resolution images, small clusters of EGFR molecules can be visualized. The heterogeneity of labeled membrane systems can be quantified using mathematical tools. One of the most

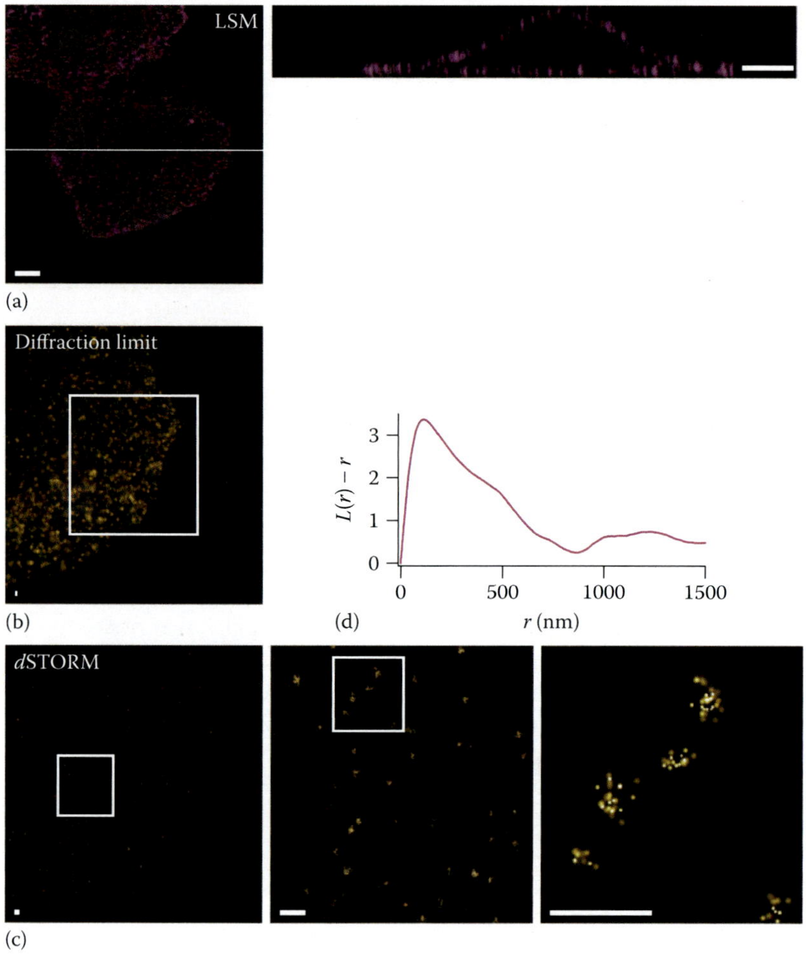

Figure 17.1

Clustering of EGFR in HeLa cells under resting conditions. HeLa cells were fixed without EGF stimulation and immunostained using mouse anti-human EGFR antibody (Ab-5) as a primary antibody and goat anti-mouse IgG antibody labeled with Alexa 647 as a secondary antibody. (a) Confocal 3D image of a cell. The left panel is a z-projection and the right panel is an xz-plane at the position indicated with the white line in the left panel. (b) Diffraction-limited image of the cell under TIRF illumination. (c) *d*STORM images of the region indicated with the white box in B. Expansion of a region indicated with the white box is shown on the right side of respective panels. (d) Ripley's K-function analysis of the clusters found in the whole region shown in the middle image of (c). Scale bars indicate 10 μm (a) or 500 nm (b and c).

commonly used methods to quantify spatial distribution heterogeneity is the Ripley's K-function (Kiskowski et al. 2009). The K-function ($K(r)$) is defined as

$$K(r)=\frac{1}{\lambda \cdot n}\sum_{i\neq j} I(d_{ij}<r) \tag{17.3}$$

where d_{ij} is the distance between the ith and jth point in a data set of n points, r is the search radius, λ is the average density of the points, and I is the indicator function, which is 1 if the operand is true and 0 if otherwise. The expected value of $K(r)$ for a random Poisson distribution is πr^2. Deviations from this value indicate some degree of clustering. The K-function can be normalized, so that its expected value for a random distribution is r (linear):

$$L(r) = \sqrt{K(r)/\pi} \tag{17.4}$$

The K-function can be further normalized, so that the expected value is 0, yielding the so-called H-function:

$$H(r) = L(r) - r \tag{17.5}$$

A positive value of $H(r)$ indicates clustering, whereas a negative value indicates dispersion. Furthermore, the maximum of $H(r)$ represents the radius of maximal clustering: the radius of a disk in which a centred test point on average contains the most points per area; therefore, it approximates the domain radius (Kiskowski et al. 2009).

Ripley's analysis was performed on simulated data that displayed clustering with different radius, from 50 to 400 nm (Figure 17.2). As depicted in Figure 17.2c, the maximum of the H-function increases with the radius of the cluster.

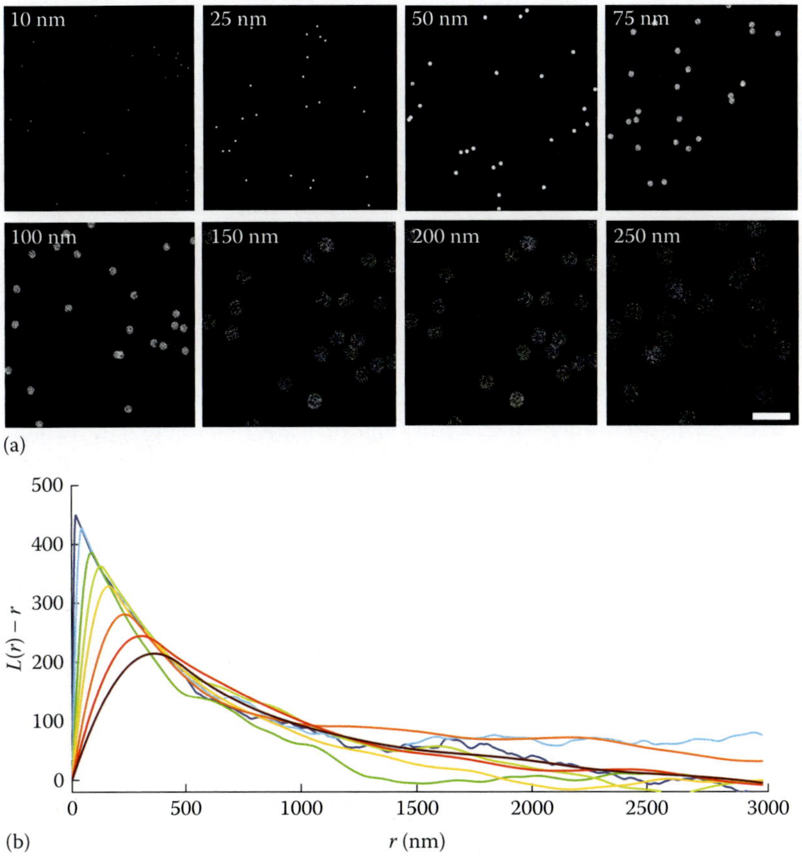

Figure 17.2

Ripley's analysis. (a) Simulated fields with 25 circular clusters (10–250 nm in diameter). Each cluster contains 250 particles. (b) Ripley's K-function analysis of the simulated fields: note the decrease of peak intensities and shift of maxima toward larger r at larger cluster sizes.

Ripley's analysis of *d*STORM images of immunostained EGFR molecules at resting conditions revealed the presence of clusters with an average size of 112 nm (Figure 17.1d). Although the antibody size (ca. 15 nm) can affect spatial resolution of the measurement, we believe that the calculated size of the observed structures is large enough to neglect the influence of the size of the antibody.

17.2.2 Labeled Ligand to Visualize Activated EGFR

Another way of labeling EGFR molecules is by using labeled agonist. Biotinylated EGF complexed to Alexa Fluor 647 (EGF-Alexa647; Life Technologies) is commercially available (Wilde et al. 1999). Labeling EGFR with EGF-Alexa647 might induce a cellular response and only liganded EGFR molecules are visualized, whereas the immunostaining method labeled all EGFR molecules and is therefore applicable to both in the presence and in the absence of the ligand. Since the agonist is smaller than the antibody, the effect of label size is minimized. Here, we used 800 nM EGF-Alexa 647 to stimulate HeLa cells, which is more than two orders of magnitude higher than the dissociation constant of EGF to EGFR (two classes of EGF-binding sites have been reported with dissociation constants [K_d] in the order of subnanomolar and nanomolar range [Berkers et al. 1991; Johannessen et al. 2001]). Under such conditions, all EGFR molecules on the plasma membrane are expected to bind to EGF-Alexa647. During longer incubation, EGF-bound receptors are likely to be removed from the plasma membrane by endocytosis; we observed by confocal microscopy that all EGFR molecules binding to EGF-Alexa 647 were endocytosed within 20 min (Figure 17.7d). The issue of endocytosis can be avoided by fixing the cells with 4% formaldehyde within 5 min after EGF-Alexa 647 addition. At this time point, all labeled EGFRs are still present in the plasma membrane, forming clusters with an average radius of 99 nm (Figure 17.3a–c). Interestingly, some of the clusters appeared to be composed of smaller clusters (cluster-of-clusters, right panel of Figure 17.3b, arrowhead). The diameter of these small clusters was around 50 nm. If the cells are fixed 20 min after EGF-Alexa647 addition, endosomes containing receptor molecules could be visualized. The radii of EGF clusters in the endosomes ranged from 50 nm to 1.5 μm, with an average of 277 nm (Figure 17.3e, f), larger than the clusters observed on the plasma membrane after 5 min of stimulation. These observations indicate that multiple clusters of EGFR on the plasma membrane were gathered and incorporated into an endosome upon prolonged stimulation with EGF.

The similarity in size, shape, and density of EGFR clusters in resting conditions and after 5 min of stimulation (Figures 17.1b, c and 17.3c, d) suggests that the clusters are already formed at rest and that 5-min incubation with EGF does not evoke any structural change. We assumed that the clusters serve as platforms for EGFR signaling and are present before EGF binding. In these platforms, EGFR molecules at rest are concentrated together with downstream molecules ready to trigger the signaling pathway.

17.2.3 Membrane Raft Probes for PALM

Fluorescent probes for the visualization of membrane rafts were recently reported, based on protein toxins that have a high affinity toward lipid components (Yamaji et al. 1998; Shimada et al. 2002; Ohno-Iwashita et al. 2004; Kiyokawa et al. 2005). We developed two different probes for PALM to visualize cholesterol- or sphingomyelin-enriched domains, based on θ-toxin and lysenin, respectively.

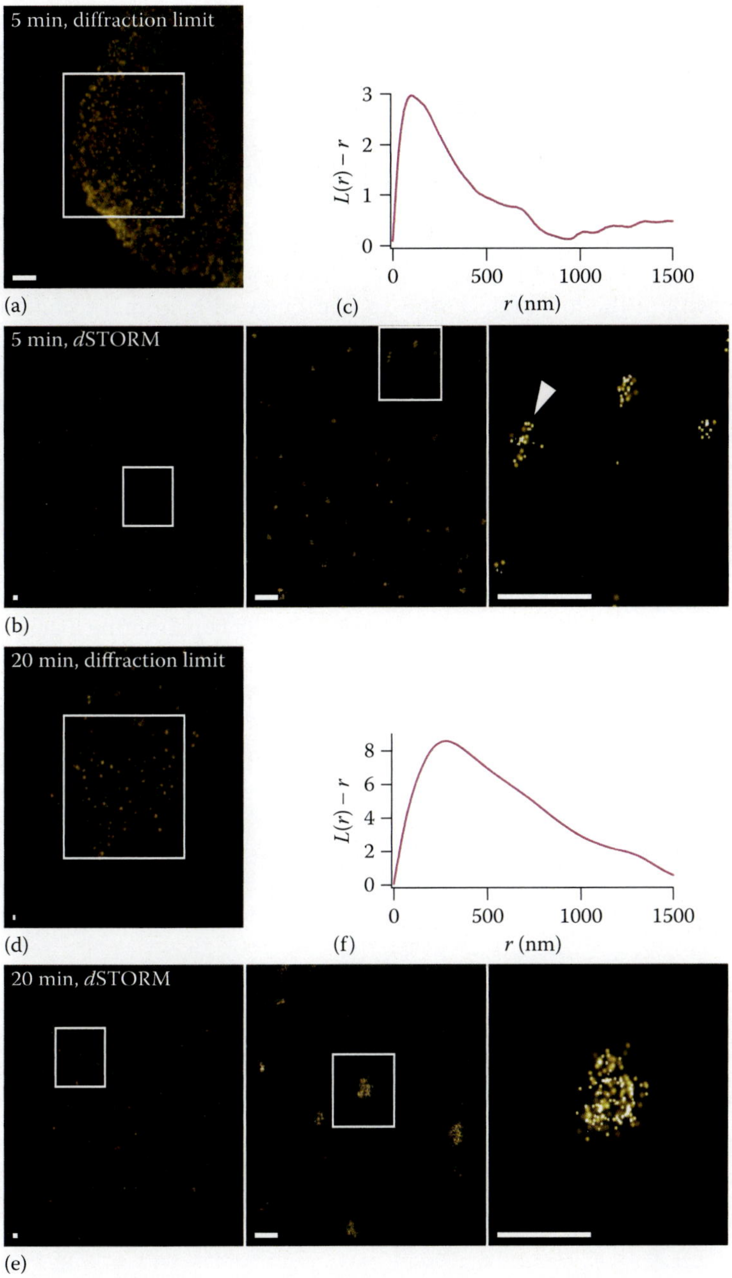

Figure 17.3

Clustering of EGFR in HeLa cells after the EGF stimulation. Cells were stimulated with 800 nM of EGF-Alexa647 for 5 min (a–c) or 20 min (d–f) and fixed with 4% ice-cold formaldehyde. Note that signals were acquired from the upper surface of the cells after 5 min stimulation (a–c), whereas the focal plane was set to inside of the cells after 20 min stimulation to visualize endosomes (d–f). (a, d) Diffraction-limited images. Boxes indicate the regions subjected to *d*STORM imaging. (b, e) *d*STORM images. Expansion of the region indicated within the box is shown on the right side of respective panels. The arrowhead on the right panel of B indicates an example of a cluster of small clusters. (c, f) Ripley's K-function analyses of the domains found over the whole regions of the middle panels of (b) and (e). Scale bars indicate 500 nm.

The cytolytic θ-toxin secreted by *Clostridium perfringens* (perfringolysin O) binds to cholesterol, where it oligomerizes to form pores in the plasma membrane. θ-toxin is a protein toxin composed of four domains, of which the C-terminal domain (θ-D4) is sufficient for specific binding to cholesterol, without inducing cellular damage (Shimada et al. 2002). The binding capacity of the lipid bilayer for the θ-toxin varied with cholesterol contents (Ohno-Iwashita et al. 1992). Liposomes with 41 mol% cholesterol contain at least two classes of θ-toxin binding sites, with K_d values of 2.7 and 170 nM. The high-affinity binding sites appear only in liposomes with 36 mol% or higher cholesterol. The binding capacity of the low-affinity sites in liposomes with 10 mol% cholesterol is two order of magnitude less than that with 40 mol% cholesterol. Hence, binding of the θ-toxin is negligible when the membrane contains less than 25 mol% of cholesterol (Ohno-Iwashita et al. 2004). This implies that θ-toxin binds only to regions enriched with cholesterol. A similar strategy was applied to design a probe for sphingomyelin-enriched membrane rafts by using a protein toxin, lysenin, derived from coelomic fluid of the earthworm *Eisenia fetida*. Lysenin specifically binds to sphingomyelin, oligomerizes, and forms pores on the membrane (Yamaji et al. 1998). The N-terminal truncated version of lysenin (NT-Lys) binds to sphingomyelin but fails to cause cell death (Kiyokawa et al. 2005). The membrane raft probes for PALM were developed by constructing a gene encoding the lipid-binding sites of the protein toxins fused to a photochromic fluorescent protein, Dronpa (Ando et al. 2004; Habuchi et al. 2005; Flors et al. 2007; Mizuno et al. 2008) on a bacterial expression vector. By using this plasmid, the probes (Dronpa-θ-D4, Dronpa-NT-Lys) can be produced in *E. coli* as recombinant proteins (Mizuno et al. 2011). It was confirmed by enzyme-linked immunosorbent assay (ELISA) that the protein toxins labeled with Dronpa conserve the specific binding to cholesterol or sphingomyelin (Figure 17.4).

Incubation of cells with 0.1 μM of purified probes for 5–10 min stains the cholesterol- or sphingomyelin-enriched membrane rafts. This concentration is close

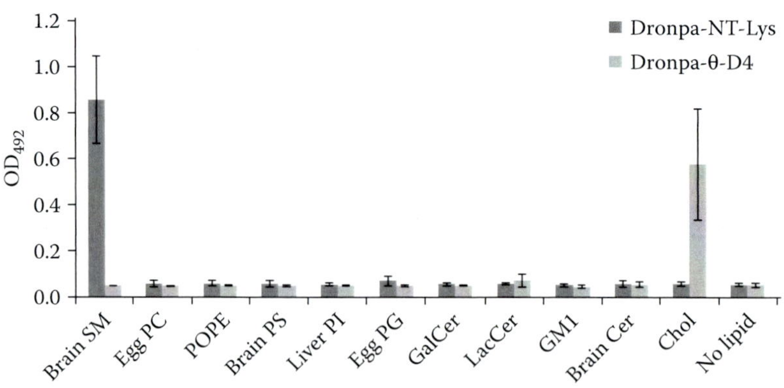

Figure 17.4

Probe specificity. Specific binding of Dronpa-θ-D4 (light gray) and Dronpa-NT-Lys (dark gray) to cholesterol and sphingomyelin, respectively, analyzed by ELISA. SM, sphingomyelin; PC, phosphatidylcholine; PE, phosphatidylethanolamine; PS, phosphatidylserine; PI phosphatidylinositol; PG, phosphatidylglycerol; GalCer, galactosylceramide; LacCer, lactosylceramide; GM1, monosialotetrahexosylganglioside; Cer, ceramide; Chol, cholesterol.

to the K_d value of cholesterol- or sphingomyelin-enriched membranes to θ-D4 and NT-Lys, respectively. 1%–3.5% of bovine serum albumin is added to the staining solution to reduce nonspecific binding of the probes. In the case of NT-Lys staining, we sometimes observed incorporation of the probe to endosomes after longer incubation (~60 min), but this was not observed when shorter incubation times were used (5–10 min). In contrast, no incorporation of θ-D4 was observed even after longer incubation times. After staining, the cells were fixed by incubation in ice-cold 4% formaldehyde for 1 h. Although it is reported that this protocol is unsuccessful in fixing the lipid components of the membrane (Tanaka et al. 2010), the proteinous nature of the probes used allows crosslinking with other proteins, and therefore, the probes are subjected to fixation by formaldehyde. After the fixation, the probes are still slightly mobile, with a diffusion coefficient of 0.013 $\mu m^2/s$, which is 2 orders of magnitude slower than that of lipid molecules in the plasma membrane (Figure 17.5). This diffusion coefficient corresponds to the mean displacement of 46 nm between consecutive frames (frame rate of

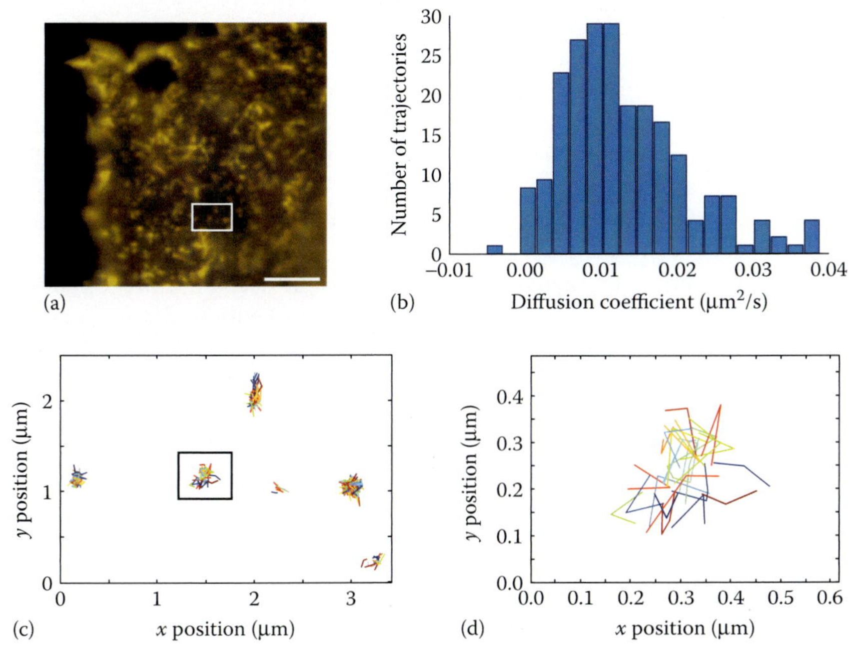

Figure 17.5

Mobility of Dronpa-θ-D4 molecules on the plasma membrane after fixation. (a) Diffraction-limited image of cholesterol-enriched domains. Scale bar: 5 μm. (b) Distribution of the diffusion coefficients of trajectories described by Dronpa-θ-D4 molecules that remain fluorescent for more than 4 consecutive frames (corresponding to 124 ms). The diffusion coefficient is calculated using the first 3 points of the MSD for each individual trajectory (for a total of 220 trajectories). The mean value of the diffusion coefficient (0.013 $\mu m^2/s$) corresponds to a mean displacement of 46 nm between consecutive frames. (c) Trajectories described by Dronpa-θ-D4 molecules in the area indicated in (a). Each color indicates a different track. Although mobile, the molecules are confined within the cholesterol-enriched domains. (d) Ten randomly chosen trajectories present in the domain, indicated by the black square in (c). Each Dronpa-θ-D4 molecule covers an area smaller than the total area of the domain. (Mizuno, H. et al., *Chem. Sci.*, 2, 1548–1553, 2011. Reproduced by permission of The Royal Society of Chemistry.)

35 ms, 2D diffusion model). The movement is confined in an area with a diameter of ≈200 nm, which is consistent with the size of membrane rafts (Mizuno et al. 2011). The probes stain the apical membranes but are not detected on the basal membranes of HeLa cells. We interpreted this as an inherent property in the cell membrane. The effect of probe accessibility is not completely excluded, but the clearance between the basal membrane and the coverslip is large enough for the probes to penetrate; Alexa647-EGF (6 kDa) can access the bottom membrane, and assuming that both EGF and the membrane probes are globular, Dronpa-NT-Lys (41 kDa) and Dronpa-θ-D4 (38 kDa) are only 1.9- and 1.8-fold larger in diameter than Alexa647-EGF. Since TIRF illumination restricts the excitation volume to a few hundred nanometers above the coverslip, it is inadequate to observe the apical membrane. Instead, we used highly inclined laminated optical sheet (HILO) illumination (Konopka and Bednarek 2008; Tokunaga et al. 2008), which illuminates a higher region with a lower background fluorescence level than epi-illumination, providing a signal-to-noise ratio high enough for single-molecule imaging.

Figure 17.6 shows cholesterol- and sphingomyelin-enriched membrane rafts observed by super-resolution microscopy. HeLa cells stained with Dronpa-θ-D4 revealed two types of clustering or nanodomains. One type of nanodomains displays a line shape, with a width of around 150 nm and length of about 0.7–5.5 μm, attributed to microvilli. This is consistent with the fact that the membrane in microvilli is enriched with cholesterol (Röper et al. 2000). In addition to the microvilli, round-shaped regions with an average radius of 118 nm (calculated using the Ripley's K-function analysis) are detected and are attributed to cholesterol-enriched membrane rafts. Using Dronpa-NT-Lys, round-shaped sphingomyelin-enriched regions could be observed, with an average radius of 124 nm. The shape and density of the regions detected with both probes are different, indicating heterogeneity of membrane rafts.

17.2.4 Dual-color Super-resolution Imaging of EFGR with Membrane Rafts

Super-resolution imaging revealed that even under resting conditions, EGFR molecules form clusters on the plasma membrane. The question remains whether the organization of EGFR microdomains is dependent on the distribution of other membrane nanodomains or not. By adding the Dronpa-labeled membrane probes and Alexa-647-EGF sequentially or simultaneously, cells can be efficiently stained with both probes, allowing visualization of both the distribution of cholesterol- or sphingomyelin-enriched membrane rafts and liganded EGFR in the same cells. The liganded EGFRs spread on both the apical and basal plasma membranes. Since the membrane rafts appeared only on the apical membrane, it is clear that the clusters of EGFR on the basal membrane are neither related to the cholesterol- nor to the sphingomyelin-enriched membrane rafts detectable with the probes (Figure 17.7a, d). Super-resolution imaging of the apical membrane showed that clusters of liganded EGFRs are distributed neither on cholesterol- nor on sphingomyelin-enriched membrane domains, except in a few regions overlapping apparently by coincidence (less than 2% of the EGFR clusters) (Figure 17.8c, d). It might be possible that the binding of the probes shields the EGF binding site of EGFR, but since the density of the EGFR clusters is similar to the one observed in the absence of the membrane raft probes, we think that the observed spatial separation reflects a physical segregation

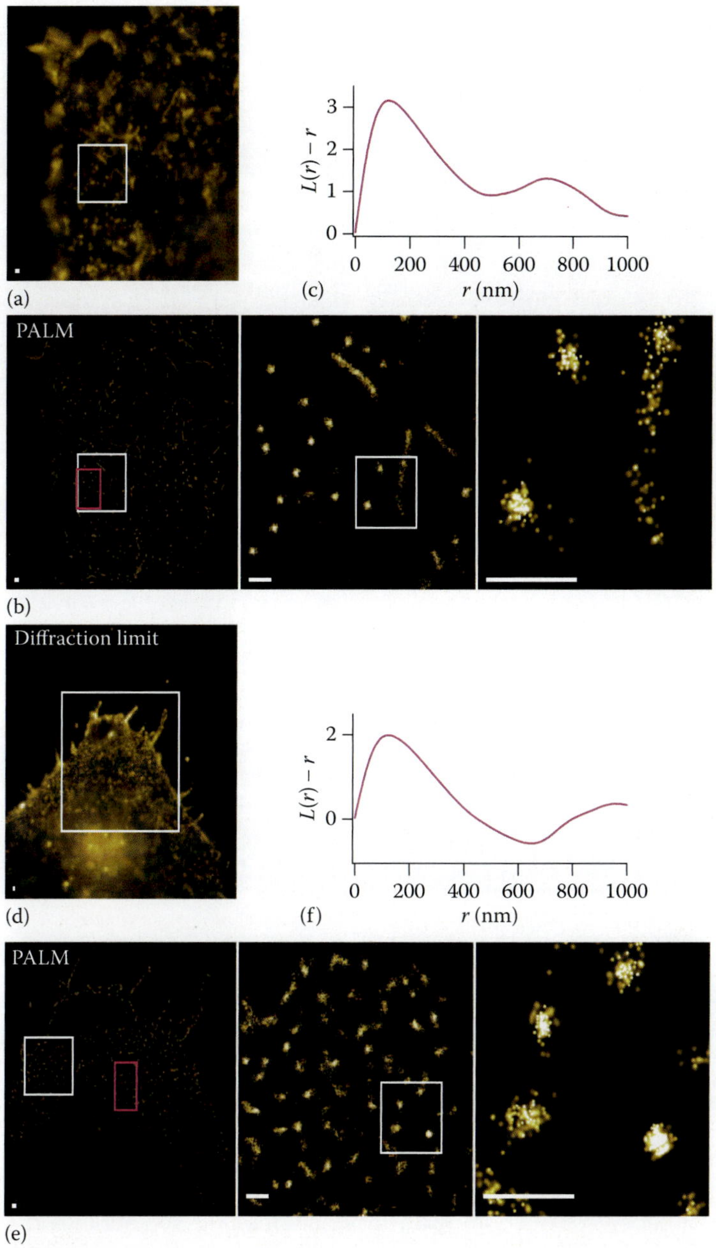

Figure 17.6

Cholesterol- and sphingomyelin-enriched domains on the plasma membrane of HeLa cells. Cells were stained with Dronpa-θ-D4 (a, b) or Dronpa-NT-Lys (d, e) for 10 min and then fixed with 4% formaldehyde. (a, d) Diffraction-limited image under HILO illumination. (b, e) PALM image of the same region as a and d. Expansion of the region indicated with the box is shown on the right side of the respective panels. The spots represent detected single molecules, and the size of the respective spot corresponds to the precision of the fitting. (c, f) Representative Ripley's K-function analysis. The analysis was applied at the regions indicated with the magenta box on the PALM image. These regions contain only round-shaped clusters. Scale bars indicate 500 nm. (Mizuno, H. et al., *Chem. Sci.*, 2, 1548–1553, 2011. Reproduced by permission of The Royal Society of Chemistry.)

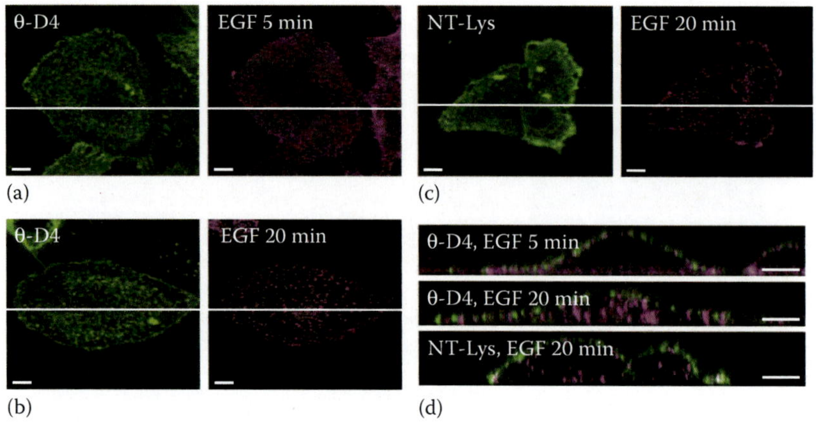

Figure 17.7

Membrane microdomains in HeLa cells stimulated with EGF-Alexa647. Cells were stained with Dronpa-θ-D4 (a, b, d *top* and *middle*) or Dronpa-NT-Lys (c, d *bottom*) for 10 min and stimulated with EGF-Alexa647 for 5 min (a, d *top*) or 20 min (b, c, d *middle* and *bottom*). (a–c) z-projections of 3D confocal images for Dronpa (green) and Alexa 647 (magenta). (d) xz-sections at the positions indicated with white lines in (a–c). After stimulation with EGF for 20 min, the labeled receptors were endocytosed, while the toxin-based probes remained on the membrane. Scale bars indicate 10 μm.

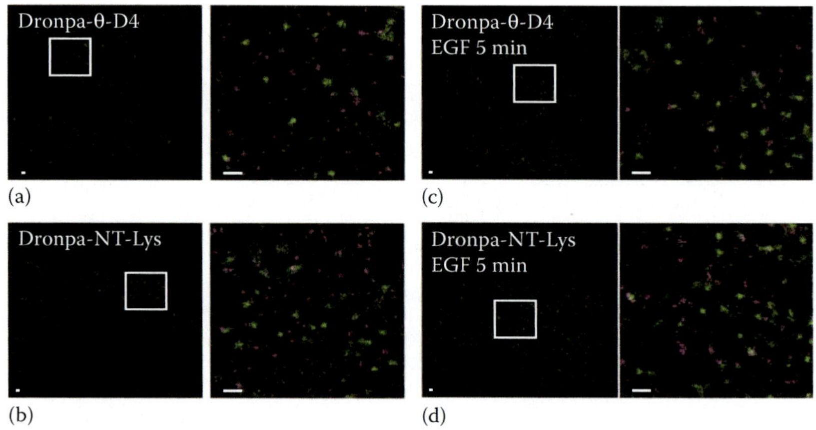

Figure 17.8

Dual-color super-resolution imaging for cholesterol- or sphingomyelin-enriched membrane microdomains with EGFR clusters. Cells were stained with Dronpa-θ-D4 (a, c) or Dronpa-NT-Lys (b, d) for 10 min, followed by immunostaining (using mouse anti-human EGFR antibody as a primary antibody and goat anti-mouse IgG antibody labeled with Alexa647) (a, b) or stimulation with EGF-Alexa647 for 5 min (c, d). The dual-color sub-diffraction images show the fluorescent signals from Dronpa (green) and from Alexa 647 (magenta). The right panels show expansions of the region indicated within the boxes. Scale bars indicate 500 nm.

between the membrane rafts and EGFR clusters. During 20 min of incubation with Alexa647-EGFR, most of the liganded EGFR molecules were endocytosed; however, the cholesterol- and sphingomyelin-enriched regions remained on the surface (Figure 17.7b–d), consistent with the hypothesis that EGFR molecules are distributed outside of the membrane rafts detectable with these probes.

As an alternative way of the dual-color super-resolution imaging, cells stained with the membrane raft probes can be subjected to indirect immunostaining of EGFR. The distributions of EGFR and the membrane rafts under resting conditions are essentially the same as that 5 min after the EGF addition (Figure 17.8a, b). The lack of substantial overlap between EGFR clusters and lipid-enriched nanodomains allows us to conclude that EGFR molecules do not form clusters at the cholesterol- or sphingomyelin-enriched domains detectable with Dronpa-θ-D4 and Dronpa-NT-Lys. Nevertheless, regions containing EGFR clusters might incorporate a small amount of cholesterol and/or sphingomyelin, with a concentration below the detectable level of our probes.

Despite that EGFR clusters are not formed on the cholesterol-enriched membrane rafts and hence the membrane rafts do not serve as platforms for EGFR clustering, there have been several reports showing that depletion of cholesterol enhances EGF signals (Furuchi and Anderson 1998; Pike and Casey 2002; Ringerike et al. 2002). It has been reported that methyl-β-cyclodextrin (MβCD) extracts cholesterol from cultured cells. Interestingly, no sphingomyelin-enriched regions could be observed after cholesterol depletion (Mizuno et al. 2011). A small amount of cholesterol might be required for the formation of these domains. In contrast, cholesterol is not required for the clustering of EGFRs; the EGFR clusters were observed after 5 min incubation with Alexa647-EGF even on the cholesterol-depleted plasma membrane of HeLa cells (Figure 17.9a, b). However, in the cholesterol-depleted HeLa cells, the EGFR clusters were not endocytosed during 20 min incubation and remained on the plasma membrane (Figure 17.9a, b, e). The absence of endocytosis probably results from perturbing the formation of clathrin-coated endocytic vesicles by cholesterol extraction with MβCD (Rodal et al. 1999). It is known that the clathrin-mediated endocytosis influences the EGF signaling pathway (Vieira et al. 1996). The perturbation of EGFR endocytosis might cause the reported enhancement of EGF signals by cholesterol depletion. In contrast, the depletion of sphingomyelin by

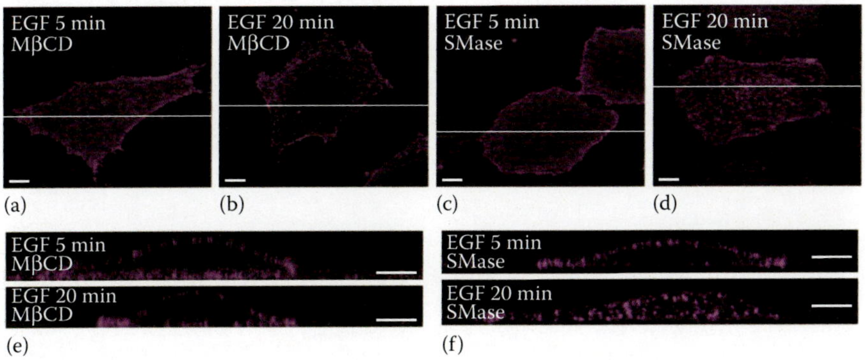

Figure 17.9

HeLa cells stimulated with EGF-Alexa647 after depletion of cholesterol or sphingomyelin. Cells were stimulated with EGF-Alexa647 for 5 min (a, c, e *top*, f *top*) or 20 min (b, d, e *bottom*, f *bottom*). (a–d) z-projections of 3D confocal images for EGF-Alexa647 (magenta). (e, f) xz-sections at the positions indicated with white lines in (a–d). Cholesterol was depleted from the cells by MβCD treatment in a, b, and e, and sphingomyelin was depleted by sphingomyelinase treatment in (c), (d), and (f). Scale bars indicate 10 μm.

sphingomyelinase treatment perturbed neither the EGFR cluster formation on the plasma membrane nor the endocytosis of EGFR (Figure 17.9c, d, f).

17.3 Single-Particle Tracking of Receptor Molecules

17.3.1 Dynamics of EGFR Molecules

In addition to acquiring super-resolution images in fixed samples, detection of single molecules allows to analyze the dynamics of EGFR in living cells. For this, Alex 647-EGF is added to living HeLa cells and observed with a wide-field fluorescence microscope under TIRF illumination. Thanks to the TIRF illumination, observation is restricted to the bottom membrane. Free Alexa 647-EGF molecules emit fluorescence but diffuse much faster than the integration time of respective images (46 ms), and as a result, they are detected as weak blurred signals. On binding to the receptor, the movement of the molecules slows down enough to appear as single spots during the recording. Each single molecule is tracked, until it bleaches or dissociates from the receptor. As for PALM/*d*STORM imaging, the precise determination of molecular localization is achieved by fitting a 2D Gaussian function to the point spread function of the detected molecules (Equation 17.1). Images of Alexa-EGF-bound EGFR were acquired, with an acquisition rate of 10 frame/s, and time trajectories of each molecule were calculated by connecting molecules in successive frames that were separated within 0.5 μm from each other. Figure 17.10 shows some example trajectories described by single Alexa647-EGF molecules. From a first look of the trajectories, one can detect different modes of diffusion. While some molecules are stationary, others are mobile, diffusing randomly.

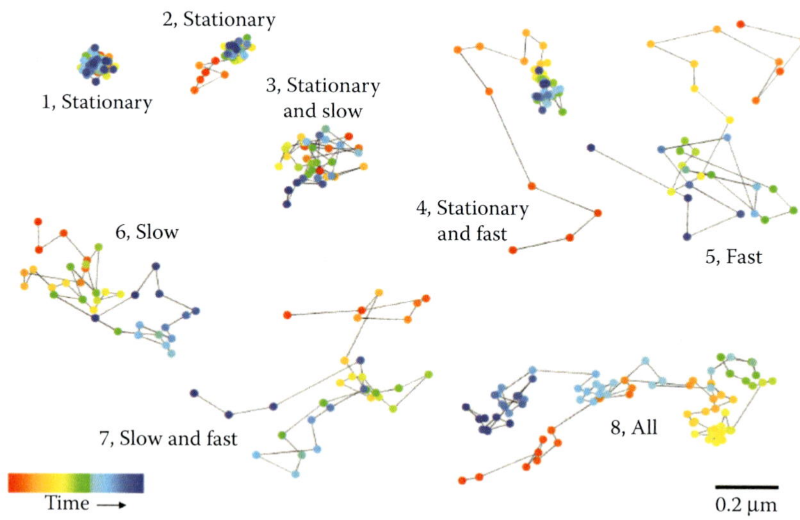

Figure 17.10

Dynamics of EGFR in HeLa cells. For SPT imaging, HeLa cells were stimulated with 1.6 nM of Alexa 647-EGF. Some examples of trajectories described by EGF receptor, showing one, two, or three different modes of motion (as indicated in the figure). (Adapted from De Keersmaecker, H. et al., *Biophys. Rev. Lett.*, 8, 229–242, Copyright at 2013 World Scientific Publishing Company.)

It is reported that some of the EGFR molecules are stationary and are thought to be immobilized by the binding to F-actin, as the EGFR has a binding site to F-actin on its C-terminal (Tang and Gross 2003). Trajectories are considered immobile if the $\langle r^2(\tau)\rangle < 4\sigma^2$ for any time lag (τ) or if $\langle r^2(\tau = \Delta t)\rangle > \langle r^2(\tau > \Delta t)\rangle$. In our experiment, 37.0 ± 10.6% of the EGFR trajectories were considered stationary (Figure 17.10, trajectories 1 and 2) and were therefore excluded from further analysis.

The mobile trajectories can be further analyzed as their motion provides information on the surroundings and interactions of the molecule. The most common analysis starts with calculation of the mean-square displacement (*MSD*). The *MSD* for single-molecule trajectories is the average displacement over the complete trajectory of all measurement intervals, $n\Delta t$:

$$MSD(\tau \equiv n\Delta t) = \langle r^2(\tau)\rangle = \frac{1}{(N-n)}\sum_{i=1}^{N-n}\left[(x_{i+n} - x_i)^2 + (y_{i+n} - y_i)^2\right] \quad (17.6)$$

where Δt is the frame interval (0.1 s in the case of the analysis of Alexa647-EGF) and N is the total number of frames for the trajectory (Konopka and Weisshaar 2004). x_i and y_i give the position of the detected particle at $t = i$. For each n, the sum runs over all possible intervals of length $\tau = n\Delta t$. For random lateral diffusion, the diffusion coefficient D can be calculated using:

$$\langle r^2(\tau)\rangle = 4D\tau + 4\sigma^2 \quad (17.7)$$

where σ is the localization precision of the experiment. The localization precision can be determined from Equation 17.2 or experimentally by tracking immobilized fluorescent molecules on the glass surface. In the case of Alexa647-EGF, 20 immobilized molecules were tracked for at least 50 consecutive frames, yielding a mean displacement of $\sigma = 28 \pm 5$ nm.

The major drawback of single-molecule SMD analysis is the assumption that each of the individual molecules exhibits homogeneous motion. For trajectories in which molecules change behavior, as in the case of EGFR, simply extracting D from the slope of the MSD curves will provide a mode of diffusion that reflects the average behavior of the molecule (Haas et al. 2014). To overcome this limitation of MSD analysis and include models for heterogeneous motion, an analytical tool that considers all steps independently of their original trajectories is required. One such method is to use cumulative distribution functions (CDFs) of the step sizes. For 2D motion, CDFs will describe the probability of a molecule staying within an area defined by a radius, r, during a given time-lag, τ. Analysis of the CDF of the step size by fitting the CDF to a multicomponent model (Equation 17.8) provides a framework for considering heterogeneous mixtures of molecular populations (Schütz et al. 1997). This model allows multiple molecule populations, each with a different diffusion coefficient, by including one exponential term per molecular population:

$$P(r^2,\tau) = 1 - \alpha_1 \cdot \exp\left(\frac{r^2}{\langle r_1^2(\tau)\rangle}\right) - \alpha_2 \cdot \exp\left(\frac{r^2}{\langle r_2^2(\tau)\rangle}\right) - \cdots \quad (17.8)$$

Fitting the CDFs to this model gives the fraction of molecules in each population (α_1, α_2, etc., where $\alpha_1 + \alpha_2 + \ldots = 1$), as well as the MSD for each population at each time lag ($\langle r_1^2(\tau)\rangle, \langle r_2^2(\tau)\rangle$, etc.). Thereafter, the MSD values are plotted as in single-molecule MSD analysis, and the diffusion coefficient for each population is calculated from the slopes of these curves, as described above (Equation 17.7). Since short trajectories contain a large number of false positives (background signals), only molecules that appeared in more than 10 successive frames were considered for the CDF analysis of EGFR diffusion in HeLa cells.

The cumulative distribution function (CDF) analysis of mobile EGF trajectories gave three diffusion coefficients and their fractions: $D_{fast} = 0.081 \pm 0.009\ \mu m^2/s$, $\alpha_{fast\_CDF} = 49 \pm 6\%$; $D_{slow} = 0.020 \pm 0.005\ \mu m^2/s$, $\alpha_{slow\_CDF} = 38 \pm 6\%$; and $D_{stationary} = 0.0015 \pm 0.0007\ \mu m^2/s$, $\alpha_{stationary\_CDF} = 13 \pm 2\%$. Note that each molecule would not always diffuse in one mode but can switch between modes. In the majority of the cases, the trajectories described by EGF were composed of 2 or more modes (Figure 17.10). Despite excluding stationary molecules from the analysis, part of the trajectories described by the mobile EGF molecules presented a diffusion coefficient lower than $4\sigma^2$ ($D_{stationary}$, considered stationary). This was due to the presence of molecules that switched from the stationary to mobile mode or *vice versa* during the observation (Figure 17.10, trajectories 3, 4, and 8). Taking the excluded molecules from the CDF analysis as stationary trajectories into account, the fractions of fast (α_{fast}), slow (α_{slow}), and stationary ($\alpha_{stationary}$) modes were calculated to be $29 \pm 9\%$, $22 \pm 6\%$, and $49 \pm 13\%$, respectively.

17.3.2 Dynamics of EGFR and Cellular Stimulation

It is proposed that the affinity of EGF toward its receptor depends on the occupancy of the receptor dimer: the high affinity is related to the binding of EGF to the EGF receptor monomer or to the unoccupied dimer, whereas the low affinity is related to the binding of EGF to the second site of the receptor dimer (Macdonald and Pike 2008; Macdonald-Obermann and Pike 2009). In the case of HeLa cells, the reported K_d values are 0.12 and 9.2 nM for the high- and low-affinity binding sites, respectively (Berkers et al. 1991). Submaximal stimulation is crucial to investigate the physiological regulation of the EGF binding, and under such conditions, the response to the stimulation can be different cell by cell. According to the reported affinity values, subnanomolar to nanomolar concentration of EGF can stimulate the cell submaximally. To evaluate the cellular response to EGF stimulation, we followed the cytosolic Ca^{2+} concentration with a genetically encoded Ca^{2+} indicator, cameleon YC2.60 (Miyawaki et al. 1997; Nagai et al. 2004). On EGF binding, the EGFR forms a dimer, which catalyzes the transautophosphorylation of several tyrosine residues of the C-terminal tail of the receptor. Among these residues, the phosphotyrosines at positions 1143, 1173, and 992 interact with PLCγ, triggering its activation (Rotin et al. 1992; Chattopadhyay et al. 1999; Bogdan and Klämbt 2001; Jorissen et al. 2003). Activated PLCγ produces IP_3 by hydrolyzing phosphatidylinositol-4,5-bisphosphate on the plasma membrane. IP_3 evokes Ca^{2+} release from the endoplasmic reticulum to the cytosol through the IP_3 receptor, leading to a rise in the concentration of cytosolic Ca^{2+}. We found that only around half of HeLa cells show Ca^{2+} response on the stimulation with 1.6 nM EGF (Figure 17.11) (De Keersmaecker et al. 2013).

With this result, we became aware of the importance of monitoring the response of the cells while investigating the dynamics of the receptor under submaximal stimulation. In order to perform functional imaging during the image acquisition for SPT of Alexa 647-EGF, we developed a new system (De Keersmaecker et al. 2013). Fluorescence images of the Alexa647 channel were

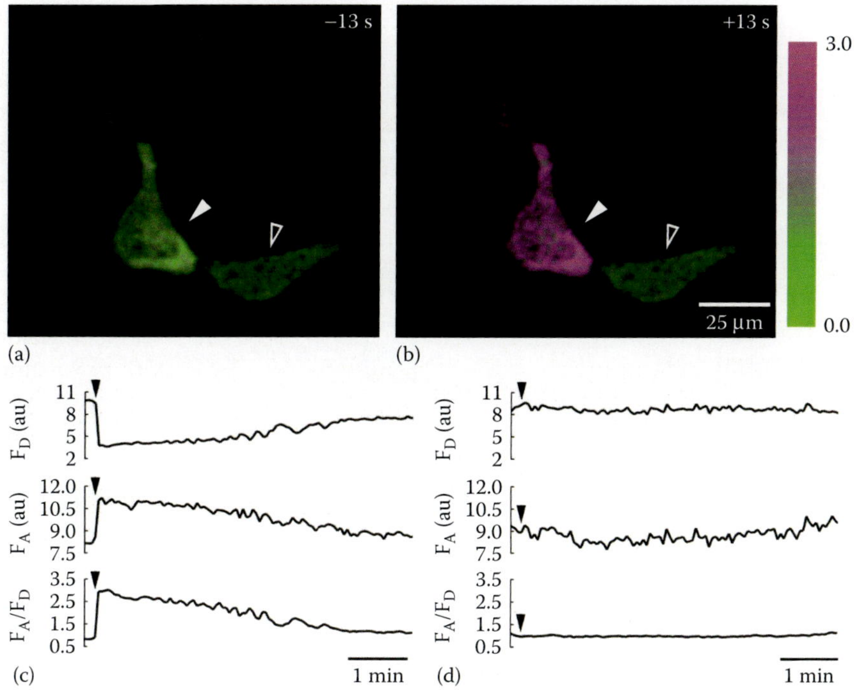

Figure 17.11

Heterogeneous Ca^{2+} response on submaximal EGF stimulation in HeLa cells. HeLa cells expressing YC2.60 were subjected to fluorescence imaging by confocal microscopy. The FRET donor, eCFP, was selectively activated and direct fluorescence from eCFP (donor channel) and fluorescence from CP173Venus (acceptor channel) were recorded. (a, b) Ratiometric images of the cells 13 s before (a) and after (b) the addition of EGF. Images shown are the ratio of the acceptor channel over the donor channel in intensity-modified pseudocolor display mode. (c, d) Time traces of the fluorescence intensities of both donor and acceptor channels (top and middle), and the intensity ratio of the acceptor channel over donor channel (bottom). Two cells that responded (c) and did not respond (d) are shown as examples. White and black arrowheads indicate the responding and nonresponding cells in (b). (Adapted from De Keersmaecker, H. et al., *Biophys. Rev. Lett.*, 8, 229–242, Copyright at 2013 World Scientific Publishing Company.)

continuously acquired every 100 ms (46 ms exposure time with 54 ms interval), whereas the images of both donor and acceptor channels of the YC2.60 were acquired every 5 s during the interval of the image acquisition of the Alexa647 channel (Figure 17.12a and b). The signal for the donor channel of YC2.60 was segregated from other channels by reflection by a long-pass dichroic mirror. Despite being sent to the same camera, the signals for the Alexa647 channel and the acceptor channel of YC2.60 could be distinguished by the image acquisition synchronized with selective laser illumination to excite the respective fluorophores: the 644-nm laser for Alexa647 and the 448-nm laser for YC2.60. Figure 17.12 shows the excitation and acquisition schemes, as well as some typical fluorescence images from the different channels. At the time point indicated with the arrow in Figure 17.12d, Alexa647-EGF was added to the sample to a final concentration of 1.6 nM. On the Alexa647 channel, signals from Alexa647-EGF trapped on the surface of the cell appeared after addition of Alexa647-EGF.

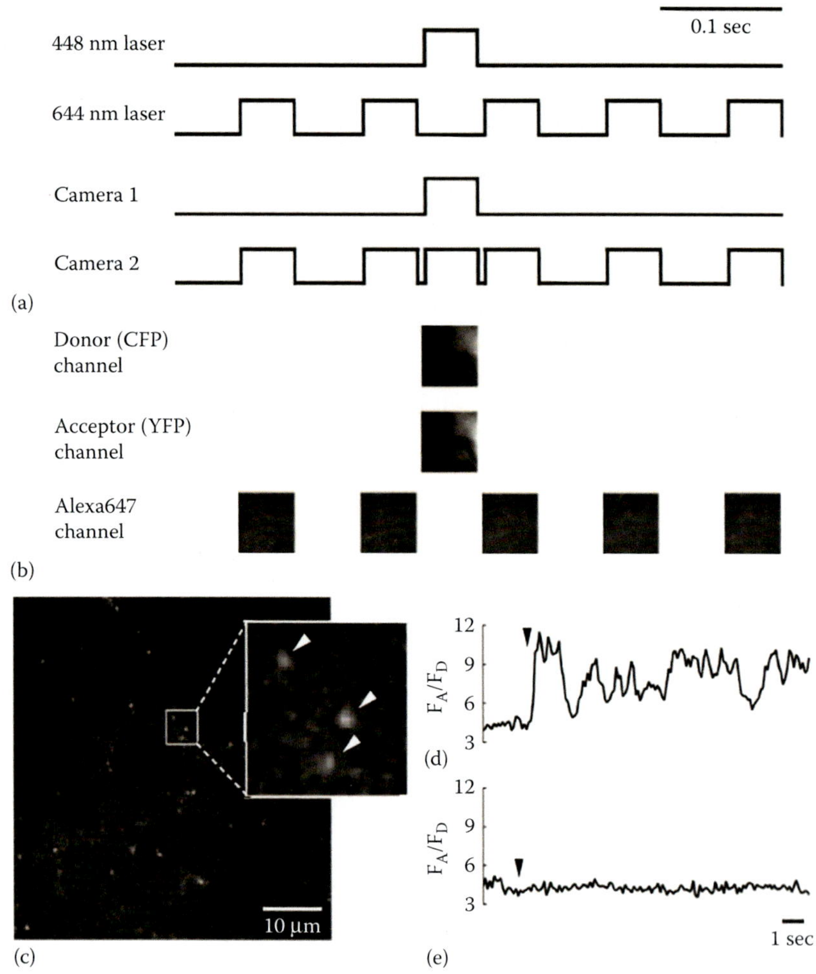

Figure 17.12

Functional imaging during SPT data acquisition using wide-field microscopy. (a) Excitation and acquisition schemes. The FRET sensor is excited using 448 nm in between the excitation of Alexa647-EGF (with 644 nm laser). The fluorescence signal from the donor is acquired with camera 1, whereas the signal from the acceptor is recorded with camera 2, as is the fluorescence coming from the Alexa647-EGF. (b) Typical fluorescence images acquired in the different channels. (c) Fluorescence image of Alexa647-EGF. Singl-molecule signals are clearly discriminated from the background (single molecules indicated by white arrow heads in the magnified area). (d, e) Examples of time traces of the fluorescence ratio F_A/F_D for cells responding to EGF stimulation (d) and nonresponding (e). EGF was added to a final concentration of 1.6 nM on the time indicated by the black arrowhead. (Adapted from De Keersmaecker, H. et al., *Biophys. Rev. Lett.*, 8, 229–242, Copyright at 2013 World Scientific Publishing Company.)

The fluorescence images of the acceptor and donor channels of YC2.60 were used to calculate the F_A/F_D ratio (Figure 17.12c, d). As expected from the previous measurements, the cytosolic Ca^{2+} increased on the EGF addition in some of the cells, whereas it remained at the basal level in other cells (Figure 17.12c and d, respectively). Based on the response in cytosolic Ca^{2+}, the cells could be separated into responding and nonresponding to EGF stimulation.

Table 17.1 Summary of the Diffusion Coefficients and Fraction of the 3 Detected Motions of EGFR in HeLa Cells Responding and Nonresponding to EGF Stimulation. The Values are Provided with Mean ± Standard Deviation

	Stationary	Slow	Fast
Responding cells	–	$D = 0.020 \pm 0.005\ \mu m^2/s$	$D = 0.081 \pm 0.009\ \mu m^2/s$
	$\alpha = 49 \pm 13\%$	$\alpha = 22 \pm 6\%$	$\alpha = 29 \pm 9\%$
Nonresponding cells	–	$D = 0.014 \pm 0.006\ \mu m^2/s$	$D = 0.069 \pm 0.009\ \mu m^2/s$
	$\alpha = 58 \pm 1\%$	$\alpha = 20 \pm 8\%$	$\alpha = 22 \pm 9\%$

With this system, we revealed that the molecular dynamics of the EGF receptor in the responding cells was different from that of nonresponding cells. The values provided in the previous section describe the dynamics of the EGF-bound receptor in responding cells. In contrast, where no change in the cytosolic Ca^{2+} concentration occurred, the same analysis yielded the following values: $D_{fast} = 0.069 \pm 0.009\ \mu m^2/s$, $\alpha_{fast} = 22 \pm 9\%$ and $D_{slow} = 0.014 \pm 0.006\ \mu m^2/s$, $\alpha_{slow} = 20 \pm 8\%$; $\alpha_{stationary} = 58 \pm 1\%$ (Table 17.1). From the calculated values, D_{fast} for nonresponding cells was smaller than that for responding cells, with statistical significance ($p < 0.05$), whereas no significant difference was observed for other values.

It has been reported that the diffusion coefficient decreased from 0.17 to 0.08 $\mu m^2/s$ on the dimer formation of EGFR (Chung et al. 2010). The larger D_{fast} of EGFR in the responding cells observed here was interpreted as a higher tendency to stay in monomeric form. Since the dimer formation is the first step of the EGFR signaling, at first glance, the higher monomer fraction apparently contradicts to the cell response. The result can be explained by considering a feedback mechanism in which EGFR is monomerized. Indeed, the $PLC\gamma/IP_3/Ca^{2+}$ signaling pathway is known to compose a feedback loop, which inhibits the kinase activity of EGF receptor (Davis 1988) and abolishes high-affinity EGF binding to EGFR (Livneh et al. 1987). Further investigation by PALM/(d)STORM and SPT will shed a light on the molecular mechanism of the feedback regulation.

17.4 Closing Remarks

Super-resolution microscopy has been successfully applied to monolayers of cultured cells to the study of several membrane proteins, namely tetherin (Lehmann et al. 2011), arrestin (Truan et al. 2013), ryanodine receptors and caveolin (Wong et al. 2013), Her2/neu receptors (Kaufmann et al. 2011), G-protein-coupled receptors (Scarselli et al. 2013), P-glycoproteins (Huber et al. 2012), SNAP receptors (Pertsinidis et al. 2013), cytokine receptors (Gabor et al. 2013), and glycine receptors (Notelaers et al. 2014). These reports and the visualization of EGFR molecules shown in this chapter demonstrate the potential of optical super-resolution microscopy in revealing protein distribution and membrane organization at the nanometer scale. The super-resolution imaging techniques are not restricted to monolayers of cultured cells but are now steadily being implemented to study more complex biological systems, including whole-animal models (Berning et al. 2012; Gao et al. 2012; York et al. 2012; Willig et al. 2014). In addition, the labeling strategy and optical setup used for PALM can be applied to investigate molecular dynamics by tracking single molecules/particles. Furthermore, combination of SPT with functional imaging provides additional information regarding the triggering of a specific signaling pathway. SPT provides access to nanometer accuracy on millisecond timescales, which allows monitoring of fast dynamics of a transmembrane receptor with high spatial resolution in a complex biological system. Super-resolution microscopy is expected to be an indispensable tool for further exploring spatiotemporal regulation of cell signaling, especially the formation and function of microdomains on the plasma membrane serving signaling platforms (Simons and Toomre 2000; Cambi and Lidke 2012).

Acknowledgments

The authors thank Dr. Y. Ohno-Iwashita for fruitful discussion and technical assistance. This work was partly supported by "Interdisciplinaire Onderzoeksprogramma (IDO) KU Leuven" (IDO/12/020). S.R. is a recipient of a "postdoctoraal mandaat van het Bijzonder Onderzoeksfonds (BOF) KU Leuven" (PDMK/14/108). H. D. K. is supported by a PhD grant Agentschap voor Innovatie door Wetenschap en Technologie (IWT number 121165) of the Institute for the Promotion of Innovation through Science and Technology in Flanders (IWT-Vlaanderen).

References

Ando, Ryoko, Hideaki Mizuno, and Atsushi Miyawaki. 2004. Regulated Fast Nucleocytoplasmic Shuttling Observed by Reversible Protein Highlighting. *Science* 306(5700):1370–1373. doi:10.1126/science.1102506.

Berkers, Jos A, Paul M van Bergen en Henegouwen, and Johannes Boonstra. 1991. Three Classes of Epidermal Growth Factor Receptors on HeLa Cells. *The Journal of Biological Chemistry* 266(2):922–927.

Berning, Sebastian, Katrin I Willig, Heinz Steffens, Payam Dibaj, and Stefan W Hell. 2012. Nanoscopy in a Living Mouse Brain. *Science* 335(6068):551. doi:10.1126/science.1215369.

Betzig, Eric, George H Patterson, Rachid Sougrat, O Wolf Lindwasser, Scott Olenych, Juan S Bonifacino, Michael W Davidson, Jennifer Lippincott-Schwartz, and Harald F Hess. 2006. Imaging Intracellular Fluorescent Proteins at Nanometer Resolution. *Science* 313(5793):1642–1645. doi:10.1126/science.1127344.

Bogdan, Sven, and Christian Klämbt. 2001. Epidermal Growth Factor Receptor Signaling. *Current Biology* 11(8):R292–R295.

Cambi, Alessandra, and Diane S Lidke. 2012. Nanoscale Membrane Organization: Where Biochemistry Meets Advanced Microscopy. *ACS Chemical Biology* 7(1):139–149. doi:10.1021/cb200326g.

Chattopadhyay, Ansuman, Manuela Vecchi, Qun-sheng Ji, Raymond Mernaugh, and Graham Carpenter. 1999. The Role of Individual SH2 Domains in Mediating Association of Phospholipase C-Gamma1 with the Activated EGF Receptor. *The Journal of Biological Chemistry* 274(37):26091–26097.

Chung, Inhee, Robert Akita, Richard Vandlen, Derek Toomre, Joseph Schlessinger, and Ira Mellman. 2010. Spatial Control of EGF Receptor Activation by Reversible Dimerization on Living Cells. *Nature* 464(7289):783–787. doi:10.1038/nature08827.

Citri, Ami, and Yosef Yarden. 2006. EGF–ERBB Signalling: Towards the Systems Level. *Nature Reviews Molecular Cell Biology* 7(7):505–516 (Nature Publishing Group). doi:10.1038/nrm1962.

Davis, Roger J. 1988. Independent Mechanisms Account for the Regulation by Protein Kinase-C of the Epidermal Growth-Factor Receptor Affinity and Tyrosine-Protein Kinase-Activity. *The Journal of Biological Chemistry* 263(19):9462–9469.

De Keersmaecker, Herlinde, Susana Rocha, Eduard Fron, Hiroshi Uji-i, Johan Hofkens, and Hideaki Mizuno. 2013. EGF Receptor Dynamics in Egf-Responding Cells Revealed by Functional Imaging During Single Particle Tracking. *Biophysical Reviews and Letters* 8(03n04):229–242 (World Scientific Publishing Company). doi:10.1142/S1793048013500070.

Flors, Cristina, Jun-ichi Hotta, Hiroshi Uji-i, Peter Dedecker, Ryoko Ando, Hideaki Mizuno, Atsushi Miyawaki, and Johan Hofkens. 2007. A Stroboscopic Approach for Fast Photoactivation-Localization Microscopy with Dronpa Mutants. *Journal of the American Chemical Society* 129(45):13970–13977. doi:10.1021/ja074704l.

Furuchi, Takemitsu, and Richard GW Anderson. 1998. Cholesterol Depletion of Caveolae Causes Hyperactivation of Extracellular Signal-Related Kinase (ERK). *The Journal of Biological Chemistry* 273(33):21099–21104.

Gabor, Kristin A, Chad R Stevens, Matthew J Pietraszewski, Travis J Gould, Juyoung Shim, Jeffrey A Yoder, Siew Hong Lam, Zhiyuan Gong, Samuel T Hess, and Carol H Kim. 2013. Super Resolution Microscopy Reveals That Caveolin-1 Is Required for Spatial Organization of CRFB1 and Subsequent Antiviral Signaling in Zebrafish. *PLoS One* 8(7):e68759. doi:10.1371/journal.pone.0068759.

Gao, Liang, Lin Shao, Christopher D Higgins, John S Poulton, Mark Peifer, Michael W Davidson, Xufeng Wu, Bob Goldstein, and Eric Betzig. 2012. Noninvasive Imaging Beyond the Diffraction Limit of 3D Dynamics in Thickly Fluorescent Specimens. *Cell* 151(6):1370–1385. doi:10.1016/j.cell.2012.10.008.

Gaus, Katharina, Enrico Gratton, Eleanor P W Kable, Allan S Jones, Ingrid Gelissen, Leonard Kritharides, and Wendy Jessup. 2003. Visualizing Lipid Structure and Raft Domains in Living Cells with Two-Photon Microscopy. *Proceedings of the National Academy of Sciences of the United States of America* 100(26):15554–15559. doi:10.1073/pnas.2534386100.

Haas, Beth L, Jyl S Matson, Victor J DiRita, and Julie S Biteen. 2014. Imaging Live Cells at the Nanometer-Scale with Single-Molecule Microscopy: Obstacles and Achievements in Experiment Optimization for Microbiology. *Molecules (Basel, Switzerland)* 19(8):12116–12149. doi:10.3390/molecules190812116.

Habuchi, Satoshi, Ryoko Ando, Peter Dedecker, Wendy Verheijen, Hideaki Mizuno, Atsushi Miyawaki, and Johan Hofkens. 2005. Reversible Single-Molecule Photoswitching in the GFP-Like Fluorescent Protein Dronpa. *Proceedings of the National Academy of Sciences of the United States of America* 102(27):9511–9516. doi:10.1073/pnas.0500489102.

Hancock, John F. 2006. Lipid Rafts: Contentious Only From Simplistic Standpoints. *Nature Reviews Molecular Cell Biology* 7(6):456–462. doi:10.1038/nrm1925.

Heilemann, Mike, Sebastian van de Linde, Mark Schüttpelz, Robert Kasper, Britta Seefeldt, Anindita Mukherjee, Philip Tinnefeld, and Markus Sauer. 2008. Subdiffraction-Resolution Fluorescence Imaging with Conventional Fluorescent Probes. *Angewandte Chemie International Edition* 47(33):6172–6176. doi:10.1002/anie.200802376.

Huber, Olga, Alexander Brunner, Patrick Maier, Rainer Kaufmann, Pierre-Olivier Couraud, Christoph Cremer, and Gert Fricker. 2012. Localization Microscopy (SPDM) Reveals Clustered Formations of P-Glycoprotein in a Human Blood-Brain Barrier Model. *PLoS One* 7(9):e44776. doi:10.1371/journal.pone.0044776.

Ishitsuka, Reiko, Akiko Yamaji-Hasegawa, Asami Makino, Yoshio Hirabayashi, and Toshihide Kobayashi. 2004. A Lipid-Specific Toxin Reveals Heterogeneity of Sphingomyelin-Containing Membranes. *Biophysical Journal* 86(1 Pt 1):296–307. doi:10.1016/S0006-3495(04)74105-3.

Johannessen, Lene E, Karianne E Haugen, Anne Carine Østvold, Espen Stang, and Inger Helene Madshus. 2001. Heterodimerization of the Epidermal-Growth-Factor (EGF) Receptor and ErbB2 and the Affinity of EGF Binding Are Regulated by Different Mechanisms. *The Biochemical Journal* 356(Pt 1):87–96.

Jorissen, Robert N, Francesca Walker, Normand Pouliot, Thomas P J Garrett, Colin W Ward, and Antony W Burgess. 2003. Epidermal Growth Factor Receptor: Mechanisms of Activation and Signalling. *Experimental Cell Research* 284(1):31–53.

Kaufmann, Rainer, Patrick Müller, Georg Hildenbrand, Michael Hausmann, and Christoph Cremer. 2011. Analysis of Her2/Neu Membrane Protein Clusters in Different Types of Breast Cancer Cells Using Localization Microscopy. *Journal of Microscopy* 242(1):46–54. doi:10.1111/j.1365-2818.2010.03436.x.

Kiskowski, Maria A, John F Hancock, and Anne K Kenworthy. 2009. On the Use of Ripley's K-Function and Its Derivatives to Analyze Domain Size. *Biophysical Journal* 97(4):1095–1103. doi:10.1016/j.bpj.2009.05.039.

Kiyokawa, Etsuko, Takeshi Baba, Naomi Otsuka, Asami Makino, Shinichi Ohno, and Toshihide Kobayashi. 2005. Spatial and Functional Heterogeneity of Sphingolipid-Rich Membrane Domains. *The Journal of Biological Chemistry* 280(25):24072–24084. doi:10.1074/jbc. M502244200.

Konopka, Catherine A, and Sebastian Y Bednarek. 2008. Variable-Angle Epifluorescence Microscopy: A New Way to Look at Protein Dynamics in the Plant Cell Cortex. *The Plant Journal: For Cell and Molecular Biology* 53(1):186–196. doi:10.1111/j.1365-313X.2007.03306.x.

Konopka, Michael C, and James C Weisshaar. 2004. Heterogeneous Motion of Secretory Vesicles in the Actin Cortex of Live Cells: 3D Tracking to 5-Nm Accuracy. *The Journal of Physical Chemistry A* 108(45):9814–9826 (American Chemical Society). doi:10.1021/jp048162v.

Kusumi, Akihiro, Taka A Tsunoyama, Kohichiro M Hirosawa, Rinshi S Kasai, and Takahiro K Fujiwara. 2014. Tracking Single Molecules at Work in Living Cells. *Nature Chemical Biology* 10(7):524–532. doi:10.1038/NCHEMBIO.1558.

Lehmann, Martin, Susana Rocha, Bastien Mangeat, Fabien Blanchet, Hiroshi Uji-i, Johan Hofkens, and Vincent Piguet. 2011. Quantitative Multicolor Super-Resolution Microscopy Reveals Tetherin HIV-1 Interaction. *PLoS Pathogens* 7(12):e1002456. doi:10.1371/journal.ppat.1002456.

Lemmon, Mark A. 2009. Ligand-Induced ErbB Receptor Dimerization. *Experimental Cell Research* 315(4):638–648. doi:10.1016/j.yexcr.2008.10.024.

Linggi, Bryan, and Graham Carpenter. 2006. ErbB Receptors: New Insights on Mechanisms and Biology. *Trends in Cell Biology* 16(12):649–656. doi:10.1016/j.tcb.2006.10.008.

Livneh, Etta, Nachum Reiss, Eva Berent, Axel Ullrich, and Joseph Schlessinger. 1987. An Insertional Mutant of Epidermal Growth-Factor Receptor Allows Dissection of Diverse Receptor Functions. *The EMBO Journal* 6(9):2669–2676.

Macdonald, Jennifer L, and Linda J Pike. 2008. Heterogeneity in EGF-Binding Affinities Arises From Negative Cooperativity in an Aggregating System. *Proceedings of the National Academy of Sciences* 105(1):112–117. doi:10.1073/pnas.0707080105.

Macdonald-Obermann, Jennifer L, and Linda J Pike. 2009. The Intracellular Juxtamembrane Domain of the Epidermal Growth Factor (EGF) Receptor Is Responsible for the Allosteric Regulation of EGF Binding. *The Journal of Biological Chemistry* 284(20):13570–13576. doi:10.1074/jbc. M109.001487.

Miyawaki, Atsushi, Juan Llopis, Roger Heim, J Michael McCaffery, Joseph A Adams, Mitsuhiko Ikura, and Roger Y Tsien. 1997. Fluorescent Indicators for Ca^{2+} Based on Green Fluorescent Proteins and Calmodulin. *Nature* 388(6645):882–887. doi:10.1038/42264.

Mizuno, Hideaki, Mitsuhiro Abe, Peter Dedecker, Asami Makino, Susana Rocha, Yoshiko Ohno-Iwashita, Johan Hofkens, Toshihide Kobayashi, and Atsushi Miyawaki. 2011. Fluorescent Probes for Superresolution Imaging of Lipid Domains on the Plasma Membrane. *Chemical Science* 2(8):1548–1553 (Royal Society of Chemistry). doi:10.1039/C1SC00169H.

Mizuno, Hideaki, Tapas Kumar Mal, Markus Wälchli, Akihiro Kikuchi, Takashi Fukano, Ryoko Ando, Jeyaraman Jeyakanthan, et al. 2008. Light-Dependent Regulation of Structural Flexibility in a Photochromic Fluorescent Protein. *Proceedings of the National Academy of Sciences of the United States of America* 105(27):9227–9232. doi:10.1073/pnas.0709599105.

Nagai, Takeharu, Shuichi Yamada, Takashi Tominaga, Michinori Ichikawa, and Atsushi Miyawaki. 2004. Expanded Dynamic Range of Fluorescent Indicators for Ca(2+) by Circularly Permuted Yellow Fluorescent Proteins. *Proceedings of the National Academy of Sciences of the United States of America* 101(29):10554–10559. doi:10.1073/pnas.0400417101.

Notelaers, Kristof, Susana Rocha, Rik Paesen, Nina Swinnen, Jeroen Vangindertael, Jochen C Meier, Jean-Michel Rigo, Marcel Ameloot, and Johan Hofkens. 2014. Membrane Distribution of the Glycine Receptor A3 Studied by Optical Super-Resolution Microscopy. *Histochemistry and Cell Biology* 142(1):79–90. doi:10.1007/s00418-014-1197-y.

Ohno-Iwashita, Yoshiko, Machiko Iwamoto, Susumu Ando, and Shintaro Iwashita. 1992. Effect of Lipidic Factors on Membrane Cholesterol Topology—Mode of Binding of Theta-Toxin to Cholesterol in Liposomes. *Biochimica et Biophysica Acta* 1109(1):81–90.

Ohno-Iwashita, Yoshiko, Yukiko Shimada, A Abdul Waheed, Masami Hayashi, Mitsushi Inomata, Megumi Nakamura, Mikako Maruya, and Shintaro Iwashita. 2004. Perfringolysin O, a Cholesterol-Binding Cytolysin, as a Probe for Lipid Rafts. *Anaerobe* 10(2):125–134. doi:10.1016/j.anaerobe.2003.09.003.

Pertsinidis, Alexandros, Konark Mukherjee, Manu Sharma, Zhiping P Pang, Sang Ryul Park, Yunxiang Zhang, Axel T Brunger, Thomas C Südhof, and Steven Chu. 2013. Ultrahigh-Resolution Imaging Reveals Formation of Neuronal SNARE/Munc18 Complexes *in Situ*. *Proceedings of the National Academy of Sciences* 110(30):E2812–E2820. doi:10.1073/pnas.1310654110.

Pike, Linda J. 2006. Rafts Defined: A Report on the Keystone Symposium on Lipid Rafts and Cell Function. *The Journal of Lipid Research* 47(7):1597–1598. doi:10.1194/jlr. E600002-JLR200.

Pike, Linda J and Laurieann Casey. 2002. Cholesterol Levels Modulate EGF Receptor-Mediated Signaling by Altering Receptor Function and Trafficking. *Biochemistry* 41(32):10315–10322.

Pike, Linda J, Xianlin Han, and Richard W Gross. 2005. Epidermal Growth Factor Receptors Are Localized to Lipid Rafts That Contain a Balance of Inner and Outer Leaflet Lipids: A Shotgun Lipidomics Study. *The Journal of Biological Chemistry* 280(29):26796–26804. doi:10.1074/jbc. M503805200.

Reid, Patrick C, Naomi Sakashita, Shigeki Sugii, Yoshiko Ohno-Iwashita, Yukiko Shimada, William F Hickey, and Ta-Yuan Chang. 2004. A Novel Cholesterol Stain Reveals Early Neuronal Cholesterol Accumulation in the Niemann-Pick Type C1 Mouse Brain. *The Journal of Lipid Research* 45(3):582–591. doi:10.1194/jlr. D300032-JLR200.

Ringerike, Tove, Frøydis D Blystad, Finn O Levy, Inger H Madshus, and Espen Stang. 2002. Cholesterol Is Important in Control of EGF Receptor Kinase Activity but EGF Receptors Are Not Concentrated in Caveolae. *Journal of Cell Science* 115(Pt 6):1331–1340.

Rodal, Siv Kjersti, Grethe Skretting, Øystein Garred, Frederik Vilhardt, Bo van Deurs, and Kirsten Sandvig. 1999. Extraction of Cholesterol with Methyl-Beta-Cyclodextrin Perturbs Formation of Clathrin-Coated Endocytic Vesicles. *Molecular Biology of the Cell* 10(4):961–974.

Röper, Katja, Denis Corbeil, and Wieland B Huttner. 2000. Retention of Prominin in Microvilli Reveals Distinct Cholesterol-Based Lipid Micro-Domains in the Apical Plasma Membrane. *Nature Cell Biology* 2(9):582–592. doi:10.1038/35023524.

Rotin, Daniela, Benjamin Margolis, Moosa Mohammadi, Roger J Daly, Guenter Daum, Nanxin Li, Edmond H Fischer, Wilson H Burgess, Axel Ullrich, and Joseph Schlessinger. 1992. SH2 Domains Prevent Tyrosine Dephosphorylation of the EGF Receptor: Identification of Tyr992 as the High-Affinity Binding Site for SH2 Domains of Phospholipase C Gamma. *The EMBO Journal* 11(2):559–567.

Rust, Michael J, Mark Bates, and Xiaowei Zhuang. 2006. Sub-Diffraction-Limit Imaging by Stochastic Optical Reconstruction Microscopy (STORM). *Nature Methods* 3(10):793–795. doi:10.1038/nmeth929.

Scarselli, Marco, Paolo Annibale, Claudio Gerace, and Aleksandra Radenovic. 2013. Enlightening G-Protein-Coupled Receptors on the Plasma Membrane Using Super-Resolution Photoactivated Localization Microscopy. *Biochemical Society Transactions* 41(1):191–196. doi:10.1042/BST20120250.

Schütz, Gerhard J, Hansgeorg Schindler, and Thomas Schmidt. 1997. Single-Molecule Microscopy on Model Membranes Reveals Anomalous Diffusion. *Biophysical Journal* 73(2):1073–1080. doi:10.1016/S0006-3495(97)78139-6.

Shimada, Yukiko, Mikako Maruya, Shintaro Iwashita, and Yoshiko Ohno-Iwashita. 2002. The C-Terminal Domain of Perfringolysin O Is an Essential Cholesterol-Binding Unit Targeting to Cholesterol-Rich Microdomains. *European Journal of Biochemistry/FEBS* 269(24):6195–6203.

Simons, Kai and Elina Ikonen. 1997. Functional Rafts in Cell Membranes. *Nature* 387(6633):569–572. doi:10.1038/42408.

Simons, Kai and Derek Toomre. 2000. Lipid Rafts and Signal Transduction. *Nature Reviews Molecular Cell Biology* 1(1):31–39. doi:10.1038/35036052.

Sorkin, Alexander and Lai Kuan Goh. 2009. Endocytosis and Intracellular Trafficking of ErbBs. *Experimental Cell Research* 315(4):683–696.

Sugii, Shigeki, Patrick C Reid, Nobutaka Ohgami, Yukiko Shimada, Robert A Maue, Haruaki Ninomiya, Yoshiko Ohno-Iwashita, and Ta-Yuan Chang. 2003. Biotinylated Theta-Toxin Derivative as a Probe to Examine Intracellular

Cholesterol-Rich Domains in Normal and Niemann-Pick Type C1 Cells. *The Journal of Lipid Research* 44(5):1033–1041. doi:10.1194/jlr.D200036-JLR200.

Tanaka, Kenji A K, Kenichi G N Suzuki, Yuki M Shirai, Shusaku T Shibutani, Manami S H Miyahara, Hisae Tsuboi, Miyako Yahara, et al. 2010. Membrane Molecules Mobile Even After Chemical Fixation. *Nature Methods* 7(11):865–866. doi:10.1038/nmeth.f.314.

Tang, Jun and David J Gross. 2003. Regulated EGF Receptor Binding to F-Actin Modulates Receptor Phosphorylation. *Biochemical and Biophysical Research Communications* 312(4):930–936.

Thompson, Russell E, Daniel R Larson, and Watt W Webb. 2002. Precise Nanometer Localization Analysis for Individual Fluorescent Probes. *Biophysical Journal* 82(5):2775–2783. doi:10.1016/S0006-3495(02)75618-X.

Tokunaga, Makio, Naoko Imamoto, and Kumiko Sakata-Sogawa. 2008. Highly Inclined Thin Illumination Enables Clear Single-Molecule Imaging in Cells. *Nature Publishing Group* 5(2):159–161. doi:10.1038/nmeth1171.

Truan, Zinnia, Laura Tarancón Díez, Claudia Bönsch, Sebastian Malkusch, Ulrike Endesfelder, Mihaela Munteanu, Oliver Hartley, Mike Heilemann, and Alexandre Fürstenberg. 2013. Quantitative Morphological Analysis of Arrestin2 Clustering Upon G Protein-Coupled Receptor Stimulation by Super-Resolution Microscopy. *Journal of Structural Biology* 184(2):329–334. doi:10.1016/j.jsb.2013.09.019.

Vieira, Amandio V, Christophe Lamaze, and Sandra L Schmid. 1996. Control of EGF Receptor Signaling by Clathrin-Mediated Endocytosis. *Science* 274(5295):2086–2089.

Wilde, Andrew, Eric C Beattie, Lawrence Lem, David A Riethof, Shu-Hui Liu, William C Mobley, Philippe Soriano, and Frances M Brodsky. 1999. EGF Receptor Signaling Stimulates SRC Kinase Phosphorylation of Clathrin, Influencing Clathrin Redistribution and EGF Uptake. *Cell* 96(5):677–687.

Willig, Katrin I, Heinz Steffens, Carola Gregor, Alexander Herholt, Moritz J Rossner, and Stefan W Hell. 2014. Nanoscopy of Filamentous Actin in Cortical Dendrites of a Living Mouse. *Biophysical Journal* 106(1):L01–L03. doi:10.1016/j.bpj.2013.11.1119.

Wong, Joseph, David Baddeley, Eric A Bushong, Zeyun Yu, Mark H Ellisman, Masahiko Hoshijima, and Christian Soeller. 2013. Nanoscale Distribution of Ryanodine Receptors and Caveolin-3 in Mouse Ventricular Myocytes: Dilation of T-Tubules Near Junctions. *Biophysical Journal* 104(11):L22–L24. doi:10.1016/j.bpj.2013.02.059.

Yamaji, Akiko, Yoshiyuki Sekizawa, Kazuo Emoto, Hitoshi Sakuraba, Keizo Inoue, Hideshi Kobayashi, and Masato Umeda. 1998. Lysenin, a Novel Sphingomyelin-Specific Binding Protein. *The Journal of Biological Chemistry* 273(9):5300–5306.

Yarden, Yosef. 2001. The EGFR Family and Its Ligands in Human Cancer. Signalling Mechanisms and Therapeutic Opportunities. *European Journal of Cancer (Oxford, England: 1990)* 37(Suppl 4):S3–S8.

York, Andrew G, Sapun H Parekh, Damian Dalle Nogare, Robert S Fischer, Kelsey Temprine, Marina Mione, Ajay B Chitnis, Christian A Combs, and Hari Shroff. 2012. Resolution Doubling in Live, Multicellular Organisms via Multifocal Structured Illumination Microscopy. *Nature Methods* 9(7): 749–754. doi:10.1038/nmeth.2025.

18

Super-Resolution Microscopy

The Dawn of a New Era in Cardiovascular Research

Marc A.M.J. van Zandvoort

18.1 Introduction
18.2 Super-Resolution Microscopy in Cardiac Research
18.3 Super-Resolution Microscopy in Vascular Research
18.4 Future Perspective
References

18.1 Introduction

Cardiovascular diseases have an enormous impact on society, and, together with cancer, they are the main cause of morbidity and death in western society. This has resulted in a vast amount of literature, both on (interventions in) pathways underlying the diseases and on defining the changes in structure and function of diseased tissues, such as atherosclerotic lesions and the infarcted or failing heart. Most studies have been carried out by applying either basic or advanced imaging techniques at various levels of resolution in cell cultures (using transillumination, widefield fluorescence, or confocal fluorescence microscopy), in *ex vivo* tissue (using either histology or widefield fluorescence microscopy on fixed tissue sections or two-photon imaging of tissue slabs), in living animals (using intravital transillumination, widefield fluorescence, or two-photon microscopy), or in humans (using either histology or widefield fluorescence on tissue sections or noninvasive imaging techniques such as magnetic resonance imaging (MRI), positron emission tomography (PET), X-ray computed tomography (CT), and ultrasound (US) on patients). All techniques have their specific niche of use. Interestingly, however, the cardiovascular application of the new kid in town, super-resolution fluorescence microscopy, in either of its forms, has so far been very limited. In this review, an overview will be given of the few piloting studies that have been carried out, ending with a future perspective. Although the review distinguishes two main areas of research, namely cardiac imaging and vascular imaging, it should be noted that cardiac and vascular diseases are often closely related. For example, atherosclerosis can cause myocardial infarction (MI), possibly followed by heart failure (HF), and/or arrhythmia. In none of the imaging studies described below, this relation has been the topic of investigation.

18.2 Super-Resolution Microscopy in Cardiac Research

Although in the past 4 years, several reviews appeared (Crossman et al. 2015b, Kohl et al. 2013, Soeller and Baddeley 2013) that highlight the enormous potential of super-resolution microscopy in cardiac imaging in general, experimental studies have been limited and centered around structural imaging in fixed cells or tissue sections. Most attention has been devoted to structural imaging of Ca^{2+}-channel-related structures and proteins (including one live-cell study), such as ryanodine receptor isoform 2 (RyR2), T-tubuli (TT [membrane invaginations]), caveolin-3 (CAV3), and junctophilin-2 (JPH2). Several fixed-cell studies investigated the Na^+ and ATP-sensitive K^+ channel ($Na_V1.5$, K_{ATP}) and their relation with structures such as connexin-43 (CX43), N-cadherin, and plakophilin-2 (PKP2) in the intercalated disc (ID) of the cardiomyocyte. Remaining work focuses on imaging a diversity of cardiomyocyte structures, such as mitochondria, CX43 and PKP2, α-actinin, actin, titin/myosin, and TT/CAV3/JPH2 in fixed cardiomyocytes. The latter study also partly uses live-cell imaging. All these papers will be discussed below.

18.2.1 Imaging of Ca^{2+} Channel-Related Structures

RyR2 is the major isoform of the ryanodine receptor in (both atrial and ventricular) cardiomyocytes, the contractile elements of the heart. This massive protein is located on the membrane of the sarcoplasmic reticulum (SR) within the cardiomyocyte and plays an important role in calcium regulation therein (for explanatory videos on this topic, you may watch numerous YouTube videos such as the very instructive ones from Ben Garside). Depolarization of the cardiomyocyte membrane due to spreading of the depolarization wave from neighboring cardiomyocytes activates membrane Na^+ channels to open and to allow extracellular Na^+ to enter into the cell (see Section 18.2.2). This further depolarizes the membrane potential and activates the voltage-gated L-type Ca^{2+} channels on the cardiomyocyte membrane in the TT network to allow small amounts of extracellular Ca^{2+} to enter into the cell and pass the dyadic cleft between the membrane and SR. There, the Ca^{2+} stimulates the Ca^{2+}-sensitive RyR2 on the SR membrane to open, and as a result, large amounts of Ca^{2+} are released from the Ca^{2+}-storing protein calsequestrin in the SR, via RyR2, into the cytoplasm. This ion current is counterbalanced by K^+ current into the SR to avoid negative potential in the SR. The so-called Ca^{2+}-induced Ca^{2+} release (CICR) synchronized among many clefts results in a sufficient rise in overall cytoplasmic Ca^{2+} and causes Ca^{2+} interaction with the troponin/tropomyosin complex that covers actin of the contractile elements throughout the cardiomyocyte. Normally, this covering prevents actin-myosin interaction, but owing to the Ca^{2+}-mediated conformational change, the interaction of myosin with actin becomes possible, and as a result, actin-myosin cross-bridges are formed, where under uncoupling of ADP and recycling of ATP, a (contractile) power stroke is developed. The coordinated combination of power strokes of many connected contractile elements, ending at the ID at the end of the cardiomyocyte, results in contraction of the cardiomyocyte as a whole and thus in the development of contractile force. To summarize, the initial membrane depolarization causes cardiomyocyte contraction, a phenomenon called excitation-contraction coupling (ECC). In view of their important function, RyR2s play a crucial role in cardiac health and disease (Blayney and Lai 2009), and malfunctioning of RyR2 has been implicated in arrhythmias and HF. Furthermore, it is important to realize that, next to the L-type Ca^{2+} channels and the RyR2s, several other ion pumps play crucial roles in cardiomyocytes, for example, in normalizing intracellular ion (Ca^{2+}, Na^{+}, and K^{+}) concentrations or as counter-ion pumps. It is beyond the scope of this review to discuss them here, but some of these channels will be briefly introduced below. Importantly, Ca^{2+} concentration in both cytoplasm and SR is carefully controlled. Therefore, Ca^{2+} has to be transported back into the SR (via the SERCA2a pump) and into the extracellular space (via the Na^{+}/Ca^{2+} exchanger NCX1) to promote relaxation of the heart during diastole. Furthermore, structural proteins in the cardiomyocyte, such as JPH2, ensure proper dyadic cleft structure to enable optimal interaction between the various ion channels (Beavers et al. 2014, Landstrom et al. 2014, Wang et al. 2014).

In the review from Kohl et al. (2013), a clear description of the various super-resolution principles and methods is given. Interestingly, there is a paragraph dedicated to probe characteristics for the various super-resolution techniques and a perspective on live-cell imaging. Finally, a summary is given of the group's own research on the TT network and the research of the group of Christian Soeller on RyR2. Both research areas will be discussed more extensively below. From the translational point of view, the 2015 review from the group of Christian Soeller (Crossman et al. 2015b) deserves special attention. It discusses the potential relevance of super-resolution microscopy in clinical decision making. First, standard clinical procedures for imaging of endomyocardial biopsies from patients suffering from various cardiac diseases, such as myocarditis, allograft rejection, infiltrative diseases, storage diseases, cardiac tumors, and arrhythmias, are explained. Next, the application of confocal and *d*STORM microscopy to achieve unprecedented information that can influence diagnosis and, hence, treatment, is discussed. Finally, an overview is given of applications of these techniques in diagnosis of the various diseases. The 2013 review from the Soeller group (Soeller and Baddeley 2013) summarizes the use of super-resolution microscopy in the imaging of RyR2s and related proteins, such as JPH2, calsequestrin, and CAV3. The latter is a protein present in cardiac caveolae, which are membrane invaginations that are important in a diversity of metabolic processes, but its association with ion channels is also thought to be related to the functioning of various ion channels, such as L-type Ca^{2+} channels, NCX, and K_{ATP}.

RyR2s and related proteins are the main focus of Christian Soeller's work in seven experimental publications since 2009 (Baddeley et al. 2009, 2011, Crossman et al. 2015a, Hou et al. 2015, Jayasinghe et al. 2012a, 2012b, Wong et al. 2013). Using *d*STORM, in these studies, the distribution of RyR2 in ventricular cardiomyocytes not only from mice, rats, and rabbits but also from humans is investigated. Interestingly, many of the studies also use histological slices (see e.g., Figure 18.1). Notable are both their clear writing style and their extensive methodological descriptions, such as the crucial effects of labeling density, fixation protocols, and imaging procedures (Crossman et al. 2015a). This kind of information makes the studies very useful from a didactical point of view. In addition, in some of the studies (Baddeley et al. 2011, Crossman et al. 2015a), the explicit use of correlative confocal and super-resolution microscopy demonstrates convincingly the gain in the information obtained from super-resolution imaging (Figure 18.1). In the study from (Wong et al. 2013), the relation between CAV3 and RyR2 is further investigated in mouse ventricular cardiomyocytes. It is found that there is only a small (around 9%) overlap between both, which is remarkable, given the role of CAV3 in the formation of TT and the close proximity of RyR2 to TT. In (Jayasinghe et al. 2012a), the overlap between RyR2 and JPH2 is investigated in rat ventricular cardiomyocytes and is found to be very high (60%–80%), suggesting at least partly interacting proteins, in line with biochemical evidence from co-immunoprecipitation experiments (Beavers et al. 2014). The work of the members of this group culminates in their most recent study (Hou et al. 2015) with tables of and discussions on the distribution of clusters of RyR2 in relation to JPH2, TT, and α-actinin. From this group's papers, it is immediately obvious that new Ca^{2+} imaging data on live cardiomyocytes

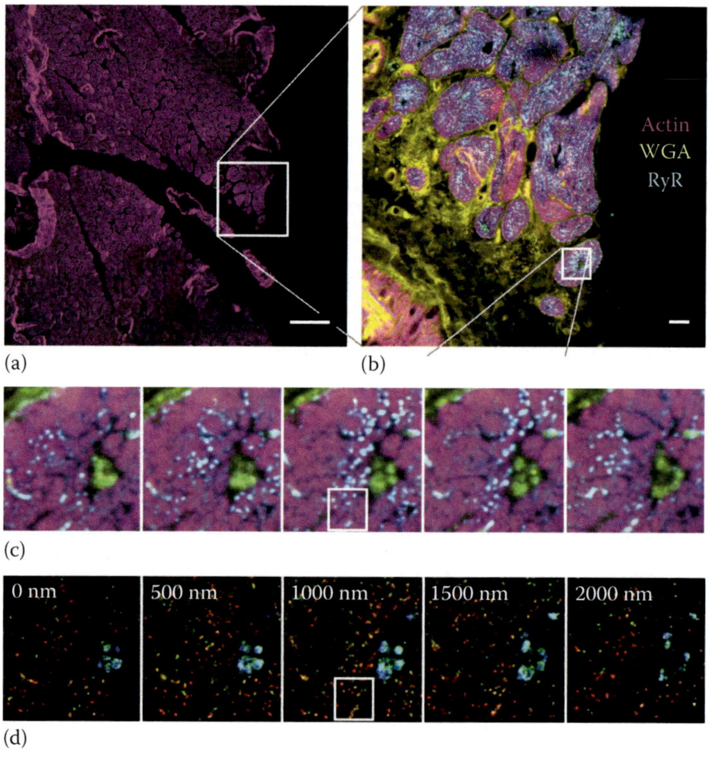

Figure 18.1

Images of 10-μm-thick human cardiac tissue section. Visualized are f-actin (phalloidin-Alexa488), cell membrane and extracellular matrix (WGA-Alexa594), ryanodine receptor (anti-RyR-Alexa680), and calsequestrin (anti-CSQ-Alexa750). Lipofuscin is furthermore visible owing to its strong autofluorescence. First, actin, WGA, and RyR were imaged on a confocal microscope and then the sample was taken to the super-resolution microscope for imaging of RyR, lipofuscin, and CSQ. (a) Actin labeling of large tissue area, scale bar: 100 μm; (b) confocal stack of area indicated in A, scale bar: 10 μm; (c) confocal stack of area indicated in (b); (d) optically sectioned super-resolution stack of area indicated in (c). (Reproduced with permission from Baddeley, D. et al., *PLoS One* 6, e20645, 2011.) *(Continued)*

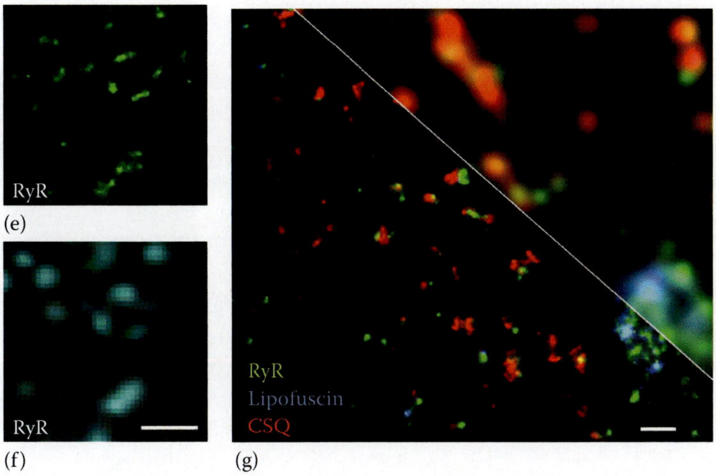

Figure 18.1 (Continued)

Images of 10-μm-thick human cardiac tissue section. Visualized are f-actin (phalloidin-Alexa488), cell membrane and extracellular matrix (WGA-Alexa594), ryanodine receptor (anti-RyR-Alexa680), and calsequestrin (anti-CSQ-Alexa750). Lipofuscin is furthermore visible owing to its strong autofluorescence. First, actin, WGA, and RyR were imaged on a confocal microscope and then the sample was taken to the super-resolution microscope for imaging of RyR, lipofuscin, and CSQ. (e) super-resolution and (f) confocal images of RyR-Alexa647 signal, scale bar: 1 μm; (g) 3-color super-resolution image of small area in the tissue sample, scale bar: 500 nm. (Reproduced with permission from Baddeley, D. et al., *PLoS One* 6, e20645, 2011.)

are needed to understand the functional role of these clusters, underpinning the need for live-cell super-resolution microscopy.

Constant SR Ca^{2+} leakage contributes to HF pathophysiology by SR Ca^{2+} depletion and loss of ECC gain, as well as by activation of Ca^{2+}-dependent signaling pathways. A study from the group of Xander Wehrens (Wang et al. 2014) demonstrates multiple independent pathways for SR Ca^{2+} leakage in ventricular cardiomyocytes isolated from tamoxifen-inducible conditional JPH2 knock-out mice. JPH2 knock-out resulted in (1) increased variation in dyadic cleft width, with the average width remaining the same; (2) increased open probability of RyR2 channels, causing increased frequency and width of Ca^{2+} sparks; and (3) reduced activity and mutual proximity of NCX1 and RyR2 channels in the dyadic cleft, leading to a net increase in local Ca^{2+} concentrations in the dyadic cleft. Although most of the study involves confocal microscopy and quantitative experiments, the reduced proximity of NCX and RyR2 is demonstrated using stochastic super-resolution imaging in fixed cells.

Two studies from the group of Raimond Winslow (Walker et al. 2014, 2015) address the relevance of the distribution of individual RyR2s within RyR2 clusters on spontaneous Ca^{2+} spark generation. These spontaneous sparks can arise on opening of a single RyR2 during diastole. This opening, in some cases, results in opening of additional RyR2s and a consequent rise in cytoplasmic Ca^{2+}, the latter being relevant in some disease conditions such as arrhythmias. Theoretical modeling is, to a limited extent, coupled to STED imaging of RyR2 clusters to show that "function" follows "structure" in the sense that specific RyR2 cluster distributions are more prone to spontaneous Ca^{2+} spark formation. These studies elegantly show how super-resolution microscopy can go hand in hand with modeling.

To the best of our knowledge, only two super-resolution studies on living cardiomyocytes exist; one of them uses SIM on Ca^{2+} channels (Jian et al. 2014). It uses a cell-in-gel system to study the effect of afterload on Ca^{2+} handling of mouse ventricular cardiomyocytes. The cells are cultured in a PVA gel, with boronic acid as the PVA-cell intermediate, linking the cell glycans to the PVA. The stiffness of the gel can be easily modified using different amounts of boronic acid cross-linker. The gel is porous, nontoxic (live-cell experiments were carried out over a period of 2 h), and transparent. The study demonstrates that afterload increases Ca^{2+} dynamics and spontaneous spark formation owing to increased RyR2 sensitivity. Among other factors influencing this process, SIM is used to determine the crucial role of NO production by NOS1, but not by NOS3. It appears that NOS1, as compared with NOS3, has a larger overlap with (0.438 and 0.125, respectively) and closer proximity to (125 nm and 370 nm, respectively) RyR2.

18.2.2 Imaging of Na^+ and K^+ Channel-Related Structures

Another ion channel that has obtained some attention in super-resolution microscopy is the voltage-gated sodium (Na^+) channel (VGSC). Arrival of the depolarization wave on the cardiomyocyte membrane causes the VGSC to open and to allow strong cardiac sodium current (I_{Na}) into the cell, responsible for further rapid depolarization of the cardiomyocyte and activation of the L-type Ca^{2+} channel to start ECC. The VGSC is also involved in regulating the duration of the cardiac action potential (AP) and propagation of the impulse throughout the myocardium via its localization at the ID (Rook et al. 2012). $Na_V1.5$ (Rook et al. 2012, Shy et al. 2013) is the pore-forming subunit of the VGSC, which plays a crucial role in arrhythmogenic cardiomyopathy (Gillet et al. 2013) and takes central stage in the studies below.

In the classical view of cardiomyocytes, intercellular adhesion and electrical excitability are separate and unrelated cellular properties, taken care of by independent protein complexes (AJ and desmosomes vs. GJ) at the ID that forms the site of connection of cardiomyocytes (Hong et al. 2012, Leo-Macias et al. 2016). All four studies on the Na^+ channel (Agullo-Pascual et al. 2014, Cerrone et al. 2014, Leo-Macias et al. 2016, Veeraraghavan et al. 2015), and also the later-discussed K^+ channel study (Hong et al. 2012), challenge this concept. The 2015 study by (Veeraraghavan et al. 2015) uses gSTED on fixed ventricular tissue slices from guinea pigs. Traditionally, GJ are considered the only possible site for conduction of sodium potentials between cells, but they show that within the perinexus (200 nm around a GJ), zones exist, where membranes of respective cells are in very close contact (EM, around 10 nm), with a high density of $Na_V1.5$ in these regions (gSTED on overlap of $Na_V1.5$ and CX43). This combination of factors is subsequently modeled to study ephaptic coupling of Na^+.

The study by (Cerrone et al. 2014) looks at it from a patient's perspective. From 200 samples of Brugada syndrome (BrS) patients without clinical signs of PKP2 (a desmosomal, thus structural, ID protein) deficiency-caused arrhythmogenic cardiomyopathy (AC), 2.5% showed single-nucleotide mutants in PKP2 genes, with five different mutations in total. Then, a set of mostly functional experiments is set up to test whether these mutations alone can cause VGSC malfunctioning. Measurements of Na^+ currents in PKP2-deficient, but endogenously $Na_V1.5$-expressing HL-1 cells and human (patient)-induced pluripotent stem cell cardiomyocytes, revealed that in both cell types, Na^+ current and $Na_V1.5$ expression at the ID are decreased. Both can be restored by transient transfection of WT-PKP2 but not by transfection with the patient's mutant forms (even in the presence of WT-PKP2). *d*STORM is then applied to visualize microtubules (EB1) and N-cadherin (as ID marker) in PKP2 dependence. The distance of microtubules to the ID in the PKP2-deficient cells appears to be increased. The complex of studies suggests a relation between PKP2, microtubules, and $Na_V1.5$.

A different study from the same year (Agullo-Pascual et al. 2014) uses *d*STORM on adult mouse ventricular cardiomyocytes (Figure 18.2) to demonstrate a second function of CX43 in the excitability of cardiomyocytes, independent of its

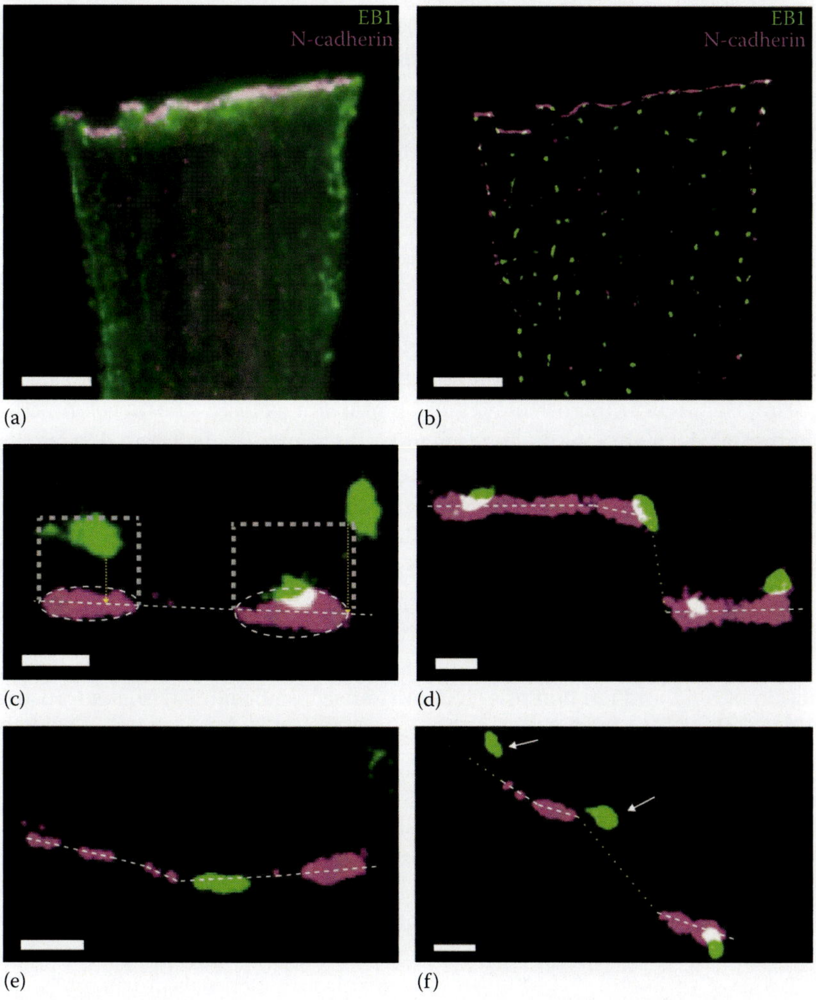

Figure 18.2

Super-resolution images of EB1 (green) and N-cadherin (purple) in adult murine ventricular cardiomyocytes. (a) TIRF image; (b) super-resolution image; (c–f) white dotted lines along the major axis of the N-cadherin plaque. Rectangular areas indicate ROI; yellow dotted lines indicate the distance to the cell end for two distal clusters. Scale bar: 500 nm. (Reproduced with permission from Agullo-Pascual, E. et al., *Cardiovasc Res* 104, 371–381, 2014.)

primary and classical role as a GJ protein at the ID. Truncation of the last five amino acids of the CX43 C-terminus causes lethal arrhythmias in mice, while CX43 GJ distribution appears to be normal. In cells of mice with this deficiency, however, the Na^+ current at the ID decreases, owing to impaired capture of cytoskeletal microtubules at the ID (*d*STORM of microtubule end-binding protein 1-EB1 and desmosomal protein N-Cadherin) and consequently results in reduced $Na_V1.5$ delivery at the ID (*d*STORM of N-cadherin and $Na_V1.5$).

The most recent study (Leo-Macias et al. 2016) applies a combination of experimental (among which SMLM) and simulation (Monte Carlo) techniques to look at the relation between $Na_V1.5$ distribution and N-cadherin, a desmosomal protein that serves as anchor for myofibril cell-cell contacts (Zuppinger et al. 2000); however, in this study, it is merely used as a signpost of the ID. The study reveals nonrandom clustering of both proteins at the ID. Of the total number of $Na_V1.5$ clusters (as seen with SMLM) that overlaid with the EM-detected ID (this correlative approach being crucial to exclude non-ID-related channel proteins), close to 50% was within 100 nm distance of N-cadherin (SMLM on mouse ventricular cardiomyocytes). Further functional studies ("angle-view" scanning patch clamp) reveal the interdependence of ID structures and channel proteins, to conclude on the mutual relation between ID and $Na_V1.5$ and the resultant crosstalk between the electrical and contractile molecular apparatus.

Finally, there is one study (Hong et al. 2012) that applies super-resolution microscopy (*d*STORM) to imaging of the ATP-sensitive K^+ channel (K_{ATP}). The K_{ATP} channel is known to open on a decrease of ATP availability and to regulate action potential duration adaptation on an increase of heart rate during exercise. It is thought to play an important role in the development of arrhythmias development during ischemia. It is generally believed that the K_{ATP} channel is statically located on the sarcolemmal surface of the cardiomyocyte, but the researchers focus on the ID population that is also clearly detected in rat ventricular tissue and isolated cardiomyocytes. The contribution of super-resolution microscopy in this paper is very limited, as shown in Figure 18.1, and the study does not provide any information on quantification of the distribution of the K_{ATP} subunit Kir6.2 at the ID in isolated rat ventricular cardiomyocyte. I do mention the study here because of its clear-written introduction on both the function of K_{ATP} and the potential relation between mechanical (via desmosomes and AJ) and electrical (via GJ) coupling.

18.2.3 Imaging of Subcellular Structures in Cardiomyocytes

Different papers focus on different cardiomyocyte structures. In (Singh et al. 2012), the group of Ligia Toro isolated mitochondria, the energy factories of all cells, from whole mouse hearts and convincingly showed that a significant part of these mitochondria remains fully functional and contains no impurities, such as nuclear proteins, SR or plasma. They then used STED imaging (with depletion at 780 nm) of this subset of 200- to 1000-nm-diameter mitochondria to visualize the outer and inner membrane distributions of specific proteins (VDAC1 and COX2, respectively) in clusters of 20–90 nm sizes.

Using an elegant combination of stochastic super-resolution microscopy and Monte Carlo simulations (Agullo-Pascual et al. 2013) studies the co-localization of CX43 and PKP2 in fixed neonatal rat ventricular cardiomyocytes and mouse cardiac tissue sections. EM has suggested that these two proteins belong to completely distinct cellular microdomains, namely GJ and desmosomal cell-cell adhesion sites, respectively. Therefore, according to this idea, an overlap between these proteins is not possible. Triggered by the effect of loss of ankyrin-G expression on both intercellular adhesion strength and electrical coupling of cardiomyocytes observed in earlier studies, they visualized and quantified possible cluster overlap. Indeed, up to 50% of CX43 and PKP2 showed such overlap, whereas siRNA-mediated loss of ankyrin-G expression caused not only larger CX43 clusters but also more CX43-PKP2 overlapping subdomains. Note that these findings are in line with the rejection of the independency concept of various ID junctions, as discussed in Section 18.2.2.

Soeller's group in another very explanatory study (Hou et al. 2014) used *d*STORM on 5- 10-μm-thick rat ventricular slices to study α-actinin (a protein in the Z-disk that connects actin to the Z-disk). Using co-staining of SERCA2a as SR marker, they demonstrated the potential of super-resolution imaging for obtaining not only Ca^{2+} channel structure but also other structural details.

A live-cell imaging gSTED study (Wagner et al. 2012) visualized and modeled changes in the TT network after MI in isolated (but living) ventricular mouse cardiomyocytes by using di-8-ANEPPS membrane staining of the TT. The TT network at 4 and 8 weeks post-MI appeared to have enlarged cross sections and to be more complex and heterogeneous, with longer segments and increased branching. Co-staining with CAV3 or JPH2 in fixed cells demonstrated proliferative remodeling. The study is very extensive, and it is beyond the scope of this chapter to go into more detail. However, in my opinion, it is a piloting study that definitely deserves thorough studying. An extensive description of the methods and models used is given in (Wagner et al. 2014).

Finally, two other papers only touch on super-resolution imaging, one using STED (Tian et al. 2012) and the other using SIM (Granzier et al. 2014). The first images actin in rat ventricular cardiomyocytes, using STED solely for obtaining higher magnification images on the effect of cytochalasin D on cellular cytoskeleton. The second study uses the combination of EM and SIM to visualize titin length, considered as a ruler protein that determines the length of the thick (myosin) filament, in control and $Ttn^{\Delta Iajxn}$ mice. In these mice, the structure of titin is changed exactly at the I-band/A-band junction, but this does not change thick filament length. However, differences are observed in mechanical behavior of isolated ventricular cardiomyocytes and in left ventricular stiffness and hypertrophy in the modified mice. Interestingly, the study demonstrated significant differences, up to 30 nm, in titin length between EM and SIM, most likely due to preparation methods.

18.3 Super-Resolution Microscopy in Vascular Research

Even more than in the cardiac field, vascular super-resolution imaging is virtually unexplored. In general, one can discriminate between imaging of cells of the vascular wall (Ecs, smooth muscle cells, and vascular fibroblasts) and of cells that interact with the vascular wall (platelets, neutrophils, natural killer cells, B cells, and T cells).

18.3.1 Imaging of the Vascular Wall: Endothelial and Smooth Muscle Cells

Endothelial cells (ECs), covered with the protective endothelial glycocalyx (Reitsma et al. 2007, 2011a, 2011b), are the gatekeepers of the vascular wall.

They form a continuous, single-cell thick layer called the tunica intima. The ECs in this layer are interconnected via so-called tight junctions (TJ), adherens junctions (AJ), and gap junctions (GJ) (Bazzoni and Dejana 2004). Each junction type consists of specific proteins, such as JAM-A for TJ, VE-cadherin and complexed catenin for AJ, and connexins for GJ (Mehta and Malik 2006). In health, the tightly connected and flexibly covered ECs protect the blood from contact with the immediate subendothelial layer, consisting of collagen IV and lamina elastica interna (LEI). In diseases such as inflammation, however, the ECs can attract leukocytes by exposing chemokines and cytokines. The leukocytes can then be captured and internalized in a carefully orchestrated process called the leukocyte adhesion cascade of capture, rolling, arrest, crawling, and internalization, using a range of expression proteins such as selectins and integrins on the luminal surface of the ECs (Ley et al. 2007). Similarly, in case of EC damage, leukocytes and platelets will immediately adhere to the site of damage to stop contact between blood and the vessel wall and to restore the vessel wall integrity.

After the LEI, the contractile vascular smooth muscle cells, embedded between elastin layers and fibers (made by them during development), are crucial in maintaining vascular tone (Lacolley et al. 2012). This layer is called the tunica media, the thickness of which depends on animal species, and within each species, its thickness varies with vessel type (artery vs. vein) and location of the vessel (e.g., carotid vs. aorta). The smallest vessels in the body, capillaries and venules, completely lack the media and instead have the basal membrane around the intima covered with contractile pericytes. The media is bounded by the lamina elastica externa (LEE). Vascular smooth muscle cells have an intense communication with the ECs, either directly via the fenestrae in the LEI or indirectly via molecules such as nitric oxide. Furthermore, the media is innervated by the sympathetic nervous system, which consists of the nerves entering from the outer layer of the vessel.

After the LEE, the tunica adventitia or tunica externa begins, consisting of mainly vascular fibroblasts that produce their own extensive extracellular matrix of various types of collagen (mainly type I and III). The adventitia embeds the vessel in surrounding tissue, gives structure, and contains smaller blood vessels, nerves, and, potentially, inflammatory cells.

All three cell types play their own important roles in vascular health and disease (such as atherosclerosis) and have various interactions, not only with each other but also with other cells. It is beyond the scope of this review to discuss it in more details; therefore, I refer the interested reader to the numerous reviews on the function of each cell type and to various papers from our own group on details of the structure of the vascular wall (Megens et al. 2007a, 2007b, 2008, 2010).

Two recent reviews indicate the (enormous) potential of super-resolution imaging for the vascular field. In the short overview (Lavina and Gaengel 2015), the various techniques are summarized and a very useful list of genetically available mouse models for obtaining vascular cells for super-resolution imaging, such as ECs, pericytes, and lymphatic ECs, is given. In (Megens et al. 2015) also, the various possible techniques are described and hints for future research are given. However, research papers on cells of the vascular wall are, surprisingly so, very limited. As far as we could find, no studies are available on histological slices of vascular tissue. This is most likely related to the fact that histological studies on vascular tissue so far mostly have been using standard widefield histology, whereas confocal microscopy is only seldom applied. Probably, clinical practice forms too big a stepping stone to apply more advanced imaging techniques. The results on super-resolution imaging of cardiac histological slices, as discussed above, should be a stimulus to initiate vascular histological studies. *In vitro* cell studies on specific vascular cell types (ECs and vascular smooth muscle cells, but not vascular fibroblasts) have been the topic of a few studies using super-resolution microscopy, and we will summarize these studies below.

In a study dating back to 2007 (Seebach et al. 2007), Jochen Seebach observed elongation of HUVECs under acute and chronic flows, accompanied by actin reorganization along the EC junctions, AJ VE-cadherin/catenin clustering (Bazzoni and Dejana 2004), increase (acute flow increase) and slight decrease (prolonged flow) in electrical EC resistance, and transient VE-cadherin/catenin tyrosine phosphorylation. Although the paper used various nonimaging techniques and most imaging was carried out using traditional confocal microscopy, VE-cadherin clustering was quantified using a home-built STED microscope. Under low-flow conditions, VE-cadherin clusters had an average diameter of around 65 nm, while the clusters were randomly distributed at 12 clusters/μm. After 15 min of high flow, the clusters maintained their diameter, but after 24 h, the clusters' diameter increased to around 80 nm, whereas cluster density decreased to 6 clusters/μm^2.

In Fernandez-Martin et al. (2012), the distribution of AJ VE-cadherin in HUVECs was studied in relation to that of the TJ PECAM-1 (Bazzoni and Dejana 2004). It goes beyond the scope of this chapter to summarize the very detailed results, but to increase curiosity, it was shown that VE-cadherin exhibited a characteristic reticular distribution at cell-cell overlay regions, while its distribution was more linear at cell-cell contacts, without association to actin. Then, in these reticular regions, the relation between VE-cadherin and PECAM-1 distribution was studied in more detail using STED microscopy. Although PECAM-1 does not exhibit this reticular structure, it appears to be crucial for the reticular distribution of VE-cadherin. The effect of various stimuli, such as TNF-α, histamine, and thrombin, was then studied.

Another paper (Huber et al. 2012) studied the distribution of P-glycoprotein (also known as PgP, MDR1, or ABCB1). This transmembrane glycoprotein is expressed in several cell types throughout the body, for example, in liver and kidney, and also in the ECs of brain capillaries. There, it is the most abundantly expressed efflux transporter, which means that it actively (via ATP) removes many substances from the ECs into the capillary lumen, and as such, it is an important component of the blood-brain barrier (BBB) (Loscher and Potschka 2005). By applying the combination of SPDM, simulation corrections, and viral expression of PgP-GFP in immortalized human cerebral microvascular ECs (hCMEC/D3), a clustered organization of PgP on the luminal membrane was found. The clusters were uniformly distributed over the cell membrane, with around 5500 PgP molecules per cell.

Sybelle Goedicke-Fritz et al. (2015) studied the clustering of the Ca^{2+}-activated K-channel (K_{Ca}) (Feletou 2009) in human microvascular ECs (HMEC-1) in relation to caveolin-1 (as an important protein of EC caveolae, CAV1) and the transient receptor potential vanilloid subtype 4 (a Ca^{2+} permeable cation channel that can activate K_{Ca}, TRPV4). K_{Ca} is present in many flavors, among which $K_{Ca2.3}$ and $K_{Ca3.1}$ (Wulff and Kohler 2013) belong to the same functional family. Both type of channels are distributed throughout the organism in distinct locations but both are found in vascular ECs, where they are known to play crucial, but clearly distinct, roles in endothelial-dependent hyperpolarization (EDH) and are thus important in the regulation of vascular tone and blood pressure (Grgic et al. 2009), independent from NO and prostacyclin. The paper found, using GSDIM, a clear association between CAV1, TRPV4, and small conductance $K_{Ca2.3}$, but not intermediate conductance $K_{Ca3.1}$ in ECs under static condition. Mechanical stimulation of cells via exposure to shear stress resulted in a partial (from 1%–5%, changing into 22%) co-localization of $K_{Ca3.1}$ with CAV1 and TRPV4. The imaging experiments were confirmed by patch-clamping on ECs from the carotid artery of Cav-$1^{-/-}$ mice. The paper thus demonstrated that K_{Ca} and TRPV form a functional unit together with caveolae.

To finish the vascular cell studies, Balint et al. (2015) used GSDIM on human vascular smooth muscle cells (HITC6 SMCs) to study their collectivization process under serum withdrawal. When the cells are deprived from serum, a change in cell-cell contacts, from transient to stable, is observed. This collectivization goes hand in hand with cell-cell strapping over distances of 60 μm. Expression studies revealed the importance of cadherin-11 and N-cadherin in this process. Using GSDIM, they found that these straps surprisingly consisted of paired, opposed, and parallel-oriented N-cadherin and cadherin-11 proteins, with a distance between the opposing proteins of around 50 nm. The main determinant in serum appeared to be TGF-β1; however, the formation of the straps had a major influence on coordination of Ca^{2+} signaling.

18.3.2 Super-Resolution Microscopy of Cells of the Immune System

In vertebrates, the defense system against infections can be divided in the, evolutionary oldest, innate (nonspecific) immune system and the adaptive immune system. The cells of the innate system consist of, on the one hand, early defense cells such as natural killer cells, mast cells, dendritic cells, eosinophils, basophils, and macrophages, and on the other hand late, cells such as neutrophils and macrophages. All these cells recognize pathogens and subsequently create an immediate response in a generic way. However, they do not confer long-lasting or protective immunity to the organism. Indeed, the major functions of the innate immune system are, on the one hand, the (immediate) identification and removal of foreign cells and on the other hand, the recruitment and activation of adaptive immune cells ($CD8^+$ T cells, $CD4^+$ T cells, γδ T cells, and B cells) at sites of infection. This is done via cytokine production and release and antigen presentation by antigen-presenting cells (APCs), respectively. Each cell type has its specialization, with various subclasses present in T cells and B cells.

Most super-resolution microscopy has been carried out on T cells and natural killer cells, whereas B cells and neutrophils are studied less. We have found no super-resolution studies on the other immune cells. One very readable review (Balagopalan et al. 2011) gives a very extensive overview of the optical techniques available for imaging lymphocytic activation. It also shortly touches on super-resolution techniques. All studies mentioned in this review (dating back to 2011 and earlier) will be discussed below.

18.3.2.1 T cells

A few recent reviews have highlighted the wealth of data existing on the relevance of (specific types of) T cells in cardiovascular diseases, such as atherosclerosis and cardiac repair after infarction (Gregersen et al. 2015, Meng et al. 2015, Ramos et al. 2016, Sattler and Rosenthal 2016). Importantly, the studies below are not specifically focused on super-resolution imaging of the role of T cells in cardiovascular diseases but study more general aspects of T cell biology. They are focused on various receptor proteins in the T cell receptor clusters (TCRs) at the interface of T cell and APC (Purtic et al. 2005). As such, they do clarify the potential impact of super-resolution imaging in T cells. A few reviews show several aspects of T cell biology that can be visualized, the first being the short but very readable paper from Shannon et al. (2015). An overview is given of the role and functioning of T cells, while the role of various protein clusters, such as LFA-1 (lymphocyte function-associated antigen-1, the first protein to bind T cells weakly to ICAM-1 on the APC), Lck (lymphocyte-specific protein tyrosine kinase, which phosphorylates the CD3 and ζ chain of the TCR, after which ZAP-70 can bind to them; ZAP-70, in turn, is phosphorylated by Lck), LAT (linker of activated T-cells, which is phosphorylated by the phosphorylated ZAP-70, after which it activates a range of other proteins that mobilize Ca^{2+} and activates the necessary signaling cascades to produce cytokines) is described (Koretzky and Myung 2001). Then, the focus turns to the various super-resolution techniques and the groups that have applied these techniques to T cells (all of which will be discussed below). The paper ends with a future perspective, concluding that three main issues that need to be solved to answer the open questions in T cell biology are speed, labeling, and 3D imaging. Two older reviews from the group of Katherina Gaus (Rossy et al. 2012, 2013b) describe in more details the findings of several super-resolution studies. It is beyond the scope of this chapter to summarize all these findings; the reviews are easy to read and anyone specifically interested in T cell biology can be advised to refer to these reviews. Interestingly, the review by Rossy et al. (2013b) also looks at the possible application of super-resolution microscopy to other leukocyte subsets, such as B cell and NK cells, the details of which will be discussed in other sections below. The review by Rossy et al. (2012) has beautiful schemes on the signaling cascades of TCR and discusses the super-resolution results within this framework. In a third review from this group (Benzing et al. 2013), however, the main focus is not on protein clustering, but on the role of (actin) cytoskeleton and endosomes in T cell signaling. This potential role is extensively discussed, and the relevance of super-resolution microscopy in unraveling that role is described. The paper is very detailed and a must-read for those who are specifically interested in signaling cascades. For this chapter, a very important statement in that paper suffices, since it forms the basis for application of super-resolution imaging in cell studies in general and TCR/BCR-APC specifically: "In receptor signaling, it is no longer sufficient to identify the components of a signaling network, because the spatial organization of signaling proteins can also determine signaling outcome. The distribution of cell surface receptors and signaling proteins into protein islands or lipid microdomains compartmentalizes the plasma membrane, but can also facilitate the sorting of proteins and protein complexes to or from intracellular compartments."

Experimental super-resolution studies of T cells are confined to a few groups. Lillemeier et al. (2010) applied high-speed PALM in 2010 to image the size of nanoclusters of LAT and TCR in T cell membranes (after removal of the cell bodies) and living T cells. They looked at nanoclusters of both LAT and CD3 (as a marker for TCR), called protein islands, and observed how the distribution of these islands changes after T cell activation. Various activating substrates were used to exclude artifacts. It appeared the islands remained distinct after activation but came closer. A number of other techniques (EM, FCCS, and so on) were used to further determine details of protein island distribution and dynamics. A study from 2014 (Roh et al. 2015) focused on the co-distribution of nanoclusters of CD4, the CD3 component of TCR (TCR-CD3 complex), and Lck. CD4 is a crucial protein in the initial activation cascade of T cells, whereas Lck is probably the initial phosphorylation kinase of TCR in the consequent signaling cascade. In nonactivated cells, they demonstrated, interestingly using dual-color PALM on living cells with image acquisition times of 10 s, the distribution of independent nanoclusters of CD4 (4–8 molecules/cluster) and TCR (13–18 molecules/cluster). On activation, the clusters of CD4 become larger (6–15 molecules/cluster), whereas the TCR clusters remain similar in size. Modeling reveals a slight increase in overlap between CD4 and TCR but no increase in interaction. The data are confirmed in dual-color *d*STORM on fixed T cells on nonimmobilizing ICAM-1-coated lipid surfaces, to exclude motional artifacts and influence of the fluorescent proteins, PSCFP2 and PAmCherry, used in PALM.

Sherman et al. (2011) applied PALM to visualization of (the interactions between) ZAP70 (zeta-chain-associated protein kinase 70), LAT, SLP-76 (lymphocyte cytosolic protein 2 gene, a gene for specific adaptor proteins), and actin. The same topic in the same year was addressed by Hsu and Baumgart (2011). In 2013, the Sherman group wrote two reviews (Sherman et al. 2013a, 2013b), including their significant own data, where they described the visualization of TCR cluster size. In (Sherman et al. 2013a), they described their own method, PALM, thoroughly, including labeling and analysis protocols, which makes the paper very useful and practical for biologists who want to apply PALM to this topic.

The group of Katerina Gaus has applied PALM microscopy to the imaging of protein cluster formation during immune signaling in three papers on Lck and LAT (Owen et al. 2010, Rossy et al. 2013a, Williamson et al. 2011).

George Ashdown et al. (2014) in a 2014 paper quantified the retrograde flow velocity and directionality of actin at the immune synapse (the contact between the T cell and the APC). This was done by using SIM on $CD4^+$ T cells on antibody-coated coverslips.

The most recent paper from the group of Roybal et al. (2015) used a very interesting and extensive systems approach and studied the distribution of more than 60 signaling cascade intermediates in T cells. The $CD4^+$ T cells were stimulated by CH27 B cell lymphoma APCs, pulsed with 100 μM moth cytochrome C antigen peptide. It appeared that during the first 3 min after activation, various intermediate distribution patterns arose, each specific for a certain intermediate. Thirty-seven percent of the intermediates exhibited a lamellar distribution at one point in time during those 3 min. They focused on this lamellar structure, seen, for example, for SLP-76, PIP_2 (membrane phospholipid) and Vav1, a gene that is part of the signaling cascade after TCR activation (Koretzky and Myung 2001). Using STED on fixed cells, they determined the relation with the actin network during the crucial 3 min after activation and found a significant, albeit not complete, overlap. Using a comparison with EM data on the undulations of the interface, it was concluded that lamellum and actin largely overlap, whereas FRAP demonstrated that the actin structures and the signaling intermediates on the lamellum diffuse on similar time scales and over equal micrometer distances. The relation of these findings with the remaining distribution profiles (central, peripheral, diffuse, invaginated, and asymmetric) are shortly mentioned in the discussion but not extended on. It is obvious that the chosen approach is a large step toward understanding the interrelation in complex signaling cascades, partly made possible by the availability of super-resolution microscopy.

18.3.2.2 B cells

Over the years, the role of the adaptive immune system in cardiovascular diseases such as atherosclerosis has become undisputed (Le Borgne et al. 2015). Some B and T cell populations are pro-atherogenic, whereas others are atheroprotective. Although the mechanisms of action become better understood, their impact on progression remains unclear. To date, successful preclinical studies with protective vaccination against atherosclerotic antigens or inhibiting pro-atherogenic B cells have even raised hope on clinical interventions. In addition, in cardiac diseases, such as HF and fibrosis, B cells seem to play a role (Cordero-Reyes et al. 2016). Again, the studies below are not focused on the cardiovascular aspects of B cell biology, but they do highlight the importance of super-resolution microscopy.

In naïve B cells, signaling through the B cell receptor cluster (BCR) *in vivo* is initiated by binding of APC, resulting in elevation of intracellular calcium and subsequent B cell activation. In response, the B cells start forming antibody-secreting plasma cells and long-lasting memory cells. Not only can APC binding initiate this cascade, but also mere actin cytoskeleton disruption alone can, implying a fundamental role for the cytoskeleton network in B cell signaling (Gasparrini et al. 2015).

We report three super-resolution studies on B cell imaging, all on BCR signaling events. The 2013 study from Mattila et al. (2013) focuses on two B cell receptors, IgD and IgM. In resting B cells, *d*STORM revealed IgD and IgM nanoclusters, the first being significantly more densely packed (70% vs. 38%), while cluster radii were similar at around 60–80 nm. The IgD clusters contain approximately 30–120 molecule, whereas the IgM resides in nanoclusters, with around 20–50 molecules per cluster. Disruption of the actin cytoskeleton (as a crucial step during early signaling cascades) does not change cluster size and density, but merely changes the mobility of the clusters to reach co-receptors such as CD19. In turn, CD19 is present in similar clusters, located in between the IgD and IgM clusters. The CD19 clusters are held in place, not by the actin network, but by a tetraspanin (e.g., CD81) network, which corresponds with the scaffolding function of the membrane-bound tetraspanin network in general.

The study of Gasparrini et al. (2015) demonstrated the role of the nanoscale organization, dynamics, and cytoskeletal cooperation of CD22 (a lectin expressed on the surface of B cells) in dampening B cell signaling. Indeed, by using SIM and *d*STORM on various fixed mouse cells, they demonstrated the existence of 100-nm-diameter nanoclusters on the membrane of naïve B cells; their dampening capability seemingly determined by their highly dynamic (2–6 times larger diffusion coefficients than other receptors) membrane behavior, as determined using single-particle tracking. The paper then extends on factors that influence CD22 organization (glycoprotein CD45 and sialic acid), dynamics (actin, CD45, and sialic acid), and dampening function (sialic acid), all using impressive super-resolution data.

In the more recent study of Zuidscherwoude et al. (2015), various tetraspanin proteins were studied using dual-color STED, to unravel the more complete web on the cell surface. It was demonstrated that CD37, CD53, CD81, and CD82 form individual clusters on the plasma membrane, with a size of 90–170 nm.

These small nanoclusters are distributed on the plasma membrane at densities of 1–5 domains per μm^2. Whereas CD53 or CD81 are in proximity to their interaction partners MHC class II or CD19; they show surprisingly little overlap with their tetraspanin colleagues CD37 and CD82. All these finding resulted in a new view of the tetraspanin network.

18.3.2.3 Natural Killer Cells

NK cells are mostly involved in defense against viral infection and malignancy, and patients with NK cell deficiencies suffer from recurrent and often fatal viral infections and malignancies. Until recently, they were not so much in the picture of cardiovascular researchers. Recent reviews (Bonaccorsi et al. 2015, Zuo et al. 2015), however, have described their possible role in atherogenesis by interaction with macrophages and dendritic cells and the mechanisms that could be involved. None of this is, however, studied in the various super-resolution papers on NK cells.

Two reviews (Mace and Orange 2012b, Pageon et al. 2012) discusses, among other techniques, the relevance of super-resolution microscopy for imaging the immune synapse (Sanborn et al. 2010) of NK cells with contacting cells. All studies they referred to are described below. Three experimental studies from the group of Daniel Davis looked at various aspects of the NK cell synapse activating receptor NKG2D, a transmembrane receptor recognizing induced-self proteins. These proteins are absent on healthy cells, but expressed abundantly on transformed, stressed, or infected cells. Recognition of these cells activates the NK cells. In the first study (Brown et al. 2011), a combination of optical tweezers and live-cell confocal microscopy first revealed that microclusters of NKG2D assemble into a ring-shaped structure at the center of intercellular synapses. Then, 3DSIM demonstrated that lytic granules, released from the immune synapse during activation, accumulate in the center of the ring at the location of a thin cortical actin network, a network that cannot be seen using standard confocal microscopy and appears to be present both during activating and inhibitory signaling. During activation, however, the periodicity of the cortical actin mesh increases in specific domains at the synapse. Two-color super-resolution imaging reveals that lytic granules dock precisely in these domains. The paper is very detailed but easy to read.

In a sequel (Brown et al. 2012), again 3DSIM was used to further study the changes in the actin network at the immune synapse, but now in response not only to activation of the NKG2D receptor. Activation of other receptors, such as CD2 (or LFA-2, adhesion and co-stimulatory molecule expressed on the surface of NK cells) and NKp46 (receptor on NK cell membrane), or triggering by influenza virus, is not sufficient to remodel the actin mesh, but this needs co-ligation of LFA-1 by ICAM-1.

In the third paper from this group (Pageon et al. 2013), the relation between NKG2D and another, inhibitory, receptor KIR2DL1 was studied using PALM and GSDIM. This receptor recognizes MHC-I and the cells lacking MHC-1 are killed. KIR2DL1 appears to form quasicircular clusters of 110–120 nm in size at the surface of the cell. The cluster density is around 3.5 clusters/μm^2, each cluster containing 1100–1900 molecules. On direct activation of KIR2DL1 (by a specific antibody), the clusters become 20% smaller, their number increases around 10%, and there density increases by 60%. Interestingly, however, also on activation of NKG2D, the same changes in KIR2DL1 clustering arise; this is a rather surprising result, since both receptors have distinct signaling function (inhibitory vs. activating).

A different group, that of Jordan Orange, actually discussed the same topic as the 2011 paper from Daniels team and found similar results (Mace and Orange 2012a,

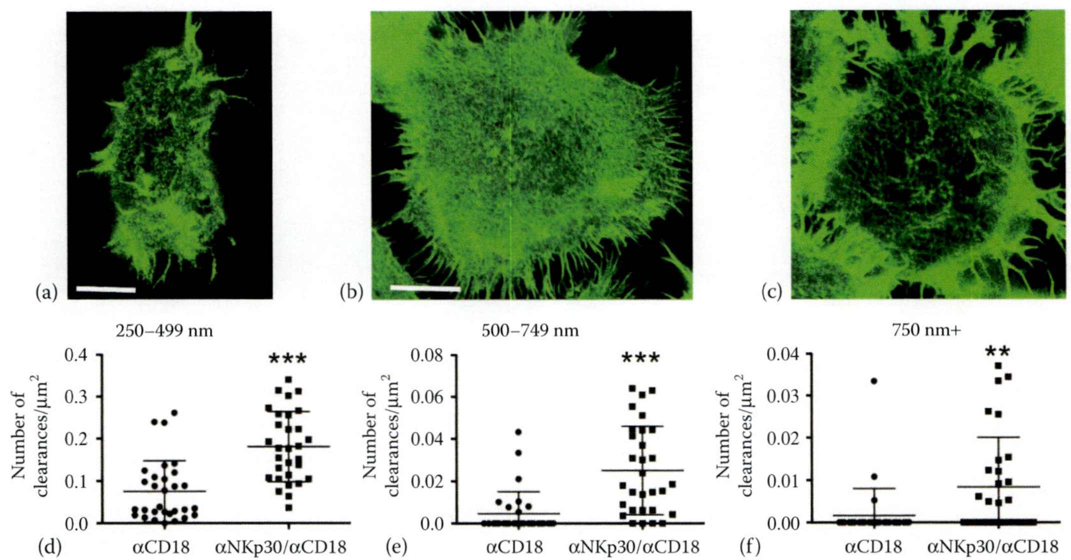

Figure 18.3

STED of NK cell. Only legends for images are given; for other details, see corresponding paper. (a) Synapse between Citrine-actin expressing NK-92 and mel1190 target in X–Y plane; (b and c) NK-92 cells were stimulated for 30 min on glass coated with either antibody to CD18 (b) or to CD18 and NKp30 (c) and stained for F-actin with phalloidin. (Reproduced with permission from Rak, G. D. et al., *PLoS Biol* 9, e1001151, 2011.) (*Continued*)

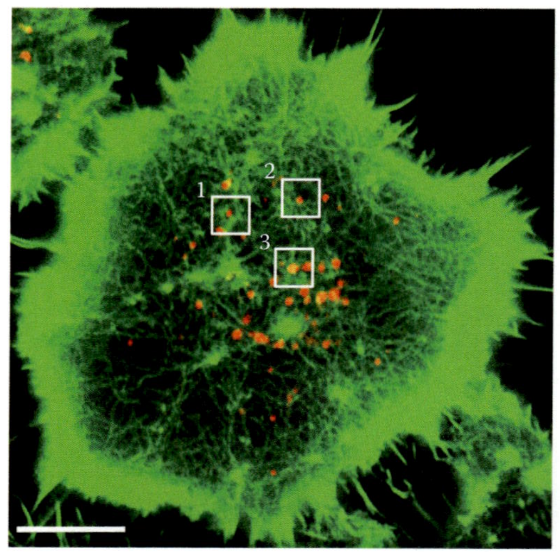

(g)

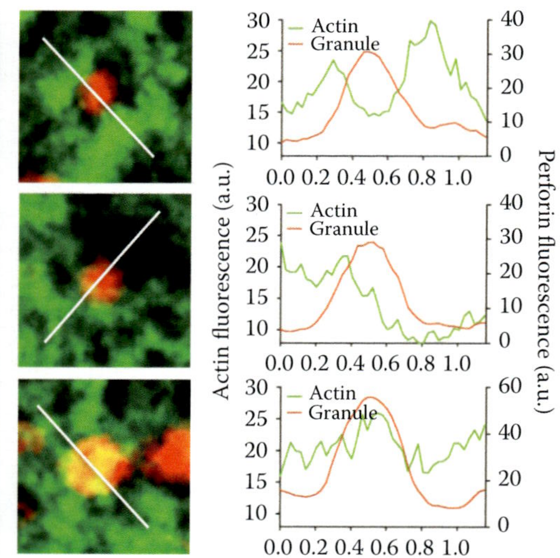

(h)

Figure 18.3 (Continued)

STED of NK cell. Only legends for images are given; for other details, see corresponding paper. (g) activated NK-92 cell stained for F-actin (green, STED) and perforin (red, confocal); (h) Insets from G1 (top), G2 (middle), and G3 (bottom). Line profile indicates pixel intensities of actin (green) and perforin (red) along white line. Scale bars: 5 mm. (Reproduced with permission from Rak, G. D. et al., *PLoS Biol* 9, e1001151, 2011.) (Reproduced with permission from Rak, G. D. et al., *PLoS Biol* 9, e1001151, 2011.)

Rak et al. 2011). The papers are well written and very explanatory, with beautiful images (Figure 18.3). The researchers used STED microscopy to study the formation of lytic granules on stimulation of NK cells using cross-linking of NKp30, a natural cytotoxicity receptor whose ligand is expressed on tumor cells, and CD18, a subunit of the heterodimeric integrin lymphocyte function-associated antigen-1 (LFA-1). The lytic granules were visualized using a home-built detector of lytic pH, pHfluorin-LAMP1. This fluorescent probe targets to lysosomes via LAMP-1, while the GFP-like protein pHfluorin increases fluorescence on increase of pH, and thus on degranulation. A very recent methodological paper (Mace and Orange 2014b) describes the sample preparation, acquisition procedures, and analysis methods needed for gSTED on NK cells; it also discusses possible sources of error in interpretation of STED data.

A 2013 paper from this group (Mace and Orange 2014a) demonstrates the essential role of coronin 1A (Coro1A, actin-binding protein) in the deconstruction of synaptic cortical filamentous F-actin, thus enabling lytic granule secretion and thus NK cytotoxicity. They applied confocal and STED microscopy on fixed cells of human origin and also extended their research to cells derived from a patient with a rare Coro1A mutation. This patient had severe NK cell function impairment and they showed a strong overlap of deficiency in cell cultures and patient cells. This clinical link is a beautiful demonstration of translational research that deserves more succession.

18.3.2.4 Neutrophils

Neutrophils are immune cells from the myeloid lineage and play an important role in innate immunity as specialized phagocytes. To be able to respond to infectious conditions, phagocytes use pattern recognition receptors (PRRs) to detect pathogen-associated molecular patterns (PAMPs) expressed on the surface of microorganisms, such as lipopolysaccharides, glycolipids, and glycoproteins. The formation of PRR-PAMP complexes activates signaling cascades, ultimately resulting in oxidative and lysosomal enzyme bursts, killing the attacker. Until recent years, neutrophils did not gain much attention for their role in, for example, atherosclerosis. A recent review (Pende et al. 2015) discusses this topic and concluded that neutrophils are important in both the early phases of atherogenesis and in plaque stability or rupture, the latter being a crucial step in the formation of clinical symptoms. Another paper (Chistiakov et al. 2015) highlighted neutrophils mechanisms in host defense against pathogens, such as neutrophil extracellular traps (NET) formation, and indicated how both in the very early and late phases of atherogenesis this mechanism might be misdirected and lead to damaging consequences. It is obvious how super-resolution imaging could yield important information on neutrophil defense mechanisms. The studies below, although not describing these aspects, can be considered illustrative for the potential of super-resolution microscopy.

The first paper by Ekyalongo et al. (2015) is actually a rather chemical review on the organization and function of glycolipid-enriched microdomains. It is mentioned here because it gives a figure on the application of STED to visualize the clustering of PRRs on the membrane of neutrophils.

The surprising study by Maurer et al. (2015) described the application of SIM to the visualization of the bitter receptor T2R38 in neutrophils. This bitter receptor is known to be important in taste buds, epithelial cells, and colon, where it has an obvious function. Here, it was demonstrated that this bitter receptor is also expressed in various leukocytes, whereas in neutrophils, it appears to be present on the membrane and in lipid droplets. They convincingly showed that this receptor not only is sensitive for AHL-12 [*N*-(3-oxododecanoyl)-L-homoserine lactone], a communication molecule among gram-negative bacteria, but is also known to activate neutrophils. SIM was used to demonstrate that the receptor co-localizes to the membrane of lipid droplets, whereas confocal microscopy was applied to show that AHL-12 is taken up by the neutrophils and co-localizes with T2R38.

18.3.3 Super-Resolution Microscopy of Platelets

The role of platelets in cardiovascular disease, such as atherosclerosis, is undisputed, and consequently, a vast amount of experimental and review papers (Ahmadsei et al. 2015, Duchene and von Hundelshausen 2015, Ed Rainger et al. 2015, Nording et al. 2015, Patzelt et al. 2015, Vajen et al. 2015) exists on this topic. It is beyond the scope of this paper to further extend on this, but the papers below, together with the readers' personal interest, should trigger enough momentum to embark on super-resolution imaging of platelets in the area of disease involved.

Platelets, also called thrombocytes, are 2- to 3-μm-diameter, small, nucleus-free, and biconvex blood particles, whose main task is to contribute to the process of stopping bleeding at the site of interrupted vessel wall endothelium. To that end, they gather at the site of injury and attach to the edge (*adhesion*), followed by a change in shape, a turn on of receptors, and secretion of chemical messengers (*activation*). Then they form receptor bridges for connection to each other (*aggregation*). The so-formed platelet plug subsequently results in activation of the coagulation cascade, with resultant fibrin deposition and linking and thus clot formation. In addition to their role in hemostasis, platelets act as modulators of inflammatory processes. They do so by interacting with leukocytes and by secretion of chemokines and cytokines. Although the general function of platelets is well described and experimentally researched, their exact structure and function are difficult to study owing to their very small size. Since their diameter is around 2 μm, studying (changes in) substructures within platelets means approaching the limit of resolution in traditional confocal microscopy. Therefore, the advance of super-resolution imaging is especially promising in this field. Simultaneously, one has to conclude that so far, only limited studies can be found on this topic.

Two studies from Daniel Rönnlund discuss the application of STED to the imaging of the substructure of platelets. Specifically, the first study (Ronnlund et al. 2012) describes the distribution of pro-angiogenic VEGF, anti-angiogenic PF-4, and fibrinogen in regional clusters significantly smaller than the size of an α-granule (Figure 18.4). The size of the VEGF clusters is furthermore significantly smaller than that of PF-4 and fibrinogen (70 and 100 nm, respectively). No co-localization between the different proteins is observed. On platelet activation by either thrombin or ADP, each of the proteins undergoes typical alterations not only in size and number of clusters but also in localization of the clusters. Based on these observations, state characterization of the platelet seems feasible. The second paper (Ronnlund et al. 2014) beautifully describes the development of four-color STED to study simultaneously VEGF, PF-4, fibrinogen, and actin. This is achieved by measurement and unmixing of various images of the four probes.

In the paper by Aslan et al. (2013), the platelet tubulin cytoskeletal network takes central stage. It is known that platelet tubulin plays a role in maintaining the typical discoid shape of resting platelets, but its functioning in the change of platelet structure on activation is not well studied. Here, using SIM, they demonstrated

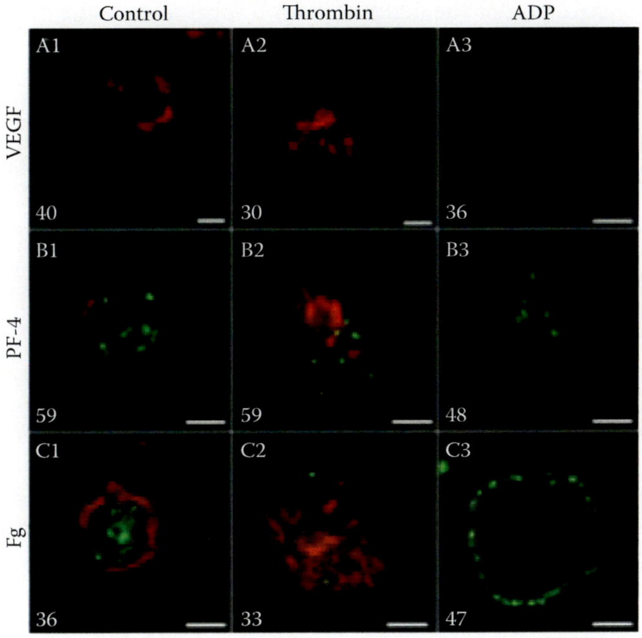

Figure 18.4

STED images of A1–A3: VEGF (green), B1–B3: PF4 (green), and C1–C3: fibrinogen (green). Actin is indicated in red. Left: inactivated control, middle: thrombin-activated platelets, right: ADP-activated platelets. Scale bar: 1 μm. (Reproduced with permission from Ronnlund, D. et al., *Adv Healthc Mater* 1, 707–713, 2012.)

that on activation, the resting platelet microtubule marginal band, connected with a network of finer fibers, collapses toward the platelet center, as platelets condense in volume. This centripetal redistribution of the marginal band appears to be associated with the deacetylation of tubulin (also shown with SIM), as inhibitors of tubulin deactetylase HDAC6 change resting marginal band structure, alter marginal band collapse, and inhibit the efficiency of platelet aggregation.

Recently, Cheepala et al. (2015) visualized the localization of the ATP-binding cassette ABCC4 in the plasma membrane of both human and mouse platelets. Although, traditionally, it is thought that this protein is localized in dense granules, where it imports ADP, recent studies have raised ambiguity. They now unambiguously demonstrated, using SIM, the localization in the membrane, where it regulates aggregation by exporting cAMP and antithrombotic drugs.

Finally, Poulter et al. (2015) in an extensive but well-written study applied a combination of SIM, *d*STORM, and TIRF microscopy to study actin nodules in mouse platelets activated on fibrinogen-coated coverslips. Using time-lapse TIRF with platelets from Lifeact-GFP mice, the nodules appeared to be podosome-like structures (see below), but without extracellular matrix degrading power, that are formed rapidly on platelet activation, are tightly controlled with respect to size (0.2–0.3 μm^2) and lifetime (22 s), and are immobile in plane, moving toward the substrate in axial direction. Alexa488-phalloidin staining was applied in combination with SIM and *d*STORM to show the interconnected network-like structure of the actin nodules in fibrinogen-activated mouse and human platelets. Co-staining of actin and vinculin (integrin-actin connecting protein), talin (integrin-actin connecting probe via vinculin), and integrin αIIbβ3 were then applied to show that all three form a network around the nodules, with variations in time after activation. Co-staining with phosphotyrosine demonstrated that the nodules are sites of phosphorylation and thus signaling. In the final part of the paper, the significant role of Wiskott–Aldrich syndrome protein (WASP) in all these processes was investigated. It appeared that WASP is required for nodule formation in both mouse and human and recruits and activates ARP2/3 complex. This underscores the role of the actin nodules in platelet aggregate formation under flow.

18.3.4 Super-Resolution Microscopy of Podosomes

Both immune cells and platelets are highly motile and therefore should have the ability to crawl along various surfaces such as ECs and extracellular matrix. Podosomes are small (0.5–2 μm) actin cytoskeletal structures in motile cells; they give to motile cells adhesional, migrational, and (cellular or matrix) transmigrational properties. They generally exist for several minutes and have the power to degrade the matrix with which they connect. It is important to note that, at least *in vitro*, ECs, smooth muscle cells, and some fibroblasts can also form podosomes. Podosomes can appear in various structures, such as rosettes, belts, and clusters.

Two reviews discuss everything you always wanted to know about podosomes. The review from Schachtner et al. (2013) discusses various aspects of podosomes: the regulation of and signaling cascades behind podosome formation, their role in cellular dynamics and how that is regulated, the (network) structure of the podosome ring and podosomes' consequent mechano-sensitivity, and their protease-release function. The review refers to the few existing research papers on super-resolution imaging of podosomes. The other review (Meddens et al. 2014) also refers to those studies and adds significant data (using 3D STORM) that further stress the need for super-resolution images in advancing the knowledge on the tiny podosomes. Importantly, this paper also stresses the urgent need for investigating podosomes in intact tissue, since the relevance of podosomes in real life is not yet known.

We found three experimental studies that focus on the podosome's structure per se. The oldest study (Cox et al. 2012) uses 3B (Bayesian analysis of blinking and bleaching) widefield microscopy to convincingly show the ring-shaped structure of the protein vinculin around the central actin podosome nucleus. The researchers performed their studies in both fixed and living cells, also determining various ways in which the podosomes form and disappear. The paper ends with a very honest comparison of the advantages (widefield, easy, and fast) and disadvantages (analysis time and effort, and analysis artifacts) of their method.

In van den Dries et al. (2013), dual-color *d*STORM is applied to study the nanostructure of podosomes in dendritic cells on coverslips with various coatings. Various proteins are considered; integrin αMBβ2, talin, and vinculin and their differences in distribution are discussed. Talin and integrin αMBβ2 are diffusely distributed across the podosome but are excluded from its core. Vinculin, in contrast, has a punctuate distribution pattern. Using destruction of the F-actin network and a mutant of vinculin that has no actin-binding site, the researchers showed that actin is crucial for this punctuate vinculin pattern. They also studied the size and shape of the podosomes and find two different patterns: circular and small (around 300 nm diameter) podosomes versus larger (around 550 nm diameter) and elongated ones. Since the cells under study were fixed, they could only speculate on the cause of the existence of these two populations, that is, the elongated podosomes are in the process of formation during fixation. Finally, the actin podosomes form a network via actin fibers, the podosomes at an interdistance around 1 μm, actin radiating with length of 0.43 μm, and vinculin oriented along the actin fibers.

Using combined SIM and STED, the vinculin ring structure was extended on by Walde et al. (2014), showing that it is not really a ring but consists of a polygonal of 400- to 600-nm-long straight vinculin strands, the strands extending from the corner points at definite angles. In addition, the distribution of talin and paxillin (a protein needed to adhere cells to the extracellular matrix) in podosomes was quantified.

18.4 Future Perspective

Clearly, super-resolution in cardiovascular research is in its infancy. The studies so far have nicely demonstrated its potential in various areas, but all are limited to structural characteristics of cardiac, vascular, or immune cells, mostly in isolated fixed cells. Without any doubt, this important type of study will continue and extend, being crucial for understanding cellular function.

Histological slices are only scarcely used, and only in cardiac imaging, but could yield significant information on cellular context and relations between various cell types and their surroundings. Here, I just want to mention a few possibilities in the vascular area of research, such as contact (1) between ECs and SMCs via elastin fenestrae, (2) of vascular fibroblasts with collagen, (3) of smooth muscle cells with elastin, (4) of ECs with the subendothelial collagen IV layer, (5) of ECs with the glycocalyx, (6) of nerve innervation hubs in the vascular wall, and (7) of NETs in the vascular wall. Indeed, personal recent studies have shown the feasibility of resonant-scanning gSTED microscopy (with 592 nm depletion laser) on histological slices of, among others, the vascular wall. This type of study can be carried out, of course, on slices of vascular tissue and also on organs such as heart (as already shown by the Soeller group), bladder, spleen, and brain. Finally, I would be very interested to see the first data on super-resolution imaging of cleared, thicker slabs of vascular or cardiac tissue. Numerous clearing methods are currently being developed, mostly for two-photon or light-sheet imaging of the brain. However, this kind of samples undoubtedly will also be available for super-resolution imaging, especially STED and SIM, of cardiovascular tissue, possibly even in combined super-resolution and multiphoton microscopy.

So far, only very limited cardiovascular live-cell studies have been carried out, all of them looking at structural, not functional, properties. Indeed, such studies require vital super-resolution imaging at frame frequencies higher than process frequency. Furthermore, the imaging should not disturb the process. Both prerequisites can only be fulfilled with a limited number of super-resolution techniques, one of them being resonant-scanning gSTED. However, the potential biological relevance and impact of such experiments are enormous. Exemplary, I mention the imaging of NO production by ECs and its trafficking to smooth muscle cells. Again, own studies using resonant-scanning gSTED have demonstrated nicely the feasibility of this technique in various cell types of the vascular wall. Other examples of functional live-cell imaging could be the visualization and quantification of (changes in) endothelial glycocalyx, thrombus formation *in vitro*, cellular cytoskeleton dynamics, nuclear envelope dynamics, or the dynamics of intracellular structures such as mitochondria, caveolae, DNA, and lysosomes.

Finally, fast super-resolution microscopy also opens up avenues for intravital super-resolution microscopy in thin tissues, such as visualization of the microcirculation or thrombus formation in action, for example, in mouse cremaster or mesentery.

References

Agullo-Pascual, E., X. Lin, A. Leo-Macias, M. Zhang, F. X. Liang, Z. Li, A. Pfenniger et al., 2014. Super-resolution imaging reveals that loss of the C-terminus of connexin43 limits microtubule plus-end capture and NaV1.5 localization at the intercalated disc. *Cardiovasc Res* 104(2):371–381. doi:10.1093/cvr/cvu195.

Agullo-Pascual, E., D. A. Reid, S. Keegan, M. Sidhu, D. Fenyo, E. Rothenberg, and M. Delmar. 2013. Super-resolution fluorescence microscopy of the cardiac connexome reveals plakophilin-2 inside the connexin43 plaque. *Cardiovasc Res* 100(2):231–240. doi:10.1093/cvr/cvt191.

Ahmadsei, M., D. Lievens, C. Weber, P. von Hundelshausen, and N. Gerdes. 2015. Immune-mediated and lipid-mediated platelet function in atherosclerosis. *Curr Opin Lipidol* 26(5):438–448. doi:10.1097/MOL.0000000000000212.

Ashdown, G. W., A. Cope, P. W. Wiseman, and D. M. Owen. 2014. Molecular flow quantified beyond the diffraction limit by spatiotemporal image correlation of structured illumination microscopy data. *Biophys J* 107(9): L21–L23. doi:10.1016/j.bpj.2014.09.018.

Aslan, J. E., K. G. Phillips, L. D. Healy, A. Itakura, J. Pang, and O. J. McCarty. 2013. Histone deacetylase 6-mediated deacetylation of alpha-tubulin coordinates cytoskeletal and signaling events during platelet activation. *Am J Physiol Cell Physiol* 305(12):C1230–C1239. doi:10.1152/ajpcell.00053.2013.

Baddeley, D., D. Crossman, S. Rossberger, J. E. Cheyne, J. M. Montgomery, I. D. Jayasinghe, C. Cremer, M. B. Cannell, and C. Soeller. 2011. 4D super-resolution microscopy with conventional fluorophores and single wavelength excitation in optically thick cells and tissues. *PLoS One* 6(5):e20645. doi:10.1371/journal.pone.0020645.

Baddeley, D., I. D. Jayasinghe, L. Lam, S. Rossberger, M. B. Cannell, and C. Soeller. 2009. Optical single-channel resolution imaging of the ryanodine receptor distribution in rat cardiac myocytes. *Proc Natl Acad Sci USA* 106(52):22275–22280. doi:10.1073/pnas.0908971106.

Balagopalan, L., E. Sherman, V. A. Barr, and L. E. Samelson. 2011. Imaging techniques for assaying lymphocyte activation in action. *Nat Rev Immunol* 11(1):21–33. doi:10.1038/nri2903.

Balint, B., H. Yin, S. Chakrabarti, M. W. Chu, S. M. Sims, and J. G. Pickering. 2015. Collectivization of vascular smooth muscle cells via TGF-beta-Cadherin-11-dependent adhesive switching. *Arterioscler Thromb Vasc Biol* 35(5):1254–1264. doi:10.1161/ATVBAHA.115.305310.

Bazzoni, G. and E. Dejana. 2004. Endothelial cell-to-cell junctions: Molecular organization and role in vascular homeostasis. *Physiol Rev* 84(3):869–901. doi:10.1152/physrev.00035.2003.

Beavers, D. L., A. P. Landstrom, D. Y. Chiang, and X. H. Wehrens. 2014. Emerging roles of junctophilin-2 in the heart and implications for cardiac diseases. *Cardiovasc Res* 103(2):198–205. doi:10.1093/cvr/cvu151.

Benzing, C., J. Rossy, and K. Gaus. 2013. Do signalling endosomes play a role in T cell activation? *FEBS J* 280(21):5164–5176. doi:10.1111/febs.12427.

Blayney, L. M. and F. A. Lai. 2009. Ryanodine receptor-mediated arrhythmias and sudden cardiac death. *Pharmacol Ther* 123(2):151–177. doi:10.1016/j.pharmthera.2009.03.006.

Bonaccorsi, I., C. De Pasquale, S. Campana, C. Barberi, R. Cavaliere, F. Benedetto, and G. Ferlazzo. 2015. Natural killer cells in the innate immunity network of atherosclerosis. *Immunol Lett* 168(1):51–57. doi:10.1016/j.imlet.2015.09.006.

Brown, A. C., I. M. Dobbie, J. M. Alakoskela, I. Davis, and D. M. Davis. 2012. Super-resolution imaging of remodeled synaptic actin reveals different synergies between NK cell receptors and integrins. *Blood* 120(18):3729–3740. doi:10.1182/blood-2012-05-429977.

Brown, A. C., S. Oddos, I. M. Dobbie, J. M. Alakoskela, R. M. Parton, P. Eissmann, M. A. Neil et al., 2011. Remodelling of cortical actin where lytic granules dock at natural killer cell immune synapses revealed by super-resolution microscopy. *PLoS Biol* 9(9):e1001152. doi:10.1371/journal.pbio.1001152.

Cerrone, M., X. Lin, M. Zhang, E. Agullo-Pascual, A. Pfenniger, H. Chkourko Gusky, V. Novelli et al., 2014. Missense mutations in plakophilin-2 cause sodium current deficit and associate with a Brugada syndrome phenotype. *Circulation* 129(10):1092–1103. doi:10.1161/CIRCULATIONAHA.113.003077.

Cheepala, S. B., A. Pitre, Y. Fukuda, K. Takenaka, Y. Zhang, Y. Wang, S. Frase et al., 2015. The ABCC4 membrane transporter modulates platelet aggregation. *Blood* 126(20):2307–2319. doi:10.1182/blood-2014-08-595942.

Chistiakov, D. A., Y. V. Bobryshev, and A. N. Orekhov. 2015. Neutrophil's weapons in atherosclerosis. *Exp Mol Pathol* 99(3):663–671. doi:10.1016/j.yexmp.2015.11.011.

Cordero-Reyes, A. M., K. A. Youker, A. R. Trevino, R. Celis, D. J. Hamilton, J. H. Flores-Arredondo, C. M. Orrego, A. Bhimaraj, J. D. Estep, and G. Torre-Amione. 2016. Full expression of cardiomyopathy is partly dependent on B-cells: A pathway that involves cytokine activation, immunoglobulin deposition, and activation of apoptosis. *J Am Heart Assoc* 5(1). doi:10.1161/JAHA.115.002484.

Cox, S., E. Rosten, J. Monypenny, T. Jovanovic-Talisman, D. T. Burnette, J. Lippincott-Schwartz, G. E. Jones, and R. Heintzmann. 2012. Bayesian localization microscopy reveals nanoscale podosome dynamics. *Nat Methods* 9(2):195–200. doi:10.1038/nmeth.1812.

Crossman, D. J., Y. Hou, I. Jayasinghe, D. Baddeley, and C. Soeller. 2015a. Combining confocal and single molecule localisation microscopy: A correlative approach to multi-scale tissue imaging. *Methods* 88:98–108. doi:10.1016/j.ymeth.2015.03.011.

Crossman, D. J., P. N. Ruygrok, Y. F. Hou, and C. Soeller. 2015b. Next-generation endomyocardial biopsy: The potential of confocal and super-resolution microscopy. *Heart Fail Rev* 20(2):203–214. doi:10.1007/s10741-014-9455-6.

Duchene, J. and P. von Hundelshausen. 2015. Platelet-derived chemokines in atherosclerosis. *Hamostaseologie* 35(2):137–141. doi:10.5482/HAMO-14-11-0058.

Ed Rainger, G., M. Chimen, M. J. Harrison, C. M. Yates, P. Harrison, S. P. Watson, M. Lordkipanidze, and G. B. Nash. 2015. The role of platelets in the recruitment of leukocytes during vascular disease. *Platelets* 26(6):507–520. doi:10.3109/09537104.2015.1064881.

Ekyalongo, R. C., H. Nakayama, K. Kina, N. Kaga, and K. Iwabuchi. 2015. Organization and functions of glycolipid-enriched microdomains in phagocytes. *Biochim Biophys Acta* 1851(1):90–97. doi:10.1016/j.bbalip.2014.06.009.

Feletou, M. 2009. Calcium-activated potassium channels and endothelial dysfunction: Therapeutic options? *Br J Pharmacol* 156(4):545–562. doi:10.1111/j.1476-5381.2009.00052.x.

Fernandez-Martin, L., B. Marcos-Ramiro, C. L. Bigarella, M. Graupera, R. J. Cain, N. Reglero-Real, A. Jimenez et al., 2012. Crosstalk between reticular adherens junctions and platelet endothelial cell adhesion molecule-1 regulates endothelial barrier function. *Arterioscler Thromb Vasc Biol* 32(8): e90–e102. doi:10.1161/ATVBAHA.112.252080.

Gasparrini, F., C. Feest, A. Bruckbauer, P. K. Mattila, J. Muller, L. Nitschke, D. Bray, and F. D. Batista. 2015. Nanoscale organization and dynamics of the siglec CD22 cooperate with the cytoskeleton in restraining BCR signalling. *EMBO J.* doi:10.15252/embj.201593027.

Gillet, L., D. Shy, and H. Abriel. 2013. NaV1.5 and interacting proteins in human arrhythmogenic cardiomyopathy. *Future Cardiol* 9(4):467–470. doi:10.2217/fca.13.38.

Goedicke-Fritz, S., A. Kaistha, M. Kacik, S. Markert, A. Hofmeister, C. Busch, S. Banfer, R. Jacob, I. Grgic, and J. Hoyer. 2015. Evidence for functional and dynamic microcompartmentation of Cav-1/TRPV4/K(Ca) in caveolae of endothelial cells. *Eur J Cell Biol* 94(7–9):391–400. doi:10.1016/j.ejcb.2015.06.002.

Granzier, H. L., K. R. Hutchinson, P. Tonino, M. Methawasin, F. W. Li, R. E. Slater, M. M. Bull et al., 2014. Deleting titin's I-band/A-band junction reveals critical roles for titin in biomechanical sensing and cardiac function. *Proc Natl Acad Sci USA* 111(40):14589–14594. doi:10.1073/pnas.1411493111.

Gregersen, I., S. Holm, T. B. Dahl, B. Halvorsen, and P. Aukrust. 2015. A focus on inflammation as a major risk factor for atherosclerotic cardiovascular diseases. *Expert Rev Cardiovasc Ther* 1–13. doi:10.1586/14779072.2016.1128828.

Grgic, I., B. P. Kaistha, J. Hoyer, and R. Kohler. 2009. Endothelial Ca^{+}-activated K^{+} channels in normal and impaired EDHF-dilator responses—Relevance to cardiovascular pathologies and drug discovery. *Br J Pharmacol* 157(4): 509–526. doi:10.1111/j.1476-5381.2009.00132.x.

Hong, M., L. Bao, E. Kefaloyianni, E. Agullo-Pascual, H. Chkourko, M. Foster, E. Taskin et al., 2012. Heterogeneity of ATP-sensitive K^{+} channels in cardiac myocytes: Enrichment at the intercalated disk. *J Biol Chem* 287(49): 41258–41267. doi:10.1074/jbc. M112.412122.

Hou, Y., D. J. Crossman, V. Rajagopal, D. Baddeley, I. Jayasinghe, and C. Soeller. 2014. Super-resolution fluorescence imaging to study cardiac biophysics: Alpha-actinin distribution and Z-disk topologies in optically thick cardiac tissue slices. *Prog Biophys Mol Biol* 115(2–3):328–339. doi:10.1016/j.pbiomolbio.2014.07.003.

Hou, Y., I. Jayasinghe, D. J. Crossman, D. Baddeley, and C. Soeller. 2015. Nanoscale analysis of ryanodine receptor clusters in dyadic couplings of rat cardiac myocytes. *J Mol Cell Cardiol* 80:45–55. doi:10.1016/j.yjmcc.2014.12.013.

Hsu, C. J. and T. Baumgart. 2011. Spatial association of signaling proteins and F-actin effects on cluster assembly analyzed via photoactivation localization microscopy in T cells. *PLoS One* 6(8):e23586. doi:10.1371/journal.pone.0023586.

Huber, O., A. Brunner, P. Maier, R. Kaufmann, P. O. Couraud, C. Cremer, and G. Fricker. 2012. Localization microscopy (SPDM) reveals clustered formations of P-glycoprotein in a human blood-brain barrier model. *PLoS One* 7(9):e44776. doi:10.1371/journal.pone.0044776.

Jayasinghe, I. D., D. Baddeley, C. H. Kong, X. H. Wehrens, M. B. Cannell, and C. Soeller. 2012a. Nanoscale organization of junctophilin-2 and ryanodine receptors within peripheral couplings of rat ventricular cardiomyocytes. *Biophys J* 102(5):L19–L21. doi:10.1016/j.bpj.2012.01.034.

Jayasinghe, I. D., D. J. Crossman, C. Soeller, and M. Cannell. 2012b. Comparison of the organization of T-tubules, sarcoplasmic reticulum and ryanodine receptors in rat and human ventricular myocardium. *Clin Exp Pharmacol Physiol* 39(5):469–476. doi:10.1111/j.1440-1681.2011.05578.x.

Jian, Z., H. Han, T. Zhang, J. Puglisi, L. T. Izu, J. A. Shaw, E. Onofiok et al., 2014. Mechanochemotransduction during cardiomyocyte contraction is mediated by localized nitric oxide signaling. *Sci Signal* 7(317):ra27. doi:10.1126/scisignal.2005046.

Kohl, T., V. Westphal, S. W. Hell, and S. E. Lehnart. 2013. Superresolution microscopy in heart—Cardiac nanoscopy. *J Mol Cell Cardiol* 58:13–21. doi:10.1016/j.yjmcc.2012.11.016.

Koretzky, G. A. and P. S. Myung. 2001. Positive and negative regulation of T-cell activation by adaptor proteins. *Nat Rev Immunol* 1(2):95–107. doi:10.1038/35100523.

Lacolley, P., V. Regnault, A. Nicoletti, Z. Li, and J. B. Michel. 2012. The vascular smooth muscle cell in arterial pathology: A cell that can take on multiple roles. *Cardiovasc Res* 95(2):194–204. doi:10.1093/cvr/cvs135.

Landstrom, A. P., D. L. Beavers, and X. H. Wehrens. 2014. The junctophilin family of proteins: From bench to bedside. *Trends Mol Med* 20(6):353–362. doi:10.1016/j.molmed.2014.02.004.

Lavina, B. and K. Gaengel. 2015. New imaging methods and tools to study vascular biology. *Curr Opin Hematol* 22(3):258–266. doi:10.1097/MOH.0000000000000141.

Le Borgne, M., G. Caligiuri, and A. Nicoletti. 2015. Once upon a time: The adaptive immune response in atherosclerosis-a fairy tale no more. *Mol Med* 21(Suppl 1):S13–S18. doi:10.2119/molmed.2015.00027.

Leo-Macias, A., E. Agullo-Pascual, J. L. Sanchez-Alonso, S. Keegan, X. Lin, T. Arcos, F. X. Liang et al., 2016. Nanoscale visualization of functional adhesion/excitability nodes at the intercalated disc. *Nat Commun* 7:10342. doi:10.1038/ncomms10342.

Ley, K., C. Laudanna, M. I. Cybulsky, and S. Nourshargh. 2007. Getting to the site of inflammation: The leukocyte adhesion cascade updated. *Nat Rev Immunol* 7(9):678–689. doi:10.1038/nri2156.

Lillemeier, B. F., M. A. Mortelmaier, M. B. Forstner, J. B. Huppa, J. T. Groves, and M. M. Davis. 2010. TCR and Lat are expressed on separate protein islands on T cell membranes and concatenate during activation. *Nat Immunol* 11(1):90–96. doi:10.1038/ni.1832.

Loscher, W. and H. Potschka. 2005. Blood-brain barrier active efflux transporters: ATP-binding cassette gene family. *NeuroRx* 2(1):86–98. doi:10.1602/neurorx.2.1.86.

Mace, E. M. and J. S. Orange. 2012a. Dual channel STED nanoscopy of lytic granules on actin filaments in natural killer cells. *Commun Integr Biol* 5(2): 184–186. doi:10.4161/cib.18818.

Mace, E. M. and J. S. Orange. 2012b. New views of the human NK cell immunological synapse: Recent advances enabled by super- and high-resolution imaging techniques. *Front Immunol* 3:421. doi:10.3389/fimmu.2012.00421.

Mace, E. M. and J. S. Orange. 2014a. Lytic immune synapse function requires filamentous actin deconstruction by Coronin 1A. *Proc Natl Acad Sci USA* 111(18):6708–6713. doi:10.1073/pnas.1314975111.

Mace, E. M. and J. S. Orange. 2014b. Visualization of the immunological synapse by dual color time-gated stimulated emission depletion (STED) nanoscopy. *J Vis Exp* (85). doi:10.3791/51100.

Mattila, P. K., C. Feest, D. Depoil, B. Treanor, B. Montaner, K. L. Otipoby, R. Carter, L. B. Justement, A. Bruckbauer, and F. D. Batista. 2013. The actin and tetraspanin networks organize receptor nanoclusters to regulate B cell receptor-mediated signaling. *Immunity* 38(3):461–474. doi:10.1016/j.immuni.2012.11.019.

Maurer, S., G. H. Wabnitz, N. A. Kahle, S. Stegmaier, B. Prior, T. Giese, M. M. Gaida, Y. Samstag, and G. M. Hansch. 2015. Tasting Pseudomonas aeruginosa biofilms: Human neutrophils express the bitter receptor T2R38 as sensor for the quorum sensing molecule N-(3-Oxododecanoyl)-l-Homoserine Lactone. *Front Immunol* 6:369. doi:10.3389/fimmu.2015.00369.

Meddens, M. B., K. van den Dries, and A. Cambi. 2014. Podosomes revealed by advanced bioimaging: What did we learn? *Eur J Cell Biol* 93(10–12):380–387. doi:10.1016/j.ejcb.2014.09.002.

Megens, R. T., M. Bianchini, M. M. Schmitt, and C. Weber. 2015. Optical imaging innovations for atherosclerosis research: Multiphoton microscopy and optical nanoscopy. *Arterioscler Thromb Vasc Biol* 35(6):1339–1346. doi:10.1161/ATVBAHA.115.304875.

Megens, R. T., M. G. oude Egbrink, J. P. Cleutjens, M. J. Kuijpers, P. H. Schiffers, M. Merkx, D. W. Slaaf, and M. A. van Zandvoort. 2007a. Imaging collagen in intact viable healthy and atherosclerotic arteries using fluorescently labeled CNA35 and two-photon laser scanning microscopy. *Mol Imaging* 6(4):247–260.

Megens, R. T., M. G. oude Egbrink, M. Merkx, D. W. Slaaf, and M. A. van Zandvoort. 2008. Two-photon microscopy on vital carotid arteries: Imaging the relationship between collagen and inflammatory cells in atherosclerotic plaques. *J Biomed Opt* 13(4):044022. doi:10.1117/1.2965542.

Megens, R. T., S. Reitsma, L. Prinzen, M. G. oude Egbrink, W. Engels, P. J. Leenders, E. J. Brunenberg et al., 2010. In vivo high-resolution structural imaging of large arteries in small rodents using two-photon laser scanning microscopy. *J Biomed Opt* 15(1):011108. doi:10.1117/1.3281672.

Megens, R. T., S. Reitsma, P. H. Schiffers, R. H. Hilgers, J. G. De Mey, D. W. Slaaf, M. G. oude Egbrink, and M. A. van Zandvoort. 2007b. Two-photon microscopy of vital murine elastic and muscular arteries. Combined structural and functional imaging with subcellular resolution. *J Vasc Res* 44(2):87–98. doi:10.1159/000098259.

Mehta, D. and A. B. Malik. 2006. Signaling mechanisms regulating endothelial permeability. *Physiol Rev* 86(1):279–367. doi:10.1152/physrev.00012.2005.

Meng, X., J. Yang, M. Dong, K. Zhang, E. Tu, Q. Gao, W. Chen, C. Zhang, and Y. Zhang. 2015. Regulatory T cells in cardiovascular diseases. *Nat Rev Cardiol*. doi:10.1038/nrcardio.2015.169.

Nording, H. M., P. Seizer, and H. F. Langer. 2015. Platelets in inflammation and atherogenesis. *Front Immunol* 6:98. doi:10.3389/fimmu.2015.00098.

Owen, D. M., C. Rentero, J. Rossy, A. Magenau, D. Williamson, M. Rodriguez, and K. Gaus. 2010. PALM imaging and cluster analysis of protein heterogeneity at the cell surface. *J Biophotonics* 3(7):446–454. doi:10.1002/jbio.200900089.

Pageon, S. V., S. P. Cordoba, D. M. Owen, S. M. Rothery, A. Oszmiana, and D. M. Davis. 2013. Superresolution microscopy reveals nanometer-scale reorganization of inhibitory natural killer cell receptors upon activation of NKG2D. *Sci Signal* 6(285):ra62. doi:10.1126/scisignal.2003947.

Pageon, S. V., D. Rudnicka, and D. M. Davis. 2012. Illuminating the dynamics of signal integration in Natural Killer cells. *Front Immunol* 3:308. doi:10.3389/fimmu.2012.00308.

Patzelt, J., A. Verschoor, and H. F. Langer. 2015. Platelets and the complement cascade in atherosclerosis. *Front Physiol* 6:49. doi:10.3389/fphys.2015.00049.

Pende, A., N. Artom, M. Bertolotto, F. Montecucco, and F. Dallegri. 2015. Role of neutrophils in atherogenesis: An update. *Eur J Clin Invest*. doi:10.1111/eci.12566.

Poulter, N. S., A. Y. Pollitt, A. Davies, D. Malinova, G. B. Nash, M. J. Hannon, Z. Pikramenou et al., 2015. Platelet actin nodules are podosome-like structures dependent on Wiskott-Aldrich syndrome protein and ARP2/3 complex. *Nat Commun* 6:7254. doi:10.1038/ncomms8254.

Purtic, B., L. A. Pitcher, N. S. van Oers, and C. Wulfing. 2005. T cell receptor (TCR) clustering in the immunological synapse integrates TCR and costimulatory signaling in selected T cells. *Proc Natl Acad Sci USA* 102(8): 2904–2909. doi:10.1073/pnas.0406867102.

Rak, G. D., E. M. Mace, P. P. Banerjee, T. Svitkina, and J. S. Orange. 2011. Natural killer cell lytic granule secretion occurs through a pervasive actin network at the immune synapse. *PLoS Biol* 9(9):e1001151. doi:10.1371/journal.pbio.1001151.

Ramos, G., U. Hofmann, and S. Frantz. 2016. Myocardial fibrosis seen through the lenses of T-cell biology. *J Mol Cell Cardiol*. doi:10.1016/j.yjmcc.2016.01.018.

Reitsma, S., M. G. oude Egbrink, V. V. Heijnen, R. T. Megens, W. Engels, H. Vink, D. W. Slaaf, and M. A. van Zandvoort. 2011a. Endothelial glycocalyx thickness and platelet-vessel wall interactions during atherogenesis. *Thromb Haemost* 106(5):939–946. doi:10.1160/TH11-02-0133.

Reitsma, S., M. G. oude Egbrink, H. Vink, B. M. van den Berg, V. L. Passos, W. Engels, D. W. Slaaf, and M. A. van Zandvoort. 2011b. Endothelial glycocalyx structure in the intact carotid artery: A two-photon laser scanning microscopy study. *J Vasc Res* 48(4):297–306. doi:10.1159/000322176.

Reitsma, S., D. W. Slaaf, H. Vink, M. A. van Zandvoort, and M. G. oude Egbrink. 2007. The endothelial glycocalyx: Composition, functions, and visualization. *Pflugers Arch* 454(3):345–359. doi:10.1007/s00424-007-0212-8.

Roh, K. H., B. F. Lillemeier, F. Wang, and M. M. Davis. 2015. The coreceptor CD4 is expressed in distinct nanoclusters and does not co-localize with T-cell receptor and active protein tyrosine kinase p56lck. *Proc Natl Acad Sci USA* 112(13):E1604–E1613. doi:10.1073/pnas.1503532112.

Ronnlund, D., L. Xu, A. Perols, A. K. Gad, A. Eriksson Karlstrom, G. Auer, and J. Widengren. 2014. Multicolor fluorescence nanoscopy by photobleaching: Concept, verification, and its application to resolve selective storage of proteins in platelets. *ACS Nano* 8(5):4358–4365. doi:10.1021/nn406113m.

Ronnlund, D., Y. Yang, H. Blom, G. Auer, and J. Widengren. 2012. Fluorescence nanoscopy of platelets resolves platelet-state specific storage, release and uptake of proteins, opening up future diagnostic applications. *Adv Healthc Mater* 1(6):707–713. doi:10.1002/adhm.201200172.

Rook, M. B., M. M. Evers, M. A. Vos, and M. F. Bierhuizen. 2012. Biology of cardiac sodium channel Nav1.5 expression. *Cardiovasc Res* 93(1):12–23. doi:10.1093/cvr/cvr252.

Rossy, J., D. M. Owen, D. J. Williamson, Z. Yang, and K. Gaus. 2013. Conformational states of the kinase Lck regulate clustering in early T cell signaling. *Nat Immunol* 14(1):82–89. doi:10.1038/ni.2488.

Rossy, J., S. V. Pageon, D. M. Davis, and K. Gaus. 2013. Super-resolution microscopy of the immunological synapse. *Curr Opin Immunol* 25(3):307–312. doi:10.1016/j.coi.2013.04.002.

Rossy, J., D. J. Williamson, C. Benzing, and K. Gaus. 2012. The integration of signaling and the spatial organization of the T cell synapse. *Front Immunol* 3:352. doi:10.3389/fimmu.2012.00352.

Roybal, K. T., E. M. Mace, J. M. Mantell, P. Verkade, J. S. Orange, and C. Wulfing. 2015. Early signaling in primary T cells activated by antigen presenting cells is associated with a deep and transient lamellal actin network. *PLoS One* 10(8):e0133299. doi:10.1371/journal.pone.0133299.

Sanborn, K. B., G. D. Rak, A. N. Mentlik, P. P. Banerjee, and J. S. Orange. 2010. Analysis of the NK cell immunological synapse. *Methods Mol Biol* 612: 127–148. doi:10.1007/978-1-60761-362-6_9.

Sattler, S. and N. Rosenthal. 2016. The neonate versus adult mammalian immune system in cardiac repair and regeneration. *Biochim Biophys Acta*. doi:10.1016/j.bbamcr.2016.01.011.

Schachtner, H., S. D. Calaminus, S. G. Thomas, and L. M. Machesky. 2013. Podosomes in adhesion, migration, mechanosensing and matrix remodeling. *Cytoskeleton (Hoboken)* 70(10):572–589. doi:10.1002/cm.21119.

Seebach, J., G. Donnert, R. Kronstein, S. Werth, B. Wojciak-Stothard, D. Falzarano, C. Mrowietz, S. W. Hell, and H. J. Schnittler. 2007. Regulation of endothelial barrier function during flow-induced conversion to an arterial phenotype. *Cardiovasc Res* 75(3):596–607. doi:10.1016/j.cardiores.2007.04.017.

Shannon, M. J., G. Burn, A. Cope, G. Cornish, and D. M. Owen. 2015. Protein clustering and spatial organization in T-cells. *Biochem Soc Trans* 43(3): 315–321. doi:10.1042/BST20140316.

Sherman, E., V. Barr, S. Manley, G. Patterson, L. Balagopalan, I. Akpan, C. K. Regan et al., 2011. Functional nanoscale organization of signaling molecules downstream of the T cell antigen receptor. *Immunity* 35(5):705–720. doi:10.1016/j.immuni.2011.10.004.

Sherman, E., V. A. Barr, and L. E. Samelson. 2013a. Resolving multi-molecular protein interactions by photoactivated localization microscopy. *Methods* 59(3):261–269. doi:10.1016/j.ymeth.2012.12.002.

Sherman, E., V. Barr, and L. E. Samelson. 2013b. Super-resolution characterization of TCR-dependent signaling clusters. *Immunol Rev* 251(1):21–35. doi:10.1111/imr.12010.

Shy, D., L. Gillet, and H. Abriel. 2013. Cardiac sodium channel NaV1.5 distribution in myocytes via interacting proteins: The multiple pool model. *Biochim Biophys Acta* 1833(4):886–894. doi:10.1016/j.bbamcr.2012.10.026.

Singh, H., R. Lu, P. F. Rodriguez, Y. Wu, J. C. Bopassa, E. Stefani, and L. Toro. 2012. Visualization and quantification of cardiac mitochondrial protein clusters with STED microscopy. *Mitochondrion* 12(2):230–236. doi:10.1016/j.mito.2011.09.004.

Soeller, C. and D. Baddeley. 2013. Super-resolution imaging of EC coupling protein distribution in the heart. *J Mol Cell Cardiol* 58:32–40. doi:10.1016/j.yjmcc.2012.11.004.

Tian, Q., S. Pahlavan, K. Oleinikow, J. Jung, S. Ruppenthal, A. Scholz, C. Schumann et al., 2012. Functional and morphological preservation of adult ventricular myocytes in culture by sub-micromolar cytochalasin D supplement. *J Mol Cell Cardiol* 52(1):113–124. doi:10.1016/j.yjmcc.2011.09.001.

Vajen, T., S. F. Mause, and R. R. Koenen. 2015. Microvesicles from platelets: Novel drivers of vascular inflammation. *Thromb Haemost* 114(2):228–236. doi:10.1160/TH14-11-0962.

van den Dries, K., S. L. Schwartz, J. Byars, M. B. Meddens, M. Bolomini-Vittori, D. S. Lidke, C. G. Figdor, K. A. Lidke, and A. Cambi. 2013. Dual-color super-resolution microscopy reveals nanoscale organization of mechanosensory podosomes. *Mol Biol Cell* 24(13):2112–2123. doi:10.1091/mbc. E12-12-0856.

Veeraraghavan, R., J. Lin, G. S. Hoeker, J. P. Keener, R. G. Gourdie, and S. Poelzing. 2015. Sodium channels in the Cx43 gap junction perinexus may constitute a cardiac ephapse: An experimental and modeling study. *Pflugers Arch* 467(10):2093–2105. doi:10.1007/s00424-014-1675-z.

Wagner, E., S. Brandenburg, T. Kohl, and S. E. Lehnart. 2014. Analysis of tubular membrane networks in cardiac myocytes from atria and ventricles. *J Vis Exp* (92):e51823. doi:10.3791/51823.

Wagner, E., M. A. Lauterbach, T. Kohl, V. Westphal, G. S. Williams, J. H. Steinbrecher, J. H. Streich et al., 2012. Stimulated emission depletion live-cell super-resolution imaging shows proliferative remodeling of T-tubule membrane structures after myocardial infarction. *Circ Res* 111(4):402–414. doi:10.1161/CIRCRESAHA.112.274530.

Walde, M., J. Monypenny, R. Heintzmann, G. E. Jones, and S. Cox. 2014. Vinculin binding angle in podosomes revealed by high resolution microscopy. *PLoS One* 9(2):e88251. doi:10.1371/journal.pone.0088251.

Walker, M. A., T. Kohl, S. E. Lehnart, J. L. Greenstein, W. J. Lederer, and R. L. Winslow. 2015. On the adjacency matrix of RyR2 cluster structures. *PLoS Comput Biol* 11(11):e1004521. doi:10.1371/journal.pcbi.1004521.

Walker, M. A., G. S. Williams, T. Kohl, S. E. Lehnart, M. S. Jafri, J. L. Greenstein, W. J. Lederer, and R. L. Winslow. 2014. Superresolution modeling of calcium release in the heart. *Biophys J* 107(12):3018–3029. doi:10.1016/j.bpj.2014.11.003.

Wang, W., A. P. Landstrom, Q. Wang, M. L. Munro, D. Beavers, M. J. Ackerman, C. Soeller, and X. H. Wehrens. 2014. Reduced junctional Na+/Ca2+-exchanger activity contributes to sarcoplasmic reticulum Ca2+ leak in junctophilin-2-deficient mice. *Am J Physiol Heart Circ Physiol* 307(9): H1317–H1326. doi:10.1152/ajpheart.00413.2014.

Williamson, D. J., D. M. Owen, J. Rossy, A. Magenau, M. Wehrmann, J. J. Gooding, and K. Gaus. 2011. Pre-existing clusters of the adaptor Lat do not participate in early T cell signaling events. *Nat Immunol* 12(7):655–662. doi:10.1038/ni.2049.

Wong, J., D. Baddeley, E. A. Bushong, Z. Yu, M. H. Ellisman, M. Hoshijima, and C. Soeller. 2013. Nanoscale distribution of ryanodine receptors and caveolin-3 in mouse ventricular myocytes: Dilation of t-tubules near junctions. *Biophys J* 104(11):L22–L24. doi:10.1016/j.bpj.2013.02.059.

Wulff, H. and R. Kohler. 2013. Endothelial small-conductance and intermediate-conductance KCa channels: An update on their pharmacology and usefulness as cardiovascular targets. *J Cardiovasc Pharmacol* 61(2):102–112. doi:10.1097/FJC.0b013e318279ba20.

Zuidscherwoude, M., F. Gottfert, V. M. Dunlock, C. G. Figdor, G. van den Bogaart, and A. B. van Spriel. 2015. The tetraspanin web revisited by super-resolution microscopy. *Sci Rep* 5:12201. doi:10.1038/srep12201.

Zuo, J., Z. Shan, L. Zhou, J. Yu, X. Liu, and Y. Gao. 2015. Increased CD160 expression on circulating natural killer cells in atherogenesis. *J Transl Med* 13:188. doi:10.1186/s12967-015-0564-3.

Zuppinger, C., M. Eppenberger-Eberhardt, and H. M. Eppenberger. 2000. N-Cadherin: Structure, function and importance in the formation of new intercalated disc-like cell contacts in cardiomyocytes. *Heart Fail Rev* 5(3):251–257. doi:10.1023/A:1009809520194.

Index

Note: Page numbers followed by f and t refer to figures and tables, respectively.

3D imaging, SD-confocal, 247f
3DSIM, 398
3D-SIM illumination pattern, 143
3D SOFI with pcDronpa, 247f
4B isoform, 322

A

A431 cell expressing mCitrine-erbB3, 80, 82f, 87, 89f, 92–93, 92f
Abbe, Ernst, 49
Abbe resolution limit, 138
Abbe's formula, 5–6
Acousto-optic modulators (AOMs), 303
Actin, 133, 160–161
  cytoskeleton, 397
  filaments, 33
  in HeLa cells, 305f
  podosomes, 405
Actin-myosin cross-bridge, 384
Adaptive shifted histogram (ASH) method, 86, 87f
AFM. *See* Atomic force microscope (AFM)
Airy pattern, 194–195
Alexa 647, 160, 359
Andor IQ software, 93
Anti-Brownian electrokinetic trap (ABEL), 228–230
Anti-fading reagents, 60
Antonie van Leeuwenhoek microscope, 6f
Atherosclerosis, 384
Atomic force microscope (AFM), 50, 149, 163, 228
  application, 150, 166
  cell probing measurement, 161
  correlative microscopy, 162f
  precise dissection, 166f
  quantification ability of STED, 157, 158f–159f
  scanning probe, 151
  static/dynamic modes, 151
  STED, STORM, 156–161, 157f, 160f, 162f
  super-resolution microscopy, 151
  topography, 159
  working principles, 151–153, 152f
ATP-sensitive K^+ channel (K_{ATP}), 390
Avalanche photodiodes (APDs), 42
"Average" single-molecule measurement, 32–33
Axial ZFS, 220

B

Bacterial chromosome, super-resolution imaging, 270–271
balanced SOFI (bSOFI), 181
Bayesian deconvolution, 109–116
  computing likelihood, 109–110
  object prior probability, 110–111, 111f
  optimization, acceleration, 114
  regularization, 112, 112f
  sample rate for deconvolution, 112–114, 113f–114f
  STED-3D deconvolution, 115–116
  STED deconvolution, 114–115, 115f

Bayesian framework, 112
B cells, 397–398
Beam modulation by LCDs, 197–198, 198f
Bimolecular fluorescence complementation (BiFC), 326, 341
configuration, 345
with PALM, 343
Binding-activated localization microscopy (BALM), 63
Biological systems, 302, 348
with NV centers in nanodiamonds
nanoscale imaging of, 221–222, 221f
nanoscale sensing of, 226–228, 227f
Blob detection, 80
β-mercaptoethanol (BME), 55
Broadening effect, 31

C

Ca^{2+}-activated K-channel (K_{Ca}), 393
Ca^{2+} channel-related structure imaging, 384–388
Ca^{2+}-induced Ca^{2+} release (CICR), 384
Caged fluorophore systems, 62
Cantilever deflections, 151
Cardiac caveolae, 385
Cardiac super-resolution imaging, 384–391
Ca^{2+} channel-related structures, 384–388
Na^+ and K^+ channel-related structures, 388–390
subcellular structures in cardiomyocytes, 390–391
Cardiovascular diseases, 383
platelets role in, 402
Caveats, 292
CD19 cluster, 397
cDNA (complementary DNA), 50
Cell
3T3, 10f
B, 397–398
culturing, 312
fluorescence microscopy in biology, 4, 5f
lines and reagents, 93
live, 253
membranes, lipids diffusion, 38, 40f
NK, 398–401, 399f–400f
probing measurement, 161
T, 394–397
Cellular process, 259
Cholesterol-depleted HeLa cells, 369
Chromosomal DNA
fluorescent labeling via click chemistry, 275–276
structural study in *E. coli,* 268–270, 268f–269f
visualization using SMLM, 264–265
CLSM. *See* Confocal laser scanning microscope (CLSM)
Colocalization analysis, 86, 275
Color center, 216–217
Complementary DNA (cDNA), 50
Compound microscope, 5, 49
Computation algorithm, SOFI, 177–179, 178f
Computer-controlled fluid injection system, 290f
Confinement strength parameter, 26, 28
Confocal laser scanning microscope (CLSM), 50, 190, 190f
Confocal microscopy, 50, 125, 246
lateral resolution, 126–128, 127f
optical sectioning, 126
RCM compared, 134, 135f
standard, 130, 131f
Convolution theorem, 100–101, 104
efficient computation, 103
fast computation, 103f
hybrid FFT/spatial, 118–119, 118f
steps, 103
Coronin 1A (Coro1A), 401
Correlation amplitude $g(\tau)$, 30, 30f
Correlation function, 27, 30
dynamic partitioning systems, 27
GFP-GPI, 37, 37f
spatiotemporal image, 23
temporal, 30f
Correlative imaging
in cargo transport
and 3d microtubule architecture, 288–289
and cellular roadblocks, 285–286
at microtubule–microtubule intersections, 286–288
technical developments in
automated correlative imaging with microfluidics, 289–291
multicolor imaging using correlative approach, 291–294
Correlative live-cell and super-resolution microscopy, 284–285
workflow, 284f

Correlative nanoscopy, 14, 149
AFM STED, AFM STORM, 156–161, 157f, 162f
cytoskeleton, 160f
quantification, 158f–159f
nanomanipulation, 163–167, 164f, 165f, 166f
overview, 149–156
optical, 153–156, 154f–155f
working principles, AFM, 151–153, 151f
Cramér–Rao lower bound, 84
Cumulative distribution function (CDF), 371–372
Cut-off frequency, 6

D

Data acquisition sequence, SIM, 140, 140f
Data reconstruction, SIM, 141–142
fourier space, 141, 141f
SIM reconstruction, 141–142, 142f
Deconvolution algorithms, 103–105
Bayesian, 109–116
iterative, 105–109, 106f–107f
linear inversion, 103–104
sample rate, 112–114, 113f–114f
super-resolution, 104–105
Defense system against infections, 394
Density-based spatial clustering analysis with noise (DBSCAN), 262, 330
Desmosomal protein, 390
Deterministic emitter switch microscopy (DESM), 226
Diffraction barrier, circumventing, 7–8, 8f
Diffuse-assisted FRET SOFI (dSOFI), 252, 253f
Diffusion law, 20–22
Dippen nanolithography (DNP), 163
Direct stochastic optical reconstruction microscopy (*d*STORM), 173, 265, 385, 389–390
dual-color, 404
imaging, 276–277
E. coli cells, 268
indirect immunostaining for, 359–362
DNA
cell cycle dependent, 265–267, 266f
chromosomal. *See* Chromosomal DNA in *E. coli*
replication and segregation, spatial organization, 264–267
super-resolution imaging, 267–268
Drift map, 114
Dronpa, 59, 245, 341, 364
*d*STORM. *See* Direct stochastic optical reconstruction microscopy (*d*STORM)
Dynamic partitioning, 37
Dynamic saturation optical microscopy (DSOM), 173–174

E

EGF-Alexa 647 addition, 362
EGFR. *See* Epidermal growth factor (EGF) receptor (EGFR)
Electron multiplied charge coupled device (EMCCD) camera, 34, 42, 273–274
Electron tomography, 33
EMCCD camera. *See* Electron multiplied charge coupled device (EMCCD) camera
Emitting molecule density, 250
Endocytosis, 362, 369
Endogenous fluorescent proteins, 58
Endothelial cells (ECs), 391–394
Epidermal growth factor (EGF), 372
signaling, 358
Epidermal growth factor (EGF) receptor (EGFR), 358
cellular stimulation, dynamics, 372–375, 373f–374f, 375t
dual-color super-resolution imaging, 366, 368–370, 368f–369f
dynamics
and cellular stimulation, 372–375, 373f–375f
molecules, 370–372, 370f
indirect immunostaining, 359–362
labeled ligand to visualize activated, 362, 363f
liganded, 366
Escherichia coli
SMLM studies in
biomolecule tagging, 271–272, 272f
imaging procedure, 273–274, 274f
post-processing, 275
sample preparation, 272–273, 273f
Spatial Organization
of DNA replication and segregation in, 264–267
of transcription, 260–264
super-resolution imaging, 267–268
Extracellular matrix (ECM), 276
EZ-rich defined medium (EZ-RDM), 276–277

F

F-actin protien, 160, 203
FALCON alogrithm, 90
Fast SOFI approaches, 250–251
FCS. *See* Fluorescence correlation spectroscopy (FCS)
Fibroblast cells, 166–167
Fitting methods, 83
FLIM. *See* Fluorescence lifetime imaging microscopy (FLIM)
Fluorescence
  fluctuation microscopy, 19
  intensity, 23
  microscope, 8
  microscopy, 189
  super-resolution microscopy techniques, 51–53, 52t
Fluorescence correlation spectroscopy (FCS), 11, 21, 32, 177–178, 251–254
Fluorescence emission, frequency response, 309–311, 310f–311f
Fluorescence lifetime imaging microscopy (FLIM), 51, 153
  SOFI for, 215–216
Fluorescence PALM (FPALM), 13, 39
Fluorescence recovery after photobleaching (FRAP), 51, 153
Fluorescence resonance energy transfer (FRET), 38, 51, 216
Fluorescent beads, 165
Fluorescent biomarkers, 215
Fluorescent labeling technique, 99
Fluorescent labels in SOFI, 243–245
  fluorescent proteins, 244–245, 245f
  QDs, 243–244
Fluorescent nanodiamonds (FNDs), 216
  general properties, 216–221, 217f, 218t–219t
  manipulation techniques
    ABEL trap, 229–230, 230f–231f
    optical tweezers, 229, 230f
    scanning probe, 228–229, 228f
  super-resolution microscopy with, 222–226, 223f–225f
Fluorescent protein (FP), 244–245
Fluorescent repressor–operator systems (FROS), 265
Fluorophore, 53–56, 54f, 302
  absorption and emission spectra, 54
  fluorescence emission, 57
  nonfluorescent dark state, 57–58, 58f–59f
  pairs, 60
  photoconvertible, 55
  photophysics, 55, 57, 60
  reversible photoswitching, 58–61, 61f
  types, 53, 53f
FNDs. *See* Fluorescent nanodiamonds (FNDs)
Force distance (FD) curves analysis, 153
Förster resonance energy transfer (FRET), 251–254
Fourier interpolation, 179–181
Fourier reweighing, SOFI, 181–183, 182f–183f
Fourier-SOFI (fSOFI), 181
Fourier space, 141, 141f
  SIM reconstruction, 141–142, 142f
fPALM (fluorescence-PALM), 13
Full width at half maximum (FWHM), 104, 132, 134

G

Gamma adjustment, 242–243, 242f
gated pulsed STED (gP-STED), 11f
Gaussian function, 83
GFP. *See* Green fluorescent protein (GFP)
GFP-GPI
  correlation function, 37f
  dynamic partitioning, 36–38, 37f
  TIRF image, 37
Gold algorithm, 120
Green fluorescent protein (GFP), 13, 50–51, 138
Green-to-red photoconversion, 61
Gustafsson's experiment, 143–144

H

HeLa cells, 366, 369, 372
  3D projection in 3D culture, 313, 314f–315f
  cytoskeletal structures, 12f
  dynamics of EGFR in, 370–371, 370f
  EGF-Alexa647, 369f
  fluorescence images, 305f, 314f
  vimentin-Dreiklang, 249f
Hermite–Gaussian (HG) mode, 193
H-function, 361
High density labeling, 55, 260, 265
Hybrid FFT/spatial convolution, 118–119, 118f
Hypervariable region (HVR), 322

I

Ideal model scheme, 58, 61
image-derived MSD (iMSD), 20, 24, 30, 38
analysis on 2D diffusion, 28, 29f
GFP-TfR, 34
plot interpretation, 25–28
from STICS to, 22f
time plot, 22, 35f
vs SPT, 32–33, 32f
Image formation process, 4, 5f, 6
Image scanning microscopy (ISM), 128–130
drawback, 130
"pinhole-image" series, 129
possibilities and limitations, 129–130
Immune system cells imaging, 394–402
B cells, 397–398
neutrophils, 401–402
NK cells, 398–401, 399f–400f
T cells, 394–397
Immunoelectron microscopy (immuno-EM), 326
Immunogold labeling, 161
Immunoprecipitation (IP), 325–326
iMSD. *See* Image-derived MSD (iMSD)
Incoherent imaging system, 100
Indirect immunostaining, 359–362
instant SIM (ISIM), 130
Irreversible photoactivation/-conversion for SMLM, 61–62
ISM. *See* Image scanning microscopy (ISM)
Isoforms
Raf kinases, 324
Ras, 322
Iterative deconvolution methods, 105–109, 106f–107f
penalty function, 108–109, 108f

J

Jablonski diagrams, 190f
Joint tagging SOFI (JT-SOFI), 250–251, 251f
JPH2 knock-out, 387

K

K_{ATP} channel, 390
K-function ($K(r)$), 360–361
Köhler illumination, 50

L

Labeling density, 250–251, 290
Laguerre–Gaussian (LG) mode, 193–194
Lamina elastica externa (LEE), 392
Laplace operator, 112, 112f
Laser scanning microscopy (LSM), 190, 194
Layard lens, 5
Liganded EGFRs, 366
Light microscopy, 281–282
Linear 2D-SIM, 142–143
Linear 3D-SIM, 143
Lipid nanodomains, 33
Lipid rafts, 33, 36–37
Liquid crystal (LC) on silicon (LCOS) devices, 198
Live cell imaging, 253f
Live-cell super-resolution microscopy, challenges, 282–283
Localization, 327
algorithms, 83
microscopy, 301
precision, 359, 371
subpixel, 82–83
uncertainty, 84–85
Localization-based techniques, 153
Lysenin, 364
Lytic granules, 398, 401

M

Magnetotactic bacteria (MTB), 227
Manipulation techniques, FND
ABEL trap, 229–230, 230f–231f
optical tweezers, 229, 230f
scanning probe, 228–229, 228f
Maximum likelihood estimation (MLE) method, 83, 110, 111f
Mechanical magnification, 131–132
Membrane
components, super-resolution imaging, 359–370
lipids diffusion on cell, 38
nanodomains, 37
raft, 357
dual-color super-resolution imaging of EFGR with, 366–370, 368f–369f
probes for PALM, 362–366, 364f, 365f, 367f
Methyl-β-cyclodextrin (MβCD), 369
Microscope detecting system, 12
Microscopic image formation, 100–103, 100f
bandwidth, 101–103, 102f
convolution, efficient computation, 103
spatial frequencies, 101, 102f
Microscopium nominare libuit, 5

Missing cone, 102f, 182
Moerner's lab, 11
Moiré effect, 139f
Monte Carlo simulations, SMLM, 90–92, 91f–92f
Motor proteins, 285
Multiemitter fitting analysis, 84
Multifork replication, 265

N

NA. *See* Numerical aperture (NA)
Na^+ and K^+ channel-related structure imaging, 388–390
Naïve B cells, 397
Nanomanipulation, 163–167, 164f, 165f, 166f
Nanoscale imaging, biological systems, 221–222, 221f
Nanoscale sensing, biological systems, 226–228, 227f
Nanoscopic images, deconvolution
  Bayesian deconvolution, 109–116
  deconvolution algorithms, 103–105
  iterative deconvolution methods, 105–109, 106f–107f
  microscopic image formation, 100–103, 100f
  overview, 99–100
  space variant PSFs, 116–119, 117f
Nanoscopium nominare libuit, 13–14
Natural killer (NK) cells, 398–401, 399f–400f
Near-field scanning optical microscopy (NSOM), 51, 222–223
Neutrophils, 401–402
Nimrud lens, 5
NK cells. *See* Natural killer (NK) cells
NKG2D receptor, 398
Nonfitting methods, 83
Nonlinear structured-illumination microscopy (nlSIM), 173–174
Nonnegative matrix factorization, 90
NSOM. *See* Near-field scanning optical microscopy (NSOM)
Numerical aperture (NA), 49, 51, 138
Nyquist criterion, 106, 283
Nyquist–Shannon theorem, 112

O

Oncogenic mutations, 324
One-component Gaussian model, 27, 37, 37f
Optically detected magnetic resonance (ODMR), 219, 226
Optical magnification ($M_{optical}$), 131
Optical microscopy, 210, 312
  physical limitation, 4
  quantitative leap, 5
  and spatial resolution, 5–6, 6f
Optical nanoscopy, 151
  with SOFI
    applications, 246–252
    fluorescent labels in, 243–245
Optical photon re-assignment (OPRA), 130
Optical sectioning, 126
Optical transfer function (OTF), 101, 103–104, 180–181, 309
Optimization scheme, 109
Organic dyes, 56, 62
  chromophore backbones, 53
  engineering-improved, 50
  fluorophore switching, reversible, 60–61, 61f
  reversible photoswitching, 61f
Organic fluorophores, 283
OTF. *See* Optical transfer function (OTF)

P

Pair correlation analysis, 86–87
PALM. *See* Photoactivated localization microscopy (PALM)
P-glycoprotein, 393
The Philosopher of the Arabs, 5
Photoactivated localization microscopy (PALM), 13, 79, 173, 226, 301, 326–327
  membrane raft probes for, 362, 364–366
  quantitative, 330–332, 331f
  sequential, 268f, 276
Photoactivation, 343
  irreversible, 61–62
Photobleaching, 54, 165, 215, 310–311
Photoswitchable fluorescent proteins, 59, 174
Photoswitching, 57, 343
  dronpa, fluorescent protein, 59
  fluorophore reversible, 58–61, 61f
  super-resolution fluorescence microscopy, 56–63, 57f
    irreversible photoactivation/-conversion for SMLM, 61–62
    nonfluorescent dark state, 57–58, 58f–59f
    reversible, 58–61, 61f
    super-resolution applications, 62–63
  super-resolution microscopy without, 63
Phototoxicity, 282

"Pinhole-image" series, ISM, 129
Platelets, super-resolution microscopy of, 402–404
Podosomes, super-resolution microscopy of, 404–405
Points accumulation for imaging in nanoscale topography (PAINT) technique, 63, 185, 268f
Point spread function (PSF), 5–6, 83, 132, 174, 196, 241, 306
  fluctuation-based, 183–185, 185f
  Gaussian, 24
  space variant, 116–119, 117f
    hybrid FFT/spatial convolution, 118–119, 118f
  widefield, 117f
Poisson noise model, 83, 90
Post-processing algorithm, 129
Posttranslational modifications, 323f, 337
Practical resolution gain, 114–115
Protein islands, 395
Protein-protein interactions (PPIs), 326, 341
Protein psCFP2, 62
Protocols
  fixed RNAP samples preparation, 275
  fluorescent labeling via click chemistry, 275–276
  sample preparation, correlative sequential approaches, 276–277
    multicolor imaging, 276
    sptPALM and *d*STORM Imaging, 276–277
PSF. *See* Point spread function (PSF)
Pupil plane, 141

Q

QDs. *See* Quantum dots (QDs)
θ-toxin, 362, 364
Quantitative PALM
  Raf multimer formation, 332–339
    induced by Raf kinase inhibitors, 337–339
    in presence of mutant Ras, 332–336, 333f–335f
    upon forced membrane localization, 337
  resolving protein multimers using PALM, 330–332, 331f
Quantitative super-resolution fluorescence microscopy
  considerations for SRM, 327–330, 328f, 330f
  resolving protein multimers using PALM, 330–332
Quantum dots (QDs), 54, 90, 174, 243–244
  fluorescent emission features, 244f

R

Radial symmetry, 83
Raf multimer formation, 332–339
  induced by Raf kinase inhibitors, 337–339, 338f
  in presence of mutant Ras, 332–336
  upon forced membrane localization, 337, 338f
Ramsey pulse sequence scheme, 219–220
Ras and Raf signal regulation, 324–326
Rasbinding domain (RBD), 324
Ras isoforms, 322
Ras-Raf BiFC diffusion trajectories, 347
Ras-Raf interactions imaging at nanoscale, 339–348
  BiFC-PALM
    based on split PAmCherryl, 339–343, 340f
    nanoscopic imaging, 343–346, 344f, 345f
    tracking of individual, 346–348, 347f
Ras-Raf-MAPK signaling pathway in cancer, 321–324, 322f
Rayleigh criterion, 51
RCM. *See* Re-scan confocal microscopy (RCM)
Reactive oxygen species (ROS), 54
Reducing and oxidizing system (ROXS), 55
Registration error, 285
Re-scan confocal microscopy (RCM), 130–133
  biological applications, 133, 133f
  confocal microscope, 134, 135f
  resolution improvement, 132–133, 133f
  scanning mirrors, 130
  set-up and principle, 130–131, 131f
  sweep factor M, 131–132
  working, 126–128
    lateral resolution, 126–128, 127f
    optical sectioning, 126
    super-resolution information, 128
RESOLFT. *See* Reversible saturable optical linear fluorescence transitions (RESOLFT)
Restricted super-resolution, 7
Reversible saturable optical linear fluorescence transitions (RESOLFT), 7–8, 10, 225, 283
Rhodamine 6G, 302, 306–307

Richardson–Lucy (RL) algorithm, 110–111, 111f, 119–120, 183
Ripley's analysis, 360–362, 361f
Ripley's K test, 329, 330, 332
RL algorithm. *See* Richardson–Lucy (RL) algorithm
RNA polymerase (RNAP), 260
  clustering at nanoscale, 262–264, 263f–264f
  E. coli
    single-molecule tracking, 270–271, 270f
    structural study, 268–270, 268f–269f
    super-resolution imaging in, 267–268
  single-molecule super-resolution microscopy, 260–262, 262f
Rotation angle, 293
Ryanodine receptor isoform 2 (RyR2), 384–385

S

SAD. *See* Simulation-aided DBSCAN (SAD)
Saturable SIM (SSIM), 8
Saturated excitation (SAX) microscopy, 302
  imaging of 3D-cultured cells, 312–317
  optical setup, 304f
  principle, 302–306
  resolution improvement, 303f
  spatial resolution, 306–309
Scalar diffraction theory, 194
Scanning near-field optical microscopy (SNOM), 151
Scanning probe microscopy (SPM), 149, 151
Scanning tunneling microscope (STM), 50, 149
Second-order cumulant image, 175
Seebach, Jochen, 393
SIM. *See* Structured illumination microscopy (SIM)
"SIMcheck" software, 146, 147f
simulation-aided DBSCAN (SAD), 330–332, 333f
Single cell force spectroscopy (SCFS), 150
Single-molecule localization-based methods, 283
Single-molecule localization microscopy (SMLM), 53, 56, 79, 260
  materials and methods, 93
  overview, 79–80
  processing using Monte Carlo simulations, 90–92
  software, 92–93
  study in *E. coli*
    biomolecule tagging, 271–272, 272f
    imaging procedure, 273–274, 274f
    post-processing, 275
    sample preparation, 272–273, 273f
  temporal resolution, 88–90
  traditional approach, 80–88, 81f
Single-molecule SMD analysis, 371
Single-molecule-tracking (smt)-PALM, 346–348
Single-particle tracking (SPT), 20, 221–222, 358
  characteristics, 20–21
  iMSD approach, 32f
  of receptor molecules
    dynamics of EGFR and cellular stimulation, 372–375
    EGFR molecule dynamics, 370–372
Single-photon counter detector, 42
Single plane illumination microscopy (SPIM), 42
Sky lantern for SOFI (Skylan-S), 245, 245f
Smearing effect, 132
SMLM. *See* Single-molecule localization microscopy (SMLM)
SOFI. *See* Stochastic optical fluctuation imaging (SOFI)
Spatial frequency, 101, 141
Spatial light modulators (SLMs), 197, 209–210
Spatial resolution, SAX microscopy, 306–309, 307f–309f
Spatial/temporal analysis, 19
Spatiotemporal image, 23–27
  iMSD plot interpretation, 25–27
  particle size contribution, 25
Spatiotemporal image correlation spectroscopy (STICS), 21
Spot
  density, 275
  finding, 80
SPT. *See* Single-particle tracking (SPT)
SPT PALM (sptPALM), 39, 274
  correlative, 276–277
Standard deviation (σ), 27
Standard microscope, super-resolution, 38–43
  biological applications, 33–38
    cell membranes, lipids diffusion, 38, 40f
    GFP-GPI dynamic partitioning, 36–38, 37f
    TfR transient confinement, 34–36, 35f–36f
    TRPV1, 38, 39f

experimental requirements, 41–42
overview, 20–22, 22f
potential impact, 42–43
spatiotemporal image, 23–27
STICS-iMSD method, 27–33
STED. *See* Stimulated emission depletion (STED)
STED-3D deconvolution, 115–116, 116f
STED-3X system, 115, 116f
Steepest descent algorithm, 108–109, 108f
STICS-iMSD method, 27–33
"average" single-molecule measurement, 32–33
correlation amplitude $g(\tau)$, 30, 30f
σ_0, particle size, and sampling speed, 31–32, 31f
simulations, 27–30, 29f
Stimulated emission depletion (STED), 7–8, 21, 51, 56, 100, 173–174, 216, 282
beam, 9
confocal and gated pulsed, 11f
deconvolution, 114–115, 115f
-FCS measurement, 40
microscope, 113–114
microscopy, 8, 149, 151, 153, 156, 164–165, 203–208
AFM, 158, 165f
improvement, 166
key aspect, 153
quantification, 159f
principle, 10
Stochastic optical fluctuation imaging (SOFI), 174
brightness values, 179
challenges, 175
computation algorithm, 177–179, 178f
detector pixel size and Fourier interpolation, 179–181, 179f
fast, 250–251
fluctuation-based PSF estimation, 183–185, 185f
fluorescent labels in, 243–245
fluorescent proteins, 244–245, 245f
QDs, 243–244
Fourier reweighing, 181–183, 182f–183f
for FRET and FCS, 251–252
hierarchical, 178f
issues, 179
*n*th-order cumulant, 178, 181
optical nanoscopy with
applications, 246–252
fluorescent labels in, 243–245
overview, 173–174, 241–243
second-order, 175–176
signal-to-noise ratio, 183
theoretical basis, 174–177
three-dimensional techniques, 246–249
Stochastic optical reconstruction microscopy (STORM), 13, 149, 156, 173, 326
AFM, 156–161
Stochastic "read-out" approach, 11–13, 12f
Stokes shift, 50
STORM. *See* Stochastic optical reconstruction microscopy (STORM)
Structured illumination microscopy (SIM), 7, 52, 56, 137, 388
acquisition sequence, 139, 140f
data reconstruction, 142f
data set, 3D-, 138f, 140, 140f
different flavors, 142–145
linear 2D-SIM, 142–143
linear 3D-SIM, 143
nonlinear, 143–145, 144f–145f
imaging, 146
data reconstruction, 146, 147f
sample preparation, 146
mathematical treatment, 142
nonlinear, 139
overview, 137
super-resolution imaging, 138–142
Abbe diffraction limit, 138–139, 138f
data acquisition sequence, 140, 140f
data reconstruction, 141–142, 141f–142f
working, 139, 139f
Submaximal stimulation, 372
Subpixel localization, 82–84
Superconducting quantum interference devices (SQUIDs), 220
Super-resolution, 4, 6–7, 104–105
fluorescence microscope, 149
fluorescence microscopy
microscope, history, 49–53
photoswitching mechanisms. *See* Photoswitching, super-resolution fluorescence microscopy
without photoswitching, 63
imaging, 366
imaging of membrane components
dual-color super-resolution imaging, 366–370, 368f–369f
indirect immunostaining, 359–362

Super-resolution (*Continued*)
labeled ligand to visualize activated, 362, 363f
membrane raft probes for PALM, 362–366, 364f–365f, 367f
microscopy, 259–260
in cardiac research. *See* Cardiac super-resolution imaging
EB1 and N-cadherin, 389f
future perspective, 405–406
of platelets, 402–404
of podosomes, 404–405
regimes, 156
techniques, 138
as tool for biological imaging, 326–327
in vascular research, 391–405
microscopy with FNDs, 222–226, 223f–225f
techniques, 241
Super-resolved ensemble fluorophore microscopy, 7–11, 9f, 10f–11f
Super-resolved fluorescence microscopy, 4
diffraction barrier, 7–8
nanoscopium nominare libuit, 13–14
optical microscopy and spatial resolution, 5–6
stochastic "read-out" approach, 11–13
super-resolution, 6–7
super-resolved ensemble fluorophore microscopy, 8–11
Super-resolved single fluorophore microscopy, 7, 11–13, 12f

T

Talbot effect, 143
Targeted read-out approach, 8–11
T cells, 394–397
Technical developments in correlative imaging
automated imaging with microfluidics, 289–291
multicolor imaging, 291–294
Temporal resolution, SMLM, 88–90
TfR. *See* Transferrin receptor (TfR)
Third-order cumulant, 176
Three-dimensional SOFI techniques, 246–249
Three-photon excitation, 189
Thrombocytes, 402
ThunderSTORM, 92–93
Tikhonov–Miller method, 104
Tip-enhanced Raman spectroscopy (TERS), 150
TIRFMs. *See* Total internal reflection fluorescence microscopes (TIRFMs)
Titanium:sapphire (Ti:Sa) laser techniques, 190
tLCDs. *See* Transmissive liquid crystal devices (tLCDs)
Total internal reflection (TIR), 34
Total internal reflection fluorescence microscopes (TIRFMs), 34, 51, 137, 359
image of GFP-GPI, 37f
TPLSM. *See* Two-photon excitation LSM (TPLSM)
Traditional approach, SMLM, 80–88, 81f
analysis, 86–88, 89f
crowded field problem, 84
multiemitter fitting analysis, 84
image filtering, 80, 82f
post processing, 84–86
drift correction, 85
filtering, 85–86
merging, 85
subpixel localization, 82–84
fitting methods, 83
localization uncertainty, 84
nonfitting methods, 83
thresholding and detection of molecules, 80–82
visualization methods, 86, 87f, 88f
Transferrin receptor (TfR), 22
transient confinement, 34–36, 35f–36f
Transient Receptor Potential Vanilloid 1 (TRPV1), 38, 39f
Transmembrane proteins, 33, 38
Transmissive liquid crystal devices (tLCDs), 192, 198–199, 203
types, 205
Transverse mode, 192–193
Trapping time (τ_T), 30
Tunica
intima, 392
media, 392
Two-photon excitation, 4, 189–190, 190f, 301–302
approach, 10
laser scanning microscopy, 7–8
regime, 13
STED microscopy, 203–208, 204f–205f, 207f
Two-photon excitation LSM (TPLSM), 190–191, 190f, 199–203
utilizing an HRP beam, 199–203

U

Unrestricted super-resolution, 7

V

van Cittert algorithm, 105, 106f–107f
van Leeuwenhoek, Antonie, 5
Variable-shaped mirrors, 197
Variational Bayesian singleparticle tracking (vbSPT), 346–348
Vascular smooth muscle cells, 391–394
Vascular super-resolution imaging, 391–405
  immune system cells, 394–402
  vascular wall imaging, 391–394
Vascular wall imaging, 391–394
VE-cadherin cluster, 393
Visualization methods, 86, 87f, 88f
Voltage-gated sodium (Na^+) channel (VGSC), 388

W

Wavelet transform, 80
Widefield (WF), 80
  A431 cell, 82f, 87, 89f, 92, 92f
  microscope, 181–182
Wiskott–Aldrich syndrome protein (WASP), 404

Z

Zero phonon line (ZPL), 216–217, 219
Zhuang's method, 84